Characterization of Lignocellulosic Materials

To Xuan, Nicholas and Lucas

Characterization of Lignocellulosic Materials

Edited by

Thomas Q. Hu
Principal Scientist, FPInnovations (Paprican Division)
Adjunct Professor, Department of Chemistry
University of British Columbia

Blackwell
Publishing

© 2008 by Blackwell Publishing Ltd

Blackwell Publishing editorial offices:
Blackwell Publishing Ltd, 9600 Garsington Road, Oxford OX4 2DQ, UK
  Tel: +44 (0)1865 776868
Blackwell Publishing Professional, 2121 State Avenue, Ames, Iowa 50014-8300, USA
  Tel: +1 515 292 0140
Blackwell Publishing Asia Pty Ltd, 550 Swanston Street, Carlton, Victoria 3053, Australia
  Tel: +61 (0)3 8359 1011

The right of the Author to be identified as the Author of this Work has been asserted in accordance with the
Copyright, Designs and Patents Act 1988.

All rights reserved. No part of this publication may be reproduced, stored in a retrieval system, or transmitted, in
any form or by any means, electronic, mechanical, photocopying, recording or otherwise, except as permitted by
the UK Copyright, Designs and Patents Act 1988, without the prior permission of the publisher.

Designations used by companies to distinguish their products are often claimed as trademarks. All brand names
and product names used in this book are trade names, service marks, trademarks or registered trademarks of
their respective owners. The Publisher is not associated with any product or vendor mentioned in this book.

This publication is designed to provide accurate and authoritative information in regard to the subject matter
covered. It is sold on the understanding that the Publisher is not engaged in rendering professional services. If
professional advice or other expert assistance is required, the services of a competent professional should be
sought.

First published 2008 by Blackwell Publishing Ltd

ISBN: 978-1-4051-5880-0

Library of Congress Cataloging-in-Publication Data
Characterization of lignocellulosic materials/edited by Thomas Q. Hu.
    p. cm.
  Includes bibliographical references and index.
  ISBN-13: 978-1-4051-5880-0 (hardback : alk. paper)
  ISBN-10: 1-4051-5880-8 (hardback : alk. paper)
  1. Lignocellulose–Analysis. 2. Lignocellulose–Biotechnology. 3. Wood-pulp–Testing.
I. Hu, Thomas Q.
TP248.65.L54C45 2008
620.1'97–dc22
                                                                                            2007025833

A catalogue record for this title is available from the British Library

Set in 10/12 pt Minion
by Newgen Imaging Systems (P) Ltd., Chennai, India
Printed and bound in Singapore by C.O.S. Printers Pte Ltd

The publisher's policy is to use permanent paper from mills that operate a sustainable forestry policy, and which
has been manufactured from pulp processed using acid-free and elementary chlorine-free practices.
Furthermore, the publisher ensures that the text paper and cover board used have met acceptable environmental
accreditation standards.

For further information on Blackwell Publishing, visit our website:
www.blackwellpublishing.com

Contents

Preface xiii
Acknowledgments xv
Editor Biography xvii
List of Contributors xix

Part I Novel or Improved Methods for the Characterization of Wood, Pulp Fibers, and Paper 1

1 2D Heteronuclear (^{1}H–^{13}C) Single Quantum Correlation (HSQC) NMR Analysis of Norway Spruce Bark Components 3
   Liming Zhang and Göran Gellerstedt
   Abstract 3
   1.1 Introduction 3
   1.2 Experimental 4
       1.2.1 Materials 4
       1.2.2 Solvent extraction 4
       1.2.3 Purification of the condensed tannin 4
       1.2.4 Milled bark tannin–lignin 5
       1.2.5 Analyses 5
   1.3 Results and discussion 5
       1.3.1 Solvent extractions and separation of bark components 5
       1.3.2 Stilbenes and tannins 7
       1.3.3 Identification by NMR 8
   Conclusions 14
   Acknowledgments 15
   References 15

2 Raman Spectroscopic Characterization of Wood and Pulp Fibers 17
   Umesh Prasad Agarwal
   Abstract 17
   2.1 Introduction 17
   2.2 Wood and components 18
       2.2.1 Woods 18

		2.2.2	Extraneous compounds	19
		2.2.3	Wood components	19
		2.2.4	Band assignments	20
		2.2.5	Microscope studies	22
		2.2.6	Fiber cell wall organization and compositional heterogeneity	22
		2.2.7	Molecular changes during tensile deformation	23
	2.3	Mechanical pulp		25
	2.4	Chemical pulp		27
	2.5	Modified/treated wood		28
		2.5.1	Heat treated	28
		2.5.2	Chemically treated	28
	2.6	Cellulose I crystallinity of wood fibers		29
	2.7	"Self-Absorption" phenomenon in near-IR FT-Raman spectroscopy		30
	Conclusions			30
	References			31
3	Surface Characterization of Mechanical Pulp Fibers by Contact Angle Measurement, Polyelectrolyte Adsorption, XPS, and AFM			36
	Per Stenius and Krista Koljonen			
	Abstract			36
	3.1	Introduction		36
	3.2	Mechanical pulps		37
	3.3	Morphology of mechanical pulp surfaces		39
	3.4	Surface chemistry of mechanical pulps		41
		3.4.1	X-ray photoelectron spectroscopy	41
		3.4.2	Time-of-flight secondary ion mass spectrometry	45
		3.4.3	Polyelectrolyte adsorption, anionic groups, and surface charge	46
	3.5	Surface energy of mechanical pulp fibers		48
	Conclusions			52
	References			52
4	Assessing Substrate Accessibility to Enzymatic Hydrolysis by Cellulases			60
	Richard P. Chandra, Ali R. Esteghlalian, and John N. Saddler			
	Abstract			60
	4.1	Introduction		60
	4.2	Enzymatic hydrolysis by cellulases		61
	4.3	Exterior surface area		65
		4.3.1	Particle size	65
	4.4	Combination measurements: Interior and exterior surface area		67
		4.4.1	Swelling and water retention value	67
		4.4.2	Simons' stain	68
		4.4.3	Nitrogen adsorption technique	69
	4.5	Interior surface area		71
		4.5.1	Mercury porosimetry	71
		4.5.2	Solute exclusion	72

Conclusions		75
References		76

5 Characterization of AKD-Sized Pilot Papers by XPS and Dynamic Contact Angle
   Measurements 81
   Rauni Seppänen
   Abstract 81
   5.1 Introduction 81
   5.2 Materials and methods 84
       5.2.1 Materials 84
       5.2.2 Methods 84
   5.3 Results and discussion 86
       5.3.1 Surface chemical composition 86
       5.3.2 Wettability 93
   Conclusions 98
   Acknowledgments 98
   References 98

6 Chemical Microscopy of Extractives on Fiber and Paper Surfaces 101
   Pedro Fardim and Bjarne Holmbom
   Abstract 101
   6.1 Introduction 101
   6.2 Surface analysis of extractives by XPS 102
   6.3 Surface analysis of extractives by ToF-SIMS 104
   6.4 Extractives on fiber and paper surfaces as investigated by XPS and
       ToF-SIMS 105
       6.4.1 Surface coverage by extractives on mechanical, chemical, and
             recycled fibers 106
       6.4.2 Surface coverage by extractives on newsprint papers 107
       6.4.3 Surface composition and distribution of extractives on fiber and
             paper surfaces 108
   6.5 Role of extractives on fiber and paper surfaces 114
   Acknowledgment 116
   References 116

Part II Characterization of Cellulose, Lignin, and Modified Cellulose Fibers 119

7 Studies of Deformation Processes in Cellulosics Using Raman Microscopy 121
   Wadood Y. Hamad
   Abstract 121
   7.1 Introduction 121
   7.2 Principles of Raman microscopy 123
       7.2.1 Instrumentation for Raman microscopy 124
       7.2.2 Raman deformation studies of polymers 125

7.3	Deformation processes in cellulosics		125
	7.3.1	The structure and solid-state properties of cellulose fibers	126
	7.3.2	Molecular deformation processes in celluloses	127
Concluding remarks and future applications			133
Dedication			134
Notes			134
References			134

8 Lifetime Prediction of Cellulosics by Thermal and Mechanical Analysis — 138
Tatsuko Hatakeyama and Hyoe Hatakeyama

Abstract			138
8.1	Introduction		138
8.2	Materials and methods		139
	8.2.1	Samples	139
	8.2.2	Thermogravimetry	139
	8.2.3	Mechanical measurements	139
8.3	Prediction of durability		140
	8.3.1	Prediction by TG	140
	8.3.2	Prediction of mechanical properties	143
References			147

9 Recent Advances in the Isolation and Analysis of Lignins and Lignin–Carbohydrate Complexes — 148
Mikhail Yurievich Balakshin, Ewellyn Augsten Capanema, and Hou-min Chang

Abstract			148
9.1	Introduction		148
9.2	Isolation of lignin and LCC preparations/samples from wood and pulps		149
	9.2.1	Isolation of lignin from wood	149
	9.2.2	Isolation of lignin from pulps	153
9.3	Lignin and LCC analysis		154
	9.3.1	Methods in lignin and LCC analysis	154
	9.3.2	Coupling of wet chemistry and NMR methods	157
	9.3.3	Strategy in the analysis of lignocellulosics	157
	9.3.4	Current understanding of the structure of lignin	159
	9.3.5	Analysis of technical lignins	162
	9.3.6	Linkages between lignin and carbohydrates	163
Conclusions			166
Acknowledgment			166
References			166

10 Chemical Composition and Lignin Structural Features of Banana Plant Leaf Sheath and Rachis — 171
Lúcia Oliveira, Dmitry Victorovitch Evtuguin, Nereida Cordeiro, Armando Jorge Domingues Silvestre, and Artur Manuel Soares da Silva

Abstract			171
10.1	Introduction		171
10.2	Materials and methods		172
	10.2.1	Preparation of plant material	172

		10.2.2	Chemical analysis	172
		10.2.3	Isolation and characterization of lignins	173
	10.3	Result discussion		175
		10.3.1	Chemical composition of leaf sheath and rachis	175
		10.3.2	Non-(hemi)cellulosic polysaccharides	176
		10.3.3	Lignin characterization	176
	Conclusions			186
	References			186
11	Recent Advances in the Characterization of Lignosulfonates			189
	Stuart E. Lebo, Svein Magne Bråten, Guro Elise Fredheim, Bjart Frode Lutnaes, Rolf Andreas Lauten, Bernt O. Myrvold, and Timothy John McNally			
	Abstract			189
	11.1	Molecular weight determination of lignosulfonates by size exclusion chromatography and multi-angle laser light scattering		189
		11.1.1	Background	189
		11.1.2	Experimental and materials	190
		11.1.3	Results and discussion	190
	11.2	Structure of lignosulfonates by two-dimensional nuclear magnetic resonance spectroscopy		192
		11.2.1	Introduction	192
		11.2.2	NMR of lignosulfonates	193
		11.2.3	Sample preparation	194
		11.2.4	Equipment	194
		11.2.5	Results and discussion	194
	11.3	Hydrophobicity of lignosulfonates by HIC		197
		11.3.1	Background	197
		11.3.2	Methods	197
		11.3.3	Results and discussion	197
	11.4	Elemental analysis of lignosulfonates by x-ray fluorescence spectroscopy		199
		11.4.1	Background	199
		11.4.2	Experimental and materials	201
		11.4.3	Results and discussion	201
	References			203
12	Integrated Size-Exclusion Chromatography (SEC) Analysis of Cellulose and its Derivatives			206
	Akira Isogai and Masahiro Yanagisawa			
	Abstract			206
	12.1	Introduction		206
	12.2	Background of MALS analysis		207
	12.3	Dissolution of cellulose in solvents for SEC analysis		209
	12.4	SEC-MALS analysis of celluloses		210
	12.5	Determination of dn/dc of cellulose in LiCl/amide solvents		212
	12.6	Determination of MM values of celluloses and pulps by SEC-MALS		212
	12.7	Conformation analysis of cellulose molecules in LiCl/DMI		214
	12.8	SEC-MALS analysis of cellulose in softwood kraft pulps using LiCl/DMI		216

12.9	SEC-PDA-MALS analysis of unbleached chemical pulps	218
12.10	SEC-MALS analysis of cellouronic acid and other water-soluble cellulose derivatives	218
12.11	SEC-MALS analysis of CMC used as wet-end additive in papermaking	222
12.12	SEC-MALS analysis of polymer-brush-type cellulose β-ketoesters	224
	References	224

13 ^{13}C CPMAS NMR Studies of Wood, Cellulose Fibers, and Derivatives   227
Sirkka Liisa Maunu

	Abstract	227
13.1	Introduction	227
13.2	Experimental	228
13.3	Special characteristics of ^{13}C CPMAS NMR	228
13.4	Applications	230
	13.4.1 Wood	230
	13.4.2 Cellulose fibers	235
	13.4.3 Derivatives	243
	Concluding remarks	246
	References	246

Part III   Characterization of Lignocellulose-Based Composites and Polymer Blends   249

14 Advances in the Characterization of Interfaces of Lignocellulosic Fiber Reinforced Composites   251
Sherly Annie Paul, Laly A. Pothan, and Sabu Thomas

	Abstract	251
14.1	Introduction	251
	14.1.1 Lignocellulosic fibers and their composites	251
	14.1.2 Interfaces of lignocellulosic fiber reinforced composites	255
14.2	Characterization of the interfaces of lignocellulosic fiber reinforced composites	258
	14.2.1 Micromechanical techniques	258
	14.2.2 Microscopic techniques	261
	14.2.3 Spectroscopic techniques	263
	14.2.4 Thermodynamic and other techniques	266
	Conclusion	269
	References	270

15 Thermal and Mechanical Analysis of Lignocellulose-Based Biocomposites   275
Hyoe Hatakeyama, Takashi Nanbo, and Tatsuko Hatakeyama

	Abstract	275
15.1	Introduction	275
15.2	Sample preparation and measurements	276
15.3	Thermal and mechanical properties of PU composites with wood meal as the bio-filler	277

15.4		Thermal and mechanical properties of PU composites with inorganic fillers	280
References			285

16 New Insights into the Mechanisms of Compatibilization in Wood–Plastic Composites — 288
Kaichang Li and John Ziqiang Lu

- Abstract — 288
- 16.1 Introduction — 288
- 16.2 Compatibilization mechanisms for WPCs without a compatibilizer — 289
- 16.3 Increasing the strength and stiffness of WPCs through strengthening and stiffening wood with wood adhesives — 290
- 16.4 Effect of hydrophobic interaction on the WPC strengths — 291
- 16.5 Increasing the strength and stiffness of WPCs with a compatibilizer — 292
- Conclusions — 296
- References — 297

17 X-Ray Powder Diffraction Analyses of Kraft Lignin-Based Thermoplastic Polymer Blends — 301
Yi-ru Chen and Simo Sarkanen

- Abstract — 301
- 17.1 Introduction — 301
  - 17.1.1 X-Ray diffraction patterns from lignin preparations — 301
  - 17.1.2 Amorphous polymeric materials — 302
- 17.2 Kraft lignin-based polymeric materials — 304
  - 17.2.1 X-Ray powder diffraction patterns from paucidisperse kraft lignin fractions — 304
  - 17.2.2 Alkylated kraft lignin-based thermoplastic blends with low-T_g polymers — 306
  - 17.2.3 X-Ray powder diffraction patterns from methylated kraft lignin-based plastics — 308
- Concluding remarks — 313
- Acknowledgments — 314
- References — 314

18 DSC and DMA of ECs and EC–MC Blends — 316
Shizuka Horita, Tatsuko Hatakeyama, and Hyoe Hatakeyama

- Abstract — 316
- 18.1 Introduction — 316
- 18.2 Experimental — 317
  - 18.2.1 Preparation of diurethane cross-linked EC films with various NCO/OH ratios — 317
  - 18.2.2 Preparation of films from EC–MC blends — 318
  - 18.2.3 Measurements — 318
- 18.3 Effect of inter-molecular hexamethylene diurethane linkage on molecular motion of EC — 320

18.4	Effect of MC blending on molecular motion of EC	325
References		327

19 DSC and AFM Studies of Chemically Cross-Linked Sodium Cellulose Sulfate Hydrogels — 329
Toru Onishi, Hyoe Hatakeyama, and Tatsuko Hatakeyama

Abstract		329
19.1	Introduction	329
19.2	Experimental	330
	19.2.1 Preparation of hydrogels from cross-linked NaCS	330
	19.2.2 Differential scanning calorimetry	330
	19.2.3 Atomic force microscopy	332
19.3	Phase diagram of NaCS–water and NaCS–PU–water systems	332
19.4	Atomic force micrographs of NaCS and NaCS–PU hydrogel	336
References		338

20 Microscopic Examination of Cellulose Whiskers and Their Nanocomposites — 340
Ingvild Kvien and Kristiina Oksman Niska

Abstract		340
20.1	Introduction	340
20.2	Characterization methods for nanostructures	342
20.3	Field emission scanning electron microscope	343
	20.3.1 Sample preparation	343
	20.3.2 The structure of CNW and nanocomposites	344
20.4	Transmission electron microscope	346
	20.4.1 Sample preparation	347
	20.4.2 The structure of CNW and nanocomposites	348
20.5	Atomic force microscope	350
	20.5.1 Sample preparation	351
	20.5.2 The structure of CNW and nanocomposites	352
Conclusions		354
Acknowledgments		354
References		354

Index — 357

Color plate section appears between pages 202 and 203

Preface

Lignocellulosic materials are a natural, abundant, and renewable resource essential to the functioning of industrial societies and critical to the development of a sustainable global economy. As wood and paper products, they have played an important role in the evolution of our civilization. Improvement of the qualities of such products and the efficiency of their manufacturing processes have often been hampered by the lack of understanding of the complex physical states, morphological features, and chemical compositions of the materials. Novel or improved methods for the characterization of lignocellulosic materials are needed.

As serious economic, sociopolitical, and environmental issues build up with the use of petrochemicals, lignocellulosic materials will be relied upon as the feedstock for the production of chemicals, fuels, and biocompatible materials. Significant progress has been made to use lignocellulosic materials as a feedstock for the production of fuel ethanol and as a reinforcing component in polymer composites. Effective and economical methods for such uses, however, remain to be developed, partly due to the difficulty encountered in the characterization of the structures of native lignocelluloses and lignocellulose-based materials.

This book was developed based on 8 presentations selected from the 2005 Pacifichem Symposium on Characterization, Photostabilization, and Usage of Lignocellulosic Materials, and 12 invited contributions from researchers renowned in the field of lignocellulosics' characterization and usage. It covers the recent advances in the characterization of wood, pulp fibers, and cellulose networks (papers). It also describes the analyses of native and modified lignocellulosic fibers and materials using advanced techniques such as time-of-flight secondary ion mass spectrometry, 2D heteronuclear single quantum correlation NMR, and Raman microscopy. Furthermore, it presents useful methods for the characterization of lignocellulose-reinforced composites and polymer blends.

It is anticipated that this book will provide references on the state-of-the-art characterization of lignocellulosic materials to both academic and industrial researchers who work in the fields of wood and paper, lignocellulose-based composites and polymer blends, and bio-based fuels and materials. It is also anticipated that this book will stimulate further efforts in the development of new processes and technologies to use lignocellulosic materials for the production of chemicals, fuels, and bio-based materials for years to come.

Acknowledgments

I would like to thank all the contributors of this book for their time, effort, and enthusiasm in writing their chapters; without their active participation and support, this book would not have been possible. More importantly, I would like to thank my wife, Xuan, for her patience and support during the preparation of this book and our two sons, Nicholas and Lucas, for being a source of entertainment and inspiration.

Dr. Thomas Q. Hu
FPInnovations – Paprican Division
Vancouver, British Columbia, Canada

Editor Biography

Dr. Thomas Q. Hu is a Principal Scientist at FPInnovations – Paprican Division and an Adjunct Professor in the Chemistry Department at the University of British Columbia (UBC). He received his B.Sc. (1983) in Polymer Science and Engineering from South China Institute of Technology, and his M.Sc. (1988) and Ph.D. (1993) in Synthetic Organic Chemistry from UBC. He joined Paprican in 1994 after a 2-year tenure at Paprican as an NSERC Canada Industrial Postdoctoral Fellow.

His area of expertise is in the application of advanced, next-generation chemistry to solve various long-standing technological problems in the pulp and paper industry. He has pioneered the work in the novel modification of lignin functional groups, the development of fiber-reactive radical scavengers for the photostabilization of lignocellulosic materials, and the bleaching of lignin-rich wood pulps with phosphorus-based chemicals. He has developed a number of novel processes for the bleaching and brightness stabilization of lignin-rich wood pulps. He has over 70 publications including one edited book on *Chemical Modification, Properties and Usage of Lignin*, five issued international patents and several pending US and Canadian patents. He has won a number of prestigious awards including the 2004 *Journal of Pulp and Paper Science* Best Paper Award and the 2005 Pulp and Paper Technical Association of Canada Douglas Atack Award for Best Mechanical Pulping Paper.

List of Contributors

Umesh Prasad Agarwal	United States Department of Agriculture, Forest Service, Forest Products Laboratory, Madison, WI, USA
Mikhail Yurievich Balakshin	Department of Wood and Paper Science, North Carolina State University, Raleigh, NC, USA
Svein Magne Bråten	Borregaard Lignotech Corporate R&D, 1701 Sarpsborg, Norway
Ewellyn Augsten Capanema	Department of Wood and Paper Science, North Carolina State University, Raleigh, NC, USA
Richard P. Chandra	Department of Wood Science, University of British Columbia, Forest Sciences Centre, Vancouver, Canada
Hou-min Chang	Department of Wood and Paper Science, North Carolina State University, Raleigh, NC, USA
Yi-ru Chen	Department of Bioproducts and Biosystems Engineering, University of Minnesota, St. Paul, MN, USA
Nereida Cordeiro	CEM and Department of Chemistry, University of Madeira, 9000-390 Funchal, Portugal
Ali R. Esteghlalian	Diversa Corporation, San Diego, CA, USA
Dmitry Victorovitch Evtuguin	CICECO and Department of Chemistry, University of Aveiro, 3810-193 Aveiro, Portugal
Pedro Fardim	Department of Chemical Engineering, Åbo Akademi University, FI-20500 Turku/Åbo, Finland
Guro Elise Fredheim	Borregaard Lignotech Corporate R&D, 1701 Sarpsborg, Norway
Göran Gellerstedt	Department of Fibre and Polymer Technology, KTH Royal Institute of Technology, SE-100 44 Stockholm, Sweden

Wadood Y. Hamad	FPInnovations – Paprican Division, Vancouver, BC, Canada
Hyoe Hatakeyama	Graduate School of Engineering, Fukui University of Technology, Fukui, Japan
Tatsuko Hatakeyama	Lignocellulose Research, Bunkyo-ku, Tokyo, Japan
Bjarne Holmbom	Department of Chemical Engineering, Åbo Akademi University, FI-20500 Turku/Åbo, Finland
Shizuka Horita	Graduate School of Engineering, Fukui University of Technology, Fukui, Japan
Akira Isogai	Department of Biomaterial Science, Graduate School of Agricultural and Life Science, The University of Tokyo, Tokyo, Japan
Krista Koljonen	VTT Technical Research Centre of Finland, Pulp and Furnish Chemistry, VTT, Finland
Ingvild Kvien	Engineering Design and Materials, Norwegian University of Science and Technology, Trondheim, Norway
Rolf Andreas Lauten	Borregaard Lignotech Corporate R&D, 1701 Sarpsborg, Norway
Stuart E. Lebo	Lignotech USA, Rothschild, WI, USA
Kaichang Li	Department of Wood Science and Engineering, Oregon State University, Corvallis, OR, USA
John Ziqiang Lu	Department of Wood Science and Engineering, Oregon State University, Corvallis, OR, USA
Bjart Frode Lutnaes	Department of Chemistry, Norwegian University of Science and Technology, 7491 Trondheim, Norway
Sirkka Liisa Maunu	Laboratory of Polymer Chemistry, University of Helsinki, Finland
Timothy John McNally	Lignotech USA, Rothschild, WI, USA
Bernt O. Myrvold	Borregaard Lignotech Corporate R&D, 1701 Sarpsborg, Norway
Takashi Nanbo	Graduate School of Engineering, Fukui University of Technology, Fukui, Japan
Kristiina Oksman Niska	Engineering Design and Materials, Norwegian University of Science and Technology, Trondheim, Norway and Division of Manufacturing and Design of Wood and Bionanocomposites, Luleå University of Technology, Skellefteå, Sweden

Lúcia Oliveira	CEM and Department of Chemistry, University of Madeira, 9000-390 Funchal, Portugal
Toru Onishi	Graduate School of Engineering, Fukui University of Technology, Fukui, Japan
Sherly Annie Paul	School of Chemical Sciences, Mahatma Gandhi University, Kottayam, Kerala, India
Laly A. Pothan	Bishop Moore College, Mavelikara, Kerala, India
John N. Saddler	Department of Wood Science, University of British Columbia, Forest Sciences Centre, Vancouver, Canada
Simo Sarkanen	Department of Bioproducts and Biosystems Engineering, University of Minnesota, St. Paul, MN, USA
Rauni Seppänen	YKI, Institute for Surface Chemistry, SE-114 86 Stockholm, Sweden
Artur Manuel Soares da Silva	CICECO and Department of Chemistry, University of Aveiro, 3810-193 Aveiro, Portugal
Armando Jorge Domingues Silvestre	CICECO and Department of Chemistry, University of Aveiro, 3810-193 Aveiro, Portugal
Per Stenius	Laboratory of Forest Products Chemistry, Helsinki University of Technology, TKK, Finland
Sabu Thomas	School of Chemical Sciences, Mahatma Gandhi University, Kottayam, Kerala, India
Masahiro Yanagisawa	Department of Biomaterial Science, Graduate School of Agricultural and Life Science, The University of Tokyo, Tokyo, Japan
Liming Zhang	Department of Fibre and Polymer Technology, KTH Royal Institute of Technology, SE-100 44 Stockholm, Sweden

Part I
Novel or Improved Methods for the Characterization of Wood, Pulp Fibers, and Paper

Chapter 1
2D Heteronuclear (^{1}H–^{13}C) Single Quantum Correlation (HSQC) NMR Analysis of Norway Spruce Bark Components

Liming Zhang and Göran Gellerstedt

Abstract

Norway spruce (*Picea abies*) bark, collected during the winter and summer seasons, has been separated into inner- and outer-bark fractions followed by extraction using a five-solvent extraction procedure. The inner-bark fraction was found to contain a rather high amount of carbohydrates but only minor amounts of tannin. In addition, the glucosides of stilbenes such as astringin and isorhapontin were detected. The outer-bark fraction had a high content of high-molecular-mass tannin but was low in carbohydrates. Lignin-derived components were also detected. Further studies of the tannin fraction using advanced nuclear magnetic resonance (NMR) techniques showed that polyflavanol and polystilbene units together with glucoside units were prevalent.

1.1 Introduction

Bark comprises ~9–15% of the wood log volume depending on species. Today, large amounts of bark materials are produced as a waste product from the pulp and paper industry as well as from the wood-processing industry. These bark materials are mostly burned as low-valued fuel. Alternative uses of bark components have been investigated but, so far, without any major progress. Thus, the use of tannin from wattle as a substitute for phenols in adhesives formulations was suggested a long time ago [1]. More recently, however, the antioxidant and cancer chemoprevention activity of various tannins has attracted a considerable interest as many natural phenolic and polyphenolic compounds are considered beneficial for human health [2–8]. It has also been noted that stilbenes found in spruce bark may act as inhibitors of certain pathogens [9] and that such compounds could have a potential as precursors for medicinally interesting substances [10], act as strong antioxidants [11], or have cancer chemopreventive activity [12]. The detailed chemical composition of bark from various wood species is, however, rather poorly understood, and before any major advances in the use of bark components can be realized, a thorough group/compound separation and identification must be done followed by further characterization work on, for example, medicinal properties.

Previous studies have shown that 24% of the spruce bark material is soluble in, and can be extracted by, organic solvents and water [13]. The holocellulose content in spruce bark was found to be ~37%, and the major extractable component was a polyphenolic tannin believed to be of the polyflavonoid type with a broad-molecular-mass distribution [14]. Some lignin-like substances were also assumed to be present in the bark, but these could not be identified.

The chemical composition of the water-soluble material from spruce bark has been studied in detail. Stilbene glucosides, glucose, fructose, sucrose, and soluble tannins were found to be the main components [15]. Furthermore, the complex mixture of lipophilic extractives present in spruce bark has been characterized and found to contain fatty acids, resin acids, fatty alcohols, β-sitosterol, and terpenoids [16].

In the present work, inner and outer bark from spruce (*P. abies*) has been separated and the overall chemical composition of each type of bark has been determined. In addition, the dominant material present, a broad range of tannins, has been further characterized using modern nuclear magnetic resonance (NMR) techniques.

1.2 Experimental

1.2.1 Materials

Winter (January) and summer (June) spruce (*P. abies*) bark materials were obtained from 60- to 100-year-old trees in the Stockholm area. The inner- and outer-bark fractions were separated by hand cutting. The bark materials were air-dried in the dark at room temperature (~24°C) for 2 weeks. After that, the dried bark was ground in a Wiley mill to a particle size of 40 mesh. β-Glucosidase and catechin were purchased from Sigma-Aldrich Sweden AB, Stockholm.

1.2.2 Solvent extraction

Each bark sample (10 g) was successively extracted with petroleum ether (bp 40–60°C), methylene chloride, acetone, water, and acetone–water (2:1 v/v). All extractions were carried out with 100 mL of the solvent at room temperature overnight. After each extraction, the liquid phase was collected and the residue was subjected to an identical second extraction with the same solvent before the next solvent in the series was used. After the acetone extraction, however, the bark residue was allowed to dry before being extracted with water. The dissolved material from each solvent extraction was isolated by evaporation of the combined solvent of the two identical extractions under reduced pressure at 40°C. When water was present, the concentrated solution was freeze-dried.

1.2.3 Purification of the condensed tannin

After freeze-drying, the residue from the acetone–water (2:1 v/v) extraction that contained the condensed tannin was sequentially extracted with methylene chloride and acetone again, in the same way as mentioned above, to eliminate any residual lipophilic and low-molecular-mass tannin and stilbene fractions. The purified condensed tannin was air-dried.

1.2.4 Milled bark tannin–lignin

The insoluble residue left behind after a complete five-solvent extraction sequence was air-dried and then subjected to ball milling for 48 h. Subsequently, the milled powder was extracted with acetone–water (2:1 v/v) twice overnight (cf. Reference 17) to give an acetone–water-soluble tannin–lignin material denoted as milled bark tannin–lignin (MBTL). The yield of MBTL was ~71% based on the "Klason lignin" content of the insoluble residue.

1.2.5 Analyses

Klason lignin analysis was conducted on the insoluble residue left behind after a complete five-solvent extraction sequence according to the TAPPI Test Method [T222 om83]. Carbohydrate analysis was carried out using acid hydrolysis and gas chromatography of alditol acetates [18].

All NMR analyses were made at 24°C on a Bruker Avance 400 MHz instrument. One-dimensional (1D) quantitative ^{13}C NMR experiments were recorded with a 5-mm broadband probe and with an inverse-gated proton-decoupling sequence. A pulse angle of 90° and a delay time of 12 s between pulses were applied during data acquisition. Two-dimensional (2D) NMR experiments were performed with an inverse proton/carbon selective probe equipped with a z-gradient coil. Standard Bruker pulse programs were applied in all experiments. Chemical shift values were referenced to the TMS signal. DMSO-d_6 was used as the solvent in most cases with some experiments performed with acetone-d_6/D$_2$O (2:1). 2D heteronuclear (^{1}H–^{13}C) single quantum correlation (HSQC) and HSQC–total correlation spectroscopy (TOCSY) NMR spectra were acquired with a spectral window of 12.8 ppm in F2 and 150 ppm in F1 with increments 2K × 0.5K, giving an acquisition time of 0.2 s in F2. The spectral center was set at 5.3 ppm in the ^{1}H and at 70 ppm in the ^{13}C dimension. Typical parameters for data acquisition included a delay time of 1 s, an average coupling constant of 150 Hz, and 128 scans per increment. For HSQC–TOCSY experiments, a mixing time of 70 ms was applied. The acquired 2D NMR data sets were processed with 2K × 1K data points using the $\pi/2$-shifted sine-bell window function in both dimensions.

High-performance liquid chromatography (HPLC) was performed on a Waters system. A Nuckeosil 5, C18 column was used. The mobile phase was water–acetonitrile, with a linear gradient from 0% to 70% acetonitrile during 50 min. Gas chromatography–mass spectrometry (GC–MS) was performed on a Finnigan TRACE GC–MS, 2000 series system. The MS ionization method used was electron impact operating at 70 eV.

1.3 Results and discussion

1.3.1 Solvent extractions and separation of bark components

There is no simple separation method or any definite distinction between inner- and outer-bark sections in spruce. Therefore, the inner soft and light brown material was separated by hand cutting from the darker hard material located on the outer part of the bark.

Table 1.1 Yields (% of the original dry bark) of the extracts and the residues from the spruce bark, collected during winter and summer times, respectively, after successive extractions with five different solvents.

Solvent	Winter bark		Summer bark		Major components
	Inner	Outer	Inner	Outer	
Petroleum ether	4.0	3.1	1.8	5.2	Resin acids, fatty acids, terpenoids
CH_2Cl_2	1.2	1.4	0.6	2.6	Resin acids, fatty acids, terpenoids
Acetone	15.1	7.0	17.6	7.9	Stilbene glucosides, tannin
Water	12.3		10.0		Glucose, fructose, sucrose
		7.0		6.8	Glucose, fructose, sucrose, tannin
Acetone–water (2:1)	1.7	8.0	1.9	7.9	Condensed tannin
Residue	65.7		68.1		"Cellulose"
		73.5		69.6	Polysaccharides, polymeric tannin–lignin (MBTL)

The mass ratio between the inner- and outer-bark materials was found to range from 1:1 to 1:2, depending on the tree age and the bark location on the tree. Bark collected from the bottom of an old tree seemed to contain more outer-bark mass.

All bark samples, viz., inner and outer bark collected during winter and summer, respectively, were successively extracted with five different solvents and the yields of the extracts and the insoluble residues are listed in Table 1.1. The compositions of the winter and summer samples were found to be similar, but with only minor differences in amounts. The extractions with petroleum ether and methylene chloride produced lipophilic extractives that accounted for ~2.4–7.8% of the original bark dry weight. Similar results have been obtained before, and most of the individual components have been identified [16].

The inner bark contained higher amounts of acetone extractable material (15.1–17.6%) than the outer bark (7.0–7.9%). Predominant components in these extracts were found to be a mixture of stilbene glucosides and a low-molecular-mass tannin fraction. Both of these structural types may act as precursors to the higher-molecular-mass material present in bark, denoted as condensed tannin. Detailed analysis of the low-molecular-mass tannin is of particular importance as condensed tannin is much more difficult to study by NMR because of the presence of rotamers [19] and the fast signal transverse relaxation.

The extraction with water at room temperature resulted in mixtures of simple sugars such as glucose, fructose, and sucrose in amounts of ~10–12% and 7% from the inner and outer bark, respectively. In addition, small amounts of tannin were obtained from the outer-bark extractions. In agreement with these results, it has been shown recently that glucose, fructose, sucrose, and stilbene glucosides are the predominant components present in the effluents from the debarking operation in pulp mills [15]. In the present work, the stilbene glucoside fraction was, however, found in the acetone extract.

After the extraction with water, a condensed tannin fraction could be obtained by extraction with acetone–water (2:1) with the outer bark containing a much higher amount than the inner bark (~8% vs <2%). Altogether, ~30% of the material could be dissolved by the series of extractions with ~70% left as an insoluble residue. After air drying and ball

Table 1.2 Klason lignin and carbohydrate analysis of the insoluble residues (% of the original dry bark) from the summer bark after successive extractions with five different solvents.

Component	Inner bark	Outer bark
Klason lignin (tannin–lignin)	3.1	34.8
Glucose	40.9	17.4
Galactose	2.9	1.8
Mannose	1.8	2.9
Arabinose	6.7	3.3
Xylose	3.3	2.6
Rhamnose	0.7	0.3
Total	59.4	63.1
Theoretical	68.1	69.6

milling, all the residues were extracted with acetone–water (2:1), resulting in the dissolution and isolation of a large amount of a tannin–lignin material (denoted as MBTL) from the outer-bark residue. The inner-bark residue, on the other hand, gave only a small amount of such a tannin–lignin material with the predominant component being glucose (after acid hydrolysis) (Table 1.2).

Furthermore, in both the inner and the outer bark, glucose was found to be the dominant sugar with only small amounts of other neutral sugars. The tannin–lignin material behaved like Klason lignin and could be analyzed as a solid residue after acid hydrolysis, whereas the amount of "acid-soluble tannin–lignin" was negligible. As the total recovery of material after the acid hydrolysis (59.4% and 63.1%) was somewhat lower than theoretical (68.1% and 69.6%) (Table 1.2), some other acid-soluble materials such as uronic acids may also be present in the residues. The major difference in composition between the inner- and outer-bark residues after the five-stage extraction can thus be summarized as a strong decrease of polysaccharides and an increase in tannin when going from the inner to the outer bark. Stilbene glucosides are, however, much more prevalent in the inner bark.

1.3.2 Stilbenes and tannins

The stilbene glucosides astringin **1** and isorhapontin **2** were identified as the predominant constituents present in the acetone extract. The presence of a small amount of piceid **3** could also be observed (Figure 1.1). After HPLC separation (Figure 1.2), the relative quantity of the compounds **1–3** was estimated as 7:2:0.7 with λ_{max}-values of 325, 325, and 319 nm, respectively. By treatment with β-glucosidase in aqueous solution, the stilbene glucosides could be hydrolyzed, releasing the free stilbenes, which subsequently could be identified by GC-MS analysis. The stilbene glucosides **1–3** have been isolated before from spruce bark in yields between 1% and 6% of the bark dry weight [20]. The biological activity of such stilbenes has also been widely studied [12,21,22].

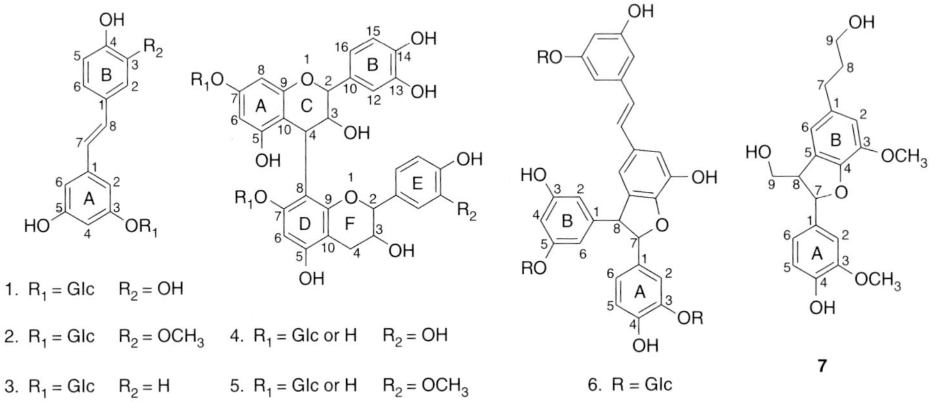

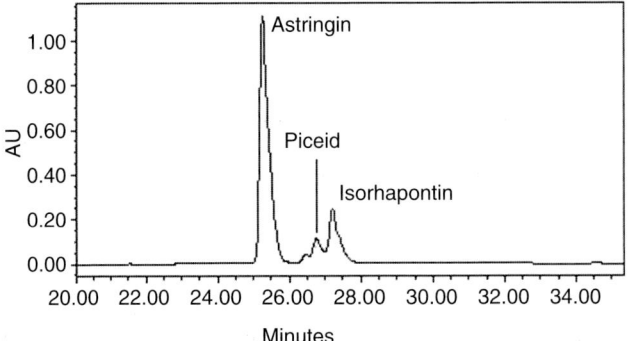

Figure 1.1 Structures identified in the acetone extract.

Figure 1.2 HPLC separation of stilbene glucosides present in the acetone extract.

In addition to the stilbene glucosides **1–3**, the presence of structures **4–7** (Figure 1.1) in the acetone extract could also be identified by applying a 2D HSQC NMR analysis on the whole mixture (Figure 1.3). Detailed ^{1}H and ^{13}C NMR shift identification and assignments for the 2D HSQC NMR spectrum in Figure 1.3 are listed in Table 1.3, which also contains HSQC–TOCSY correlations (see the following discussion). On the basis of these assignments, the origins of some of the signals found in the similar spectra of the condensed tannin and MBTL were identified (see next section).

1.3.3 Identification by NMR

Compound **4** had a typical B-type proanthocyanidin structure with a 4–8 interflavonoid linkage and with a catechin end unit [23]. Model compound studies performed in the present work showed that the catechin structure in DMSO-d_6 produced 2D HSQC NMR signals at δ values (ppm) 2.36, 2.66/27.7 (**4**, F4), 3.82/66.2 (**4**, F3), and 4.49/80.8 (**4**, F2), in agreement with literature data [24]. All these signals were also found in the 2D HSQC NMR

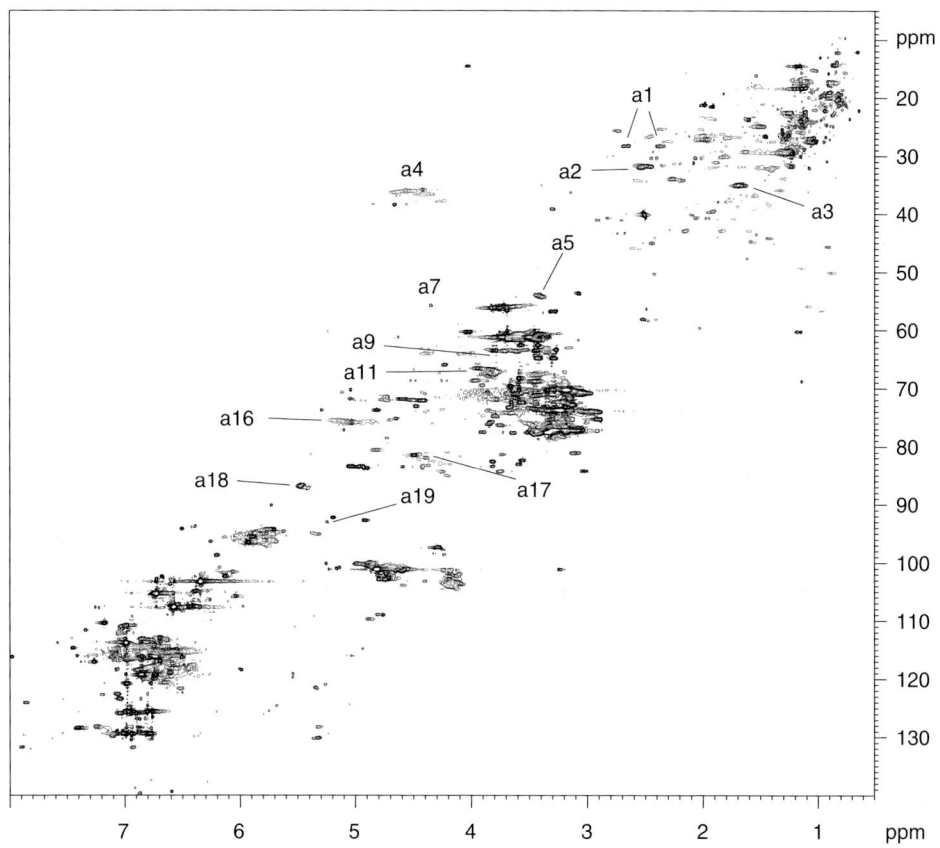

Figure 1.3 2D HSQC NMR spectrum of the winter-outer-bark acetone extract (in DMSO-d$_6$). For a complete signal assignment (see Table 1.3).

spectrum of the acetone-extracted tannin from the winter outer bark (Figure 1.3). Results from the 2D HSQC–TOCSY NMR experiment proved that all these NMR signals were correlated to each other, confirming the presence of catechin end units in the acetone-extracted tannin. No NMR signals originating from epicatechin (epimer of catechin; catechin and epicatechin have 2,3-*trans* and 2,3-*cis* stereochemistries, respectively) end units [24] could be observed, however, indicating that epicatechin end structures were either not present or present only with a very low abundance. The origin of the catechin end units can either be as a part of flavonoid oligomers or as catechin itself. It has also been shown before that various flavonoids isolated from spruce root bark all carry a catechin end unit [25].

The C4 carbon in compound **4** can give ^{13}C NMR signals at two different chemical shift values depending on the stereochemistry at the 2 and 3 positions of ring C. The 2,3-*cis* units will give a ^{13}C signal for C4 at ∼36.5 ppm (in acetone-d$_6$) [26] and the 2,3-*trans* units a ^{13}C signal for C4 at ∼38.0 ppm (in acetone-d$_6$) [27,28]. The 2D HSQC NMR analysis showed that in the winter-outer-bark acetone extract, the 2,3-*cis* units were more abundant. In DMSO-d$_6$, the 2,3-*cis* units present in the acetone extract were found to give a C4 signal

Table 1.3 Assignments for cross-peaks in the 2D HSQC NMR spectrum of the stilbene glucosides and tannin obtained from the acetone extract of the outer bark (^{1}H and ^{13}C δ-values in DMSO-d_6).

Symbol	^{1}H	^{13}C	HSQC–TOCSY correlations	Assignment	References
a1	2.36–2.66	27.7	a11–a17	**4, 5**-F4	24
a2	2.48	31.2	a3–a10	7-B7	32
a3	1.66	34.4	a2–a10	7-B8	32
a4	4.29–4.69	35.1–35.9		**4, 5**-C4	27
a5	3.40	53.4	a9–a18	7-A8	33
a6	3.62–3.82	55.2–55.6		2, 5, 7-OCH$_3$	33
a7	4.34	55.3	a19	6-A8	30,31
a8	3.48–3.72	60.7		Glc-6	35
a9	3.60–3.70	62.9		7-A9	33
a10	3.40	60.0		7-B9	32
a11	3.82	66.2	a1–a17	**4, 5**-F3	24
a12	3.16	69.8		Glc-4	35
a13	3.21	73.4		Glc-2	35
a14	3.27–3.31	76.8–77.1		Glc-3,5	35
a15	3.59–4.00	70.1–71.2		**4, 5**-C3	27
a16	4.79–5.19	75.3		**4, 5**-C2	27
a17	4.49	80.8	a1–a11	**4, 5**-F2	24
a18	5.45	86.4		7-A7	33
a19	5.24	92.6	a7	6-A7	30,31
a20	5.87–5.91	95.1–96.1		**4, 5**-A6; A8	27
a21	4.80	100.7		Glc-1	35
a22	6.32	102.8		1-A4	35
a24	6.70	104.9		1-A2	35
a23	6.55	107.2		1-A6	35
a25	6.98	113.3		1-B2	35
a26	6.73–6.69	114.4–115.0		**4, 5**-B12; B15	15,27
a27	6.72	115.5	a29	1-B5	35
a28	6.60	118.3		**4, 5**-B16	15,27
a29	6.85	118.6	a27	1-B6	35
a30	6.77	125.0	a31	1-7	35
a31	6.96	128.8	a30	1-8	35

at ~35.5 ppm (symbol a4 in Table 1.3). A much smaller signal at 37.6 ppm (Figure 1.3) could be assigned to the 2,3-*trans* units. In the 2D HSQC NMR spectrum of the condensed tannin sample dissolved in acetone-d_6/D$_2$O (Figure 1.4), the C4 signal was mainly observed at 36.3 ppm. Therefore, the internal flavonoid units in spruce tannin should be present mainly in the 2,3-*cis* (epicatechin) configuration, in contrast to the end catechin units, which have been observed to be mostly in the 2,3-*trans* form. In a previous study, the 2,3-*cis* monomer units were found to be dominant (70%) in polyflavonoids isolated from spruce root bark [25]. However, the HSQC NMR signals corresponding to the internal 2,3-*trans* units could still be observed.

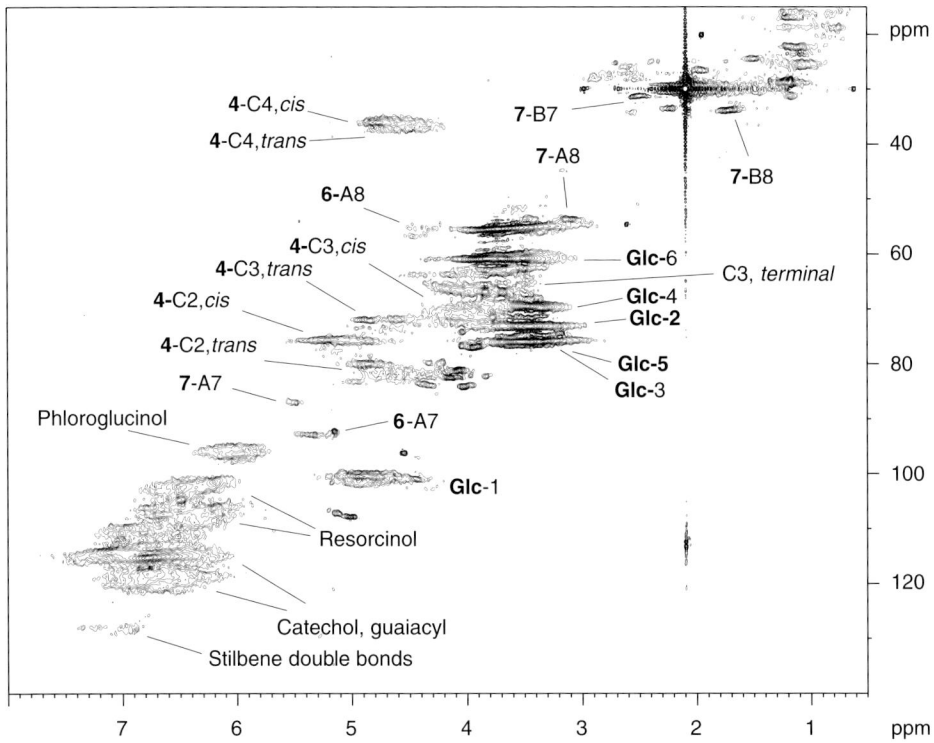

Figure 1.4 2D HSQC NMR spectrum of the condensed tannin found in the acetone–water (2:1) extract (in 2:1 acetone-d_6–D_2O).

Structure **5** represents tannin units containing a methoxy group. In the condensed tannin fractions isolated from both the outer and inner bark, the methoxy group content was found to be ~30% of the flavonoid units based on a quantitative ^{13}C NMR estimation (see later section). Part of the methoxy groups may be attributed to the presence of lignin or lignan contaminations in the tannin preparation, such as structure **7**. The actual ratio between catechol and guaiacyl moieties in spruce tannin can be estimated to be similar to that observed with the stilbene glucosides, that is, 7:2. Guaiacyl-containing flavonoids of the structure type **5** have previously been isolated from spruce root bark [25].

Traditionally, the condensed tannin from spruce bark has been regarded as having a pure polyflavonoid structure with interlinkages between the C4 and C8 (shown as linkage between the carbon 4 in ring C and the carbon 8 in ring D of structures **4** and **5** in Figure 1.1) or C4 and C6 carbon atoms [25,29]. Liquid chromatography on Sephadex LH-20 has been used as a key step in the isolation of pure tannin. In most cases, however, the yield of purified tannin has not been reported, but it has been observed that Sephadex LH-20 has a strong affinity for tannin [25]. In the present work, the condensed tannin was isolated by solvent extraction at room temperature in a yield of ~8% from the outer bark (Table 1.1).

Structure **6** is a dehydrodimer of a stilbene structure. The HSQC NMR signals at 5.24/92.6 and 4.34/55.3 (symbols a19 and a7 in Figure 1.3, respectively) were assigned to the side-chain

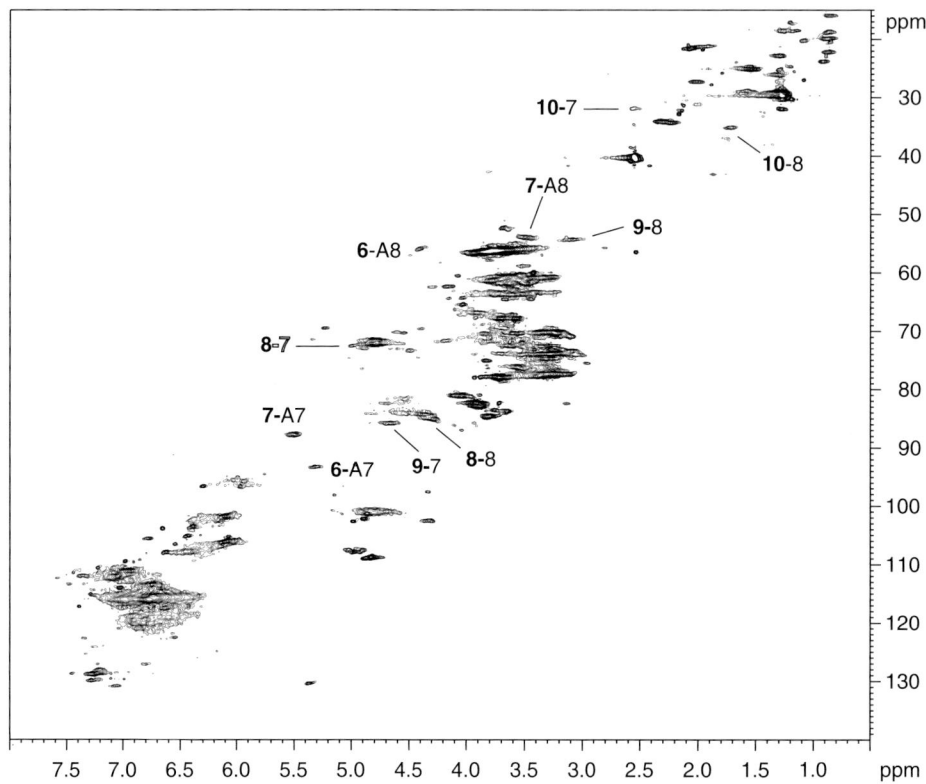

Figure 1.5 2D HSQC NMR spectrum (in DMSO-d_6) of the milled bark tannin–lignin (MBTL) sample from the winter outer bark. Structures **8**, **9**, and **10** referred to in the figure are shown in Figure 1.6.

CH-groups C7 and C8, respectively, based on NMR data from similar type of structures [30,31]. Results obtained from 2D HSQC–TOCSY analysis showed that the two ^{1}H NMR signals belonged to the same molecule and were coupled to each other. In the present work, the structural type **6** was found both in the condensed tannin and the MBTL samples, as revealed by their 2D HSQC NMR spectra (shown as signals **6**-A7 and **6**-A8) (Figures 1.4 and 1.5).

Compound **7** is related to lignin and is a dimer of coniferyl alcohol. This structure was found to be present in the acetone extract as well as being part of the condensed tannin. No other lignin-related structures were found in this tannin. The MBTL sample, on the other hand, contained not only compounds **6** and **7** but also significant amounts of other lignin structures including linkages of the β-O-4 and β–β types (structures **8** and **9**, respectively, Figure 1.6). Furthermore, dihydroconiferyl alcohol structures (**10**) were detected [32,33]. The MBTL sample had a methoxy content more than three times as high as that of the condensed tannin sample. Therefore, these results clearly demonstrate that typical lignin structures can coexist with condensed tannin in spruce outer bark.

In the quantitative ^{13}C NMR analysis of the condensed tannin (Figure 1.7), the NMR signal integration was referenced to the total, mostly aromatic region between 93 and 163 ppm.

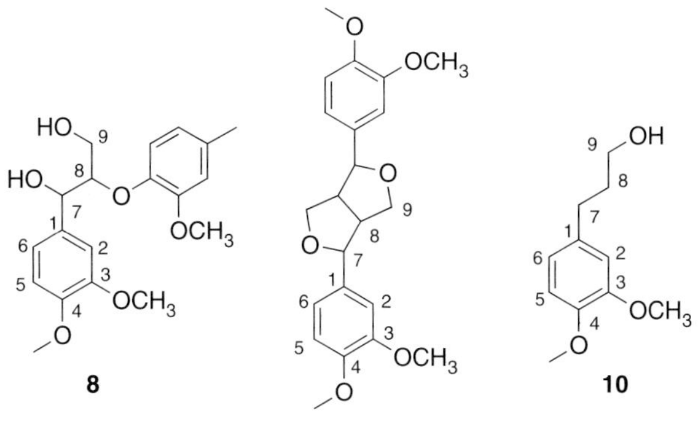

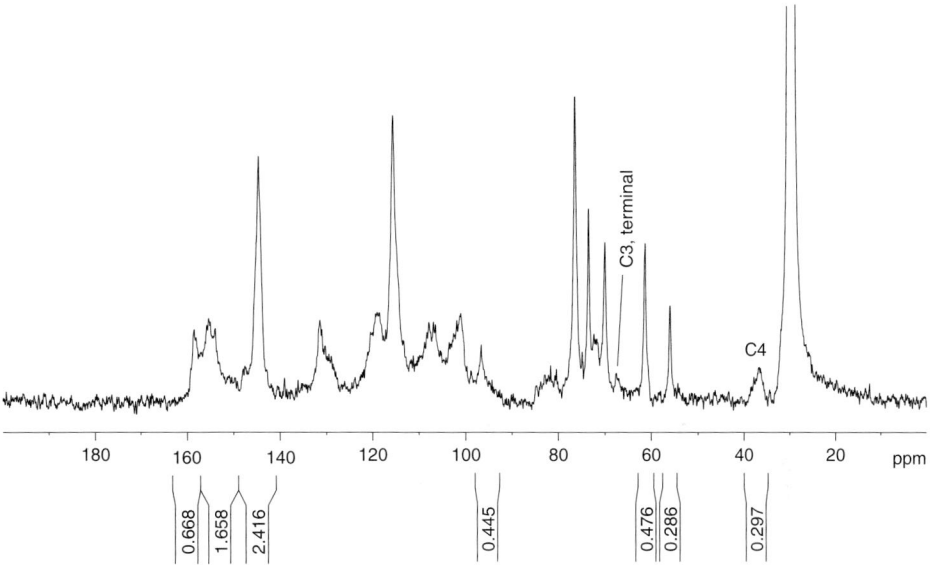

Figure 1.6 Types of lignin structures identified in the milled bark tannin–lignin (MBTL) sample.

Figure 1.7 Quantitative ^{13}C NMR spectrum of the acetone–water (2:1) extracted condensed tannin from the winter outer bark.

This range was given a total value of 13 carbon atoms based on the following calculations. Each flavanol monomer containing a total of 15 carbon atoms contributes 12 aromatic carbons. The presence of glucose units was found to contribute ~0.48 carbons to this mostly aromatic region through the anomeric carbon located at 101 ppm. Other nonaromatic carbons such as the double bond in stilbenes (~125–129 ppm) and the carbon C7 (~93 ppm) in structural units belonging to compound **6** (Figure 1.1) were estimated to contribute less than 0.5 carbons to this mostly aromatic region. Thus, the polyflavanol type of structure with C4–C8 and C4–C6 linkages is estimated to account for only ~30% of the total units

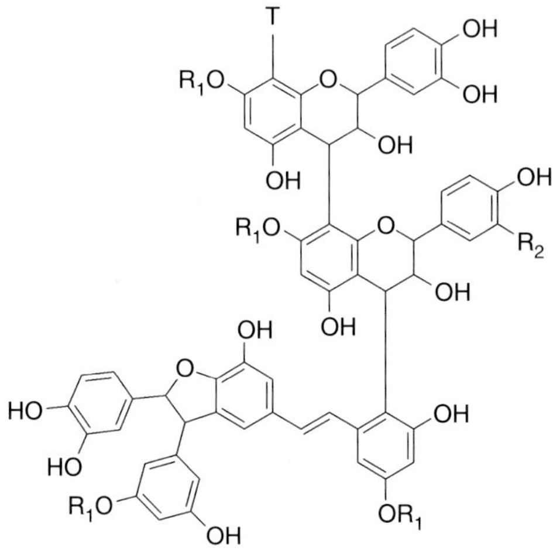

Figure 1.8 Proposed chemical structure of a segment of the condensed tannin obtained in the acetone–water (2:1) extract. R_1 = Glucose or H, R_2 = OH or OCH_3, T = tannin.

in the condensed tannin, because the C4 signal at 35–40 ppm has an integral value of 0.3. The methoxy group content is estimated to be ~29% of the flavonoid units according to the integral value of OCH_3 carbons at 56 ppm. The content of glucose can be estimated by integration of the C6 carbon at 60.7 ppm. This signal gave an integration value of 0.48, demonstrating that every second flavonoid monomer in the condensed tannin contained a glucose unit. Stilbene glucosides as well as structures derived from compound **6** were found to be one of the major components in the condensed tannin fraction.

On the basis of the carbon shifts for model phenols, such as resorcinol, the aromatic HSQC NMR signals between the ^{13}C shift values of 102 and 109 ppm were assigned to the resorcinol units present in the stilbene moieties. In the ^{13}C NMR spectrum, a resorcinol unit shows two aromatic carbons at 157–161 ppm, whereas a phloroglucinol unit usually gives three aromatic carbons between 153 and 157 ppm. Thus, by signal integration over these two regions, it could be estimated that the flavonoid units accounted for ~53% of the monomers in the condensed tannin, whereas stilbene-derived structures accounted for ~33%. The remaining aromatic units may originate from lignin-like structures. It has been observed before that stilbene moieties of the resorcinol type can participate in condensation reactions leading to the formation of condensed tannin [34]. A proposed chemical structure for spruce condensed tannin is shown in Figure 1.8.

Conclusions

By a series of extractions employing solvents of different polarities, a complete separation of spruce bark components, originating from inner and outer bark, could be achieved. It was found that, in the outer bark, a series of tannin glucosides having different solubility characteristics was predominant, whereas the inner bark mostly contained stilbene glucosides, monomeric sugars, and a large glucan residue, possibly with a cellulose structure. The differences between bark collected during winter and summer time were small.

Various 1D and 2D NMR techniques were used to obtain detailed structural information about the complex chemical structure of the tannins. Thereby, it could be shown that condensed tannin in spruce is predominantly built up by flavonoid (~53%) and stilbene (~33%) units and with a high degree of glucose substitution. A minor amount of lignin-related structures of the β-5 type was also identified. In the major, highest-molecular-mass tannin fraction, a high degree of copolymerization with lignin-derived structures was found.

Acknowledgments

This work is part of the Swedish–Finnish project "Value-added products from barks of Nordic wood species using bioconversion and chemical technology (WoodBiocon)". Thanks are due to Vinnova, the Swedish Governmental Agency for Innovation Systems, for financial support to one of us (LZ).

References

1. Pizzi, A. and Scharfetter, H.J. (1978) The chemistry and development of tannin-based adhesives for exterior plywood. *J. Appl. Polym. Sci.* 22:1745.
2. Sakagami, H., Kuribayashi, N., Iida, M. *et al.* (1995) Induction of DNA fragmentation by tannin- and lignin-related substances. *Anticancer Res.* 15:2121–2128.
3. Hagerman, A.E., Riedl, K.M., Jones, G.A. *et al.* (1998) High molecular weight plant polyphenolics (tannins) as biological antioxidants. *J. Agric. Food Chem.* 46:1887–1892.
4. Gil, M.I., Tomas-Barberan, F.A., Hess-Pierce, B., Holcroft, D.M., and Kader, A.A. (2000) Antioxidant activity of pomegranate juice and its relationship with phenolic composition and processing. *J. Agric. Food Chem.* 48:4581–4589.
5. Quideau, S., Jourdes, M., Lefeuvre, D. *et al.* (2005) The chemistry of wine polyphenolic C-glycosidic ellagitannins targeting human topoisomerase II. *Chem. Eur. J.* 11:6503–6513.
6. Afaq, F., Saleem, M., Krueger, C.G., Reed, J.D., and Mukhtar, H. (2005) Anthocyanin- and hydrolyzable tannin-rich pomegranate fruit extract modulates MAPK and NF-κB pathways and inhibits skin tumorigenesis in CD-1 mice. *Int. J. Cancer* 113:423–433.
7. Malik, A., Afaq, F., Sarfaraz, S., Adhami, V.M., Syed, D.N., and Mukhtar, H. (2005) Pomegranate fruit juice for chemoprevention and chemotherapy of prostate cancer. *PNAS* 102:14813–14818.
8. Raghuraman, A., Tiwari, V., Thakkar, J.N. *et al.* (2005) Structural characterization of a serendipitously discovered bioactive macromolecule, lignin sulfate. *Biomacromolecules* 6:2822–2832.
9. von Alcubilla, M., Diaz-Palacio, M.P., Kreutzer, K., Laatsch, W., Rehfuess, R.E., and Denzel, G. (1971) Relation between nutrition and heart rot attack of Norway spruce (*Picea abies*) and the fungistatic effect of its inner bark. *European J. For. Path.* 1:100.
10. Mannila, E., Talvitie, A., and Kolehmainen, E. (1993) Anti-leukaemic compounds derived from stilbenes in *Picea abies* bark. *Phytochemistry* 33:813–816.
11. Merillon, J.-M., Fauconneau, B., Teguo, P.W., Barrier, L., Vercauteren, J., and Huguet, F. (1997) Antioxidant activity of the stilbene astringin, newly extracted from *Vitis vinifera* cell cultures. *Clin. Chem.* 43:1092–1093.
12. Waffo-Teguo, P., Hawthorne, M.E., Cuendet, M. *et al.* (2001) Potential cancer-chemopreventive activities od wine stilbenoids and flavans extracted from grape (*Vitis vinifera*) cell cultures. *Nutrition Cancer* 40:173–179.

13. Peltonen, S. (1981) Studies on bark extracts from Scots pine (*Pinus sylvestris*) and Norway spruce (*Picea abies*), Part 1. Main chemical composition. *Pap. Puu* 63:593–595.
14. Peltonen, S. (1981) Studies on bark extracts from Scots pine (*Pinus sylvestris*) and Norway spruce (*Picea abies*), Part 2. Chromatographic analysis of hot water and alkali soluble polyflavonoids. *Pap. Puu* 63:681–687.
15. Kylliainen, O. and Holmbom, B. (2004) Chemical composition of components in spruce waters. *Pap. Puu* 86:289–292.
16. Norin, T. and Winell, B. (1972) Extractives from the bark of common spruce, *Picea abies* L. Karst. *Acta Chem. Scand.* 26:2289–2296.
17. Björkman, A. (1956) Studies on finely divided wood. Part 1. Extraction of lignin with neutral solvents. *Svensk Papperstidn.* 59:477–485.
18. Theander, O. and Westerlund, E.A. (1986) Study on dietary fiber. 3. Improved procedure for analysis of dietary fiber. *J. Agric. Food Chem.* 343:330–336.
19. Laouenan, P., Michon, V., Herve de Penhoat, C., Mila, I., and Scalbert, A. (1997) NMR structural investigations and conformational analysis of condensed tannins. A continuing challenge due to restricted rotation about the interflavanoid linkage. *Analysis* 25(8):M29–M32.
20. Mannila, E. and Talvitie, A. (1992) Stilbenes from *Picea abies* bark. *Phytochemistry* 31:3288–3289.
21. Dore, S. (2005) Unique properties of polyphenol stilbenes in the brain: more than direct antioxidant actions; gene/protein regulatory activity. *Neurosignals* 14:61–70.
22. Roupe, K.A., Remsberg, C.M., Yanez, J.A., and Davies, N.M. (2006) Pharmacometrics of stilbenes: seguing towards the clinic. *Curr. Clin. Pharmacol.* 1:81–101.
23. Ferreira, D. and Li, X.-C. (2000) Oligomeric proanthocyanidins: naturally occurring O-heterocycles. *Nat. Prod. Rep.* 17:193–212.
24. Shen, C.-C., Chang Y.-S., and Ho, L.-K. (1993) Nuclear magnetic resonance studies of 5,7-dihydroxyflavonoids, *Phytochemistry* 34:843–845.
25. Pan, H. and Lundgren, L.N. (1995) Phenolic extractives from root bark of *Picea abies. Phytochemistry* 39:1423–1428.
26. Khan, M.L., Haslam, E., and Williamson, M.P. (1997) Structure and conformation of the procyanidin B-2 dimer. *Magn. Res. Chem.* 35:854–858.
27. De Bruyne, T., Pieters, L.A.C., Dommisse, R.A. *et al.* (1996) Unambiguous assignments for free dimeric proanthocynidin phenols from 2D NMR. *Phytochemistry* 43:265–272.
28. Saito, A., Nakajima, N., Tanaka, A., and Ubukata, M. (2002) Synthesis study of proanthocyanidins. Part 2: Stereoselective gram-scale synthesis of procyanidin-B3. *Tetrahedron* 58:7829–7837.
29. Behrens, A., Maie, N., Knicker, H., and Kögel-Knabner, I. (2003) MALDI-TOF mass spectrometry and PSD fragmentation as means for the analysis of condensed tannins in plant leaves and needles. *Phytochemistry* 62:1159–1170.
30. Ohyama, M., Tanaka, T., and Iinuma, M. (1995) Five resveratrol oligomers from roots of *Sophora leachiana. Phytochemistry* 38:733–740.
31. Waffo-Teguo, P., Lee, D., Cuendet, M., Merillon J.-M., Pezzuto, J.M., and Kinghorn, D. (2001) Two new stilbene dimer glucosides from grape (*Vitis vinifera*) cell cultures. *J. Nat. Prod.* 64:136–138.
32. Zhang, L., Henriksson, G., and Gellerstedt, G. (2003) The formation of β–β structures in lignin biosynthesis – are there two different pathways? *Org. Biomol. Chem.* 1:3621–3624.
33. Ralph, J., Marita, J., Ralph, S.A. *et al.* (1999) Solution-State NMR of Lignins. In D.S. Argyropoulos (Ed.) *Advances in Lignocellulosics Characterization.* TAPPI Press, Atlanta, GA, pp. 55–108.
34. Steyberg, J.P., Ferreira, D., and Roux, D.G. (1983) The first condensed tannins based on a stilbene. *Tetrahedron Lett.* 24:4147–4150.
35. Waffo-Teguo, P., Decendit, A., Krisa, S., Deffieux, G., Vercauteren, J., and Merillon J.-M. (1996). The accumulation of stilbene glycosides in *Vitis vinifera* cell suspension culture. *J. Nat. Prod.* 59:1189–1191.

Chapter 2
Raman Spectroscopic Characterization of Wood and Pulp Fibers

Umesh Prasad Agarwal

Abstract

This chapter reviews applications of Raman spectroscopy in the field of wood and pulp fibers. Most of the literature examined was published between 1998 and 2006. In addition to introduction, this chapter contains sections on wood and components, mechanical pulp, chemical pulp, modified/treated wood, cellulose I crystallinity of wood fibers, and the "self-absorption" phenomenon in near-infrared (IR) Fourier-transform (FT)-Raman spectroscopy. When needed, the sections are further categorized into various subsections. For example, the section on wood and components contains a variety of topics ranging from Raman band assignments to molecular changes during tensile deformation. From this review it will become clear that Raman spectroscopy has become an essential analytical technique in the field of wood and pulp fibers.

2.1 Introduction

Techniques that can provide information on the chemical constitution of wood and pulp fibers and on processing-related changes of these materials are of great interest. Raman spectroscopy [1] is one such technique that provides fundamental knowledge on a molecular level and does not require chemical treatment of the sample (contrary to the case, for example, in fluorescence and electron microscopy). It has become an important analytical technique for nondestructive, qualitative, and quantitative analysis of materials. The technique is useful in a number of areas, including analytical measurements, mechanistic studies, and structural determinations. Studying a sample is very simple and usually involves sampling an area of interest by directing a laser excitation beam and analyzing the collected molecularly specific scattered light [2]. Although initially, obtaining good-quality Raman data required operator skill and training, this has started to change as a result of the commercial availability of compact, easy-to-use, integrated instruments at reasonable cost.

Nevertheless, although the specificity of Raman spectroscopy is very high, its sensitivity is somewhat poor. Considering that only a small number of the incident laser photons are inelastically scattered, the detection of analytes present in very low concentrations is limited. To overcome this problem, special Raman signal-enhancing techniques can be applied. The two most prominent approaches are the resonance Raman effect [3] and the surface-enhanced Raman scattering (SERS) [4,5]. Resonance Raman and SERS are only two of many special techniques in the field of Raman spectroscopy. Nonetheless, these two techniques open the possibility of investigating low-concentration samples. Moreover, resonance Raman and SERS spectra provide additional important information about the investigated system.

In resonance Raman spectroscopy, the resonance state arises when the wavelength used to excite the Raman effect lies within the electronic absorption band of the sample, causing the vibrational modes involved in the electronic transition to be selectively enhanced (by a factor of up to 10^6 compared with nonresonant excitation). In addition, resonance Raman spectroscopy allows the site-specific investigation of chromophores within a molecule.

In Raman literature a large number of examples can be found where SERS has been used to enhance the Raman scattering of an analyte. SERS is an effect wherein a significant increase in the intensity of Raman light scattering can be observed when molecules are brought into close proximity of certain metal surfaces (e.g., Ag and Au). The metal surface should contain roughness with sizes in the order of 1/10th of the wavelength of the excitation light to obtain the largest enhancement factors. The SERS effect entails enhancement (up to 10^{11}–10^{14} times) of the Raman scattering of a molecule located in the vicinity of nanosized metallic structures (usually Ag and Au). Electromagnetic and chemical enhancement mechanisms are responsible for this effect. SERS is important not only for Raman spectroscopy but also for surface science and nanoscience. Thus, SERS-active metal surfaces serve as model substrates to investigate the type of interactions between a molecule and a substrate, the molecule's adsorption site, and, at times, the properties of such adsorbed molecules.

In studies of wood and pulp fibers, Raman spectroscopy has become an important analytical technique because of the important technical developments [6] with respect to both new instrumentation (especially instruments that successfully limit sample fluorescence) and new interpretive advances that have taken place within the past two decades. Such progress has made the macro- and microlevel Raman investigations of wood and fibers much more valuable. For instance, in the area of microinvestigations, chemical imaging of fiber cell walls has become a reality and images of cellulose and lignin distributions in the plant cell wall have provided useful information on the compositional and organizational characteristics of woody tissues [7–9].

Raman spectroscopy and its various techniques have started to have an impact on the field of wood and pulp fiber science. This chapter presents a brief overview of the recent advances. The intent of the author is to present an overview that summarizes the various capabilities of Raman spectroscopy in the field of lignocellulosics research. The chapter focuses on the characterization work published between 1998 and 2006. An earlier review by the author on the applications of Raman spectroscopy to the field of lignocellulosic materials was published in 1999 [10].

2.2 Wood and components

2.2.1 Woods

Raman spectroscopy has been applied to study both softwoods and hardwoods in their native states. Earlier studies [11–15] consisted of distinguishing these classes of woods based on chemical composition-related Raman spectral differences. Although the chemical composition of wood is complex and varies from species to species, there are three structural polymeric components, namely, cellulose, lignin, and hemicellulose, which are common to all woods. Early on, wood spectra were assigned to various wood polymeric components [16–18]. For example, for black spruce, it was reported that observed spectral features were mostly due to cellulose and lignin [18]. Additionally, it was reported that, although present in significant quantity (20–30%), very few hemicellulose Raman contributions were detected. In more recent work [19], ultraviolet resonance Raman (UVRR) in conjunction with principal component analysis has been used to characterize woods, and using this approach, it was possible to distinguish between various aromatic and other unsaturated structures in wood.

FT-Raman spectroscopy was applied to *Eucalyptus* wood meal samples [20–28] to not only determine cell types, cell morphology, and wood constituents (α-cellulose, hemicellulose and hemicellulose sugar composition, lignin and lignin syringyl-to-guaiacyl ratio, extractives, etc.) but also to develop a rapid quantitative method for accessing wood properties for kraft pulp production. Multivariate data analysis, including cross-validation, produced highly significant correlations between conventionally measured and Raman-predicted values for a number of traits.

Nondestructive monitoring of wood decay [29] and evaluation of acid–base properties of wood [30] are two others areas of research that have taken advantage of the capabilities of FT-Raman spectroscopy. In the former case, both the C=O and the glycosidic-bond bands were used to monitor wood decay by *Coriolus versicolor*, *Dichomitus squalens*, and *Ceriporiopsis subvermispora* [29]. Next, FT-Raman spectroscopy was used to investigate the acid–base properties of bulk pine wood [30], and it was concluded that the pine was largely acidic. In the investigation, various solvents were used as probe liquids, and spectral changes at $\sim$2936 and $\sim$1657 cm^{-1} were used as measures of acid–base properties.

2.2.2 Extraneous compounds

In woods where extraneous compounds (other plant metabolites) including heartwood-resins (e.g., flavonoids and pinosylvins) were present, they have been detected using near-IR FT-Raman [31–33] and UVRR [34,35] spectroscopy. Specific bands in the spectra can be used in both wood identification and quantitative determination of specific resin compounds. For example, FT-Raman spectroscopy was used to characterize brazilwood [32] to isolate key Raman biomarker bands that could provide the basis for an identification protocol.

2.2.3 Wood components

Raman spectra of wood components – cellulose, hemicellulose, and lignin – have been obtained [18,36,37]. It is reported that lignin spectrum varies depending upon the excitation wavelength of the laser [37,38]; no such sensitivity to excitation wavelength exists in the case of cellulose and hemicelluloses. For the former, this is explicable due to the occurrence of the resonance Raman effect at shorter excitation wavelengths (UV), where different structures of lignin (*p*-hydroxyphenyl, guaiacyl, and syringyl) absorb the UV light differently. Additionally, as previously reported [39], when excited in the visible range, band intensity differences for specific modes in the spectrum of lignin arise due to pre-resonance [39] and conjugation effects [40]. Of the Raman spectra of three polymer components, only cellulose spectrum has been well characterized and assigned [36,41], although high degree of coupling exists between vibrations that involve C—C and C—O bonds. This knowledge has helped in the interpretation of wood and pulp fiber spectra. Considering that the Raman spectra of lignins (syringyl, guaiacyl, and coumaryl) are complex, significant advances have been made to interpret them. This has come about as a result of studying, with a number of approaches, not only lignins (native [18], milled wood [18,42,43], and residual lignins [37,44–47]) and their models (deuterated and normal dehydrogenation polymer [DHP] lignins [48] and a large number of lignin model compounds [38,49]) but also chemically modified (e.g., bleaching, hydrogenation, and acetylation) lignins and lignocellulosics [10,39,43,50,51]. An additional tool in the aid to interpretation has been the theoretical calculations on lignin models wherein some degree of preliminary work has been done [52,53], but a lot more remains. In recently reported research findings, an additional approach, partial least squares modeling, was successfully used to interpret the UVRR spectra of lignin model compounds [54].

Considering that lignin's heterogeneous structure consists not only of interphenylpropane-unit linkages of carbon-to-carbon and carbon-to-oxygen but also of side chains with various substituent and functional groups, a study of a large number of representative models is essential. In the work recently completed in the author's laboratory [49], 40 lignin models were investigated (Figure 2.1) in the neat state (near-IR FT-Raman and FT-IR), in solution (near-IR FT-Raman), and on cellulose (near-IR FT-Raman). These models represent a large number of substituent/functional groups and substructures in lignin–aliphatic and phenolic OH, C=O, CHO, COOH, CH_3, OCH_3, $\alpha, \beta C$=C, furan, and interunit C—O—C and C—C linkages. Raman band positions associated with various groups were identified and Raman frequencies and intensities were compared between the models. The position of the 1600 cm^{-1}-Raman-mode was found to be only minimally sensitive to substituents because in ~70% of the models, including those containing more than one phenyl group, the vibrational frequency was confined between 1594 and 1603 cm^{-1}. Sensitivity of a model's vibrational frequencies to its environment was evaluated by comparing frequency data in three different sampling states – in neat, in solution, and on cellulose. Several Raman band positions shifted compared to the value in the neat state.

2.2.4 Band assignments

The Raman spectra of wood and pulp fibers were assigned to the three polymer components and, in the case of lignin, further to its substructural units. Such interpretation of the experimental Raman data was based on pure-component spectra and their assignments. However, sometimes, a clear assignment of the measured Raman band to specific vibration is unclear. To further improve this situation, an assignment of the observed Raman bands can be assisted by a systematic comparison with the theoretically calculated spectra [52,53]. Although quantum chemical calculations and normal coordinate analysis have been available for sometime, recently it was shown that density functional theory (DFT) methods provide a powerful alternative, as they are much less computationally demanding (compared with the quantum calculations) and take account of the effects of electron correlation. Recently, a DFT method using the B3LYP functional and 6-31 + G(d) basis set was used to

	R₁	R₂	R₃	R₄
1	–COOH	–H	–OCH₃	–H
2	–COOH	–OCH₃	–OCH₃	–H
3	–CH=CHCHO	–H	–OCH₃	–H
4	–H	–OCH₃	–OH	–OCH₃
5	–CCH₃ (O)	–OCH₃	–OCH₃	–H
6	–CCH₃ (O)	–OCH₃	–OH	–H
7	figure	–H	–H	–H
8	–CH=CHCOOH	–H	–OCH₃	–H
9	–H	–OCH₃	–OCH₃	–H
10	–CCH₃ (O)	–OCH₃	–OCCH₃ (O)	–H
11	–CCH₃ (O)	–H	–H	–H
12	–CHO	–OCH₃	–OCH₃	–H
13	–CCH₃ (O)	–H	–OH	–H
14	–CH=CHCOOH	–OCH₃	–OH	–H
15	–CCH₃ (O)	–OCH₃	–OCH₃	–OCH₃
16	figure	–OCH₃	–OH	–H
17	–CH=CHCOOH	–OCH₃	–OH	–OCH₃
18	–CH₂OH	–H	–OH	–H
19	figure	–H	figure	–H
20	–CCH₃ (O)	–H	figure	–H

	R₁	R₂	R₃	R₄
21	–CCH₂CH₃ (O)	–OCH₃	–OH	–OCH₃
22	figure	–OCH₃	–OCH₃	–H
23	figure	–OCH₃	–OH	–H
24	figure	–OCH₃	–OCH₃	–H
25	–CHCH₃ (OH)	–OCH₃	–OCH₃	–H
26	–CH₃	–OCH₃	–OH	–H
27	–CH₂OH	–OCH₃	–OCH₃	–H
28	–CH₃	–OCH₃	–OCCH₃ (O)	–H
29	–CH₂OH	–H	–OCH₃	–H
30	–CH₂CH₂CH₃	–OCH₃	–OH	–H
31	–CH=CHCH₃	–OCH₃	–OH	–H
32	–COOH	–OCH₃	–OH	–H
33	–CHO	–OH	–OCH₃	–H
34	–CHO	–OCH₃	–OH	–H
35	–COOH	–OCH₃	–OH	–OCH₃
36	–CHO	–OCH₃	–OH	–OCH₃
37	–CH=CHCOOH	–H	–OH	–H
38	–COOH	–H	–OH	–H
39	–CHO	–OCH₃	–OCH₃	–OCH₃
40	–CCH₃ (O)	–OCH₃	–OH	–OCH₃

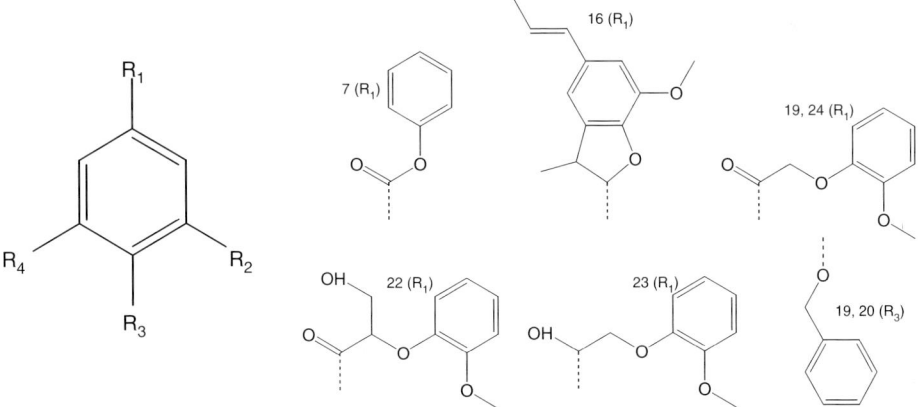

Figure 2.1 Molecular structures of 40 lignin models used in the lignin model study.

predict the vibrational frequencies of some simple lignin models [53]. The performance of this DFT method was evaluated by comparison of the computed frequencies with observed Raman spectra of the appropriate model compounds. Assignments for vibrational bands of the lignin polymer were presented. In future, this approach is expected to result in reliable assignment of those experimentally observed lignin Raman bands that have not yet been properly assigned, which in turn will lead to a detailed understanding of the geometrical and electronic structure of the investigated molecular lignin model units.

2.2.5 Microscope studies

Although Raman microscopy (lateral spatial resolution ~1.0 μm) preceded IR microscopy (resolution ~10 μm) commercially by ~5 years, infrared microscopy has been used more often in the past. Currently easier to use and increasingly available, Raman microscopy is being used more frequently. In the investigations of wood and pulp fibers, Raman studies range from investigating tensile deformation of single wood fibers [55,56] to detailed investigations, at the subcellular level, of composition and ultrastructure of native woody tissue [7–9].

Indeed, in the early years of wood investigations by Raman using the 514.5 nm argon ion laser excitation, a Raman microprobe was more successful because of its ability to circumvent the sample fluorescence, which invariably accompanied the Raman signal. Moreover, subsequently, with the availability of the 633- and 785-nm-laser-based Raman microscopy systems, the fluorescence interference has been further minimized. Consequently, it became realistic to investigate individual morphological regions in the woody tissue and spectra at high spatial resolution could be obtained. Although a resolution of 1 μm (obtained using a 100× microscope objective) in Raman spectroscopy is limited due to the diffraction effect, compared with IR microscopy, it is 10 times better. It was polarized Raman microscopy that showed, for the first time, that secondary wall lignin was ordered [16,17]. The recent technological advances [6] have benefited Raman microscopy instrumentation extensively and a state-of-the-art system is capable of providing new information at the individual cell wall morphological level [9].

2.2.6 Fiber cell wall organization and compositional heterogeneity

As mentioned earlier, with a confocal Raman microscope spectra of a very small sample-volume can be measured, and therefore, the chemical composition of very small structures can be determined. For heterogeneous samples such as wood fibers, the distributions of cellulose and lignin are important. Such information can be obtained with a Raman microscope that provides images based on a Raman band characteristic for the component of interest [18]. In cell wall Raman imaging, inherent scattering characteristics of cellulose and lignin are used to visualize their distribution and no external labels are necessary [8,9]. In confocal Raman microscopy, point-by-point imaging couples the mobility of an automated scanning stage with the high numerical aperture of the microscope objective to take the spectrum at each sample point. The technique has been used to study micron-size regions in the fiber cell wall.

Although, in the studies of wood fibers, use of Raman microscopy preceded macro-investigations, the progress towards detailed cell wall investigations was hindered due to the limitations driven by both sampling and instrumentation. Sample fluorescence and a lignin band that was less than fully stable made Raman microscopy investigations difficult. In addition, a highly efficient Raman microprobe equipped with an automated high-resolution x, y stage was not available. In the 1990s, new technologies such as the holographic notch filter and the availability of charge-coupled devices that acted as multichannel detectors decreased acquisition time by more than an order of magnitude. Rugged, air-cooled lasers (e.g., He–Ne 633 nm) simplified utility requirements and provided more beam-pointing stability compared with that of water-cooled lasers. Furthermore, sampling in confocal mode reduced fluorescence by physically blocking the signal originating from the volume of the sample not in focus. The detected Raman signal came from the illuminated spot. The latest Raman microprobe incorporates these advances and is well suited to investigate lignin and cellulose distribution in the cell walls of woody tissue. These capabilities permit compositional mapping of the woody tissue with chosen lateral and axial spatial resolutions.

Most recent applications of confocal Raman microscopy to woody cell walls consisted of chemical imaging of poplar (including the tension wood) and spruce woods [8,9] as well as topochemical studies of beech wood [7]. In the case of spruce [9], a state-of-the-art 633-nm-laser-based confocal Raman microscope was used to determine the distribution of cell wall components in the cross section of black spruce wood *in situ*. Chemical information from morphologically distinct cell wall regions was obtained and Raman images of lignin and cellulose spatial distribution were generated. While cell corner (CC) lignin concentration was the highest on average, lignin concentration in compound middle lamella (CmL) was not significantly different from that in secondary wall (S2 and S2–S3). Images generated using the 1650 cm^{-1} band showed that coniferaldehyde and coniferyl alcohol distribution followed that of lignin and no particular cell wall layer/region was therefore enriched in the ethylenic residue. In contrast, cellulose distribution showed the opposite pattern – low concentration in CC and CmL and high in S2 regions. Nevertheless, cellulose concentration varied significantly in some areas, and concentrations of both lignin and cellulose were high in other areas. Figure 2.2a, taken from Reference 9, shows the bright field image of the cell wall area that was investigated by Raman microscopy. Chemical images, indicating lignin and cellulose distributions, are shown in Figure 2.2b and c, respectively.

Besides using visible laser excitation in Raman microscopy, for the first time, use of UVRR (244 nm excitation) to study individual plant cell walls has been reported [57]. Because damage to the cell walls is expected at such high-energy wavelengths of excitation, a series of spectra were obtained along a given cell wall at low power and for short duration. These were then averaged to produce a mean spectrum with a high signal-to-noise ratio and represented that region of the cell wall.

2.2.7 Molecular changes during tensile deformation

In other studies of wood fibers [55,56], Raman microscopy was used to show that during tensile deformation the $\sim$1095 cm^{-1} Raman band shifts towards a lower wave number due to molecular deformation of cellulose. This shift has been shown to be useful in understanding the micromechanisms of deformation in wood and pulp fibers. In one study [56],

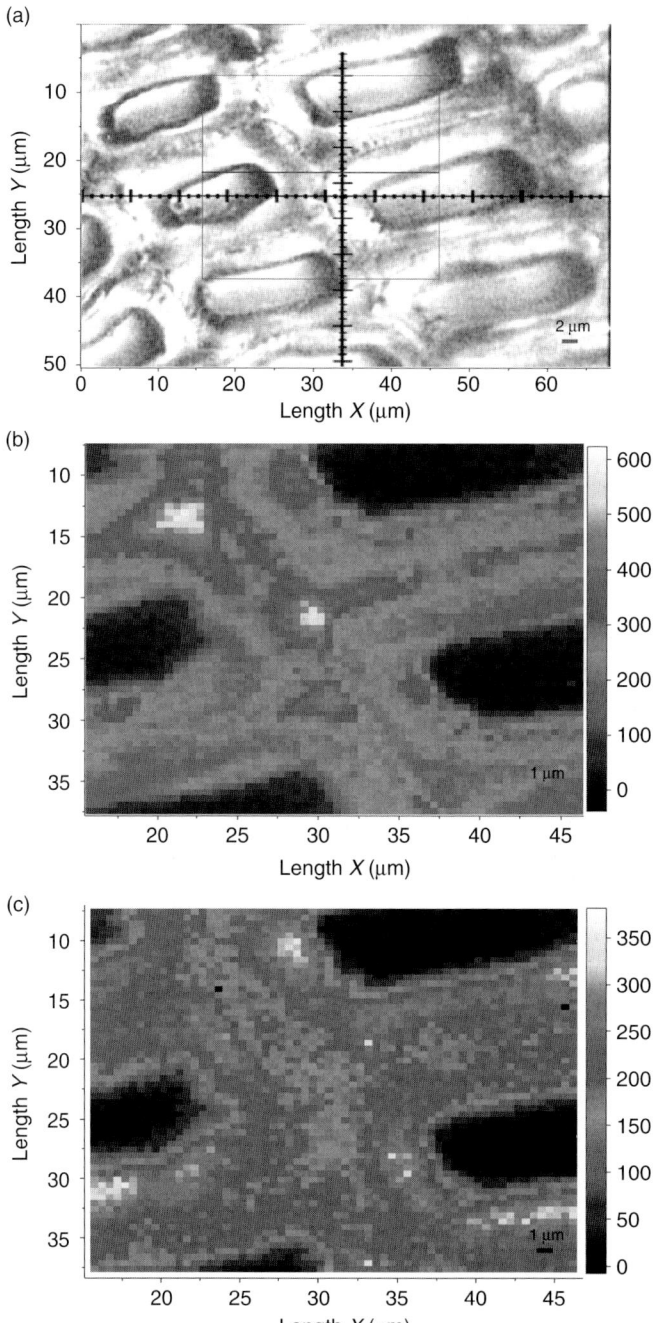

Figure 2.2 See opposite page for caption.

isolated fibers of spruce latewood were mechanically strained and molecular changes due to deformation were monitored by stress–strain curves and changes in band positions and intensities. Frequency changes for the 1097 cm^{-1} band correlated well with the applied stress and strain. Additionally, a decrease in the peak intensity ratio of the 1127 and 1097 cm^{-1} bands was related to the straining of the latewood fiber. Changes observed in the O—H stretch region were interpreted in terms of weakening of the hydrogen-bonding network as a result of straining. Stretching of the spruce fiber was found to have no influence on lignin Raman bands.

2.3 Mechanical pulp

Bleaching of spruce thermomechanical pulp by alkaline hydrogen peroxide, sodium dithionite, and sodium borohydride was studied using FT-Raman spectroscopy [50]. The Raman examination revealed that spectral differences between bleached and unbleached pulp fiber spectra were primarily due to coniferaldehyde and *p*-quinone structures in lignin. This was new direct evidence that such bleaching removes *p*-quinone structures. In the unbleached pulp, the contribution of *p*-quinones was detected in the 1665–1690 cm^{-1} region. This contribution was removed to varying degrees in the bleached pulps, depending on the extent to which the pulps were brightened. As shown in Figure 2.3, the relative change in the post-color number of the pulp correlated linearly with the Raman intensity decline in the 1665–1690 cm^{-1} region. Although the correlation was not great, it nonetheless indicated that *p*-quinone structures were largely responsible for determining pulp brightness.

FT-Raman and other spectroscopic methods were used to detect lignin oxidation products in thermomechanical pulp (TMP) [58]. It was shown that using 2,2′-azinobis-3-ethylbenzthiazoline-6-sulfonate cation radical (ABTS$^{•+}$) and FT-Raman spectroscopy, detailed information on lignin reactions can be obtained. Another approach [59] in probing lignin structure by (ABTS$^{•+}$) was to make use of the pre-resonance and resonance Raman techniques in conjunction with the Kerr gate. The Kerr gate approach permits temporary rejection of high level of fluorescence from resonance Raman spectra. Spectral information obtained depended upon both the wavelength of excitation and lignin excitation profiles.

Figure 2.2 (a) Bright-field image of spruce cross section (micrometer scale superimposed) showing selected area (red rectangle) for Raman mapping. Rectangle encloses cell walls of six adjoining mature tracheid cells and contains three CC regions. Raman images of this area are shown in (b) lignin and (c) cellulose. A Y-segment at $Y = 21.6$ µm is marked; from Reference 9. For a better visual illustration, see the colored figure in the reference. (b) Raman image (false color) of lignin spatial distribution in selected cell wall area in (a) in two-dimensional representation. Intensity scale appears on the right. Bright white/yellow locations indicate high concentration of lignin; dark blue/black regions indicate very low concentration, for example, lumen area; from Reference 9. See Plate 1 for the color image. (c) Raman images (false color) of cellulose spatial distribution in cell wall area selected in (a) in two-dimensional representation. Bright white/yellow locations indicate high cellulose concentration; dark blue/black regions indicate very low concentration, for example, lumen area; from Reference 9. See Plate 2 for the color image. (From Agarwal [9].)

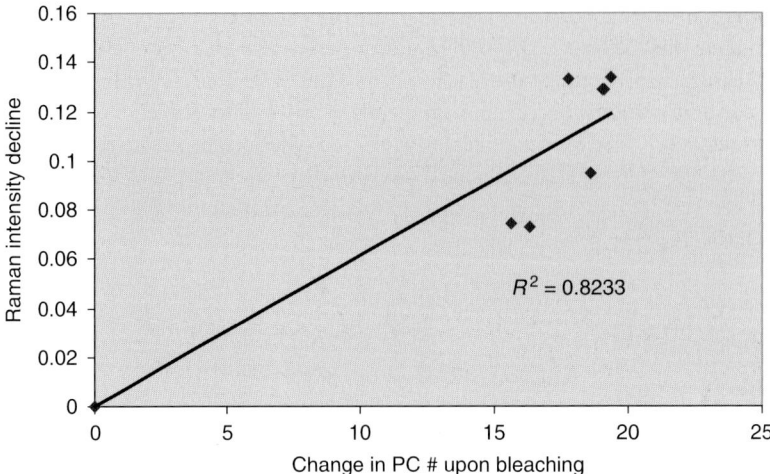

Figure 2.3 Linear regression between ΔPC and the decline in Raman intensity of the 1665–1690 cm^{-1} region (contribution primarily due to p-quinones).

Hydrothermally degraded ground-wood-fibers containing paper (75% ground wood and 25% bleached sulfite) was studied with the help of near-IR Raman spectroscopy [60]. Degradation reactions were monitored by observing the appearance of C=O stretching bands upon formation of carbonyl containing species. However, other changes that resulted in the C=C band intensity decline suggested involvement of the coniferyl alcohol groups.

Photochemically modified mechanical pulp fibers in printing and writing papers were investigated by both macro- and micro-FT-Raman spectroscopy to determine the extent to which aromatic lignin structure was destroyed [61]. In the microscopic study, 10 spots through the thickness of the paper were analyzed, and the ratio of aromatic to cellulose vibration intensity (I_{1600}/I_{1095}) was calculated. The results showed the destruction of aromatic structures on the photoexposed surface, as well as an increase in aromatic structure intensity deeper in the sheet. This increased intensity was attributed to C=O bonds formed on the aromatic structures during initial stages of photo-oxidation. Macro-Raman analysis revealed that upon photo-oxidation the band at 1654 cm^{-1} decayed and a new band appeared at 1675 cm^{-1}, indicating destruction of ring conjugated C=C structures and formation of p-quinone structures.

Bleaching and photoyellowing of TMPs was also studied using UV–vis and resonance Raman spectroscopy [62]. In the latter technique, the chosen excitation wavelength was close to the absorption bands of the light-generated chromophores to take advantage of the resonance enhancement. The results revealed that the originally present coniferaldehyde structures were partly removed by alkaline peroxide bleaching and they were further degraded upon light exposure. On the basis of the resonance Raman analysis of the photoexposed pulps, formation of quinone structures, possibly p-quinones, was found to be a plausible explanation for the brightness reversion of the pulps.

The cause of darkening in biomechanical pulp was investigated by FT-Raman and other spectroscopy techniques [63]. The results indicated that quinones (both o- and p-) were produced when wood chips were treated with the fungus. This was also the case when fungus-treated wood chips were refined at higher steaming pressures (up to 85 psi). For a given pressure, compared with control pulps, treated pulps were darker because their quinone content was higher. Within the set of biopulps that were produced at various steaming pressures, higher-pressure pulps were darker and contained more quinones.

2.4 Chemical pulp

Using conventional Raman spectroscopy fully bleached chemical pulp fibers have been studied earlier in the 1980s and differences in the pulp spectra were found to exist [64]. However, due to the problem of fluorescence associated with unbleached and partly bleached pulps, these could not be investigated and their Raman studies have to wait until near-IR-excited FT-Raman technique was developed. Fluorescence signal was minimized in FT-Raman spectroscopy and the technique was well suited to investigate residual lignin in chemical pulps. Lignin in pine kraft pulps was quantified by FT-Raman based on the ratio of the lignin and cellulose band intensities [44]. The data was found to correlate well with the kappa number of the pulps. However, because in FT-Raman only a single prominent lignin band at 1600 cm^{-1} was observed in the pulp spectra, it was still not possible to obtain information on the detailed residual lignin structural changes that were being introduced by the bleaching chemicals.

UVRR spectroscopy was used [37,46,47,65] not only for determination of lignin in unbleached and fully bleached chemical pulps, but also to detect bleaching-related changes in residual lignin. Both lignin and hexenuronic acid could be investigated. Whereas concentration of lignin was predicted fairly accurately, the standard error of prediction was higher for hexenuronic acid [65]. The UVRR method was found to be highly sensitive and selective for detecting lignin structures.

In another application of resonance Raman spectroscopy to detect highly fluorescent residual lignin in chemical pulps, optical Kerr gate approach was successfully used [47]. Sample fluorescence was effectively suppressed and much weaker bands of lignin could be observed and information on chromophore groups in pulps could be obtained.

Recently, in the author's laboratory, for the first time, SERS effect in unbleached kraft pulp fibers was induced using nano- and microparticles of silver (Figure 2.4) [66]. Further, it was shown that only residual lignin bands were enhanced in intensity. The lack of carbohydrate signal enhancement implied that SERS method can be used to selectively investigate lignin in presence of carbohydrate polymers. The structural transformations of lignin in pulps bleached with polyoxometallates were investigated using SERS [45].

In addition to studies of residual lignin in pulps, a small amount of carbonyl groups in cellulose pulps (μmol/g range), have been investigated by a combination of carbonyl-selective fluorescence labeling method and UVRR. It was demonstrated [67] that carbonyls in the pulps are not only present as a C=O structure with an sp^2-hybridized carbon, but also to a significant extent in sp^3-hybridized form as hydrates or hemiacetals.

FT-Raman technique has been applied to the area of wood pulp recycling. In one study [68], the composition of recycled pulp (including pigments such as calcium carbonate, talc, gypsum, titanium dioxide and kaolin) was analyzed whereas in the other [69] structural change in cellulose was the subject of investigation.

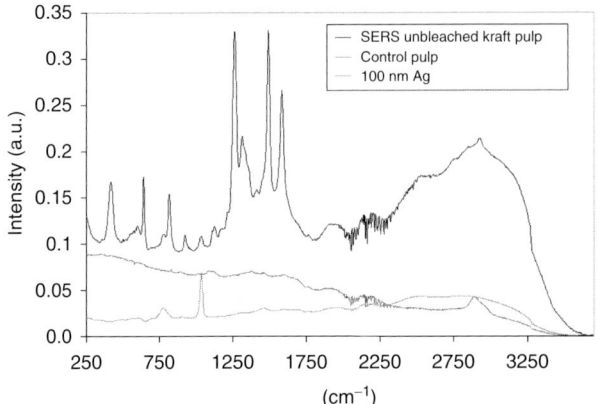

Figure 2.4 Near-IR FT-Raman spectra of various samples: top, SERS of ball-milled unbleached KP (pulp-to-Ag weight ratio 1:1); middle, normal spectrum of same pulp without Ag; bottom, spectrum of 100–150 nm Ag particles. Spectra have been vertically offset for improved visualization. In the pulp SERS spectrum, most enhanced bands were present between 250 and 1800 cm^{-1}. In normal pulp spectrum (middle) hardly any peaks were detected due to high sample fluorescence.

2.5 Modified/treated wood

2.5.1 Heat treated

Thermal modification of woods was studied by examining chemical changes in wood structure by UVRR spectroscopy [70]. It was found that upon heat treatment at 180°C or higher lignin became partly soluble in acetone. Spectra of extracted wood samples indicated that the structure of unextracted lignin remained unchanged when heated to 200°C. Additionally, formation of new carbonyl groups was reported and it was found that the lignin content of the samples was increased as a result of degradation of wood hemicelluloses.

In one study that used FT-Raman to look at the pyrolyzed Japanese cedar [71] in a nitrogen atmosphere at various temperatures (200–1000°C), observation of two characteristic bands at 1340 and 1590 cm^{-1} was reported and the spectral features were found to change markedly as a function of temperature. A second study [72] involved Raman investigations of Japanese larch heartwood and Japanese beech sapwood where the wood samples were heated for 22 h at constant temperatures in the temperature range of 50–180°C. The spectral changes, mainly detected in the C=C and C=O regions, were interpreted in terms of condensation reactions of lignin during heat treatment. At higher temperatures, spectral changes in the region 1200–1500 cm^{-1} suggested that the wood constituents were partly decomposed.

2.5.2 Chemically treated

Propiconazole, a fungistatic agent, has a Raman band in the region 647–693 cm^{-1}. Using a 785-nm-laser-excited conventional microprobe, the band intensity in the latter region was used to determine the propiconazole distribution in white spruce [73]. The Raman study of longitudinal depth distribution pattern of propiconazole correlated well with the gas chromatography–mass spectroscopy (GC–MS) profile.

A Raman microscopic investigation to measure the melamine–formaldehyde resin content within cell walls of impregnated spruce wood was carried out [74] and the findings were compared with the UV method. Not only Raman was found to be a good method for estimating the resin content, the sharpness of the Raman melamine peak compared with the not-well-defined increase in the UV absorbance was an added advantage in the Raman analysis. Previously, FT-Raman spectroscopy was used for characterization of melamine–formaldehyde resins [75].

Additionally, identification of adhesives in wood has been carried out using FT-Raman method [76]. For example, wood adhesives such as urea resins, melamine resins, phenolic resins, resorcinol resins, and vinyl acetate resins were studied. How spectral features were affected upon curing was also evaluated, and the FT-Raman method was shown to be a useful nondestructive analytical technique to investigate synthetic resins in wood.

Because of environmental benefits, boron compounds are often used for preservative treatment of wood as they are less toxic to mammals while being regarded as both an effective insecticide and fungicide. FT-Raman spectroscopy was used to understand the role of boric acid in Japanese cedar wood [77]. A symmetrical stretching vibrational band of BO_3 group at 879 cm^{-1} was used to study boric acid distribution of two different types of wood blocks treated in aqueous boric acid solution, one cross-sectional and one longitudinal. It was concluded that from the intensity and peak position of the Raman band due to $B(OH)_3$, the $B(OH)_3$ unit in wood can be classified into two groups based on the chemical species around it. One group comprises the microcrystalline state of $B(OH)_3$ precipitated in the lumens, and the other comprises the $B(OH)_3$ units penetrating the cell wall. An obvious tendency recognized from the Raman line maps in the longitudinal direction was that $B(OH)_3$ microcrystals were made significantly more abundant near the cut ends in a longitudinal wood block due to the air-drying process.

2.6 Cellulose I crystallinity of wood fibers

Using near-IR FT-Raman spectroscopy and Whatman CC31 cellulose samples that were partially crystalline, a cellulose I crystallinity estimation model based on the Raman band intensity ratio of the 378 and 1096 cm^{-1} bands was developed [78]. This model was then used to calculate cellulose I crystallinity in sour orange (a hardwood). Raman determined crystallinity values of wood cellulose were reliable and produced low standard error. To detect if cellulose crystallinity changed in trees that were grown under elevated CO_2 climate conditions, the FT-Raman model was used. No significant change in cellulose crystallinity was found (Figure 2.5) although mean values varied between 46% and 51% and one of the control samples (SOAC#7 in Figure 2.5) showed slightly higher absolute cellulose crystallinity. While "crystallinity determination by Raman" work is ongoing, it is expected that the Raman method will be applicable to fibers of softwoods, hardwoods, and pulps.

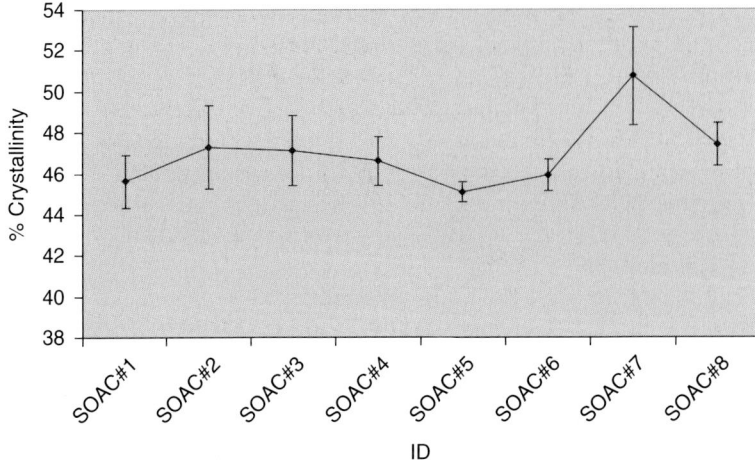

Figure 2.5 Cellulose I crystallinity of sour orange (hardwood) acid chlorite-delignified (SOAC) stem wood calculated using the FT-Raman method; Controls – #1, #2, #7 and #8, and elevated CO_2 – #3–#6; Error bars represent mean ± SD.

2.7 "Self-Absorption" phenomenon in near-IR FT-Raman spectroscopy

Cellulose and lignocellulosic materials can be conveniently studied using near-IR FT-Raman spectroscopy, and the problem of laser-induced fluorescence can be avoided or minimized. However, it was reported [79] that in the FT-Raman spectral region 2800–3500 cm^{-1} the phenomenon of self-absorption (defined as absorption of Raman scattered photons by the sample itself) occurs, and therefore, the Raman intensity of bands in this region diminishes. This was a study where Whatman cellulose paper and spruce TMP samples were studied and the self-absorption effect was found to suppress the intensity of the 2895 cm^{-1} Raman band in the C—H stretch region. Experimental observations indicated that such suppression was caused by cellulose in both the samples, and lignin, in the case of pulp, was not implicated. Furthermore, hydroxyl groups present in cellulose/hemicellulose and water were suspected to have played a role. Although certain precautions can be taken in FT-Raman quantitative work to minimize the detrimental effect of this phenomenon, it is best to choose a band in the spectrum that is not part of the C—H stretch region.

Conclusions

As a material characterization technique, Raman is complementary to IR and both must be used to obtain comprehensive information on samples. The field of Raman spectroscopy has benefited immensely from recent technological advantages in instrumentation. Because of such advances, at present, wood and pulp fiber materials can be studied using UV-, visible-, and near-IR-Raman methods. Application of Raman methods to wood and pulp fibers covers wide and diverse research areas. In fundamental and applied research, the technique has produced useful insights and generated new information. Using the traditional approaches, not only large areas of wood and fiber samples have been investigated but, with the help of a confocal Raman microscope, sample domains as small as 1.0 μm have been analyzed and chemical images generated. This review convincingly demonstrates that Raman spectroscopy has become an indispensable analytical method for studying wood and pulp fiber materials.

References

1. Schmitt, M.; Popp, J., Raman Spectroscopy at the Beginning of the Twenty-First Century; *J. Raman Spectrosc.* 2006, 37, 20–28.
2. Ferraro, J.R.; Nakamoto, K.; Ferraro, J., *Introductory Raman Spectroscopy*, 2nd Edition; Academic Press, New York, 2003.
3. Long, D.A., *Raman Spectroscopy*; McGraw Hill, New York, 1977, Chapter 7.
4. Campion, A.; Kambhampti, P., Surface-Enhanced Raman Scattering; *Chem. Soc. Rev.* 1998, 27, 241–250.
5. Moskovits, M., Surface-Enhanced Raman Spectroscopy: A Brief Retrospective; *J. Raman Spectrosc.* 2005, 36, 485–496.
6. Adar, F., Evolution and Revolution of Raman Instrumentation – Application of Available Technologies to Spectroscopy and Microscopy. In: Lewis, I.R.; Edwards, H.G. (eds); *Handbook of Raman Spectroscopy*; Marcel Dekker Inc, New York, 2002; pp. 11–40.
7. Roder, T.; Koch, G.; Sixta, H., Application of Confocal Raman Spectroscopy for the Topochemical Distribution of Lignin and Cellulose in Plant Cell Walls of Beech Wood (*Fagus sylvatica* L.) Compared to UV Microspectrophotometry; *Holzforschung* 2004, 58, 480–482.
8. Gierlinger, N.; Schwanninger, M., Chemical Imaging of Poplar Wood Cell Walls by Confocal Raman Microscopy; *Plant Physiol.* 2006, 140, 1246–1254.
9. Agarwal, U.P., Raman Imaging to Investigate Ultrastructure and Composition of Plant Cell Walls: Distribution of Lignin and Cellulose in Black Spruce Wood (*Picea mariana*); *Planta* 2006, 224, 1141–1153.
10. Agarwal, U.P., An Overview of Raman Spectroscopy as Applied to Lignocellulosic Materials. In: Argyropoulos, D.S. (ed.); *Advances in Lignocellulosics Characterization*; TAPPI Press, Atlanta, GA, 1999; pp. 201–225.
11. Kenton, R.C.; Rubinovitz, R.L., FT-Raman Investigations of Forest Products; *Appl. Spectrosc.* 1990, 44, 1377–1380.
12. Evans, P.A., Differentiating "Hard" from "Soft" Woods Using Fourier Transform Infrared and Fourier Transform Raman Spectroscopy; *Spectrochim. Acta* 1991, 47A, 1441–1447.
13. Lewis, I.R.; Daniel, Jr., N.W.; Chaffin, N.C.; Griffiths, P.R., Raman Spectrometry and Neural Networks for the Classification of Wood Types–1; *Spectrochim. Acta* 1994, 50A, 1943–1958.
14. Yang, H.; Lewis, I.R.; Griffiths, P.R., Raman Spectrometry and Neural Networks for the Classification of Wood Types. 2. Kohonen Self-Organizing Maps; *Spectrochim. Acta Mol. Biomol. Spectrosc.* 1999, 55A, 2783–2791.
15. Lavine, B.K.; Davidson, C.E.; Moores, A.J.; Griffiths, R.P., Raman Spectroscopy and Genetic Algorithms for the Classification of Wood Types; *Appl. Spectrosc.* 2001, 55, 960–966.
16. Atalla, R.H.; Agarwal, U.P., Raman Microprobe Evidence for Lignin Orientation in the Cell Walls of Native Woody Tissue; *Science* 1985, 227, 636–638.

17. Agarwal, U.P.; Atalla, R.H., *In-situ* Raman Microprobe Studies of Plant Cell Walls: Macromolecular Organization and Compositional Variability in the Secondary Wall of *Picea mariana* (Mill.) B.S.P.; *Planta* 1986, 169, 325–332.
18. Agarwal, U.P.; Ralph, S.A., FT-Raman Spectroscopy of Wood: Identifying Contributions of Lignin and Carbohydrate Polymers in the Spectrum of Black Spruce (*Picea mariana*); *Appl. Spectrosc.* 1997, 51, 1648–1655.
19. Nuopponen, M.H.; Wikberg, H.I.; Birch, G.M. et al., Characterization of 25 Tropical Hardwoods with FT-IR, Transform Infrared, Ultraviolet Resonance Raman, and 13C-NMR Cross-Polarization/Magic-Angle Spinning Spectroscopy; *J. Appl. Poly. Sci.* 2006, 102, 810–819.
20. Ona, T.; Sonoda, T.; Ito, K. *et al.*, Rapid Determination of Cell Morphology in Eucalyptus Wood by Fourier Transform Raman Spectroscopy; *Appl. Spectrosc.* 1999, 53,1078–1082.
21. Ona, T.; Sonoda, T.; Ito, K. *et al.*, *In Situ* Determination of Proportion of Cell Types in Woods by FT-Raman Spectroscopy, *Anal. Biochem.* 1999, 268, 43–48.
22. Ona, T.; Sonoda, T.; Ito, K. *et al.*, Quantitative FT-Raman Spectroscopy to Measure Wood Cell Dimensions, *Analyst* 1999, 124, 1477–1480.
23. Ona, T.; Ohshima, J.; Adachi, K.; Yokota, S.; Yoshizawa, N., Length Determination of Vessel Elements in Tree Trunks Used for Water and Nutrient Transport by Fourier Transform Raman Spectroscopy; *Anal. Bioanal. Chem.* 2004, 380, 958–963.
24. Ona, T.; Sonoda, T.; Ito, K.; Shibata, M.; Kato, T.; Ootake, Y., Non-destructive Determination of Hemicellulosic Neutral Sugar Composition in Native Wood by Fourier Transform Raman Spectroscopy; *J. Wood Chem. Technol.* 1998, 18, 43–51.
25. Ona, T.; Sonoda, T.; Ito, K. *et al.*, Non-Destructive Determination of Lignin Syringyl/Guaiacyl Monomeric Composition in Native Wood by Fourier Transform Raman Spectroscopy; *J. Wood Chem. Technol.* 1998, 18, 27–41.
26. Ona, T.; Sonoda, T.; Ito, K.; Shibata, M.; Kato, T.; Ootake, Y., Determination of Wood Basic Densities by Fourier Transform Raman Spectroscopy; *J. Wood Chem. Technol.* 1998, 18, 367–379.
27. Ona, T.; Sonoda, T.; Ito, K. *et al.*, Rapid Prediction of Pulp Properties by Fourier Transform Raman Spectroscopy of Native Wood; *J. Pulp Paper Sci.* 2000, 26, 43–47.
28. Ona, T.; Ohshima, J.; Adachi, K.; Yokota, S.; Yoshizawa, N., A Rapid Quantitative Method to Assess Eucalyptus Wood Properties for Kraft Pulp Production by FT-Raman Spectroscopy; *J. Pulp Paper Sci.* 2003, 29, 6–10.
29. Wariishi, H.; Kihara, M.; Takayama, M.; Tanaka, H., Nondestructive Monitoring of Wood Decay Process Using FT-Raman Technique; Kami Parupu Kenkyu Happyokai Koen Yoshishu 2002, 69, 8–11.
30. Shen, Q.; Rahiala, H.; Rosenholm, J.B., Evaluation of the Structure and Acid–Base Properties of Bulk Wood by FT-Raman Spectroscopy; *J. Colloid Interface Sci.* 1998, 206, 558–568.
31. Yamauchi, S.; Shibutani, S.; Doi, S., Characteristic Raman Bands for Artocarpus Heterophyllus Heartwood; *J. Wood Sci.* 2003, 49, 466–468.
32. Edwards, H.G.M.; De Oliveira, L.F.C.; Nesbitt, M., Fourier-Transform Raman Characterization of Brazilwood Trees and Substitutes; *Analyst* 2003, 128, 82–87.
33. Holmgren, A.; Bergström, B.; Gref, R.; Ericsson, A., Detection of Pinosylvins in Solid Wood of Scots Pine Using Fourier Transform Raman and Infrared Spectroscopy; *J. Wood Chem. Technol.* 1999, 19, 139–150.
34. Nuopponen, M.; Willför, S.; Jääskeläinen, A.-S.; Sundberg, A.; Vuorinen, T.; A UV Resonance Raman (UVRR) Spectroscopic Study on the Extractable Compounds of Scots Pine (*Pinus sylvestris*) Wood. Part I: Lipophilic Compounds. *Spectrochim. Acta A* 2004, 60, 2953–2961.
35. Nuopponen, M.; Willför, S.; Jääskelainen, A.S.; Vuorinen, T., A UV Resonance Raman (UVRR) Spectroscopic Study on the Extractable Compounds in Scots Pine (*Pinus sylvestris*) Wood. Part II. Hydrophilic compounds; *Spectrochim. Acta A* 2004, 60, 2963–2968.

36. Wiley, J.H.; Atalla, R.H., Raman Spectra of Celluloses. In: Atalla, R.H. (ed.); *The Structures of Cellulose*, ACS Symposium Series 340; Washington, DC, 1987; pp. 151–168.
37. Halttunen, M.; Vyörykkä, J.; Hortling, B. *et al.*, Study of Residual Lignin in Pulp by UV Resonance Raman Spectroscopy; *Holzforschung* 2001, 55, 631–638.
38. Saariaho, A.-M.; Jääskeläinen, A.-S.; Nuopponen, M.; Vuorinen, T., Ultra Violet Resonance Raman Spectroscopy in Lignin Analysis: Determination of Characteristic Vibrations of p-Hydroxyphenyl, Guaiacyl and Syringyl Structures; *Appl. Spectrosc.* 2003, 57, 58–66.
39. Agarwal, U.P.; Atalla, R.H., Raman Spectral Features Associated with Chromophores in High-Yield Pulps; *J. Wood Chem.Technol.* 1994, 14, 227–241.
40. Agarwal, U.P.; Atalla, R.H., Using Raman Spectroscopy to Identify Chromophores in Lignin-Lignocellulosics. In: Glasser, W.G.; Northey, R.A.; Schultz, T.P. (eds); *Lignin: Historical, Biological, and Materials Perspectives*; ACS symposium series 742; American Chemical Society, Washington, DC, 2000; pp. 250–264.
41. Wiley, J.H.; Atalla, R.H., Band Assignments in the Raman Spectra of Celluloses; *Carbohydr. Res.* 1987, 160, 113–129.
42. Agarwal, U.P.; Ralph, S.A.; Atalla, R.H., FT Raman Spectroscopic Study of Softwood Lignin; Proc. 9th Intern. Symp. Wood Pulping Chem., *Canadian Pulp and Paper Assn.*; Montreal, 1997; pp. 8-1–8-4.
43. Agarwal, U.P.; McSweeny, J.D.; Ralph, S.A., An FT-Raman Study of Softwood, Hardwood, and Chemically Modified Black Spruce MWLs; Proc. 10th Intern. Symp. Wood Pulping Chem., Vol. II, Japan Technical Assn. Pulp & Paper Industry; TAPPI Press, Atlanta, GA, 1999; pp. 136–140.
44. Agarwal, U.P.; Weinstock, I.A.; Atalla, R.H., FT Raman Spectroscopy for Direct Measurement of Lignin Concentrations in Kraft Pulps; *Tappi J.* 2003, 2, 22–26.
45. Bujanovic, B.; Reiner, R.S.; Ralph, S.A.; Agarwal, U.P.; Atalla, R.H., Structural Changes of Residual Lignin of Softwood and Hardwood Kraft Pulp Upon Oxidative Treatment with Polyoxometallates; Proc. 2005 TAPPI Eng. Pulping Environ. Conf., Presentation Book 2, Paper # 30–3; TAPPI, Atlanta, GA, 2005.
46. Jääskeläinen, A.-S.; Saariaho, A.-M.; Vuorinen, T., Quantification of Lignin and Hexenuronic Acids in Bleached Hardwood Kraft Pulps: A New Calibration Method for UVRR Spectroscopy and Evaluation of the Conventional Methods; *J. Wood Chem. Technol.* 2005, 25, 51–65.
47. Saariaho, A.-M.; Jääskeläinen, A.-S.; Matousek, P.; Towrie, M.; Parker, A.W.; Vuorinen, T., Resonance Raman Spectroscopy of Highly Fluorescing Lignin Containing Chemical Pulps: Suppression of Fluorescence with an Optical Kerr Gate; *Holzforschung* 2004, 58, 82–90.
48. Agarwal, U.P.; Terashima, N., FT-Raman Study of Dehydrogenation Polymer (DHP) Lignins; Proc. 12th Intern. Symp. Wood Pulping Chem., Vol. III; Dept. of Forest Ecology and Management, University of Wisconsin, Madison, 2003; pp. 123–126.
49. Agarwal, U.P; Reiner, R.S.; Pandey, A.K.; Ralph, S.A.; Hirth, K.C.; Atalla, R.H., Raman Spectra of Lignin Model Compounds; Proc. 13th Inter. Symp. Wood Fiber Pulping Chem., Vol. 2; Appita, Carlton, Australia, 2005; pp. 377–384.
50. Agarwal, U.P.; Landucci, L.L., FT-Raman Investigation of Bleaching of Spruce Thermomechanical Pulp; *J. Pulp Paper Sci.* 2004, 30, 269–274.
51. Agarwal, U.P.; McSweeny, J.D., Photoyellowing of Thermomechanical Pulps: Looking beyond α-Carbonyl and Ethylenic Groups as the Initiating Structures; *J. Wood Chem. Technol.* 1997, 17, 1–26.
52. Ehrhardt, S.M., An Investigation of the Vibrational Spectra of Lignin Model Compounds, Ph.D. Thesis Dissertation, Institute of Paper Science and Technology, Atlanta, GA, 1984.
53. Larsen, K.L.; Barsberg, S.; Agarwal, U.P.; Elder, T., Prediction and Assignments of Vibrational Bands of Lignin Moieties by DFT; 231st American Chemical Society Meeting, Cell Division, March 26–31, ACS, Washington, DC, 2006, Abstract #7.

54. Saariaho, A.-M.; Argyropoulos, D.S.; Jääskeläinen, A.-S.; Vuorinen, T., Development of the Partial Least Squares Models for the Interpretation of the UV Resonance Raman Spectra of Lignin Model Compounds; *Vib. Spectrosc.* 2005, 37, 111–121.
55. Eichhorn, S.J.; Sirichaisit, J.; Young, R.J., Deformation Mechanisms in Cellulose Fibres, Paper and Wood; *J. Mater. Sci.* 2001, 36, 3129–3135.
56. Gierlinger, N.; Schwanninger, M.; Reinecke, A.; Burgert, I., Molecular Changes during Tensile Deformation of Single Wood Fibers Followed by Raman Microscopy; *Biomacromolecules* 2006, 7, 2077–2081.
57. Czaja, A.D.; Kudryavtsev, A.B.; Schopf, J.W., New Method for the Microscopic, Nondestructive Acquisition of Ultraviolet Resonance Raman Spectra from Plant Cell Walls; *Appl. Spectrosc.* 2006, 60, 352–355.
58. Vester, J.; Felby, C.; Nielsen, O.F.; Barsberg, S., Fourier Transform Raman Difference Spectroscopy for Detection of Lignin Oxidation Products in Thermomechanical Pulp; *Appl. Spectrosc.* 2004, 58, 404–409.
59. Barsberg, S.; Matousek, P.; Towrie, M., Structural Analysis of Lignin by Resonance Raman Spectroscopy; *Macromol. Biosci.* 2005, 5, 743–752.
60. Proniewicz, L.M.; Paluszkiewicz, C.; Weselucha-Birczynska, A.; Baranski, A.; Dutka, D., FT-IR and FT-Raman Study of Hydrothermally Degraded Groundwood Containing Paper; *J. Mol. Struct.* 2002, 614, 345–353.
61. Hunt, C.; Yu, X.; Bond, J.; Agarwal, U.; Atalla, R., Aging of Printing and Writing Papers upon Exposure to Light: Part 2-Mechanical and Chemical Properties; Proc. 12th Intern. Symp. Wood Pulping Chem., Vol. III; Dept. of Forest Ecology and Management, University of Wisconsin, Madison, 2003; pp. 231–234.
62. Jääskeläinen, A.-S.; Saariaho, A.-M.; Vyörykkä, J.; Vuorinen, T.; Matousek, P.; Parker, A.W., Application of UV-Vis and Resonance Raman Spectroscopy to Study Bleaching and Photoyellowing of Thermomechanical Pulps; *Holzforschung* 2006, 60, 231–238.
63. Agarwal, U.P.; Akhtar, M., Understanding Fungus-Induced Brightness Loss of Biomechanical Pulps; Proc. 2000 TAPPI Pulping, Process, and Product Quality Conference; TAPPI, Atlanta, GA, 2000; pp. 1–12.
64. Atalla, R.H.; Ranua, J.; Malcolm, E.W., Raman Spectroscopic Studies of the Structures of Cellulose: A Comparison of Kraft and Sulfite Pulps; *Tappi J.* 1984, 67, 96–99.
65. Saariaho, A.-M.; Hortling, B.; Jaaskelainen, A.-S.; Tamminen, T.; Vuorinen, T., Simultaneous Quantification of Residual Lignin and Hexenuronic Acid from Chemical Pulps with UV Resonance Raman Spectroscopy and Multivariate Calibration; *J. Pulp Paper Sci.* 2003, 29, 363–370.
66. Agarwal, U.P.; Reiner, R.S.; Ralph, S.A.; Using Nano- and Micro-Particles of Silver in Lignin Analysis; Proc. TAPPI Intern. Conf. Nanotechnol.; TAPPI Press, Atlanta, GA, April 26–28, 2006; NanoCD-06, ISBN 1-59510-121-7.
67. Potthas, A.; Rosenau, T.; Kosma, P.; Saariaho, A.-M.; Vuorinen T., On the Nature of Carbonyl Groups in Cellulosic Pulps; *Cellulose* 2005, 12, 43–50.
68. Niemelä, P.; Hietala, E.; Ollanketo, J.; Tornberg, J.; Pirttinen, E.; Stenius, P., FT-Raman Spectroscopy as a Tool for Analyzing the Composition of Recycled Paper Pulp; *Prog. Pap. Recycling* 1999, 4, 15–24.
69. Somwang, K.; Enomae, T.; Isogai, A.; Onabe, F., Changes in Crystallinity and Re-swelling Capability of Pulp Fibers by Recycling Treatment; *Japan Tappi J.* 2002, 56, 863–869.
70. Nuopponen M.; Vuorinen, T.; Jämsä, S.; Viitaniemi, P., Thermal Modifications in Softwood Studied by FT-IR and UV Resonance Raman Spectroscopies; *J. Wood Chem. Technol.* 2004, 24, 13–26.
71. Yamauchi, S.; Kurimoto, Y., Raman Spectroscopic Study on Pyrolyzed Wood and Bark of Japanese Cedar: Temperature Dependence of Raman Parameters; *J. Wood Sci.* 2003, 49, 235–240.

72. Yamauchi, S.; Iijima, Y.; Doi, S., Spectrochemical Characterization by FT-Raman Spectroscopy at Low Temperatures: Japanese Larch and Beech; *J. Wood Sci.* 2005, 51, 498–506.
73. Kurti, E.; Heyd, D.V.; Wylie, R.S., Raman Microscopy for the Quantitation of Propiconazole in White Spruce; *Wood Sci. Technol.* 2005, 39, 618–629.
74. Gierlinger, N.; Hansmann, C.; Röder, T.; Sixta, H.; Gindl, W.; Wimmer, R., Comparison of UV and Confocal Raman Microscopy to Measure the Melamine–Formaldehyde Resin Content Within Cell Walls of Impregnated Spruce Wood; *Holzforschung* 2005, 59, 210–213.
75. Scheepers, M.L.; Gelan, J.M.; Carleer, R.A. *et al.*, Investigation of Melamine–Formaldehyde Cure by Fourier Transform Raman Spectroscopy; *Vib. Spectrosc.* 1993, 6, 55–69.
76. Yamauchi, S.; Tamura, Y.; Kurimoto, Y.; Koizumi, A., Vibrational Spectroscopic Studies on Wood and Wood Based Materials. III. Identification of Adhesives in Wood by Using FT-Raman Spectroscopy as a Nondestructive Analytical Method; *Nippon Setchaku Gakkaishi* 1997, 33, 380–388.
77. Yamauchi, S.; Doi, S., Raman Spectroscopic Study on the Behavior of Boric Acid in Wood; *J. Wood Sci.* 2003, 49, 227–234.
78. Agarwal, U.P.; Reiner, R.S.; Ralph, S.A., Dependable Cellulose I Crystallinity Determination using Near-IR FT-Raman; CELL Division, Abstract 95, 233rd ACS National Meeting; Chicago, March 25–29, 2007.
79. Agarwal, U.P.; Kawai, N., Self-Absorption Phenomenon in Near-Infrared Fourier Transform Raman Spectroscopy of Cellulosic and Lignocellulosic Materials; *Appl. Spectrosc.* 2005, 59, 385–388.

Chapter 3

Surface Characterization of Mechanical Pulp Fibers by Contact Angle Measurement, Polyelectrolyte Adsorption, XPS, and AFM

Per Stenius and Krista Koljonen

Abstract

The surfaces of pressure groundwood (PG), thermomechanical pulp (TMP), and chemithermomechanical pulp (CTMP) were studied by atomic force microscopy (AFM), x-ray photoelectron spectroscopy (XPS), contact angle measurement, and polyelectrolyte adsorption. The combination of these methods yielded a comprehensive picture of the chemistry, adhesive properties and morphology of the fiber surfaces and how they were affected by washing, peroxide bleaching, dithionite bleaching, and ozone treatment. Overall, the fiber surfaces were found to be very heterogeneous. Sulfite treatment in the production of CTMP pulp, peroxide bleaching under alkaline conditions, and ozone treatment modified especially the lignin and/or the pectins so that more acidic groups were introduced into the pulps. The adhesion between water and fibers increased when hydrophobic extractives (pitch) were removed by extraction with dichloromethane.

3.1 Introduction

In the development of theories and experiments aiming at describing the papermaking properties of mechanical pulps, fiber characteristics such as size distribution, coarseness, and surface properties are main subjects of interest [1–3]. Numerous investigations of various wood pulps have shown that the surface chemical composition of the pulps can be very different from their bulk composition of the pulps. The surfaces of fibers and fines come into contact during web forming, and hence, surface chemistry and morphology play an important role in the formation of bonds between fibers, between fibers and fines, and between fibers and different additives such as polyelectrolytes and fillers. The importance of the surface properties in coating and printing is obvious. Knowledge of the surface properties of mechanical pulp fibers will promote understanding of (1) factors affecting the different bonds in the paper network; (2) how the surface should be chemically modified to gain good strength/bonding properties of paper; (3) how to govern interactions between surfaces and different additives; and/or (4) how to tailor wetting properties of surfaces for different coatings and printing inks.

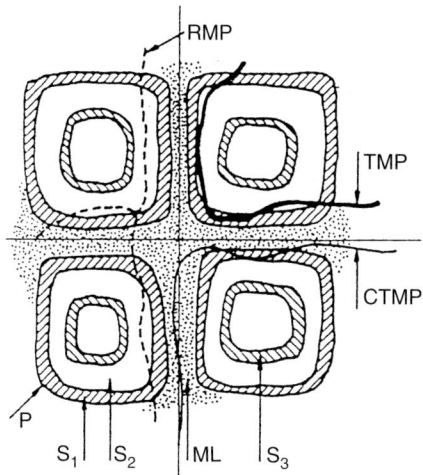

Figure 3.1 Schematic of fiber rupture in different mechanical pulping processes [8].

Mechanical pulp fiber surfaces are rough and chemically heterogeneous, due to rupture of different fiber layers during the mechanical treatment (Figure 3.1) and to changes in composition induced by chemical treatments. Accordingly, a range of methods is required to characterize the surface properties. This chapter describes the characterization of mechanical pulp fiber surfaces using AFM (morphology) [4], Wilhelmy balance (surface energy/adhesion) [5], XPS, also called electron spectroscopy for chemical analysis (ESCA) (chemistry) [4], and polyelectrolyte adsorption [6] (charge). The chemistry of mechanical pulp surfaces has also been studied by secondary ion mass spectrometry (SIMS), in particular time-of-flight SIMS (ToF-SIMS). This technique, which yields more detailed but less quantitative information of the surface chemistry than XPS, is discussed in Chapter 6 and will be only briefly considered here.

The aim of the research presented here was to evaluate how the surfaces of mechanical pulps – pressure groundwood (PGW), thermomechanical pulp (TMP), and chemithermomechanical pulp (CTMP) are modified by the following treatments: (1) washing in alkali (E), (2) peroxide (P) and dithionite (Y) bleaching, (3) ozone treatment (Z), and (4) water washing (w) in comparison with the untreated (unwashed) pulp (uw) (Table 3.1). Table 3.2 shows the approximate analysis depths of the various characterization methods. Following an introduction to mechanical pulps and mechanical fiber surfaces is given, each method is shortly described, earlier investigations are briefly reviewed and some recent results are presented. More detailed descriptions of the results and the effects of pulp surface properties on paper properties are given elsewhere [4–7].

3.2 Mechanical pulps

Depending on the manufacture process of mechanical pulps, different cell wall layers are exposed (Figure 3.1). During manufacture of TMP and PGW pulps, the separation mainly

Table 3.1 Notation and treatments of mechanical pulps prepared from Norwegian spruce (*Picea abies*) [7], chemical charges are based on oven-dried weight of the pulps and all the treatments were made in laboratory.

Notation	Description	Treatment
PGW-uw	Pressure groundwood (CSF 45 mL) from mill	None
PGW-w	Washed PGW	Three times washing with deionized water
PGW-E	Alkaline-treated PGW	pH 11.2 (NaOH), 60°C, 1.5 h, 10% consistency[a]
PGW-P	Peroxide-bleached PGW	3% H_2O_2 at pH 11.2 (NaOH), 60°C and 10% consistency for 1.5 h[a]
PGW-Y	Dithionite-bleached PGW	1.5% $Na_2S_2O_4$ at pH 6.0, 60°C and 3% consistency for 1 h[a]
PGW-Z	Ozone-treated PGW	Dispersed (in deionized water) at 1% consistency and 60°C for 3 h, centrifuged (1500 rpm, 60 min), ozone (55 min, dosage 5.2%, 12% consistency), pH adjusted to 7.0, shives removed by screening, dispersed at 1% consistency and 60°C for 2 h
TMP-uw	Thermomechanical pulp (CSF 120 mL) from mill	None
TMP-w	Washed TMP	Same treatment as PGW-w
TMP-E	Alkaline-treated TMP	Same treatment as PGW-E
TMP-P	Peroxide-bleached TMP	Same treatment as PGW-P
TMP-P2	Peroxide-bleached TMP (CSF 35 mL) from mill	None
CTMP-uw	Chemithermomechanical pulp (CSF 470 mL) from mill	None
CTMP-P-uw	Peroxide-bleached CTMP (CSF 465 mL) from mill	None

[a] After the treatments, the pH was adjusted to 5.3 and the pulps were washed three times with water.

occurs between primary and secondary walls, chiefly between the S1 and S2 layers [8–10]. The sulfonation of lignin by sodium sulfite used in CTMP pulping makes the cell wall more hydrophilic and easier to soften [11,12]. It has been suggested that the rupture of the fiber walls occurs then to a greater extent in the primary layer (P) and in the middle lamella (ML) [8].

Mechanical fibers are rather stiff, and their composition is almost the same as that of the wood. Therefore, mechanical and/or chemical modification is usually needed to render them more flexible/swelling. Acidic groups are introduced by alkaline treatment (associated with hydrogen peroxide bleaching) [13–15] and sulfonation [14,16]. Their presence results in swelling of fibers due to electro-osmotic interactions, but the nature of the counter ions of the acidic groups is also very important in determining the refining efficiency

Table 3.2 Surface analysis methods used in our studies and their analysis depths [7].

Method	1. Input radiation 2. Radiation-detected	Application	Analysis depth
ESCA/XPS	1. X-ray 2. Photoelectrons from inner orbitals	Elemental composition, chemical states	Few atom layers, 4–10 nm
ToF-SIMS	1. Primary ions 2. Secondary ions (atoms, small and larger molecules)	Elemental and molecular information	~1 nm
AFM	–	Topography	~1 nm
Wilhelmy balance	–	Contact angle between single fiber and liquid	Outermost surface

[17,18]. Plasma treatment [19–23], enzymatic modification [16,24–27], and hemicellulose adsorption [28–30] have also been used to modify the surfaces of mechanical pulps.

The fine particles formed during mechanical pulping can be roughly classified as fibrillar and flake-like materials. The fibrillar material, thought to originate mainly from the primary and secondary cell walls, has a high specific surface area and a low lignin content. The flakes consist mainly of ML fragments and have a lower specific surface area and a higher lignin content. The flakes provide strength to paper sheets less efficiently than the fibrils [31].

3.3 Morphology of mechanical pulp surfaces

Atomic force microscopy is one of the most useful techniques available for the study of the morphology of mechanical pulp surfaces. The method provides detailed information on the fibrillar and globular surface structures with dimensions of 10–100 nm [32,33] and on the structure of fines. Some results obtained are

- Mechanical pulp fiber surfaces to a large extent consist of layers with well-oriented parallel microfibrillar structures [34].
- The surface roughness of pine TMP fibers increases with increasing refiner pressure [35].
- Comparison of TMP with RTS (low retention, high temperature, high speed) processed pulp shows that the raw material from which the pulp is manufactured has a greater effect on the exposed cell wall layer than the temperature or type of process used [36,37].
- Fibrillar and granular structures present on the fiber surface after one-stage TMP refining have been identified as P/S1 cell wall layers. Two-stage refining results in much more heterogeneous surfaces [37].

AFM phase-contrast images usually reveal more details of the fiber structure than topography images. Examples of a topography image and a phase image of an unbleached PGW fiber are given in Figure 3.2. Often there is no clear phase contrast between different components, which are distinguished mainly on the basis of structural differences. However, the phase image (image on the right, Figure 3.2) shows a distinct phase contrast. Because

Figure 3.2 Topography (left) and phase (right) images of an unbleached PGW fiber surface. The image size was 3 μm × 3 μm [7]. Instruments and settings: NanoScope IIIa Multimode scanning probe microscope, commercial Si cantilevers (Digital Instruments, Inc. Santa Barbara), tapping mode in air, resonance frequency 300–360 Hz, free amplitude ∼20 nm, set-point ratio r_{sp} 0.4–0.6, no image processing.

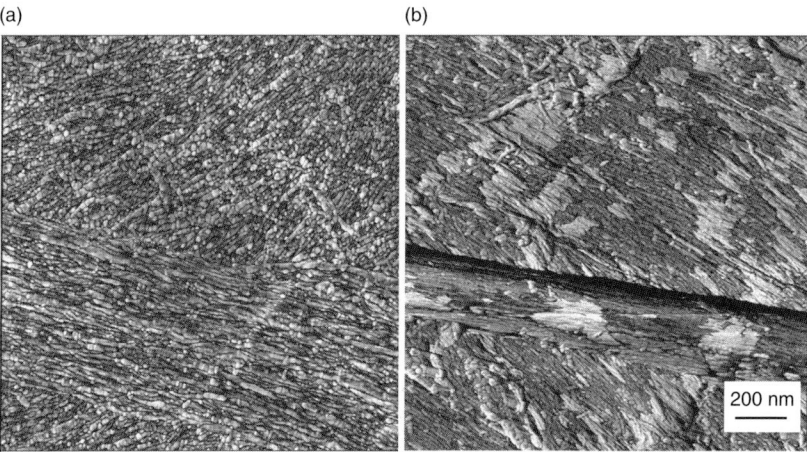

Figure 3.3 AFM phase image of the PGW w before (a) and after (b) ozone treatment. The image size is 3 μm × 3 μm [4]. Instrument and settings are the same as those described in the caption of Figure 3.2.

a rather light tapping was used, this contrast is probably due to the differences in hydrophobicity. The AFM tip adheres more strongly to hydrophilic areas that appear darker in the image.

AFM micrographs also indicated that S2 layers and a more irregular layer (S1 or P) were exposed by pressure grinding (PGW) as well as chip refining (TMP) [4]. The surfaces were not markedly different although the pulp freeness levels were quite different.

The surface morphology of PGW and TMP fibers was not changed by peroxide bleaching. However, the PGW fiber surfaces were modified by ozone treatment (Figure 3.3). Distinct

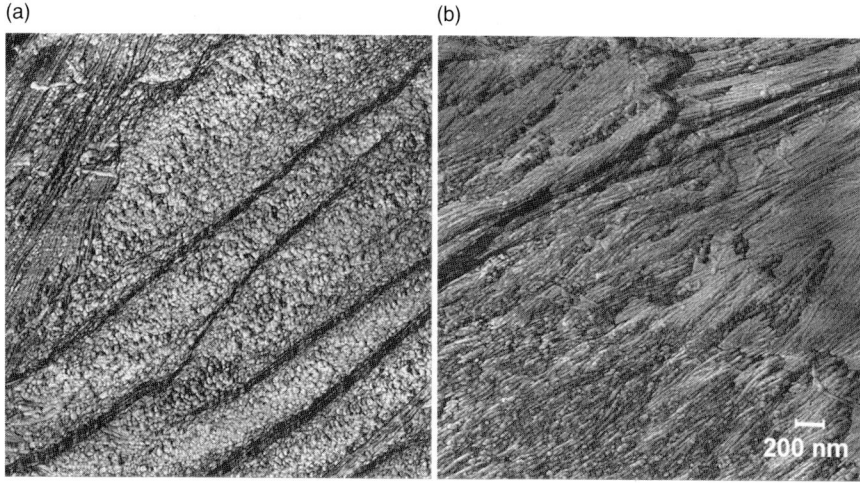

Figure 3.4 AFM phase image of the CTMP-uw before (a) and after (b) peroxide bleaching. The image size is 3 μm × 3 μm [4]. Instrument and settings are the same as those described in the caption of Figure 3.2.

granules were observed on the surface of unbleached CTMP fibers (Figure 3.4a). These granules were no longer present after peroxide bleaching (Figure 3.4b).

Despite the ability of AFM to distinguish between different cell wall structures, clarification of the nature of the granules has turned out to be difficult. They have been linked to, for example, extractives in unbleached CTMP [38], adsorbed xylan in modified CTMP [28] or lignin in kraft pulps [39–42]. Furthermore, different extractives seem to behave differently on the fiber surfaces, which does not make the interpretation any easier. Stearic acid and calcium stearate form aggregates with diameter of 100–500 nm, while oleic acid forms a more uniform layer on kraft pulp surface [43]. When hexane-extracted extractives (pitch) from TMP were precipitated on bleached kraft fibers, they formed a thin film, as verified by XPS [44].

3.4 Surface chemistry of mechanical pulps

XPS and SIMS have yielded detailed information on the surface chemistry of mechanical pulps. Infrared (IR) and Raman spectroscopies have also been utilized, but the analysis depths of these methods are orders of magnitude higher (micrometers) than those of XPS or SIMS (nanometers). Thus, they do not probe the surface layers on a molecular scale.

3.4.1 X-ray photoelectron spectroscopy

XPS was first applied to pulp fibers by Dorris and Gray [45–47] in the late 1970s, and since then it has been used in numerous investigations of the elemental and chemical composition of pulp/fiber/fines surfaces. The most common approach is to combine XPS with solvent extraction (acetone or dichloromethane, DCM) to evaluate the amount of extractives (resin)

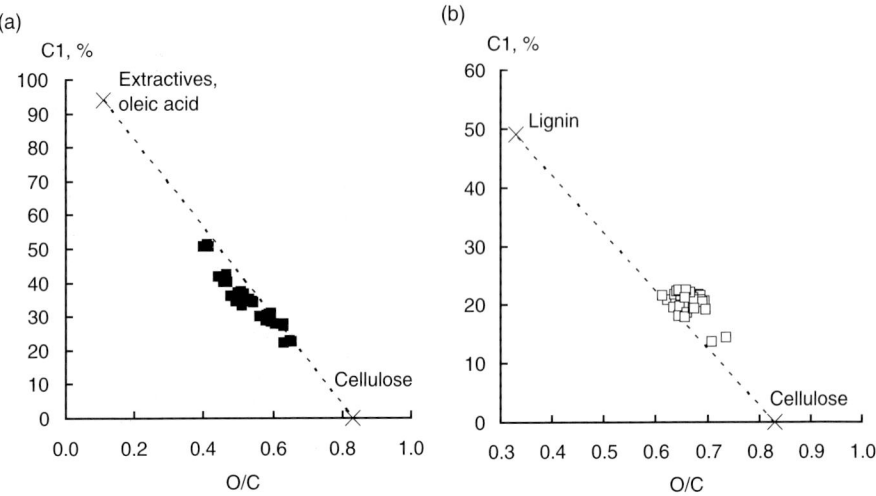

Figure 3.5 C1 as a function of O/C of the mechanical pulps investigated before (a) and after (b) DCM extraction [7]. Kratos Analytical AXIS 165 electron spectrometer, monochromated Al$_{K_\alpha}$ x-ray source (run at 100 W), survey scans (O1s, C1s) at 80 eV pass energy and 1.0 eV step, high-resolution spectra (C1s) at 20 eV pass energy and 0.1 eV step. Theoretical values for cellulose, lignin, and extractives from Reference 60.

and lignin on the surface. Using this approach, the amounts of lignin and extractives on PGW, TMP [45–52], and CTMP pulp sheet surfaces [19,48,53,54] have been evaluated. It has been shown that

- Lower amounts of lignin or higher amounts of carbohydrate on the surfaces of CTMP pulps are favored by a lower temperature in press-drying (170–140°C) [53], a lower sodium sulfite charge during sulfonation [54], and the manufacture of pulp by steam explosion [49].
- Extractives are enriched on the surface of fines [55], especially flake-like fines of TMP pulp [56].

We have conducted a detailed XPS study of the surface chemistry of different mechanical pulps [4]. The correlation of high-resolution C1s data (C1 carbon) to O/C atomic ratios from wide-scan spectra was established, as recommended by Johansson *et al.* [57–59] (Figure 3.5). Quantification of surface lignin and extractives was based on C—C (C1 carbon) percentages as these were found to give more reproducible results than O/C ratios. The larger scattering in O/C ratios especially for the extracted samples (Figure 3.5b) may be due to adsorbed water present on the hygroscopic fiber surface.

Assuming that pure milled wood lignin contains 49% C1 carbon while pure cellulose contains no C1 carbon [7], the surface content of lignin was estimated using Equation (3.1):

$$\phi_{\text{lignin}} = \frac{(C1_{\text{extracted pulp}} - a) \times 100\%}{49\%} \quad (3.1)$$

where $C1_{\text{extracted pulp}}$ is the relative amount of the C1 component in the high-resolution C1s spectrum of the extracted pulp and a is the contribution to this component from surface contamination. The value used for a was 2% because this was the lowest amount of C1 detected by the XPS instrument on the surface of a fully bleached pulp. The surface content of extractives was calculated using Equation (3.2):

$$\phi_{\text{extractives}} = C1_{\text{pulp}} - C1_{\text{extracted pulp}} \qquad (3.2)$$

It was found that the surfaces of the mechanical pulps were 50–75% covered by lignin and extractives [4]. The bulk fibers contained only 30% of these components. Lignin was slightly enriched on the surface (surface 35–40%, bulk 25–29%). The amount of lignin on the surface was not readily reduced by E, P, or Y treatments but lowered by ozone treatment (Figure 3.6).

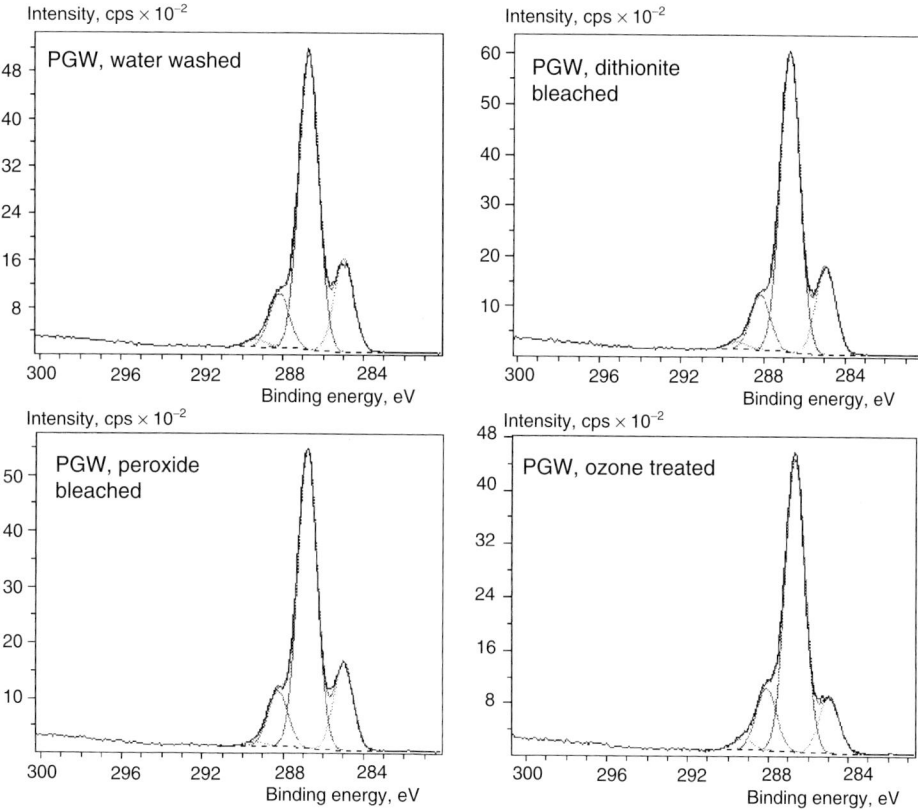

Figure 3.6 High-resolution C1s XPS spectra of water-washed (PGW-w), dithionite-bleached (PGW-Y), peroxide-bleached (PGW-P), and ozone-treated (PGW-Z) PGW pulps, respectively. Before taking the spectra, extractives were removed with DCM extraction [4]. Spectrometer and settings were the same as those described in Figure 3.5 caption, curve fitting was done with symmetric Gaussians; chemical shifts for C2, C3, and C4 relative to C1 were assumed to be 1.7, 3.1, and 4.4 eV, respectively.

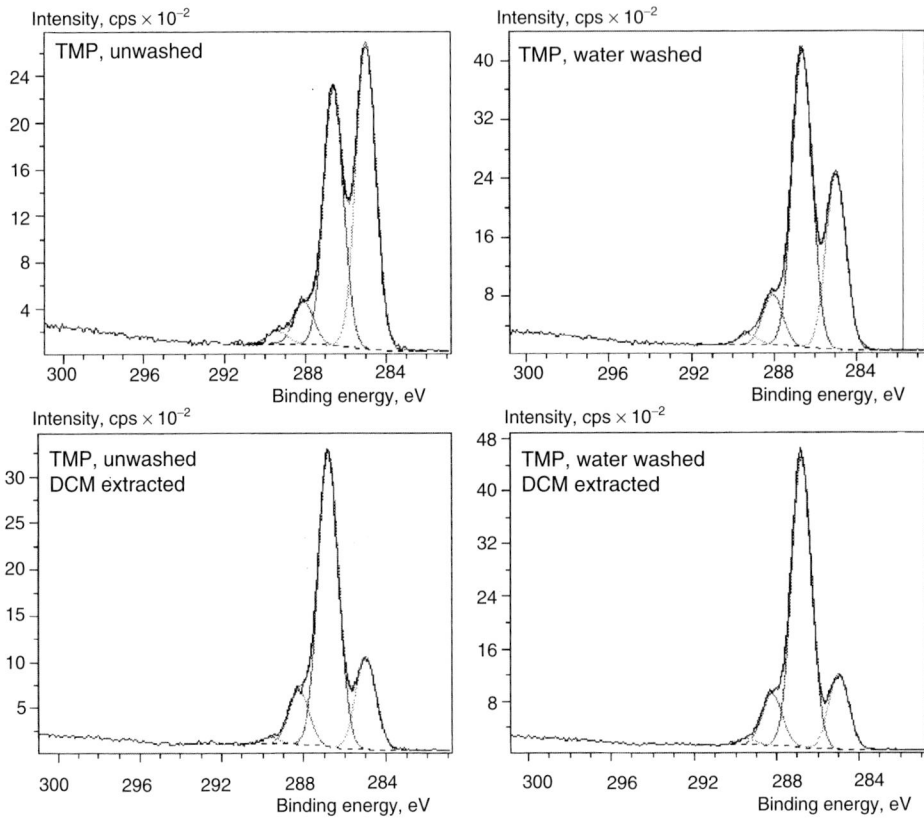

Figure 3.7 High-resolution C1s carbon spectra of unwashed (TMP-uw), water-washed (TMP-w), unwashed and DCM extracted, and water-washed and DCM extracted TMP pulps [4]. Spectrometer, settings and fitting are the same as those described in the caption of Figure 3.6.

Extractives were strongly enriched on the surface (surface 13–32%, bulk 0.2–2.3%). The effect of washing and extraction with DCM on TMP-uw is shown in Figure 3.7. Washing with water reduced the amount of surface extractives by 10–15%.

By applying the method introduced by Tougaard [61,62] to TMP, Johansson *et al.* [63] obtained quantitative information on the depth distribution of surface lignin and extractives. An ECF bleached softwood kraft pulp (BKP) was used for comparison. Peak-to-background ratios (D) were determined by dividing the area under the O1s XPS peak with the background intensity at 40 eV from the O1s peak center (Figure 3.8) [44]. A decrease in D indicates stronger inelastic scattering of the electrons and thus a longer path traveled by them in the solid. For polymeric materials, $D > 25$ eV indicates homogeneous depth distributions or clusters thicker than the XPS escape depth [63]. $D < 25$ eV suggests surface-depleted distributions such as thin surface films. Table 3.3 shows that surface coverage by extractives and lignin on the TMP-uw was much higher than on the BKP but the layers were, on average, thinner than the escape depth. After the removal of extractives, TMP was still partially covered with materials containing less oxygen than cellulose, but the covering

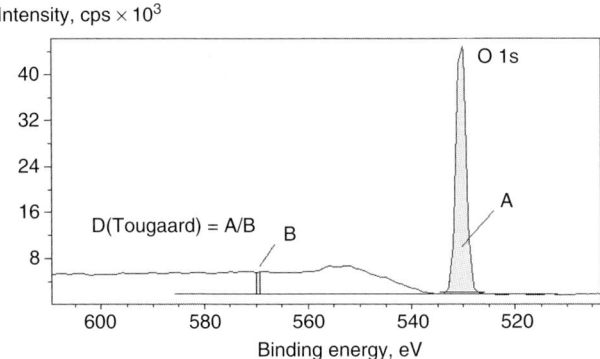

Figure 3.8 A close up from the survey spectra region of the O1s peak for TMP-uw. D values are calculated by dividing the peak area (A) by the background area (B) at 40 eV from the peak center [44]. Spectrometer and settings are the same as those described in the caption of Figure 3.6.

Table 3.3 O/C ratios, relative amounts of C1 carbon (carbon atoms bound only to other carbon atoms or hydrogen), relative surface areas (Φ) for lignin and extractives (extracted with DCM) and peak-to-background ratios (D) determined by XPS [63].

Sample	XPS analysis				D (eV)
	O/C ratio	C1 (% of total carbon)	Φ_{lignin} (atom-%)	$\Phi_{extractives}$ (atom-%)	
TMP	0.41	51	40	29	18
TMP, extracted	0.65	22	40	–	24
ECF	0.80	4	2	1	25
ECF, extracted	0.80	3	2	–	25

layer was thicker than the escape depth. Thus, on the TMP-uw the extractives formed a film that was thinner than the analysis depth of XPS ($\sim$10 nm).

Westermark [51] and Heijnesson-Hultén et al. [64,65] labeled the lignin in the fibers by mercurization and determined the amount of mercury bound to lignin in the surface by XPS. Their results confirm that the lignin content of different fibers is slightly higher on the surface than in the bulk.

3.4.2 Time-of-flight secondary ion mass spectrometry

ToF-SIMS has recently been used for the analysis of the surfaces of fibers and fines [66–72]. It has been shown that:

- The fines of TMP pulp contain more extractives, lignin, and pectin than the bulk fibers [67].

- The surfaces of fibrillar and flake-like fines of unbleached groundwood (GW), PGW, and TMP contain more lignin and extractives than the surface of ray cells [68,69]. Fibrillar fines in TMP contain more extractives, cellulose and mannan but less lignin than flakes [70]. Flake-like fines in CTMP pulp are enriched with extractives [69].
- Peroxide bleaching decreases surface coverage by extractives on different mechanical pulps. Residual extractives are adsorbed more on fibrillar than on flake-like fines [71].
- Fibrillar fines consist of outer fiber wall layers (P + S1). The amount of S1 material increases with refining energy. Flakes produced during mainline refining originate from middle lamella and P. In reject refining they are released also from the outer S1 wall layer [72].

Extractives are enriched on the surface of fines [55,67–69]. Studies of whether there are more extractives on flakes [56,69] or on fibrils [70,71] have given somewhat contradictory results. Very recently it has been shown by XPS that the surfaces of flakes from Norwegian spruce contain more extractives than fibrils [73].

3.4.3 Polyelectrolyte adsorption, anionic groups, and surface charge

The content of anionic (acidic) groups in pulps has been determined by conductometric titration [74–76], potentiometric titration [77,78], polyelectrolyte adsorption [14,79–81], magnesium or benzidinium ion exchange [82,83], and methylene blue adsorption [84,85]. Several authors have compared these methods [86–90].

Some of cations used in these methods may not have access to all anionic groups in the fibers, but only polyelectrolyte adsorption is capable of selectively detecting the outermost surface charge. In this method, adsorption isotherms of polycations with different molecular weights (M_w) are determined by equilibration of the fibers with polycation solutions. The equilibrium concentration of polycation in solution is determined by titration with a polyanion. The point of equivalence can be detected by using a cationic dye indicator (e.g., toluidine blue) or a particle charge detector (PCD). The amount of cationic groups adsorbed at the plateau level of the isotherm is assumed to be stoichiometrically equivalent to the amount of accessible anionic groups in the fibers. The accessibility depends on the M_w of the polycation [91,92]. All groups in the fibers will be accessible to a polymer with low M_w, while polymers with sufficiently high M_w ($>10^5$) will have access only to groups on the outermost surface of the fibers. It is important that the ionic strength is chosen to ensure stoichiometric reaction with the accessible anionic groups [86,93–95]. This will generally be the case as long as the charge density of the polymer is high and electrostatic interactions are not screened by the electrolyte in the solution.

The effect of different treatments on the surface charge of mechanical pulps in their sodium form was investigated using highly charged polycations [6]. Polybrene (i.e., 1,5-dimethyl-1,5-diazaundecamethylene polymethobromide) ($M_w \simeq 8.0 \times 10^3$, charge density = 5.35 mmol g^{-1}) was assumed to have access to all of the anionic groups in the fibers. Poly(diallyldimethylammonium chloride) (PDADMAC) with charge density 6.19 mmol g^{-1} and M_w 1.0–3.0 × 10^5 and M_w > 1.0–3.0 × 10^5 were used to determine the charge distribution and surface charge. The amount of nonadsorbed polycation was determined by titration with sodium polyethylenesulfonate. Typical adsorption isotherms are shown in Figure 3.9.

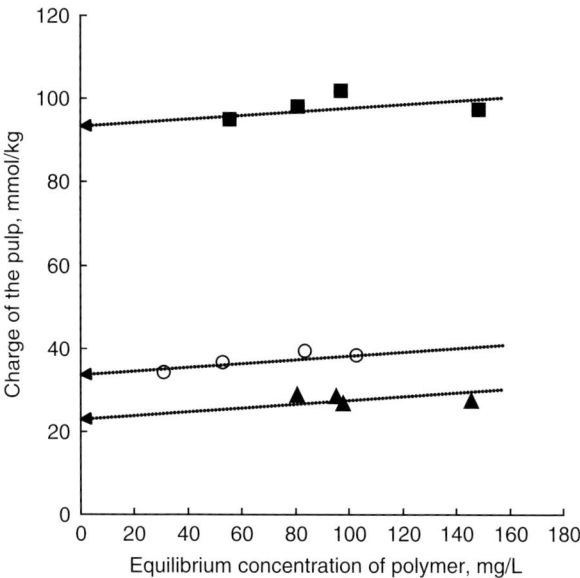

Figure 3.9 Adsorption isotherms for polybrene (■), and PDADMAC with $M_w = 1.0–3.0 \times 10^5$ (○) and $M_w > 3.0 \times 10^5$ (▲), respectively, in 0.01 M NaCl and pH $\simeq$ 7.0 on unbleached PGW pulp with its anionic groups in sodium form. The amount of polycation corresponding to neutralization of the accessible charge on the fibers was determined by extrapolation of the linear part of the isotherm to zero polymer concentration (arrows).

The charge distributions in PGW-uw and TMP-uw were found to be similar (surface charge accessible to PDADMAC with $M_w > 3.0 \times 10^5 = 22$ mmol kg^{-1}, surface charge accessible to PDADMAC with M_w 1.0 – 3.0 $\times 10^5 = 35$ mmol kg^{-1}, total charge = 80–95 mmol kg^{-1}). The total charge of CTMP-uw, which contained sulfonate groups, was higher (surface charge = 35 mmol kg^{-1}, total charge = 140–160 mmol kg^{-1}, amount of SO_3^- at pH 2.5 $\simeq$ 50 mmol kg^{-1}). Figure 3.10 shows that end-point detection with the PCD yielded charges that were consistently somewhat lower than detection with the dye indicator. Treatments of the PGW pulp caused the following changes in the charge profiles (total charge profiles shown in Figure 3.10):

Dithionite bleaching (Y). The surface and total charges increased only moderately.

Alkaline treatment (E) and peroxide bleaching (P). Both the surface and total charges increased two- to threefold over those of the untreated pulp. Other investigations [96] have shown that during treatments in alkali, free carboxyl groups are formed by demethylation of methyl ester groups in pectins. Peroxide bleaching introduced some additional carboxyl groups into the pulp.

Ozone treatment (Z). The total charge and especially the surface charge of the PGW increased markedly. The surface layer of PGW-Z was more acidic than that of PGW-uw. The increase in the charge was most probably due to the formation of carboxylic groups in lignin [97].

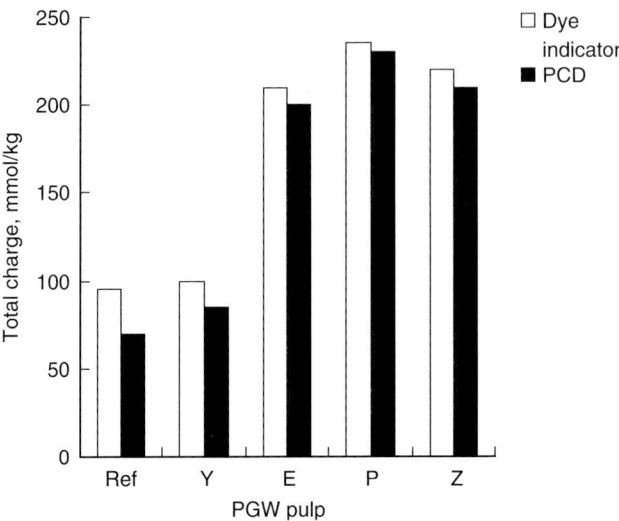

Figure 3.10 Total charge (polybrene adsorption) of the differently-treated PGW pulps measured with the dye indicator (toluidine blue, □) and particle charge detector (Mütek, ■); uncertainty in the determinations = ±5 mmol kg^{-1} [7].

Figure 3.11 shows the total charge against the surface charge for various pulps: unbleached, Y- or P-bleached, and E-treated PGW, TMP and CTMP pulps, as well as Z-treated PGW. In most pulps ∼35% of the anionic groups were accessible to the polyelectrolyte with M_w 100 000–300 000. The value for PGW-Z was ∼70%. This selective effect of ozone treatment on the surface charge is consistent with the well-known fact that penetration of ozone into fibers is relatively slow, but it may also partly be explained by changes in topography.

An independent determination of the amount of carboxyl groups on the fiber surface was obtained by XPS analysis [4]. Usually, the amount of carboxyl groups on the surface is determined by measuring the relative intensity of the peak representing carboxyl groups in the high-resolution XPS C1s spectrum (C4 carbon, see Figures 3.6 and 3.7). However, the intensity of this peak is low and it includes carboxyls in methyl esters. Therefore, the free carboxyl groups of the PGW pulps were label ed by conversion of the fibers into calcium form, followed by determination of calcium by XPS. One calcium ion is assumed to correspond to two carboxyl groups on the fiber surface. As shown in Figure 3.12, the calcium content of the surface (in atom-%) is linearly dependent on the surface charge determined by adsorption of PDADMAC. This confirms that PDADMAC adsorption does give a measure of the surface charge of the pulp.

3.5 Surface energy of mechanical pulp fibers

Contact angle measurements using the Wilhelmy balance have been widely applied to single fibers [98–102], yielding information on factors governing the surface energy

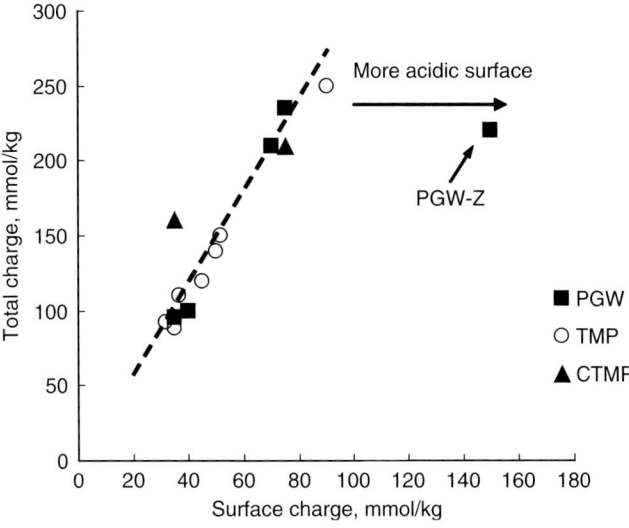

Figure 3.11 Total charge as a function of surface charge for various unbleached, Y- or P-bleached, and E-treated PGW (■), TMP (○) and CTMP (▲); the data point at the far right (■) was obtained for PGW-Z. Uncertainty in the determinations = ±5 mmol/kg [7].

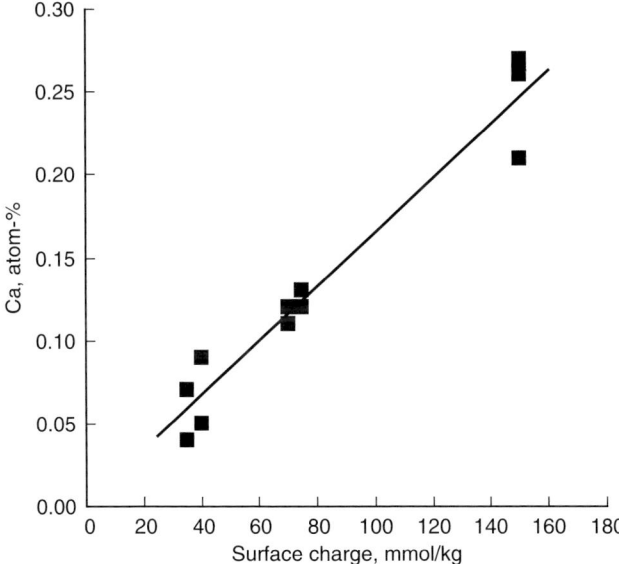

Figure 3.12 Content of calcium on the outer surface of the PGW pulps determined by XPS vs surface charge determined by polyelectrolyte titration (toluidine blue) [6]. Spectrometer is the same as that described in the caption of Figure 3.5, high resolution spectra (Ca 2p) at 20 eV pass energy and 0.1 eV step.

of kraft [103,104], recycled [105,106], and mechanical fibers [38,54,101,107–110]. This method measures the downward force acting upon a fiber suspended vertically through a liquid surface [101,111,112]. Inverse gas chromatography (IGC) has also been extensively used to evaluate the surface energy and acid/base properties of wood fibers [104,113–117] and wood surfaces [118–120]. However, single fiber wetting and IGC measurements have provided different results. According to the wetting data, most wood fibers are basic, whereas the IGC results show that they are predominantly acidic [114,117,121,122]. It has been suggested that IGC mainly assesses the high-energy sites on the surface, while the advancing contact angle mainly probes low-energy domains [114,123].

We measured the wetting properties of single pulp fibers using the Wilhelmy balance technique [5].

The contact angles θ were calculated from Equation (3.3):

$$F = p\gamma_L \cos\theta + mg - (\rho_L - \rho_v)Vg \tag{3.3}$$

where F is the force acting on the fiber, p is the wetted perimeter, γ_L is the surface tension of the liquid, m is fiber mass, V is submerged fiber volume, and ρ_L and ρ_v are the liquid and vapor densities, respectively. The buoyancy force, $mg - (\rho_L - \rho_v)Vg$, could be neglected [101,112].

The receding contact angle is assumed to be zero [99,101,124–126]. Then, by combining measurements of average advancing and receding forces (F_A and F_R), the advancing contact angle, θ_A, can be calculated from Equation (3.4):

$$\cos\theta_A = \frac{F_A}{F_R} \tag{3.4}$$

Contact angles were calculated only from the first immersion cycle of the fibers. Two additional cycles were used to confirm that no marked changes in fiber properties occurred due to the repeated immersion and withdrawal of the fibers.

The work of adhesion, W_a, between the liquid and the fiber was calculated using the Dupré equation, Equation (3.5):

$$W_a = \gamma_L(1 + \cos\theta_A) \tag{3.5}$$

The Lewis acid–base (W_a^{LW}) and Lifshitz–van der Waals (W_a) contributions to the work of adhesion (W_a) were calculated using the van Oss–Chaudhury–Good approach [127–129] (Equation [3.6]):

$$W_a = W_a^{LW} + W_a^{AB} = 2\left(\sqrt{\gamma_S^{LW}\gamma_L^{AB}}\right) + 2\left(\sqrt{\gamma_S^+\gamma_L^-} + \sqrt{\gamma_S^-\gamma_S^+}\right) \tag{3.6}$$

where the subscripts L and S refer to liquid and solid phases and γ^{LW}, γ^+, and γ^- are the Lifshitz–van der Waals, acid, and base parameters, respectively. These were determined using three solvents with known acid–base properties: water, ethylene glycol, and α-bromonaphthalene (BN). Values of γ_L^{LW}, γ_L^+, and γ_L^- proposed by Della Volpe et al. [130,131] were used. BN, for which γ_l^+ and $\gamma_L^- \approx 0$, was used to calculate $W_a^{LW,BN}$ from Equation (3.7):

$$W_a^{LW,BN} = \frac{1}{4}\gamma_L^{BN}(1 + \cos\theta^{BN})^2 \tag{3.7}$$

Table 3.4 Fiber surface free energy parameters (mJ m^{-2}) [5] calculated from Equations (3.6) and (3.7) using liquid parameters from References 130 and 131.

Sample	Surface free energy parameters (mJ m^{-2})				
	γ_S	γ_S^{LW}	γ_S^{AB}	γ_S^+	γ_S^-
PGW fibers					
Unwashed (uw)	45	40	4	0.4	12
Water washed (w)	45	43	2	0.3	4
Alkaline E (w)	45	41	4	0.3	12
Peroxide P (w)	46	43	3	0.3	8
Dithionite Y(w)	46	43	3	0.2	11
Ozone Z (w)	48	41	7	0.5	22
TMP fibers					
Unwashed (uw)	43	41	2	0.1	9
(uw) DCM extracted (e)	47	43	4	0.6	7
P (uw)	47	43	4	0.5	7
P (uw) (e)	48	43	5	0.5	13
P (w)	47	42	5	0.4	14
P (w) (e)	49	43	6	0.5	16
CTMP fibers					
Unwashed (uw)	48	43	5	0.5	15
(uw) (e)	48	41	6	0.6	18
P (uw)	47	42	4	0.3	18
P (uw) (e)	48	42	5	0.3	22
Kraft fibers					
Fully-bleached kraft	49	43	5	0.4	21

Although van Oss et al. assume that for water $\gamma_L^+ = \gamma_L^-$, according to Della Volpe et al. water is more acidic than basic ($\gamma_L^+/\gamma_L^- = 4.3/1$).

Results for the mechanical pulp fibers (Table 3.4) show that the Lewis base parameters of the fibers are substantially larger than the Lewis acid parameters. If water is more acidic than basic, the water–fiber interaction will be predominantly due to the basic groups in the fibers. However, this does not exclude the possibility that the fibers may contain some Lewis' acid groups.

Changes in surface energies were due to changes in acid–base interactions, the Lifshitz–van der Waals component remaining more or less constant. The increased hydrophilicity resulting from different treatments was due to an increase in γ_S^-, that is, the exposure of more basic groups in the fibers resulted in increased adhesion of the acid water. The removal of extractives from the fiber surface also increased the basicity/hydrophilicity, as illustrated in Figure 3.13. CTMP was the most hydrophilic of the unbleached mechanical pulps, with the lowest contact angle between water and fibers. Lignin-modifying treatments, such as ozonation and sulfonation followed by peroxide bleaching, increased the hydrophilicity of the fibers.

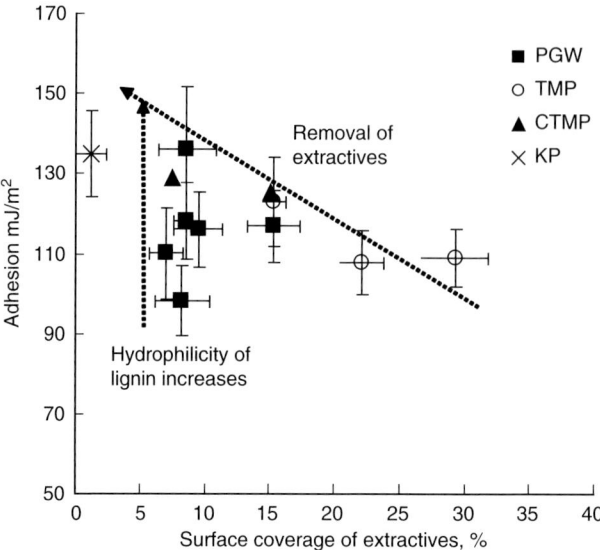

Figure 3.13 Work of adhesion between fibers and water as a function of the surface coverage of extractives for PGW (■), TMP (○), CTMP (▲) and fully bleached softwood kraft pulp (×). Arrows indicate trends [7].

Conclusions

A combination of AFM, XPS, contact angle measurement, and polyelectrolyte adsorption yields a comprehensive picture of the chemistry, adhesive properties, and morphology of mechanical pulp and fiber surfaces. Overall, the surfaces of mechanical pulps and fibers are extremely heterogeneous. The morphology, chemical composition and charge profile of the fiber surfaces vary widely, depending on manufacturing methods and chemical treatments. Water washing of the pulp after the treatments also has an influence on the surface composition of the pulps/fibers, especially on the extractives content.

From the papermaker's point of view, swelling, water removal, and strength of mechanical pulps are important factors. Generally, swelling of mechanical fibers and fines is increased when ionic groups are introduced, resulting in the formation of stronger fiber–fiber bonds. Sulfite treatment in the production of CTMP pulp, peroxide bleaching under alkaline conditions, and ozone treatment modify especially the lignin and/or the pectins so that more acidic groups are introduced into the pulps. Hydrophobic extractives (pitch) should be removed to increase the adhesion between fibers.

References

1. Karnis, A. (1994) The Mechanism of Fibre Development in Mechanical Pulping. *J. Pulp Pap. Sci.* 20, J280–J288.
2. Mohlin, U.-B. (1997) Fibre Development During Mechanical Pulping Refining. *J. Pulp Pap. Sci.* 23, J28–J33.

3. Corson, S. (2002) Process Impacts on Mechanical Pulp Fibre and Sheet Dimensions. *Pulp Pap. Can.* 103, T38–T35.
4. Koljonen, K., Österberg, M., Johansson, L.S., Stenius, P. (2003) Surface Chemistry and Morphology of Different Mechanical Pulps Determined by ESCA and AFM. *Colloids Surf. A Phys. Chem. Eng. Asp.* 228, 143–158.
5. Koljonen, K., Stenius, P. (2005) Surface Characterisation of Single Fibres from Mechanical Pulps by Contact Angle Measurements. *Nordic Pulp Pap. Res. J.* 2005, 107–113.
6. Koljonen, K., Mustranta, A., Stenius, P. (2004) Surface Characterisation of Mechanical Pulps by Polyelectrolyte Adsorption. *Nordic Pulp Paper Res. J.* 19, 500–510.
7. Koljonen, K. (2004) Effect of Surface Properties of Fibres on Some Paper Properties of Mechanical and Chemical Pulp. Ph.D. Thesis, Helsinki University of Technology, Espoo, Finland.
8. Franzén, R. (1986) General and Selective Upgrading of Mechanical Pulps. *Nordic Pulp Pap. Res. J.* 1, 4–13.
9. Cisneros, H.A., Williams, G.J., Hatton, J.V. (1995) Fibre Surface Characteristic of Hardwood Refiner Pulps. *J. Pulp Pap. Sci.* 21, J178–J184.
10. Salmén, L., Lucander, M., Härkönen, E., Sundholm, J. (1999) Fundamentals of Mechanical Pulping. In: *Mechanical Pulping*, Sundholm, J. (ed.) Fapet Oy, Helsinki: pp. 16–21.
11. Sundholm, J. (1999) What is Mechanical Pulping? In: *Mechanical Pulping*, Sundholm, J. (ed.) Fapet Oy, Helsinki: pp. 16–21.
12. Lindholm, C.-A., Kurdin, J.A. (2000) Chemimechanical Pulping. In: *Mechanical Pulping*, Sundholm, J. (ed.) Fapet Oy, Helsinki: pp. 223–249.
13. Engstrand, P., Sjögren, B., Ölander, K., Htun, M. (1991) The Significance of Carboxylic Groups for the Physical Properties of Mechanical Pulp Fibres. *Proc. 6th International Symposium on Wood and Pulping Chemistry*. Melbourne, Australia, Appita, pp. 75–79.
14. Zhang, Y., Sjögren, B., Engstrand, P., Htun, M. (1994) Determination of Charged Groups in Mechanical Pulp Fibres and Their Influence on Pulp Properties. *J. Wood Chem. Technol.* 14, 83–102.
15. Kuys, K. (1992) The Effect of Bleaching on the Surface Chemistry of High Yield Pulps. *46th Appita Annual General Conference* Launceston, Tasmania, Australia, Appita, pp. 53–62.
16. Peng, F., Johansson, L. (1996) Characterization of Mechanical Pulp Fibres. *J. Pulp Pap. Sci.* 22, J252–J257.
17. Scallan, A.M. (1983) The Effect of Acidic Groups on the Swelling of Pulps: A Review. *Tappi J.* 66, 73–75.
18. Hammar, L.-Å., Htun, M., Ottestam, C., Salmén, L., Sjögren, B. (1996) Utilizing the Ionic Properties of the Wood Polymers in Mechanical Pulping. *J. Pulp Pap. Sci.* 22, J219–J223.
19. Carlsson, G. (1996) Surface Composition of Wood Pulp Fibres. Relevance to Wettability, Sorption and Adhesion. Ph.D. Thesis, Royal Institute of Technology, Stockholm, Sweden.
20. Navarro, F., Dávalos, F., Denes, F., Cruz, L.E., Young, R.A., Ramos, J. (2003) Highly Hydrophobic Sisal Chemithermomechanical Pulp (CTMP) Paper by Fluorotrimethylsilane Plasma Treatment. *Cellulose* 13, 157–170.
21. Östenson, M., Järund, H., Toriz, G., Gatenholm, P. (2006) Determination of Surface Functional Groups in Lignocellulosic Materials by Chemical Derivatization and ESCA Analysis. *Cellulose* 13, 157–170.
22. Van der Wielen, L.C., Östenson, M., Gatenholm, P., Ragauskas, A.J. (2005) Mechanism of Dielectric-Barrier Discharge Initiated Wet-Strength Development. *J. Appl. Polym. Sci.* 98, 2219–2225.
23. Van der Wielen, L.C., Elder, T., Ragauskas, A.J. (2005) Analysis of the Topochemical Effects of Dielectric-Barrier Discharge on Cellulosic Fibres. *Cellulose* 12, 185–196.
24. Hafrén, J., Daniel, G. (2005) Chemoenzymatic Modifications of Charge in Chemithermomechanical Wood Pulp. *Nordic Pulp Pap. Res. J.* 20, 200–204.

25. Mustranta, A., Koljonen, K., Holmbom, B., Stenius, P., Buchert, J. (1998) Characterization of the Surface Chemistry of PGW Spruce Pulps. *Proc. 5th European Workshop on Lignocellulosics and Pulp.* Aveiro, Portugal, pp. 11–14.
26. Mustranta, A., Koljonen, K., Lappalainen, A. *et al.* (2000) Characterization of Mechanical Pulp Fibres with Enzymatic, Chemical and Immunological Methods. *Proc. 6th European Workshop on Lignocellulosics and Pulp.* Bordeaux, France, pp. 19–25.
27. Yang, J.-L., Pettersson, B., Eriksson, K.-E. (1988) Development of Bioassays for the Characterization of Pulp Fibre Surfaces. I. Characterization of Various Mechanical Pulps Fibre Surfaces by Specific Cellulotic Enzymes. *Nordic Pulp Pap. Res. J.* 3, 19–25.
28. Henriksson, Å., Gatenholm, P. (2002) Surface Properties of CTMP Fibres Modified with Xylans. *Cellulose* 9, 55–64.
29. Hannuksela, T., Fardim, P., Holmbom, B. (2003) Sorption of Spruce O-Acetylated Galactoglucomannans onto Different Pulp Fibres. *Cellulose* 10, 317–324.
30. Zhou, Q., Baumann, M.J., Brumer, H.I., Teeri, T. (2006) The Influence of Surface Chemical Composition on the Adsorption of Xyloglucan to Chemical and Mechanical Pulps. *Carbohydr. Polym.* 63, 449–458.
31. Luukko, K. (1999) Characterization and Properties of Mechanical Pulp Fines. Ph.D. Thesis, Helsinki University of Technology, Espoo, Finland.
32. Hanley, S.J., Gray, D.G. (1994) Atomic Force Microscopy Images of Black Spruce Woodsections and Pulp Fibres. *Holzforschung* 48, 29–34.
33. Niemi, H., Paulapuro, H., Mahlberg, R. (2002) Review: Application of Scanning Probe Microscopy to Wood, Fibre and Paper Research. *Pap. Puu* 84, 389–406.
34. Gustafsson, J. (2004) Surface Characterization of Chemical and Mechanical Pulp Fibres by AFM and XPS. Ph.D. Thesis, Åbo Akademi University, Turku (Åbo), Finland.
35. Snell, R., Groom, L.H., Rials, T.G. (2001) Characterizing the Surface Roughness of Thermomechanical Pulp Fibres with Atomic Force Microscopy. *Holzforschung* 55, 511–520.
36. Gustafsson, J., Lehto, J.H., Tienvieri, T., Ciovica, L., Peltonen, J. (2003) Surface Characteristics of Thermomechanical Pulps; the Influence of Defibration Temperature and Refining. *Colloids Surf. A Phys. Chem. Eng. Asp.* 225, 95–104.
37. Gustafsson, J., Ciovica, L., Lehto, J.H., Tienvieri, T., Peltonen, J. (2001) The Influence of Defibration Temperature and Refining on the Surface Characteristics of Mechanical Pulps. *Proc. Research Techniques for Tomorrow's Papermaking; COST Action E11 Meeting, 4–5 October.* Helsinki, Finland, pp. 67–70.
38. Börås, L., Gatenholm, P. (1999) Surface Composition and Morphology of CTMP Fibres. *Holzforschung* 53, 188–194.
39. Simola, J., Malkavaara, P., Alén, R., Peltonen, J. (2000) Scanning Probe Microscopy of Pine and Birch Kraft Pulp Fibres. *Polymer* 41, 2121–2126.
40. Pereira, D.E.D., Meuer, E., Guntherodt, H.-J. (2001) The Use of AFM to Investigate the Delignification Process. Part II: AFM Performance in Pulp Prebleached with Oxygen. *Rev. ATIP* 55, 6–9.
41. Simola-Gustafsson, J., Hortling, B., Peltonen, J. (2001) Scanning Probe Microscopy and Enhanced Data Analysis on Lignin and Elemental-Chlorine-Free or Oxygen-Delignified Pine Kraft Pulp. *Colloid. Polym. Sci.* 279, 221–231.
42. Gustafsson, J., Ciovica, L., Peltonen, J. (2002) The Ultrastructure of Spruce Kraft Pulps Studied by Atomic Force Microscopy (AFM) and X-Ray Photoelectron Spectroscopy (XPS). *Polymer* 44, 661–670.
43. Fardim, P., Gustafsson, J., von Schoultz, S., Peltonen, J., Holmbom, B. (2005) Extractives on Fibre Surfaces Investigated by XPS, ToF-SIMS and AFM. *Colloids Surf. A Phys. Chem. Eng. Asp.* 255, 91–103.

44. Österberg, M., Koljonen, K., Johansson, L.-S. (2005) Detecting Wood Extractives on Pulp Fibre Surfaces Using AFM and ESCA. *Proc. 13th International Symposium on Wood, Fibre and Pulping Chemistry.* Auckland, New Zealand, pp. 69–75.
45. Dorris, G.M., Gray, D.G. (1978) The Surface Analysis of Paper and Wood Fibres by ESCA (Electron Spectroscopy for Chemical Analysis). I. Application to Cellulose and Lignin. *Cellul. Chem. Technol.* 12, 9–23.
46. Dorris, G.M., Gray, D.G. (1978) The Surface Analysis of Paper and Wood Fibres by ESCA. II. Surface Composition of Mechanical Pulps. *Cellul. Chem. Technol.* 12, 721–734.
47. Dorris, G.M., Gray, D.G. (1978) The Surface Analysis of Paper and Wood Fibres by ESCA. III. Interpretation of Carbon (1s) Peak Shape. *Cellul. Chem. Technol.* 12, 735–743.
48. Barry, A.O., Koran, Z., Kaliaguine, S. (1990) Surface Analysis by ESCA of Sulfite Post-Treated CTMP. *J. Appl. Polym. Sci.* 39, 31–42.
49. Hua, X., Kaliaguine, S., Kokta, B.V., Adnot, A. (1993) Surface Analysis of Explosion Pulps by ESCA. Part I. Carbon (1s) Spectra and Oxygen-to-Carbon Ratios. *Wood Sci. Tech.* 27, 449–459.
50. McDonald, A.G., Clare, A.B., Dawson, B. (1999) Surface Characterisation of Radiate Pine High-Temperature TMP Fibres by X-Ray Photo-Electron Spectroscopy. *Proc. 53rd Appita Annual Conference.* Rotorua, New Zealand, Appita, pp. 51–57.
51. Westermark, U. (1999) The Content of Lignin on Pulp Fibre Surfaces. *Proc. 10th International Symposium on Wood and Pulping Chemistry 7–10 June.* Yokohama, Japan, pp. 40–42.
52. Matuana, L.M., Balatinecz, J.J., Sodhi, R.N.S., Park, C.B. (2001) Surface Characterization of Esterfied Cellulosic Fibres by XPS and FTIR Spectroscopy. *Wood Sci. Tech.* 35, 191–201.
53. Koubaa, A., Riedl, B., Koran, Z. (1996) Surface Analysis of Press Dried-CTMP Paper Samples by Electron Spectroscopy for Chemical Analysis. *J. Appl. Polym. Sci.* 61, 545–560.
54. Börås, L., Gatenholm, P. (1999) Surface Properties of Mechanical Pulps Prepared under Various Sulphonation Conditions and Preheating Time. *Holzforschung* 53, 429–434.
55. Luukko, K., Laine, J., Pere, J. (1999) Chemical Characterization of Different Mechanical Pulp Fines. *Appita J.* 52, 126–131.
56. Mosbye, J., Laine, J., Moe, S. (2003) The Effect of Dissolved Substances on the Adsorption of Colloidal Extractives to Fines in Mechanical Pulp. *Nordic Pulp Pap. Res. J.* 18, 63–68.
57. Johansson, L.-S. (2002) Monitoring the Papermaking Processes with XPS. *Microchim. Acta* 138, 217–223.
58. Johansson, L.-S., Campbell, J.M., Koljonen, K., Stenius, P. (1999) Evaluation of Surface Lignin on Cellulose Fibers with XPS. *Appl. Surf. Sci.* 92–95, 144–145.
59. Johansson, L.-S., Campbell, J.M., Fardim, P., Heijnesson-Hultén, A., Boisvert, J.-P., Ernstsson, M. (2005) An XPS Round Robin Investigation on Analysis of Wood Pulp Fibres and Filter Paper. *Surf. Sci.* 584, 126–132.
60. Laine, J., Stenius, P., Carlsson, G., Ström, G. (1994) Surface Characterization of Unbleached Kraft Pulp by Means of ESCA. *Cellulose* 1, 145–160.
61. Tougaard, S., Sigmund, P. (1983) Influence of Elastic and Inelastic Scattering on Energy Spectra of Electrons Emitted from Solids. *Phys. Rev. B* 25, 4452–4466.
62. Tougaard, S., Ignatiev, A. (1983) Concentration Depth Profiles by XPS; a New Approach. *Surf. Sci.* 129, 355–365.
63. Johansson, L.-S., Campbell, J.M., Koljonen, K., Kleen, M., Buchert, J. (2004) On Surface Distribution in Natural Cellulosic Fibres. *Surf. Interface Anal.* 36, 706–710.
64. Heijnesson-Hultén, A., Paulsson, M. (2003) Surface Characterization of Unbleached and Oxygen Delignified Kraft Pulp Fibres. *J. Wood Chem. Technol.* 23, 31–46.
65. Heijnesson-Hultén, A., Basta, J., Larsson, P., Ernstesson, M. (2006) Comparison of Different XPS Methods for Fibre Surface Analysis. *Holzforschung* 60, 14–19.

66. Kleen, M., Räsänen, H., Ohra-aho, T., Laine, C. (2001) Surface Chemistry of TMP Fibres and Fines. *Proc. 11th International Symposium on Wood and Pulping Chemistry.* Nice, France, pp. 255–258.
67. Kleen, M., Kangas, H., Laine, C. (2003) Chemical Characterization of Mechanical Pulp Fines and Fibre Surface Layers. *Nordic Pulp Pap. Res. J.* 18, 361–368.
68. Kleen, M., Kangas, H., Kristola, J. (2002) Surface and Bulk Chemical Properties of Groundwood Pulp Fibres and Fines. *Proc. 7th European Workshop on Lignocellulosics and Pulp.* Turku (Åbo), Finland, pp. 447–450.
69. Kangas, H., Salmi, J., Kleen, M. (2002) Surface Chemistry and Morphology of Thermomechanical Pulp Fibres and Fines. *Proc. 7th European Workshop on Lignocellulosics and Pulp.* Turku (Åbo), Finland, pp. 115–118.
70. Kangas, H., Kleen, M. (2004) Surface Chemical and Morphological Properties of Mechanical Pulp Fines. *Nordic Pulp Pap. Res. J.* 19, 191–199.
71. Kangas, H., Kleen, M. (2003) The Effect of Peroxide Bleaching on the Surface Chemical Composition of Mechanical Pulps. *Proc. 12th International Symposium on Wood and Pulping Chemistry.* Madison, WI, USA, pp. 381–384.
72. Kangas, H., Pöhler, T., Heikkurinen, A., Kleen, M. (2004) Development of the Mechanical Pulp Fibre Surface as a Function of Refining Energy. *J. Pulp Pap. Sci.* 30, 298–306.
73. Johnsen, I.A., Stenius, P., Tammelin, T., Österberg, M., Johansson, L.-S., Laine, J. (2006) The Influence of Dissolved Substances on Resin Adsorption to TMP Fine Material. *Nordic Pulp Pap. Res. J.* 21, 629–637.
74. Katz, S., Beatson, R.P., Scallan, A.M. (1984) The Determination of Strong and Weak Acidic Groups in Sulfite Pulps. *Svensk Papperst.* 87, R48–R53.
75. Pu, Q., Sarkanen, K. (1989) Donnan Equilibria in Wood–Alkali Interactions. Part 1. Quantitative Determination of Carboxyl-, Carboxyl Ester and Phenolic Hydroxyl Groups. *J. Wood Chem. Technol.* 9, 293–312.
76. Scallan, A.M., Katz, S., Argyropoulos, D.S. (1989) Conductometric Titration of Cellulosic Fibres. In: *Cellulose and Wood-Chemistry and Technology,* Schuerch, C. (ed.) John Wiley & Sons, New York, NY, USA, pp. 1457–1451.
77. Budd, J., Herrington, T.M. (1989) Surface Charge and Surface Area of Cellulose Fibres. *Colloids Surf. A Phys. Chem. Eng. Asp.* 88, 277–287.
78. Laine, J., Lövgren, L., Stenius, P., Sjöberg, S. (1994) Potentiometric Titration of Unbleached Kraft Cellulose Fibre Surfaces. *Colloids Surf. A Phys. Chem. Eng. Asp.* 88, 277–287.
79. Terayama, H. (1952) Method of Colloid Titration (a New Titration between Polymer Ions). *J. Polym. Sci.* 8, 243–253.
80. Wågberg, L., Winter, L., Lindström, T. (1985) Determination of Ion-Exchanged Capacity of Carboxymethylated Cellulose Fibres Using Colloid and Conductometric Titrations. In: *Papermaking Raw Materials. Proc. 8th Fundamental Research Symposium,* Punton, V.W. (ed.) Mechanical Engineering Publishers, Oxford, England, pp. 917–923.
81. Wågberg, L., Ödberg, L., Glad-Nordmark, G. (1989) Charge Determination of Porous Substrates by Polyelectrolyte Adsorption. *Nordic Pulp Pap. Res. J.* 4, 71–76.
82. Sundberg, A., Pranovich, A., Holmbom, B. (2000) Distribution of Anionic Groups in TMP Suspension. *J. Wood Chem. Technol.* 20, 71–92.
83. Sjöström, E., Janson, J., Haglund, P., Enström, P. (1965) The Acidic Groups in Wood and Pulp as Measured by Ion Exchange. *J. Polym. Sci.* C11, 221–241.
84. Fardim, P., Holmbom, B. (2005) Origin and Surface Distribution of Anionic Groups in Different Papermaking Fibres. *Colloids Surf. A Physicochem. Eng. Asp.* 252, 237–242.
85. Rohrsetzer, S., Kovacs, P., Kabai-Faix, M., Papp, J., Volgyi, P. (1995) Investigation of the Inside and Outside Surfaces of Pulp Fibres by Adsorption of Molecular, Colloidal and Coarse Size Particles. *Cellul. Chem. Technol.* 29, 65–75.

86. Fardim, P., Holmbom, B., Ivaska, A., Karhu, J., Mortha, G., Laine, J. (2002) Critical Comparison and Validation of Methods for Determination of Anionic Groups in Pulp Fibres. *Nordic Pulp Pap. Res. J.* 17, 346–351.
87. Fardim, P., Moreno, T., Holmbom, B. (2005) Anionic Groups on Cellulosic Fibre Surfaces Investigated by XPS, FTIR-ATR, and Different Sorption Methods. *J. Colloid Interface Sci.* 290, 383–391.
88. Gill, R.I.S. (1989) The Use of Potentiometric Titration and Polyelectrolyte Titration to Measure the Surface Charge of Cellulose Fibre. In: *Fundamentals of Papermaking. Proceedings of the 9th Fundamental Research Symposium*, Baker, C.F., Punton, V.W. (eds) Mechanical Engineering Publishers, Cambridge, England, pp. 437–452.
89. Stenius, P., Laine, J. (1994) Studies of Cellulose Surfaces by Titration and ESCA. *Appl. Surf. Sci.* 75, 213–219.
90. Lindgren, J., Öhman, L.-O., Gunnars, S., Wågberg, L. (2002) Charge Determinations of Cellulose Fibres of Different Origin – Comparison between Different Methods. *Nordic Pulp Pap. Res. J.* 17, 89–96.
91. Laine, J., Buchert, J., Viikari, L., Stenius, P. (1996) Characterization of Unbleached Kraft Pulps by Enzymatic Treatment, Potentiometric Titration and Polyelectrolyte Adsorption. *Holzforschung* 50, 208–214.
92. Swerin, A., Wågberg, L. (1994) Size-Exclusion Chromatography for Characterization of Cationic Polyelectrolytes Used in Papermaking. *Nordic Pulp Pap. Res. J.* 9, 18–25.
93. Lindström, T., Wågberg, L. (1983) Effects of pH and Electrolyte Concentration on the Adsorption of Cationic Polyacrylamides on Cellulose. *Tappi J.* 66, 83–85.
94. Lindström, E., Lindström, T. (2003) Polyelectrolyte Charge Titration of Cellulosic Fibres. *Proc. 5th International Paper Coating Chemistry Symposium.* Montreal, PQ, Canada, pp. 175–177.
95. Horvath, A.E. (2003) Appropriate Conditions for Polyelectrolyte Titration to Determine the Charge of Cellulosic Fibres. Lic. Thesis, Royal Institute of Technology, Stockholm, Sweden.
96. Holmbom, B., Pranovich, A., Sundberg, A., Buchert, J. (2000) Charged Groups in Wood and Mechanical Pulps. In: *Cellulosic Pulps, Fibres and Materials*, Kennedy, J.F., Phillips, G.O., Willikams, P.A. (eds) Woodhead Publishing Ltd, Cambridge, UK, pp. 109–119.
97. Robert, D.R., Szadeczki, M., Lachenal, D. (2000) Chemical Characteristics of Lignins Extracted from Softwood TMP after O_3 and ClO_2 Treatment. In: *Lignin: Historical, Biological and Materials Perspective*, Glasser, W.G., Northey, R.A., Schultz, T.P. (eds) American Chemical Society, Washington, DC, USA, pp. 520–531.
98. Young, R.A. (1976) Wettability of Wood Pulp Fibres – Applicability of Methodology. *Wood Fiber Sci.* 8, 120–128.
99. Klungness, J.H. (1981) Measuring the Wetting Angle and Perimeter of Single Wood Pulp Fibres: a Modified Method. *Tappi J.* 64, 65–66.
100. Berg, J.C. (1993) The Importance of Acid–Base Interactions in Wetting, Coating, Adhesion and Related Phenomena. *Nordic Pulp Pap. Res. J.* 8, 75–85.
101. Hodgson, K.T., Berg, J.C. (1988) Dynamic Wettability Properties of Single Wood Pulp Fibres and Their Relationship to Adsorbency. *Wood Fiber Sci.* 20, 3–15.
102. Lam, C.C.N., Lu, J.J., Neumann, A.W. (2001) Measuring Contact Angle. In: *Handbook of Applied Surface and Colloid Chemistry*, Holmberg, K., Shah, D.O., Schwuger, M.J. (eds) John Wiley & Sons, Chichester, UK, Vol. 2, pp. 251–280.
103. Laine, J., Hynynen, R., Stenius, P. (1997) The Effect of Surface Chemical Composition and Charge on the Fibre and Paper Properties of Unbleached and Bleached Kraft Pulps. In: *Fundamentals of Papermaking Materials: Transactions of the 11th Fundamental Research Symposium*, Baker, F. (ed.) Pira International, Cambridge, UK, Vol. 2, pp. 859–892.
104. Shen, W., Parker, I.H., Sheng, Y.J. (1998) The Effects of Surface Extractives and Lignin on the Surface Energy of Eucalypt Kraft Pulp Fibre. *J. Adhes. Sci. Technol.* 12, 161–174.

105. Wistara, N., Zhang, X., Young, R.A. (1999) Properties and Treatments of Pulps from Recycled Paper. Part II. Surface Properties and Crystallinity of Fibres and Fines. *Cellulose* 6, 325–348.
106. Tze, W.T., Gardner, D.J. (2001) Swelling of Recycled Wood Pulp Fibres: Effect on Hydroxyl Availability and Surface Chemistry. *Wood Fiber Sci.* 33, 364–376.
107. Jacob, P., Berg, J.C. (1993) Zisman Analysis of Three Pulp Fibre Furnishes. *Tappi J.* 76, 105–107.
108. Gellerstedt, G., Gatenholm, P. (1999) Surface Properties of Lignocellulosic Fibres Bearing Carboxylic Groups. *Cellulose* 6, 103–121.
109. Ness, J., Hodgson, K.T. (1999) The Effects of Peroxide Bleaching on Thermo-Mechanical Pulp Self Sizing. *Nordic Pulp Pap. Res. J.* 14, 111–115.
110. Östenson, M., Gatenholm, P., Gulliermo, T. (2004) Effects of Extractives on the Surface Chemistry and Wettability of High Temperature Chemithermomechanical Pulps. *Nordic Pulp Pap. Res. J.* 19, 53–58.
111. Berg, J.C. (1986) The Use and Limitations of Wetting Measurements in the Prediction of Adhesive Performance. In: *Composite Systems from Natural and Synthetic Polymers*, Salmén, L., de Ruvo, A., Seferis, J.C., Stark, E.B. (eds) Elsevier Publishing Co, Amsterdam, The Netherlands, pp. 23–46.
112. Tiberg, F., Daicic, J., Fröberg, J. (2001) Surface Chemistry of Paper. In: *Handbook of Applied Surface and Colloid Chemistry*, Holmberg, K., Shah, D.O., Schwuger, M.J. (eds) John Wiley & Sons, Chichester, UK, Vol. 1, pp. 123–173.
113. Felix, J.M., Gatenholm, P. (1993) Characterization of Cellulose Fibres Using Inverse Gas Chromatography. *Nordic Pulp Pap. Res. J.* 8, 200–203.
114. Jacob, P., Berg, J.C. (1994) Acid–Base Surface Energy Characterization of Microcrystalline Cellulose and Two Wood Pulp Fibre Types Using Inverse Gas Chromatography. *Langmuir* 10, 3086–3093.
115. Matuana, L.M., Balatinecz, J.J., Park, C.B., Woodhams, R.T. (1999) Surface Characteristics of Chemically Modified Newsprint Fibres Determined by Inverse Gas Chromatography. *Wood Fiber Sci.* 31, 116–127.
116. Shen, W., Parker, I.H. (1999) Surface Composition and Surface Energetics of Various Eucalypt Pulps. *Cellulose* 6, 41–55.
117. Belgacem, M.N. (2000) Characterisation of Polysaccharides, Lignin and Other Woody Components by Inverse Gas Chromatography. *Cellul. Chem. Technol.* 34, 357–383.
118. Wålinder, M.E.P., Gardner, D.J. (2000) Surface Energy of Extracted and Non-Extracted Norway Spruce Wood Particles Studied by Inverse Gas Chromatography. *Wood Fiber Sci.* 32, 478–488.
119. Wålinder, M.E.P., Johansson, L. (2001) Measurement of Wood Wettability by Wilhelmy Method. Part 1. Contamination of Probe Liquids by Extractives. *Holzforschung* 55, 33–41.
120. Wålinder, M.E.P., Ström, G. (2001) Measurement of Wood Wettability by Wilhelmy Method. Part 2. Determination of Apparent Contact Angles. *Holzforschung* 55, 33–41.
121. Shen, W., Sheng, Y.J., Parker, I.H. (2000) Evaluating the Surface Energy of Hardwood Fibres Using the Wilhelmy and Inverse Gas Chromatography Methods. In: *Cellulosic Pulps, Fibres and Materials*, Kennedy, J.F., Williams, P.A. (eds) Woodhead Publishing Ltd, Cambridge, UK, pp. 181–196.
122. Wålinder, M.E.P. (2002) Study of Lewis Acid–Base Properties of Wood by Contact Angle Analysis. *Holzforschung* 56, 363–371.
123. Lee, L.-H. (1996) Correlation between Lewis-Acid–Base Surface Interaction Components and Linear Solvation Energy Relationship Solvatochromic A and B Parameters. *Langmuir* 12, 1681–1687.
124. Krüger, J., Hodgson, K.T. (1994) Single Fibre Wettability of High Sized Pulp Fibre. *Tappi J.* 77, 83–87.
125. Krüger, J., Hodgson, K.T. (1995) The Relationship between Single Fibre Contact Angle and Sizing Performance. *Tappi J.* 78, 154–161.

126. Deng, Y., Abazeri, M. (1998) Contact Angle Measurement of Wood Fibres in Surfactant and Polymer Solutions. *Wood Fiber Sci.* 30, 155–164.
127. Fowkes, F.M., Mustafa, M.A. (1978) Acid–Base Interactions in Polymer Adsorption. *Ind. Eng. Chem. Prod. Res. Develop.* 17, 3–7.
128. Fowkes, F.M. (1991) Quantitative Characterization of the Acid–Base Properties of Solvents, Polymers, and Inorganic Surfaces. In: *Acid–Base Interactions: Relevance to Adhesion Science and Technology*, Mittal, K.L., Anderson, H.R. (eds) Brill Academic Publishers, Leiden, The Netherlands, pp. 93–115.
129. van Oss, C.J. (2006) *Interfacial Forces in Aqueous Media*. 2 edn; CRC Press, Boca Raton, FL, USA, p. 438.
130. Della Volpe, C., Siboni, S. (2000) Acid–Base Surface Free Energies of Solids and the Definition of Scales in the Good–van Oss–Chaudhury Theory. *J. Adhes. Sci. Technol.* 14, 235–272.
131. Della Volpe, C., Siboni, S. (2001) The Evaluation of Electron-Donor and Electron-Acceptor Properties and Their Role in the Interaction of Solid Surfaces with Water. In: *Water in Biomaterials Surface Science*, Morra, M. (ed.) John Wiley & Sons Ltd, Chichester, UK, pp. 83–214.

Chapter 4
Assessing Substrate Accessibility to Enzymatic Hydrolysis by Cellulases

Richard P. Chandra, Ali R. Esteghlalian, and John N. Saddler

Abstract

Although the structure and function of cellulase systems continue to be the subject of intense research, it is widely acknowledged that the rate and extent of the cellulolytic hydrolysis of lignocellulosic substrates is influenced not only by the effectiveness of the enzymes but also by the chemical and morphological characteristics of the heterogeneous substrate. Over the last few years we have examined various characteristics of lignocellulosic substrates at the fiber, fibril, and microfibril levels that undergo modifications during pretreatment and subsequent hydrolysis. For example, changes can be monitored at the fiber level by measuring fiber width, length, and particle size, and changes at the microfibril level can be assessed by measuring crystallinity and degree of polymerization (DP), while changes in specific surface area (SSA) affect all three levels of structural organization. Previous work in our group has shown that effective hydrolysis can occur even though little change is detected in the crystallinity and DP of the lignocellulosic substrates. In contrast we have shown a direct correlation between the SSA of lignocellulosic substrates and their ease of enzymatic hydrolysis. Assessing the "exterior surface area" of the substrate by measuring fiber length or particle size, and the "interior surface area", using techniques such as pore volume, or Simons' Stain (SS), has strongly indicated that a given substrate's accessibility to cellulases is closely tied to its overall hydrolyzability. The various techniques utilized for the measurement of both exterior and interior surface area and how the results of these techniques have been used to assess the susceptibility of various lignocellulosic substrates to subsequent hydrolysis by cellulases will be discussed.

4.1 Introduction

Cellulose is the most abundant renewable biopolymer on earth, and plants are the most plentiful source of cellulose. However, cellulose, unlike starch, was primarily "designed" by nature to fill a structural role and, as a result, it is usually associated with other components, such as lignin and hemicelluloses, which collectively provide the structural integrity of plant materials. Most of the world's ethanol that is currently produced as a transportation fuel has been derived from sugar (Brazil) or starch (USA) feedstocks. It is recognized that, if the global utilization of ethanol is ever to reach significant levels, ethanol derived from a range of agricultural and forestry derived lignocellulosic materials will have to be developed, due to their global availability and the likely lower costs associated with these feedstocks. However, it is known that natural degradation of lignocellulosic materials is a relatively slow process and that for the bioconversion process to ever be economically viable, there is a need to both increase the accessibility of the substrate to the hydrolytic enzymes while finding ways to enhance the hydrolytic potential of the overall "cellulase complex".

The lignocellulosic biomass-to-ethanol bioconversion process usually consists of three major steps. These include: pretreatment to improve the accessibility of the substrate to subsequent enzymatic hydrolysis; enzymatic hydrolysis by carbohydrate degrading enzymes to create monomeric sugars from cellulose and hemicelluloses; and fermentation of the resulting sugars to ethanol. To be fully effective, the pretreatment step should be inexpensive, simple, and allow recovery of the lignin component in a higher-value form, with maximum recovery of the hemicellulose sugars, ideally in a monomeric form while resulting in a cellulosic substrate that can be rapidly hydrolyzed by low concentrations of cellulases. Unfortunately, this ideal scenario is beyond the scope of any pretreatment processes that are currently being advocated. One of the goals of this review is to discuss the enzyme and substrate factors that influence the effective pretreatment and hydrolysis of "generic" lignocellulosic substrates. It has been shown that the source, type, and physical handling of the biomass material prior to pretreatment and enzymatic hydrolysis will all have a significant influence on the overall efficiency of conversion. For example, a more friable lignocelluolosic substrate such as corn fiber will require a significantly different pretreatment and enzyme hydrolysis regime than would a commercial batch of softwood chips. These types of variables are beyond the scope of the current review. Instead, we will initially discuss the progress that has been made in the understanding of what constitutes an "effective cellulase system" and then review the methods that can be used to determine which key substrate characteristics influence the enzymatic hydrolysis of a generic lignocellulosic substrate.

4.2 Enzymatic hydrolysis by cellulases

It is recognized that the enzymatic degradation of lignocellulosic substrates is influenced by both substrate- and enzyme-related properties [1,2]. Cellulases are multicomponent enzyme systems that depolymerize cellulose to monomeric glucose through synergistic binding and catalysis [3,4]. Cellulolytic enzymes are either associated into multienzymatic complexes, that is cellulosomes, or remain as individual multidomain proteins in a liquid cocktail [5,6]. It has been shown that the rate and extent of cellulose hydrolysis by cellulases is strongly influenced by numerous substrate characteristics, such as the cellulose index of crystallinity, DP, and the occurrence of structural irregularities, such as amorphous regions. In plant biomass, cellulose digestibility is further complicated by the presence of the lignin–hemicellulose matrix in which cellulose is tightly embedded. The numerous studies conducted with various cellulase systems and different cellulosic substrates all seem to point to the fact that it is ultimately the "accessibility" of the cellulose fraction to the enzyme system that determines how fast (reaction rate) and how far (% conversion) the hydrolysis reaction can proceed [2,7]. The term "substrate heterogeneity" has also been used to highlight the

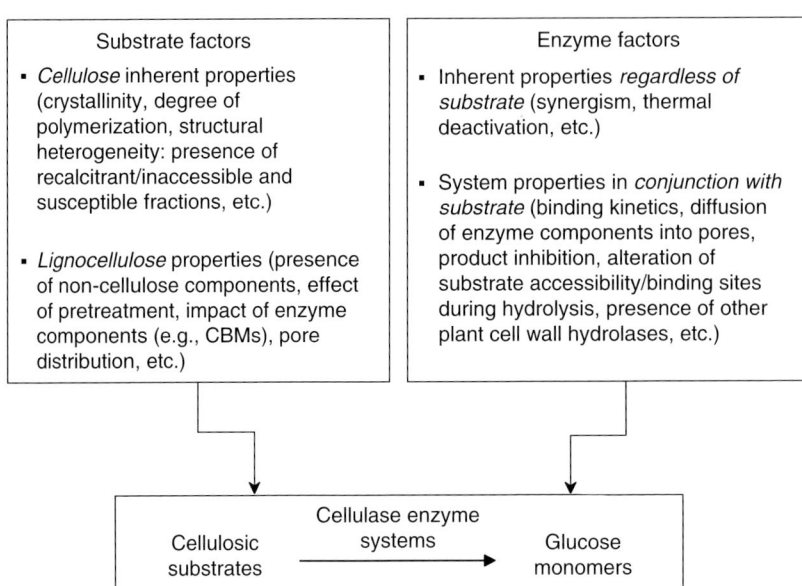

Figure 4.1 Enzyme and substrate properties both influence the rate and extent of enzymatic hydrolysis of lignocellulosic substrates [9].

fact that different portions of the substrate accommodate and react with cellulase enzymes differently, thereby producing different reaction rates and yields [8]. We and other groups have shown that the substrate accessibility is an "enzyme–substrate system" property that only becomes meaningful when a cellulase system encounters a specific substrate with a defined intrinsic pore size distribution, degree of swelling, and other gross and detailed substrate characteristics [9]. In addition to its morphological properties, the reactivity of a cellulosic substrate is also strongly influenced by the type of the physical and/or chemical treatments that it has undergone during its preparation/pretreatment (Figure 4.1).

It is generally recognized that the depolymerization of cellulose by a cellulase system occurs through the synergistic action of three enzyme components (1) endoglucanases (EGs), which cleave the cellulose along the backbone; (2) cellobiohydrolases (CBHs), which act on the mid-chain free ends created by EGs hydrolzying the cellulose chain and produce water-soluble cellobiose units; and (3) β-glucosidase that hydrolyzes dimeric cellobiose to monomeric glucose. Studies with the fungus *Trichoderma reesei* have confirmed that the plant-cell-wall-degrading enzyme system produced by this fungus includes not only cellulases (5 endo and 2 exo) but also ten different types of hemicellulases [10]. Others have also shown that the presence of hemicellulases, such as feruloyl esterase, can help uncover and degrade cellulose more efficiently [11]. The majority of plant cell wall hydrolases, including cellulases, have a multidomain structure. The catalytic domain that performs the actual hydrolytic cleavage of β-1,4 bond between two adjacent glucose units in a cellulose polymer is linked to one or more binding domains that facilitate the anchoring of the enzyme on the insoluble substrate. Relatively short and flexible polypeptide chains, known as linkers, connect the various domains of an enzyme.

The aggregated form of cellulases, cellulosomes, are perhaps the largest extracellular enzyme complexes in nature (0.65–2.5 MDa) consisting of a large noncatalytic scaffolding protein to which individual glycosyl hydrolase (GH) enzyme components are attached through a 22-amino-acid sequence called a dockerin [5]. In addition to the catalytic domain and the dockerin linker, the scaffolding-bound cellulosomal enzymes may contain other domains such as carbohydrate binding modules (CBMs), immunoglobulin-like domains and surface layer homology (SLH) domains. The protein components of the scaffolding, that is, scaffoldins, include cohesin domains (Coh), cellulose binding domains (CBDs) as well as several other domains with unidentified function. The cohesin domains act as the binding site for the dockerin domains of the GH enzymes attached to the scaffolding [5].

It has been shown that the removal of carbohydrate binding modules (CBMs) from GHs has no detrimental effect on the enzyme activity on soluble substrates. However, it appears that the deletion of CBMs reduce enzyme activity against insoluble and more complex substrates [6,12]. Therefore, it has been suggested that the functional role of cellulose and xylan binding modules (CBMs and XBMs) include attaching the protein onto the substrate to increase enzyme concentration in the proximity of cellulose, orienting the catalytic domain toward the cleavage site and, with a lesser degree of certainty, disrupting the polysaccharide structure to enhance substrate accessibility [13].

The CBMs of CBHs have been suggested to have a "plough-like" structure whereby they can peel off individual cellulose polymers from the crystalline structure and feed it through the tunnel-shaped catalytic domain for hydrolysis to cellobiose units. The "accessibility enhancing" function of CBMs has not been fully validated [7] and continues to be the subject of research [14]. It has also been observed that CBMs can impact the processivity of an enzyme; for instance, the binding module of *Thermobifidia fusca* Cel9A has been used to explain why Cel9A, despite being a EGs, act processively on filter paper producing only 13% insoluble products. More conventional endocellulases produce 30–48% insoluble products on the same substrate [8,15]. The ratio of soluble to insoluble reducing sugars produced by the cellulolytic digestion of filter paper is often used to gauge how processive the action of an enzyme is and helps distinguish the exo- and endo-activities of an enzyme [16].

Recently, some other noncatalytic proteins were shown to disrupt the cellulose structure without degrading the polysaccharide. One such protein was isolated from *T. reesei* culture supernatant and has been named, swollenin, as it has been shown to swell cotton fibers without producing detectable amounts of reducing sugars [10]. This protein has a modular structure containing an N-terminus CBM and has a significant sequence identity to plant expansins that are considered to be responsible for loosening the cell wall structure during plant cell growth [11]. It has been proposed that expansins enlarge cell wall cavities by disrupting hydrogen bonding between adjacent cellulose microfibrils or between cellulose and other cell wall polysaccharides without any hydrolytic effect [10]. In this work, the authors demonstrated that incubation with swollenin reduced the wet tensile strength of filter paper strips by 10–15%, caused swelling in cotton fibers, and disrupted Valonia cell wall fragments, without releasing any reducing sugars from any of the three substrates [10].

Although the structure and function of cellulase systems continue to be the subject of intense research, it is widely acknowledged that the rate and extent of the cellulolytic hydrolysis of lignocellulosic substrates depends not only on the enzyme properties but also on the substrate chemical, physical, and morphological properties. Approaches such as site saturation mutagenesis or directed evolution can be employed to improve cellulase properties such as binding affinity, catalytic activity, or thermostability. However, a complementary goal should be to gain a thorough understanding of the structural, physical, and chemical properties of the substrate and to develop ways to alter these properties and maximize substrate accessibility to the cellulase complex.

At this time there continues to be a lack of fundamental understanding of the substrate properties that govern effective hydrolysis. Historically, substrate characteristics such as crystallinity, DP, and accessible surface area have all been thought to influence the hydrolysis of cellulose [1,2]. Several researchers have found that, when all other substrate factors are maintained at a similar level, changes in the crystallinity of lignocellulosic substrates do not have a significant effect on the rate or extent of hydrolysis [17,18]. In a similar vein, it is also problematic to assess the effect of DP on cellulose hydrolysis as it is difficult to alter the DP exclusively, without affecting other substrate factors such as the accessible surface area.

It is intuitive to think that of all the substrate factors that potentially affect hydrolysis, an increase in the specific surface area (SSA) available for cellulases would have a significant effect on both the extent and the rate of hydrolysis, as the enzyme active site requires contact with the substrate for hydrolysis to occur. The pioneering work of Stone et al. [19] estimated that, during hydrolysis, the rate limiting pore size of a substrate lies within the 40–50 Å range, while Cowling and Kirk [20] showed that the diameter of most cellulases was in the 24–77 Å range, indicating that the 40–70 Å pore size was necessary to accommodate cellulases to catalyze hydrolysis. Since then there have been numerous reports stressing the necessity to develop adequate surface area accessible to cellulases to achieve effective hydrolysis [21]. The exposed surfaces and pores of either wood chips or even wood particles that are accessible to cellulases can be best described as scarce [22]. The very existence of a substrate pretreatment step originates from the requirement to increase the accessibility of the substrate to cellulases by altering cellulose structure and the distribution of lignin and hemicelluloses [23,24].

Frequently, especially in the case of lignocellulosic hydrolysis, investigators have attributed enhanced hydrolysis performance to changes in the proportion of the lignin, hemicellulose, and cellulose contents of the substrate during pretreatment. However, it is important to take this conclusion one step further as it is likely that decreases in lignin and hemicellulose content that occur as a result of pretreatment [21,25] also can increase the accessibility of cellulases to cellulose [26]. Several investigations have concluded that the SSA of cellulosic substrates is a controlling factor for effective enzymatic hydrolysis by cellulases [22,27,28]. A recent study relating substrate characteristics of pretreated corn stover to the initial hydrolysis rate to develop predictive models for hydrolysis concluded that the addition of surface area measurements would improve the predictive capability of their model [29]. Indeed, direct correlations have been found between the initial pore volume [22] and particle sizes [30] of cellulosic substrates and their initial rate and extent of hydrolysis. It has been proposed that the efficacy of cellulose hydrolysis is enhanced when the pores of the substrate are large enough to accommodate both large and small enzyme components to maintain the synergistic action of the cellulase enzyme system [28].

Although there have been numerous cellulose hydrolysis reviews that have stressed the importance of SSA [1,2], these papers have not been able to identify or recommend a suitable method for the measurement of SSA that is capable of assessing the susceptibility of substrates to subsequent hydrolysis by cellulases. The various methods that have been utilized to measure the SSA have differed in the type of surface area that has been measured,

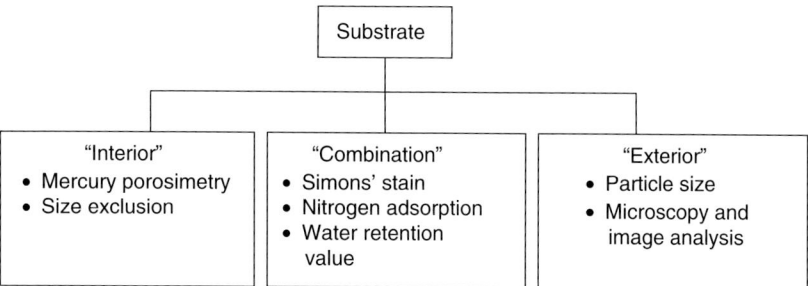

Figure 4.2 Measurement techniques that estimate interior specific surface area (SSA), exterior SSA, and gross measurements that estimate the combined interior and exterior SSAs.

and they can generally be divided into those methods that measure either the interior or the exterior surface area. Cellulose particles consist of internal pores, fissures, and microcracks which can be referred to as "interior surface area". Techniques such as mercury porosimetry and solute exclusion have been used to quantify interior surface area. The "exterior" surface area, which is mainly determined by particle dimensions, has been measured by microscopy combined with image analysis or by particle size analyzers such as the fiber quality analyzer (FQA). Techniques such as SS, nitrogen adsorption, and water retention value (WRV) measure a combination of both interior and exterior surface area. A summary of these techniques is presented in Figure 4.2. Owing to the complex structure of cellulose, and the limitations of each measurement, each of these methods has inherent advantages and disadvantages. The measurements of SSA are complicated to a greater degree by the presence of lignin and hemicelluloses, which are typically present during the hydrolysis of lignocellulosic substrates.

In the following sections we will describe the various methods used to measure the interior and exterior surface area of lignocellulosics and discuss how effective they have been in predicting the susceptibility of cellulose and pretreated lignocellulosic substrates to subsequent enzymatic hydrolysis.

4.3 Exterior surface area

4.3.1 Particle size

It is known that the SSA of a mixture of particles is inversely proportional to their average diameter, and therefore, a smaller average particle size results in an increase in surface area. Thus, it would be expected that a relationship between particle size and cellulose hydrolysis would occur. However, only a few studies have presented data to correlate these two factors and the studies that have attempted to correlate particle size to cellulosic hydrolysis have only presented circumstantial evidence with no direct measurements of the SSA of the various particle sizes. The size of cellulosic particles can be difficult to measure, as their shapes are irregular and they have a tendency to agglomerate [31]. Particle size distributions can be measured using a variety of methods, including visually by microscopy and image analysis, or using a variety of automated particle size analyzers, such as those based on the Coulter principle [32], interactions with lasers [33,34], and high-throughput image analysis. It should be noted that these types of gross estimations assume smooth particles and do not consider surface topology, cracks, and fissures, which may account for significant increases in SSA. Nevertheless, the general consensus from the limited available literature on this topic is that a smaller particle size increases the rate of hydrolysis by cellulases due to an increase in SSA.

It has been shown that ball milling of Pangola grass stems from an average particle size of 1.0–0.1 mm increased the extent of cellulose hydrolysis by cellulases from 40% to 90% during a 100-h hydrolysis period [30]. Surprisingly, decreasing the particle size was more effective than a chlorite delignification treatment of the 1.0 mm particles prior to hydrolysis. During a study assessing the effects of crystallinity on the enzymatic hydrolysis of lignocellulosic substrates subjected to various pretreatments, Puri [17] concluded that particle size and SSA were controlling factors affecting enzymatic hydrolysis rather than crystallinity. Sawada *et al.* [35,36] and others [18,37] have shown that lignocellulosics pretreated with steam explosion at a series of elevated temperatures produce substrates with progressively smaller particle sizes that result in increases in hydrolysis yields.

Further work addressing the effects of particle size on the digestibility of substrates using cellulases has mainly been performed on pulp fibers, as pulps used for the formation of paper usually contain a heterogeneous distribution of particle sizes. Consequently, the application of cellulases to pulp samples results in varying levels of performance depending on the fiber size distribution [18].

Methods for the measurement of fiber length include microscopy, the Kajaani, and more recently, the FQA, which consists of a flow cell as well as optical and image analysis systems [38]. The flow cell transports a dilute suspension of pulp fibers past the optical and imaging systems and orients fibers into the two-dimensional plane where they are measured by the image detection system. Recently, the high-resolution FQA has enabled high-throughput measurement of fiber width, in addition to length, and coarseness [39]. Cellulases have been thought to act on the surface of pulp fibers, resulting in a "peeling effect" [18], and therefore, smaller particle sizes with a greater amount of SSA would be expected to rapidly hydrolyze. In the case of kraft pulps, it has been shown that decreasing fiber size results in both increased hydrolysis rates and the ability to hydrolyze at lower cellulase loadings [40]. In an investigation assessing the hydrolysis of Douglas-fir (*Pseudotsuga menziesi*) kraft and mechanical pulps, Mooney *et al.* [26] showed that, at equal lignin contents, the "fines" of a delignified mechanical pulp were hydrolyzed faster than the longer fibers of the kraft pulp. When each fiber length fraction was hydrolyzed separately, it was seen that the isolated long-fiber fraction hydrolyzed slower and had less cellulases adsorbed than did the whole pulp [26]. The increased hydrolysis rate of the whole pulp was attributed to the greater amount of specific surface available for the adsorption of cellulases provided by the pulp fines and short fibers. Although it is apparent that particle size has a significant effect on cellulose hydrolyzability [41], it has also been shown that fiber delamination and enhanced swelling that results from mechanical treatment of kraft pulp fibers has a greater effect on hydrolysis by cellulases than does a decrease in particle size. As mentioned previously, the measurement of particle size should not be used exclusively to evaluate substrate accessibility to cellulase enzymes, as particle size does not account for the surface topology and porosity of each particle. However, particle size measurements may be effective in assessing potential substrates for cellulases when employed in combination with other measurements that assess interior surface area, such as pore size measurements.

4.4 Combination measurements: Interior and exterior surface area

4.4.1 Swelling and water retention value

A significant factor affecting the interior surface area or porosity of potential substrates for cellulases is the swelling of cellulosics/lignocellulosics that occurs in aqueous environments. The swelling of substrates is heavily influenced by the charged groups that populate the interior and exterior fiber surfaces, the pH of the medium and the presence of electrolytes [42]. As mentioned earlier, some previous work has attributed improved hydrolysis performance to increases in porosity due to fiber swelling [43,44].

A common method used to assess swelling is the centrifugal WRV measurement originally developed by Hopner et al. [45]. The WRV measurement consists of measuring a known weight of pulp and subjecting it to soaking under controlled conditions, followed by centrifugation and drying in an oven. The mass of the sample after soaking and centrifugation is compared with the sample mass after oven-drying to obtain a WRV. Although attempts have been made to standardize the WRV method [46], there have been numerous variations in the centrifugation speed and the temperature and duration of the oven-drying step. Although the basic aspects of the WRV test remain the same, the variations in methodology present a challenge when comparing the results from different investigations. In the case of highly swollen pulps, the WRV method tends to underestimate the swellability of highly swollen pulps compared with measurements obtained by the solute exclusion technique [47]. However, good correlations were obtained between WRVs and pore size measurements when performed using the nitrogen adsorption and mercury porosimetry techniques for white birch treated with ten swelling agents to create a range of WRVs and pore size values [43]. Owing to the simplicity of the WRV method and its strong correlation with pore volume measurements, it has been utilized on a handful of occasions to assess the susceptibility of substrates to enzymatic hydrolysis.

In an elaborate study, Ogiwara and Arai [44,48] used the WRV method to measure the degree of swelling of sulfite, kraft, and semimechanical pulps, and found a linear correlation between initial cellulose hydrolysis rates and WRV values. This linear relationship was maintained as increased hydrolysis rates were shown to correspond to increases in the WRV value that occurred from either a mechanical or NaOH and $ZnCl_2$ treatments of the pulps. Enhanced hydrolysis performance has also been associated with increases in WRV in the case of textile cotton wastes [49] and woven fabrics from cotton, Lyocell, modal, and viscose [50]. As recycled pulps originate from fiber sources that undergo irreversible changes in their structure upon drying [51], their swelling properties must be regenerated by employing a mechanical treatment referred to as "refining" or "beating". After beating, the pulp sample usually drains at a rate inadequate for use on a high-speed paper machine. Consequently, cellulases have been shown to improve the drainage of recycled pulps [52]. Oksanen et al. [53] applied separate EG1, EG2, and CBH1 cellulase components to pulps during each recycling run. As each pulp was beaten after recycling, the WRV value increased and the pulp became more responsive to cellulases, especially EG1 and EG2. Similar results were obtained with sulfite pulp, as the hydrolysis rate doubled after the pulp was subjected to a single beating treatment [48]. These results are not surprising as it has been established that mechanical treatment of lignocellulosics results in a cracking and delamination of the secondary cell wall layers, which increases porosity and the propensity for fibers to swell [54]. The exposure of new surfaces containing amorphous hydrophilic hemicelluloses in a process referred to as internal fibrillation [54] also increases the swelling of lignocellulosic substrates, which increases pore size. The WRV measurement has been shown to be a useful tool for indirectly assessing swelling and the potential for hydrolysis by cellulases. Methods that have been used to obtain a direct measurement of area or volume of pores in a substrate include nitrogen adsorption, mercury porosimetry, and solute exclusion, while SS provides an estimation of the pore size distribution in a substrate.

The pores in a substrate represent the true internal or interior surface area. Pores with diameters up to 2 nm are referred to as micropores, while pores in the 2–50 nm range are called "mesopores". "Macropores" have diameters >50 nm [55]. The nitrogen adsorption and mercury porosimetry methods differ from the WRV in that they require the cellulosic/lignocellulosic sample to be completely devoid of water prior to measurement. The SS [56] method uses the competitive adsorption of two direct dyes that differ in their size and affinity for cellulose in order to obtain an estimate of the pore volume distribution in a sample [57].

4.4.2 Simons' stain

Simons' stain (SS) was originally developed as a pulp fiber staining method to be used in the microscopic evaluation of mechanical damage undergone by pulp fibers during beating [56]. The SS method uses a direct blue and direct orange dye. The direct blue dye has a smaller molecular size and a weaker affinity for cellulose compared with the direct orange dye. The original method utilized the direct orange and direct blue dyes as received. However, later work revealed that the high-molecular-weight fraction of the direct orange dye was responsible for the increased affinity of the dye for cellulose, while the low-molecular-weight fraction of the dye had a similar affinity for cellulose as the direct blue dye [57]. Therefore, the method was subsequently modified to include a mixture of the direct blue and the high-molecular-weight fraction of the direct orange dye.

Initially, the direct blue dye molecules populate the smaller pores of the fiber, while the direct orange dye enters the larger substrate pores and the surface. Upon an increase in pore size of the lignocellulosic sample either by physical or fungal action [58], the direct orange dye gains access to the enlarged pores and displaces the direct blue dye due to their higher affinity for cellulose hydroxyl groups [59]. The ratio of the adsorbed orange dye to the blue dye (O/B) has been used as an estimation of substrate porosity [59]. Measurements of pore volume using SS on Avicel were shown to be directly proportional to measurements of SSA using nitrogen adsorption [59]. Therefore, the application of the mixture of direct blue and direct orange dyes can serve as a semiquantitative indicator of the pore size distribution of lignocellulosic samples.

The SS method has been used frequently to evaluate the effects of mechanical treatment on the development of lignocellulosic fibers for subsequent use in papermaking applications [60,61]. The SS method has also been used to measure energy savings resulting from biological treatment of wood chips prior to mechanical refining or "biomechanical pulping" [58]. The differences in the fiber structure as measured by SS that occurred during the fungal treatment of wood chips and subsequent refining were shown to correlate with the energy savings observed during mechanical fiberization [59].

Although the SS method has been used to characterize pulps with varying degrees of porosity as a result of mechanical or combined biomechanical treatments, there has only been a single study evaluating the ability of the SS method to assess the accessibility of substrates to cellulases. As it is well known that wood pulps undergo significant reductions in pore volume upon drying, Esteghlalian et al. [9] applied the SS method to compare the relative porosity of never-dried kraft pulp fibers with those of oven-dried, air-dried, and freeze-dried samples. Of the dried samples, the freeze-dried pulp adsorbed the greatest amount of the orange dye, and thus retained a greater amount of porosity compared to the original never-dried pulp sample. Also, the SS method was effective in predicting the susceptibility of these substrates to hydrolysis by cellulases as the extent of adsorption of the orange dye directly correlated with the extent of hydrolysis after 12 h (Figure 4.3).

4.4.3 Nitrogen adsorption technique

Unlike the SS Technique, nitrogen adsorption is a direct measure of SSA. Nitrogen or other probe gases that are unreactive with cellulose have been used extensively to characterize the pores within potential substrates for cellulases [62,63]. The method was first applied for characterizing the surface area of pulp fibers by Haselton [64]. He showed that the use of nitrogen as a probe gas in combination with the Brunauer, Emmett, and Teller (BET)

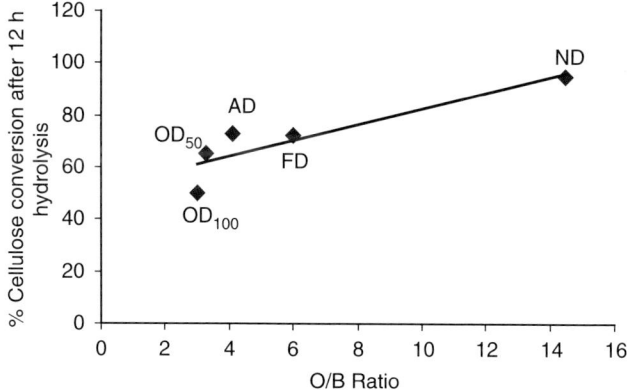

Figure 4.3 The correlation between the ratio of orange to blue dye adsorbed (O/B) on never-dried (ND), freeze-dried (FD), air-dried (AD), and oven-dried at 50°C and 100°C (OD$_{50}$ and OD$_{100}$) kraft pulps and the extent of hydrolysis after 12 h. The data was plotted from Reference 9. Hydrolysis conditions: Celluclast (40 FPU g^{-1} dry pulp) and β-glucosidase (80 CBU g^{-1} dry pulp), 50°C, pH 4.8, 2% pulp consistency.

method [65] was suitable for measuring the surface area of lignocellulosic pulps produced by the pulp and paper industry.

In the N_2 adsorption method, the surface of the cellulosic or lignocellulosic material is exposed to adhering molecules of N_2 at a series of increasing pressures. The total surface area accessible to the applied N_2 could be quantified by counting the number of molecules that filled the substrate pores. Usually these measurements are performed in an automated device at low temperatures. Nitrogen gas adsorbs on the dried sample, and using the BET equation [65], the known amount of adsorbed nitrogen (calculated from the applied pressure and the adsorption isotherm), the size of the nitrogen molecule (0.345 nm at 77 K or $-196.2°C$) [66], and the SSA of the cellulosic sample can be determined. Usually the adsorption isotherm at lower vapor pressures is used to calculate the surface area as the accuracy of the measurement is greatest at a monolayer coverage of nitrogen. The drying of the lignocellulosic samples presents a complication in the measurement of SSA using N_2 adsorption, because of the surface-tension-triggered collapse of substrate pores that can occur during the drying process [9,51]. In an attempt to circumvent this problem, researchers have used the solvent drying technique, discussed in depth in an early paper by Ingmanson *et al.* [67] and more recently by Jin *et al.* [68]. The solvent-exchange drying technique involves exposing the sample to a series of solvents of increasing hydrophobicity with a final air drying step to ensure a drying process that begins in the absence of water. The solvent drying process has been used for the preparation of cellulosic and lignocellulosic samples for subsequent SSA measurement using N_2 adsorption to evaluate the susceptibility of substrates to hydrolysis by cellulases.

Using N_2 adsorption with solvent drying using methanol followed by benzene, Fan *et al.* [69] measured the surface area of pure cellulose sulfite pulp samples (Solka Floc) in an attempt to determine if a relationship existed between the measured SSA and the initial hydrolysis rate. Substantial variations in hydrolysis rates were observed at similar measured values of SSA, whereas a linear relationship between sample crystallinity and hydrolysis rate was found. In another study [62], the same authors used gamma ray irradiation to treat pure cellulose samples to increase SSA measured by N_2 adsorption while preserving a constant level of crystallinity. Although the initial hydrolysis rate was not plotted against the SSA, it was evident from the results that the gamma ray treatment increased the SSA, which was directly proportional to the increase in the hydrolysis rate. This demonstrated the dependency of hydrolysis rate on SSA. The discrepancy between the conclusions from the two studies is difficult to ascertain. However, it is likely that the gamma ray treatment increased the SSA by directly increasing the number of pores in the substrate accessible to cellulases. In a later study, the same authors correlated the rate of enzymatic hydrolysis to both SSA and crystallinity and developed an equation to predict the hydrolysis rate of pure cellulose substrates from measurements of crystallinity and SSA [70].

Studies relating the SSA measured by N_2 adsorption to hydrolysis performed on lignocellulosic substrates have been far less frequent than studies performed on pure cellulose. One of these studies was performed by Gharpuray *et al.* [63], who applied a two-stage pretreatment of wheat straw consisting of ball milling with subsequent treatment with either peracetic acid or ethylene glycol. It was shown that the ball milling decreased crystallinity and increased SSA, while the ethylene glycol treatment removed lignin and increased SSA, thus increasing the initial rate of hydrolysis. The authors used the results to develop a formula to fit the data relating the initial rate of hydrolysis to a combination of increased

SSA and decreased crystallinity [70]. As will be discussed in further detail later, the relative infrequency of N_2 adsorption studies on lignocellulose is most likely because of the underestimation of the surface area of substrates using this technique. The N_2 molecule is ~3200 times smaller [71] than the size of cellulase enzymes, and therefore measures the area in pores that are inaccessible to cellulases. Another potential deficiency of the nitrogen adsorption method is the substantial difference in the measured results obtained when the calculations are based on either the adsorption or the desorption isotherm [66]. Further studies utilizing N_2 adsorption to determine the relationship between SSA and hydrolysis rate have combined the measurements of N_2 adsorption with mercury porosimetry as N_2 adsorption is better suited for the measurement of mesopores while mercury porosimetry can measure both meso- and macropores [72].

4.5 Interior surface area

4.5.1 Mercury porosimetry

The mercury porosimetry method strictly measures the interior surface area of substrates, whereas the N_2 adsorption measurement represents a gross surface area determination including both the pores and exterior substrate surfaces. As mercury is a nonwetting liquid for cellulose, pressure is required to facilitate the penetration of mercury into substrate pores. Therefore, when a dried cellulose or lignocellulose sample is submerged in mercury, the applied pressure for the penetration of the liquid into the pores is inversely proportional to the pore radius, as per the Washburn equation [73,74]. As the applied pressure increases, smaller pores are penetrated by the mercury, and thus a pore size distribution can be obtained. The accuracy of the test is compromised when there are large pores with smaller openings that require higher mercury pressures for penetration, resulting in an underestimation of pore volume. Applying higher pressures can also distort or collapse adjacent pores; thus mercury porosimetry is best suited for measuring macropores.

Mercury porosimetry and N_2 adsorption were applied to measure the surface area of an ethanol organosolv pulp from white birch (*Betula papyrifera*) to assess if a relationship existed between the surface area and hydrolysis of the pulp by cellulases [75]. The authors found that the increase in surface area that occurred upon increasing the ethanol concentration during pulping was directly proportional to an increase in the initial hydrolysis rate. In a later study by the same authors [43], white birch was pretreated by a host of swelling agents, followed by steam explosion. The surface area of the resulting pulps was measured by N_2 adsorption, mercury porosimetry, and solute exclusion. Similar to the results with the ethanol-pretreated pulp, the hydrolysis rate of the steam exploded samples treated with the various swelling agents was directly proportional to the median pore size and SSA estimated by N_2, mercury and solute exclusion. However, the solute exclusion technique was applied utilizing a molecular probe with a diameter of 90 Å compared to the pore size estimations of 18 and 29 969 Å determined by N_2 adsorption and mercury porosimetry, respectively. The results for mercury porosimetry were most likely obtained by low pressure application. However, the underestimation of pore size by the N_2 measurement is most likely due to the inadequate preservation of the substrate pore structure during solvent drying. The SSA measurements by N_2 have been reported to be 10–15 times lower [70] than those reported by the solute exclusion technique [76].

4.5.2   Solute exclusion

The solute exclusion technique can be regarded as an accurate measurement of surface area of pores in cellulosic and lignocellulosic substrates as it is a direct measurement of the pore volume of a wet, fully swollen sample. The solute exclusion technique has been referred to as a "more correct" method to determine the volume of pores in a substrate when compared to the WRV measurement [77]. Solute exclusion was initially used for the measurement of the pore size distribution of cotton and rayon fibers [78] and was later adapted and applied for the measurement of pulps [79]. The method has also been used frequently for the measurement of pore volume of various pretreated substrates prior to enzymatic hydrolysis for saccharification [21,27,28].

The solute exclusion method measures the volume of water accessible and inaccessible to the pore structure of the cellulosic substrate under examination. Cellulose samples are exposed to a series of solutions containing known concentrations of a molecular probe. An effective probe for the pore volume measurement should be obtainable in a range of narrowly distributed molecular weights, should be spherical in solution, and should not be an electrolyte or adsorbed on cellulose during the measurement [80]. Considering these criteria, the molecular probes that have been usually employed to study cellulose and lignocellulosic samples include polyethylene glycol (PEG) and dextran. The PEG or dextran molecules diffuse into substrate pores with a diameter greater than their molecular diameter, and thus the volume of water inaccessible to substrate pores can be calculated from the change in PEG or dextran concentration of the impregnation solution.

Batch measurements of pore volume can be performed by placing a cellulosic sample in a bottle containing solutions of the probe at a known concentration. The bottles are allowed to incubate at room temperature with intermittent shaking followed by measurement of the PEG or dextran concentration in the supernatant by viscosity measurement or via a polarimeter. Although the batch method has been employed successfully for pure cellulose and lignocellulosic substrates [19,21], usually the changes in concentration of the impregnating solutions are quite small, resulting in substantial measurement errors [71,81]. Alternatively, solute exclusion measurements on cellulosic substrates can also be performed continuously using gel permeation chromatography. In the continuous measurement, the column is packed with the cellulosic substrate, with subsequent loading of the column with solutions of known concentration and molecular weight of the probe (e.g., PEG or dextran). The probe molecules which do not penetrate the substrate in the column elute and can be analyzed using a differential refractometer, which is reasonably sensitive to small changes in solute concentration. Therefore, the pore volume measurements using the continuous method are more precise [71]. More recent work characterizing substrates have utilized the packed-column continuous method to measure changes in probe concentrations using a differential refractometer [82]. Although, the solute exclusion technique is effective, it requires a significant investment in time to obtain reproducible results. Also, it is likely that the pores in a lignocellulosic substrate will have irregular shapes, and will thus affect the accuracy of the measurement [83]. Another drawback is that the method does not measure the area in pores that are larger than the size of the probe, which would provide the easiest access for cellulases. However, despite these drawbacks, the solute exclusion method has been the most frequently applied method used to determine whether there is a relationship between cellulase hydrolysis performance and surface area accessibility.

The initial studies characterizing substrates via solute exclusion utilized molecular probes to estimate the rate-limiting pore size necessary for hydrolysis to occur. Stone et al. [19] measured pore volume on a series of cotton linter samples swollen to varying degrees using H_3PO_4 to calculate the surface area within the swollen fibers. Subsequent correlation of the surface area of the swollen cotton linters to the initial rate of hydrolysis with cellulases showed a direct linear relationship between increasing surface area and initial hydrolysis rate. By extrapolating the plot of initial hydrolysis rate vs the volume accessible to each probe of known size to the origin, Stone et al. [19] concluded that the threshold pore size for effective hydrolysis was 40 Å; thus it was hypothesized that the size of the cellulose enzyme must lie in this range. Similarly, later work investigating the hydrolysis of a set of pure cellulose samples including Sigmacell, milled cotton, Solka Floc, Avicel, and carboxymethyl cellulose by Weimer and Weston [27] also found a direct correlation between pores accessible to a probe size of 43 Å and the initial rate of hydrolysis with a coefficient of determination (r^2) of 0.83 and 0.85 for *Trichoderma* and *Clostridium* cellulases, respectively. When comparing the hydrolysis performance of free cellulases to cellulase components aggregated by cross-linking, Tanaka et al. [28] concluded that, when smaller pores predominate, the smaller cellulose components enter the pores and become deactivated due to a lack of synergism, but when the pores are large enough to accommodate all the components, hydrolysis proceeds.

In a lignocellulosic substrate, lignin and hemicellulose are intermingled with cellulose, thus adding a layer of complexity to the determination of accessible area available to cellulases. Similar to the results obtained for pure cellulose substrates, Grethlein et al. and others [84] have found that the rate-limiting pore size for the hydrolysis of acid-pretreated hardwoods and softwoods was 51 Å, thus the solute exclusion technique was effective for both cellulose and lignocellulosic substrates as originally stated by Stone et al. [19]. Although the dimensions of the pores in lignocellulosic substrates may be sufficient to accommodate cellulases, lignin can physically block access to the β-1,4 bonds of cellulose and can also bind cellulases, thus decreasing their overall hydrolytic capabilities [85,86]. However, extensive work has substantiated the relationship between pore volume measurements and hydrolysis [24] and simultaneous saccharification and fermentation [87] performance with pretreated lignocellulosic substrates. From the data of previous studies applied to pure cellulose, the accessibility of a 51 Å probe has been used as a benchmark for evaluating the accessibility of substrates to subsequent enzymatic hydrolysis. The data in Table 4.1 demonstrates the effects of various pretreatments applied to lignocellulosics on the area accessible to a 51 Å probe and the resulting hydrolysis yield after 2 h. It is evident that the development of adequate accessible surface area during substrate pretreatment is essential for effective hydrolysis by cellulases.

Grethlein and coworkers have obtained hydrolysis yields of 90–100% after acid pretreatment of corn stover, newsprint, oak, poplar, and mixed hardwood [22,88,89], while a hydrolysis yield of 65% was obtained in the case of white pine softwood. The difference in hydrolysis performance among the feedstocks was attributed to a "resistance" of the white pine, as a greater amount of hemicellulose remained in the substrate after the acid pretreatment. Even though other substrate factors, such as the recalcitrance of the guaiacyl-rich

Table 4.1 Initial (2 h) hydrolysis rate and the accessibility to a 51 Å probe of various pretreated lignocellulosic substrates.

Pretreated substrate	Pretreatment conditions	Specific surface area (m²/g) accessible to 51 Å probe	Cellulose conversion in 2 h hydrolysis (%)	Hydrolysis conditions	References
Steam-exploded poplar	Untreated	20	9.1	Cellulase: 16.2 FPU g^{-1} wood, β-glucosidase: 76 U g^{-1} wood, 50°C, pH 4.8	81
	250 psig, 5 min	60	19.1		
	350 psig, 5 min	90	34		
	450 psig, 5 min	110	30.7[a]		
Alkaline peroxide (H_2O_2) birch–maple (90:10)	50% H_2O_2, 0.5 h, 25°C	24.5	26	Cellulase: 40.8 FPU g^{-1} wood, β-glucosidase: 40.8 U g^{-1} wood, 50°C, pH 4.8	24
	50% H_2O_2, 5 h, 25°C	66.7	37		
	50% H_2O_2, 10 h, 25°C	70.8	41		
Ethylenediamine (EDA) birch–maple (90:10)	10% EDA, 90°C	30.7	31	Cellulase: 40.8 FPU g^{-1} wood, β-glucosidase: 40.8 U g^{-1} wood, 50°C, pH 4.8	24
	50% EDA, 90°C	47.3	41		
	70% EDA, 90°C	94.5	54		
Autohydrolysis birch–maple (90:10)	240°C, 8–9 s	34.7	17.5	Cellulase: 40.8 FPU g^{-1} wood, β-glucosidase: 40.8 U g^{-1} wood, 50°C, pH 4.8	24
	260°C, 8–9 s	38.0	25		
	280°C, 8–9 s	48.4	32.5		
Acid hydrolysis birch–maple (90:10)	180°C, 1% H_2SO_4, 8–9 s	25.8	16.5	Cellulase: 40.8 FPU g^{-1} wood, β-glucosidase: 40.8 U g^{-1} wood, 50°C, pH 4.8	24
	200°C, 1% H_2SO_4, 8–9 s	85.7	51		
	220°C, 1% H_2SO_4, 8–9 s	128	71		
Steam-exploded *Pinus radiata* Chips	1 min, 195°C, 1% SO_2	46	7.5	Cellulase: 40 FPU g^{-1} wood, β-glucosidase: 50 U g^{-1} wood, 50°C, pH 4.8	84
	3 min, 215°C, 0% SO_2	22	7.8		
	9 min, 195°C, 1% SO_2	96	18.9		
	9 min, 195°C, 6.5% SO_2	162	35.9		
	3 min, 215°C, 2.6% SO_2	215	36.3		
	3 min, 215°C, 9.3% SO_2	250	45.9		

[a] Considerable scatter in hydrolysis yield measurements.

lignin in the softwood, may have affected the development of porosity, the authors rationalized their results by equating the development of pores within the substrate to a removal of hemicellulose. It should be noted that lignocellulose is a heterogeneous matrix composed of lignin, hemicellulose, and cellulose; therefore, it is difficult to unequivocally associate the removal of hemicellulose to the development of pores without the consideration of lignin.

During a comparison of the hydrolysis of Avicel to *n*-butylamine treated rice straw, Tanaka et al. [28] reported that, although the rice straw contained 25% lignin, the *n*-butylamine treatment resulted in a larger pore size than the Avicel. The initial hydrolysis rate of the corn stalk was double that of the Avicel [28], thus demonstrating the significance of pore volume. Similar results were also observed previously with rice straw, but the surface area was not measured [90]. Later work by Wong and coworkers [84] has shown that, although amplifying the severity of steam explosion pretreatment increased pore volume and hydrolysis yields, washing with alkali to remove lignin after pretreatment resulted in decreased pore volume and cellulose hydrolysis. The authors concluded that the alkali extraction redistributed lignin within the substrate pores, thereby reducing the overall accessibility of the substrate to cellulases. The results described by Tanaka et al. [28] and Wong et al. [84] indicate the importance of the location of lignin rather than the lignin content. Mooney et al. [91] used the solute exclusion method to measure the pore volume of refiner mechanical pulp (RMP), sulfonated RMP, sodium chlorite delignified RMP, and kraft pulp from Douglas-fir to assess the susceptibility of these pulps to subsequent hydrolysis by cellulases. As mentioned earlier, the delignification of the RMP resulted in a greater rate and extent of hydrolysis than the kraft pulp sample, which may be attributed to the smaller particle size of the RMP. The sulfonation of the RMP dramatically increased swelling; however, unlike the delignification, this did not translate into either enhanced access to the 5.1 nm probe or hydrolysis performance. The most feasible explanation for these results is that the lignin content of the sulfonated pulp (30.9%) inhibited hydrolysis, regardless of the greater swelling of the pores. These results further support the effectiveness of the solute exclusion measurement in predicting the rate of substrate hydrolysis by cellulases. Overall, despite their various limitations, it is apparent that methods utilizing either polymeric probes or dyes to evaluate substrate accessibility in a wet state (such as solute exclusion and SS) provide some useful information and give a relatively good indication of how various lignocellulosic substrates will respond to subsequent cellulase treatments.

Conclusions

It is recognized that both substrate characteristics and the nature of the enzyme complex have a significant effect on how effectively a pretreated lignocellulosic material will be hydrolyzed by cellulase systems. Our survey of the literature shows that, although several reports appear to substantiate the view that the initial hydrolysis rate of lignocellulosics is directly proportional to substrate surface area, at this time there does not seem to be one specific, rapid, accurate, and reproducible method by which the SSA of a substrate can be measured. The methods that have been used are either indirect, tedious, or compromised by the requirement of sample drying prior to the measurement. Comparisons between data in the literature are difficult to carry out, primarily because of inconsistencies in sample preparation and variations in the methods used to perform the measurements. Additionally, any measurements that only provide estimates of surface area usually do not account for the presence of hemicelluloses and lignin that may block reactive sites for cellulases. At the core of this discussion is the assumption that the cellulase enzyme components require contact with the substrate to catalyze hydrolysis of β-1,4-glycosidic bonds; therefore, the most desirable assays would measure both the SSA required for accommodating the enzymes and the actual "reactive sites" on the cellulose or lignocellulosic substrate. A method possessing these capabilities would be invaluable in the development of effective biomass pretreatments that produce substrates amenable to subsequent enzymatic hydrolysis.

References

1. Mansfield, S.D.; Mooney, C.; Sadder, J.N., Substrate and Enzyme Characteristics that Limit Cellulose Hydrolysis; *Biotechnol.* 1999, 15, 804–816.
2. Zhang, P.; Lynd, L.R., Toward an Aggregated Understanding of Enzymatic Hydrolysis of Cellulose: Noncomplexed Cellulase Systems; *Biotech. Bioeng.* 2004, 88, 797–824.
3. Jeoh, T.; Wilson, D.B.; Walker, L.P., Cooperative and Competitive Binding in Synergistic Mixtures of *Thermobifida fusca* Cellulases Cel5A, Cel6B, and Cel9A; *Biotechnol. Progr.* 2000, 18, 760–769.
4. Jeoh, T.; Wilson, D.B.; Walker, L.P., Effect of Cellulase Mole Fraction and Cellulose Recalcitrance on Synergism in Cellulose Hydrolysis and Binding; *Biotechnol. Progr.* 2006, 22, 270–277.
5. Doi, R.H.; Kosugi, A.; Mursashima, K.; Tamaru, Y.; Han, S.O., Cellulosomes from Mesophilic Bacteria; *J. Bacteriol.* 2006, 185, 5907–5914.
6. Carrard, G.; Koivula, A.; Soderlund, H.; Beguin, P., Cellulose-binding Domains Promote Hydrolysis of Different Sites in Crystalline Cellulose; *Proc. Natl. Acad. Sci. USA* 2000, 97, 10, 342–347.
7. Esteghlalian, A.; Mansfield, S.D.; Saddler, J.N., Cellulases: Agents for Fiber Modification or Bioconversion? The Effect of Substrate Accessibility on Cellulose Enzymatic Hydrolyzability. In: Viikari, L. and Lantto, R. (eds); *Biotechnology in the Pulp and Paper Industry: Progress in Biotechnology*; Elsevier Science, Netherlands; 2002; Vol 21; pp. 21–36.
8. Wilson, D., Studies of *Thermobifida fusca* Plant Cell Wall Degrading Enzymes; *Chem. Rec.* 2004, 4, 72–82.
9. Esteghlalian, A.; Bilodeau, M.; Mansfield, S.D.; Saddler, J.N., Do Enzymatic Hydrolyzability and Simons' Stain Reflect the Changes in the Accessibility of Lignocellulosic Substrates to Cellulase Enzymes?; *Biotechnol. Progr.* 2001, 1049–1054.
10. Saloheimo, M.; Paloheimo, M.; Hakola, S. *et al.* Swollenin a *Trichoderma reesei* Protein with Sequence Similarity to the Plant Expansins, Exhibits Disruption Activity on Cellulosic Materials; *Eur. J. Biochem.* 2002, 269, 4202–4211.
11. Wong, D.W.S., Feruloyl Esterase: A Key Enzyme in Biomass Degradation; *Appl. Biochem. Biotechnol.* 2006, 133, 87–112.
12. Raghothama, S.; Simpson, P.J.; Szabo, L.; Nagy, T.; Gilbert, H.; Williamson, M.P., Solution Structure of the CBM10 Cellulose Binding Module from *Pseudomonas xylanase* A; *Biochemistry* 2000, 39, 978–984.
13. Boraston, A.B.; Bolam, D.N.; Gilbert, H.J.; Davies, G.J., Carbohydrate Binding Modules: Fine-tuning Polysaccharide Recognition; *Biochem. J.* 2004, 382, 769–781.
14. Vaaje-Kolstad, G.; Horn, S.; van Aalten, D.M.F.; Synstad, B.; Eijsink, V.G.H., The Non-catalytic Chitin-binding Protein CBP21 from *Serratia marcescens* is Essential for Chitin Degradation; *J. Biol. Chem.* 2005, 280, 28492–28497.
15. Esteghlalian, A.; Srivastava, V.; Gilkes, N.; Gregg, D.J.; Saddler, J.N., An Overview of Factors Influencing the Enzymatic Hydrolysis of Lignocellulosic Feedstocks. In: Himmel, M.; Baker, J.;

Saddler, J.N. (eds); *Glycosyl Hydrolases for Biomass Conversion* American Chemical Society Symposium Series; American Chemical Society; 2000; pp. 100–111.

16. Zhang, S.; Barr, B.K.; Wilson, D.B., Effects of Noncatalytic Residue Mutations on Substrate Specificity and Ligand Binding of *Thermobifida fusca* Endocellulase Cel6A; *Eur. J. Biochem.* 2000, 267, 244–252.
17. Puri, V.P., Effect of Crystallinity and Degree of Polymerization of Cellulose on Enzymatic Eaccharification; *Biotechnol. Bioeng.* 1984, 26, 1219–1222.
18. Ramos, L.P; Breuil, C.; Saddler, J.N., Effect of Enzymatic Hydrolysis on the Morphology and Fine Structure of Pretreated Cellulosic Residues; *Enzyme Microb. Technol.* 1993, 15, 821–831.
19. Stone, J.E.; Scallan, A.M.; Donefer, E.; Ahlgren, E., Digestibility as a Simple Function of a Molecule of a Similar Size to a Cellulose Enzyme; *Adv. Chem. Ser.* 1969, 95, 219.
20. Cowling E.B.; Kirk, T.K., Properties of Cellulose and Lignocellulosic Materials as Substrates for Enzymatic Conversion Processes; *Biotechnol. Bioeng. Symp.* 1976, 6, 95–123.
21. Grethlein, H.E., The Effect of Pore Size on the Rate of Enzymatic Hydrolysis of Cellulosic Substrates; *Biotechnology* 1985, 3, 155–160.
22. Grethlein, H.E.; Allen, D.C.; Converse, A.O., A Comparative Study of the Enzymatic Hydrolysis of Acid Pretreated White Pine and Mixed Hardwood; *Biotech. Bioeng.* 1984, 26, 1498–1505.
23. Chang, M.M.; Chou, T.Y.; Tsao, G.T., Structure, Pretreatment and Hydrolysis of Cellulose; *Adv. Biochem. Eng.* 1981, 20, 15–42.
24. Thomson, D.N.; Chen, H.C., Comparison of Pretreatment Methods on the Basis of Available Surface Area; *Bioresour. Technol.* 1992, 39, 155–163.
25. Boussaid, A.L.; Esteghlalian, A.R.; Gregg, D.J.; Lee, K.H.; Saddler, J.N., Steam Pretreatment of Douglas-fir Wood Chips: Can Conditions for Optimum Hemicellulose Recovery Still Provide Adequate Access for Efficient Enzymatic Hydrolysis?; *Appl. Biochem. Biotechnol.* 2000, 84–86, 693–705.
26. Mooney, C.M.; Mansfield, S.D.; Beatson, R.P.; Saddler, J.N, The Effect of Fiber Characteristics on Hydrolysis and Cellulase Accessibility to Softwood Substrates; *Enzyme Microb. Technol.* 1999, 25, 644–650.
27. Weimer, P.J.; Weston, W.J., Relationship between Fine Structure of Native Cellulose and Cellulose Degradability by the Cellulase Complexes of *Trichoderma reesei* and *Clostridium thermocellum*; *Biotech. Bioeng.* 1985, 27, 1540–1547.
28. Tanaka, M.; Ikesaka, M.; Matsuno, R., Effect of Pore Size and Substrate Diffusion of Enzyme on Hydrolysis of Cellulosic Materials with Cellulase; *Biotech. Bioeng.* 1988, 32, 698–706.
29. Laureano-Perez L.; Farzaneh, T.; Hasan, A.; Dale, B.E., Understanding Factors that Limit Enzymatic Hydrolysis of Biomass: Characterization of Pretreated Corn Stover; *Appl. Biochem. Biotechnol.* 2005, 121–124, 1081–1099.
30. Ford, C.W., Effect of Particle Size and Delignification on the Rate of Digestion of Hemicellulose and Cellulose by Cellulase in Mature Pangola Grass Stems; *Aust. J. Agric. Res.* 1983, 34, 241–248.
31. Ragnar E.; Alderborn, G.; Nyström, C., Particle Analysis of Microcrystalline Cellulose: Differentiation between Individual Particles and their Agglomerates; *Int. J. Pharm.* 1994, 111, 43–50.
32. Allen, T., *Particle Size Measurement*, 4th Edition; Chapman and Hall; London, 1990.
33. Aharonson, E.P.; Karasikov. N.; Rojtberg, M.; Shamir. J., Galai CIS-1. A Novel Approach to Aerosol Particle Size Analysis; *J. Aerosol Sci.* 1986, 17, 530–536.
34. Barth, H., *Modern Methods of Particle Size Analysis*; Wiley; New York, 1984.
35. Sawada, T.; Nakamura, Y.; Kobayashi, F.; Kuwahara, M.; Watanabe, T., Effects of Fungal Pretreatment and Steam Explosion Pretreatment on Enzymatic Saccharification of Plant Biomass; *Biotechnol. Bioeng.* 1995, 48, 719–724.
36. Sawada, I.; Kuwahara, M.; Nakamura, Y.; Suda, H., Effect of a Process of Explosion for the Effective Utilization of Biomass; *Int. Chem. Eng.* 1987, 27, 686–693.

37. Tanahashi, M., Characterisation and Degradation Mechanisms of Wood Components by Steam Explosion and Utilization of Exploded Wood; *Wood Res.* 1990, 77, 49–117.
38. Olson, J.A.; Robertson, A.G.; Finnigan, T.D.; Turner, R.R.H., An Analyzer for Fibre Shape and Length; *J. Pulp Pap. Sci.* 1995, 21, 367–373.
39. OpTest Equipment; *HiRes Fiber Quality Analyzer – User Manual*; OpTest, 1999.
40. Mooney, C.A.; Mansfield, S.D.; Beatson, R.P.; Saddler, J.N., The Effect of Fibre Characteristics on Accessibility and Hydrolysis of Softwood Substrates; *Proc. 7th Intern. Conf. Biotechnol. Pulp Paper Ind.* June 16–19, 1998; Vancouver, BC. Book 3: pp. 99–102.
41. Nahzad, M.M.; Ramos, L.P.; Paszner, L.; Saddler, J.N., Structural Constraints Affecting the Initial Enzymatic Hydrolysis of Recycled Paper; *Enzyme Microb. Technol.* 1995, 17, 68–74.
42. Lindstrom, T., Chemical Factors Affecting the Behaviour of Fibres During Papermaking; *Nordic Pulp Pap. Res. J.* 1992, 4, 181–191.
43. Bendzalova, M.; Pekarovicova, A.; Kokta, B.V.; Chen, R., Accessibility of Swollen Cellulosic Fibers; *Cellul. Chem. Technol.* 1996, 30, 19–32.
44. Ogiwara, Y.; Arai, K., Change in Degree of Polymerization of Wood Pulp with Cellulase Hydrolysis; *Text. Res. J.* 1969, 39, 422–427.
45. Hopner, T.; Jayme, G.; Ulrich, J.C., Determination of the Water Retention (Swelling Value) of Pulps; *Das Papier* 1955, 9, 476–482.
46. Chemical Pulp – Water Retention Value, Test Method SCAN-C 62:00, Scandinavian Pulp, Paper and Board Testing Comm., Stockholm, Sweden; 2000.
47. Scallan, A.M.; Caries, I.E., The Correlation of the Water Retention Value with the Fibre Saturation Point; *Svensk Papperstidn.* 1972, 75, 699–703.
48. Ogiwara, Y.; Arai, K., Swelling Degree of Cellulose Materials and Hydrolysis Rate with Cellulase; *Text. Res. J.* 1968, 38, 885–891.
49. Bonaventura, F.; Marzetti, A.; Sarto, V.; Beltrame, P.L.; Carniti, P., Cellulosic Materials: Structure and Enzymatic Hydrolysis Relationships; *J. Appl. Polym. Sci.* 1996, 30, 19–32.
50. Schimper, C.; Keckeis, R.; Ibanescu, C.; Burtscher, E.; Manian, A.P.; Bechtold, T., Influence of Steam and Dry Heat Pretreament on Fibre Properties and Cellulase Degradation of Cellulosic Fibres; *Biocatal. Biotransform.* 2004, 22, 383–389.
51. Laivins G.V.; Scallan A.M., The Mechanism of Hornification of Wood Pulps. In: Baker, C.F (ed.); Products of Papermaking. *Transactions of the 10th Fundamental Research Symposium*; Pira International; Oxford, 1993; pp. 1235–1260.
52. Pommier, J.; Fuentes, J.L.; Goma, G., Using Enzymes to Improve the Process and the Product Quality in the Recycled Paper Industry; *Tappi J.* 1989, 72, 187.
53. Oksanen, T.; Pere, J.; Paavlainen, L.; Buchert, J.; Viikari, L., Treatment of Recycled Kraft Pulps with *Trichoderma reesei* Hemicellulases and Cellulases; *J. Biotechnol.* 2000, 78, 39–48.
54. Page, D.H., The Beating of Chemical Pulps – the Action and Effects. In: Baker, C.F. and Punton, V.W. (eds); *Fundamentals of Papermaking Volume 1*; Mechanical Engineering Publications Ltd; London, 1989; pp. 1–38.
55. Sing, K.S.W.; Everett, D.H.; Haul, R.A.W. et al., Reporting Physisorption Data for Gas/Solid systems with Special Reference to the Determination of Surface Area and Porosity; *Pure Appl. Chem.* 1985, 57, 603–619.
56. Simons, F.L., A Stain for Use in the Microscopy of Beaten Fibers; *Tappi J.* 1950, 33, 312–314.
57. Yu, X.; Minor, J.L.; Atalla, R.H., Mechanism of Action of Simons' Stain; *Tappi J.* 1995, 78, 175–180.
58. Akhtar, M.; Blanchette, R.A.; Burnes, T.A., Using Simon's Stain to Predict Energy Savings during Biomechanical Pulping; *Wood Fiber Sci.* 1995, 27, 258–264.
59. Yu, X.; Atalla, R.H., A Staining Technique for Evaluating the Pore Structure Variations of Microcrystalline Cellulose Powders; *Powder Technol.* 1998, 98, 135–138.

60. Jayme, G.; Harders-Steinhauser, M., Determination of Differences in Density in Cellulose Fibers by Double Color Staining; *Papier* 1955, 9, 507–510.
61. Joutsimo, O.; Robertsen, L., The Effect of Mechanical Treatment on Softwood Kraft Pulp Fibers; *Paperi ja Puu.* 2005, 87, 41–45.
62. Beardmore, D.H.; Fan, L.T.; Lee, Y.H., Gamma-ray Irradiation as a Pretreatment for the Enzymatic Hydrolysis of Cellulose; *Biotechnol. Lett.* 1980, 2, 435–438.
63. Gharpuray, M.M.; Lee, Y.H.; Fan, L.T., Structural Modification of Lignocellulosics by Pretreatments to Enhance Enzymatic Hydrolysis; *Biotechnol. Bioeng.* 1983, 25, 157–172.
64. Haselton, W.R., Gas Adsorption by Wood, Pulp, and Paper. I. The low-temperature Adsorption of Nitrogen, Butane, and Carbon Dioxide by Sprucewood and its Components; *Tappi J.* 1954, 37, 404–412.
65. Brunauer, S.; Emmett, P.H.; Teller, E., Adsorption of Gases in Multimolecular Layers; *J. Am. Chem. Soc.* 1938, 60, 309–319.
66. Robens, E.; Dabrowski, A.; Kutarov, V.V., Comments on the Surface Structure Analysis by Water and Nitrogen Adsorption; *J. Therm. Anal. Calorim.* 2004, 76, 647–657.
67. Thode, E.F.; Swanson, J.W.; Becher, J.J., Nitrogen adsorption on Solvent-Exchanged Wood Cellulose Fibers: Indications of "Total" Surface Area and Pore-Size Distribution; *J. Phys. Chem.* 1958, 62, 1036–1039.
68. Jin, H.; Nishiyama, Y.; Wada, M.; Kuga, S., Nanofibrillar Cellulose Aerogels; *Colloids Surf. A* 2004, 240, 63–67.
69. Fan, L.T.; Lee, Y.H.; Beardmore, D.H., Mechanism of the Enzymatic Hydrolysis of Cellulose: Effects of Major Structural Features of Cellulose on Enzymatic Hydrolysis; *Biotechnol. Bioeng.* 1980, 22, 177–199.
70. Fan, L.T.; Lee, Y.H.; Beardmore, D.H., The Influence of Major Structural Features of Cellulose on Rate of Enzymatic Hydrolysis; *Biotechnol. Bioeng.* 1981, 23, 419–424.
71. Neuman, R.P.; Walker, L.P., Solute Exlusion from Cellulose in Packed Columns: Experimental Investigation and Pore Volume Measurements; *Biotech. Bioeng.* 1992, 40, 218–225.
72. Westermarck, S.; Juppo, A.M.; Yliruusi, J., Mercury Porosimetry of Mannitol Tablets: Effect of Scanning Speed and Moisture; *Pharm. Dev. Technol.* 2000, 5, 181–188.
73. Keller, J.U.; Staudt, R., *Gas Adsorption Equilibria: Experimental Methods and Adsorption Isotherms*; Springer; New York, 2005.
74. Gregg, S.J.; Sing, K.S.W., *Adsorption, Surface Area, and Porosity*, 2nd Edition; Academic Press; New York, 1985.
75. Bendzalova, M.; Pekarovicova, A.; Kokta, B.V., Surface Characteristics of Fibers in High Yield Pulping with Ethanol; *Cellul. Chem. Technol.* 1995, 29, 713–724.
76. Van Dyke, B.H.; Wilson, D.B., Enzymatic Hydrolysis of Cellulose. A Kinetic Study; PhD. Thesis, Massachusetts Institute of Technology, Cambridge, MA, 1972.
77. Eriksson, I.; Haglind, I.; Lidbrandt, O.; Salmen, L., Fiber Swelling Favoured by Lignin Softening; *Wood Sci. Technol.* 1991, 25, 135–144.
78. Aggenbrandt, L.; Samuellson, O., Penetration of Water Soluble Polymers into Cellulose Fibers; *J. Appl. Polym. Sci.* 1964, 8, 2801–2812.
79. Stone, I.E.; Scallan, A.M., A Structural Model for the Cell Wall of Water Swollen Pulp Fibers Based on their Accessibility to Macromolecules; *Cellul. Chem. Technol.* 1968, 2, 343–358.
80. Lin, J.K.; Ladisch, M.R.; Patterson, J.A.; Noller, C.H., Determining Pore Size Distribution in Wet Cellulose by Measuring Solute Exclusion Using a Differential Refractometer; *Biotech. Bioeng.* 1987, 29, 976–981.
81. Grethlein, H.E., Effect of Steam Explosion Pretreatment on Pore Size and Enzymatic Hydrolysis of Poplar; *Enzyme Microb. Technol.* 1986, 8, 274–280.
82. Mansfield, S.D.; de Jong, E.; Stephens, R.S.; Saddler, J.N., Physical Characterization of Enzymatically Modified Kraft Pulp Fibers; *J. Biotechnol.* 1997, 57, 205–216.

83. Converse, A.O., Substrate Factors Limiting Enzymatic Hydrolysis. In: Saddler, J.N. (ed.); *Bioconversion of Forest and Agricultural Plant Residues*; CAB International; Wollingford, UK, 1993; pp. 93–106.
84. Wong, K.Y.; Deverell, K.F.; Mackie, K.L.; Clark, T.A.; Donaldson, L., The Relationship between Fiber Porosity and Cellulose Digestibility in Steam Exploded *Pinus Radiata*; *Biotech. Bioeng.* 1988, 31, 447–456.
85. Yang, B.; Wyman, C.E., BSA Treatment to Enhance Enzymatic Hydrolysis of Cellulose in Lignin Containing Substrates; *Biotech. Bioeng.* 2006, 94, 611–617.
86. Eriksson, T.; Börjesson, J.; Tjerneld, F., Mechanism of Surfactant Effect in Enzymatic Hydrolysis of Lignocellulose; *Enzyme Microb. Technol.* 2002, 31, 353–364.
87. Meunier-Goddik, L.; Bothwell, M.; Sangseethong, K. et al., Physiochemical Properties of Pretreated Poplar Feedstocks during Simultaneous Saccharification and Fermentation; *Enzyme Microb. Technol.* 1999, 24, 667–674.
88. Knappert, D.R.; Grethlein, H.E.; Converse, A.O., Partial Acid Hydrolysis of Cellulosic Material as a Pretreatment for Enzymatic Hydrolysis; *Biotech. Bioeng.* 1980, 22, 1449–1463.
89. Knappert, D.R.; Grethlein, H.E.; Converse, A.O., Partial Acid Hydrolysis of Poplar Wood as a Pretreatement for Enzymatic Hydrolysis; *Biotech. Bioeng. Symp.* 1981, 11, 66–77.
90. Goel, S.C.; Ramachandran, K.B., Comparison of the Rates of Enzymic Hydrolysis of Pretreated Rice Straw and Bagasse with Cellulases; *Enzyme Microb. Technol.* 1983, 5, 281–284.
91. Mooney, C.A.; Mansfield, S.D.; Touhy, M.G.; Saddler, J.N., The Effect of Initial Pore Volume and Lignin Content on the Enzymatic Hydrolysis of Softwoods; *Bioresour. Technol.* 1998, 64, 113–119.

Chapter 5
Characterization of AKD-Sized Pilot Papers by XPS and Dynamic Contact Angle Measurements

Rauni Seppänen

Abstract

Alkyl ketene dimer (AKD)-sized pilot papers with varying doses of AKD were prepared and the influence of AKD on the surface chemical composition was investigated by x-ray photoelectron spectroscopy (XPS). In addition, the chemical nature of the papers was evaluated in a dynamic wetting study using two test liquids with different surface tensions. The papers were made of bleached elemental chlorine-free (ECF) pulp, unfilled and filled with 20% high opacity scalenohedral precipitated calcium carbonate (PCC), respectively. The results showed that with increasing AKD addition a higher amount of aliphatic carbon C1 originating from AKD was present on the paper surface. The C1 concentration was significantly lower on the PCC-filled paper surfaces mainly because of the higher specific surface area of the filler. The background evaluation of the XPS oxygen signal suggested that there was a thin overlayer film of AKD on the unfilled paper surface only at a high AKD addition level.

The wettability of the sized papers was observed to be greatly dependent on the sizing degree of the papers and the surface tension of the liquid. The contact angles of ethylene glycol, which has a lower surface tension than water, showed more clear and stepwise increase with the increase of AKD on the paper surfaces than those of water. On the other hand, contact angles of water above 90° were reached at a low AKD coverage. The PCC-filled papers wetted more rapidly than the corresponding unfilled ones, although the total amount of AKD in the papers was the same.

5.1 Introduction

Most paper and board grades need to be resistant to wetting and absorption by polar liquids such as water, aqueous solutions, and suspensions. Internal sizing refers to the process of introducing chemical additives to make fibers hydrophobic. In addition to improving the barrier properties of liquid packaging, sizing (hydrophobizing) also improves, for example, paper printability by controlling ink spreading and absorption. For instance, newsprint

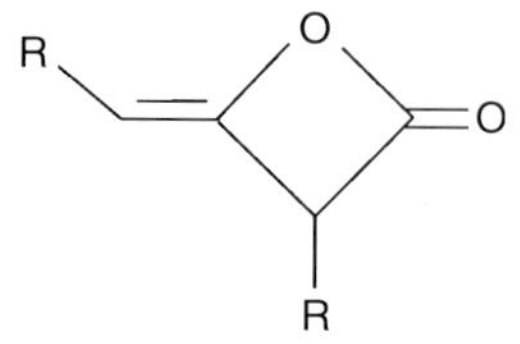

Figure 5.1 Molecular structure of AKD; R comprises C16 carbons.

grade with reduced grammage is sized to hinder print-through and absorption problems. It is also frequently necessary to size paper to retard or hinder the penetration of liquids in cartons, packaging paper and board, paper cups, and so on. For uncoated paper grades, the buildup of hydrophobicity depends on interactions between a sizing agent and different papermaking components such as fibers, fines, filler, and various chemical additives.

An effective method of sizing of paper and board at neutral or alkaline pH is to apply an internal sizing agent, for instance, AKD (see Figure 5.1 for structure) to the wet end of a paper machine (for a comprehensive overview, see Reference 1). The AKD molecule consists of a fiber-reactive group, which is expected to anchor to the cellulose surface, and a hydrocarbon part, which gives the desired hydrophobic properties. AKD is added to the furnish as an emulsion, stabilized often with cationic starch, so as to enhance its retention with negatively charged fibers. Upon sizing, polar hydroxyl groups of cellulose react or get shielded by absorbing AKD moieties leading to reduced paper wettability. Thus, the improved water resistance of sized paper can be explained by the formation of low-energy regions on fiber surfaces that inhibit spreading and capillary penetration of water into paper.

High retention, large coverage of fiber and filler surfaces by the sizing agent and well-anchored size molecules are the most evident requirements for high sizing efficiency. A large number of papers dealing with the spreading of AKD have been published. It has been proposed that the mechanism is complete spreading, which leads to a monomolecular layer of AKD on the fiber surface [2]. Contradictory observations have, however, been reported. Ström et al. [3] suggested that spreading of molten AKD on the paper surface stops at a certain film thickness and does not continue to a monomolecular layer. The spreading of AKD wax and a commercial AKD emulsion on model and cellulose surfaces has been investigated, and a complete spreading of AKD to a monolayer was not observed [4,5]. Seppänen et al. [6] concluded that AKD spread over hydrophilic cellulose and silica surfaces, but that the spreading was not complete. Further spreading was shown to occur by surface diffusion in the form of an autophobic monolayer precursor, which grows from the foot of the AKD drop or particle. The monolayer diffusion process was observed to be quite slow, the apparent surface diffusion coefficient being on the order 10^{-11} m^2 s^{-1} at 45°C. Shchukarev et al. [7] obtained a diffusion coefficient of the same order at 80°C.

The surface chemistry of paper is important because it influences paper properties such as wettability, adhesion, friction, and strength, which are critical in converting and end-use applications. The aim of this work is to investigate the influence of increased AKD amounts added to furnish on the surface chemical composition of paper. Using surface-sensitive x-ray photoelectron spectroscopy (XPS), we have analyzed the aliphatic C—C carbon peak of the high-resolution C1s signal. This is possible, as an unsized paper made from bleached pulp fibers contains only small amounts of unoxidized carbon, while this type of carbon is the major component of AKD [3]. Since the pioneering work by Dorris and Gray [8], XPS has been widely used to study lignocellulosic surfaces – for example, unbleached kraft pulp fibers [9,10], different mechanical pulp fibers [11], chemically modified lignocellulosic fibers [12–14], plasma-treated surfaces [15,16], and also AKD-sized papers [3,7,17].

In XPS, the measuring principle is that a sample, placed in high vacuum, is irradiated with well-defined x-ray energy, resulting in the emission of photoelectrons. Only those from the outermost surface layers reach the detector. By analyzing the kinetic energy of these photoelectrons, their binding energy can be calculated, thus giving their origin in relation to the element and the electron shell. XPS provides quantitative data on both the elemental composition and different chemical states of atoms in the surface layer. The chemical shifts for carbon (C1s) in pulp fibers have been classified into four categories [8]: unoxidized carbon (C—C), carbon with one bond to oxygen (C—O), carbon with two bonds to oxygen (O—C—O or C=O), and carbon with three bonds to oxygen (O=C—O).

Wettability is often characterized by measuring the contact angle formed between a liquid drop and a solid surface. Contact angle of water is a common measure of the hydrophobicity of a surface. However, paper substrates are not compliant with the assumptions of Young's equation. Paper surfaces are porous and chemically and physically heterogeneous. A liquid drop placed on a porous paper does not only spread on the surface but also penetrates into the paper, thus modifying its wetting properties. Therefore, the ability to consistently measure the contact angle for papers has long been a significant challenge because of spreading, absorption, dissolution, and swelling, which can occur on similar time scales. Despite these obstacles, there is now a relatively good qualitative understanding of the factors affecting the determination of the equilibrium contact angle on paper and the dynamics of wetting, for example, on sized paper [17–21] and unsized paper [17,18]. During the last few years there has been a renewed interest in wetting, especially on the wetting of rough surfaces [22–24] based on theoretical foundations of the Wenzel (1936) [25] and Cassie [26] relations. Briefly, these theories attempt to explain the effect of surface roughness that increases the surface area of a hydrophobic material, and the effect of geometrically enhancing hydrophobicity (homogeneous wetting). On the other hand, air can remain trapped below the liquid drop, which also leads to a superhydrophobic behavior, since the drop rests partially on air (heterogeneous wetting).

In the present work, the objective was to study the influences of internal sizing by AKD on pilot paper characteristics, such as surface chemical composition, size distribution, and wettability. To obtain a deeper understanding, papers were prepared from elemental chlorine-free (ECF) bleached kraft pulp fibers without any filler but were also filled with 20 wt% precipitated calcium carbonate (PCC). The pilot papers were prepared to represent fine paper quality or fully bleached top-ply of cartonboard, but were not surface sized or calendered.

The surface chemical composition and size coverage were determined by XPS, whereas a sessile drop technique was used to evaluate the wetting properties of the pilot papers. The wetting of the papers was studied by using the dynamic approach, since in many practical applications, such as printing, the dynamic conditions are more important than the static conditions.

5.2 Materials and methods

5.2.1 Materials

5.2.1.1 Pulp and sizing agents

Bleached ECF 70% hardwood and 30% softwood pulp refined to 27 SR (Shopper Riegler drainage evaluation) was obtained from Stora Enso Berghuizer mill, the Netherlands. Commercial AKD, AKD-C18 (melting point 55–60°C), emulsion was obtained from Hercules. The chemical structure of AKD-C18 is shown in Figure 5.1.

5.2.1.2 Filler and retention aid

Twenty wt% calcitic scalenohedral PCC based on paper from Specialty Minerals, UK, was used. The specific area of the PCC was $6.0 \, m^2 \, g^{-1}$. In case of unfilled papers 0.0125% cationic retention copolymer from Hercules was used. In case of PCC-filled papers 0.5% dry-based cationic potato starch ($N = 0.35\%$) from Roquette Frères, France, and 0.015% of the above cationic retention polymer together with 0.2% refined bentonite from Ciba Specialty Chemicals, UK, were used.

5.2.2 Methods

5.2.2.1 Preparation of pilot papers

Pilot papers with a grammage of 80 $g \, m^{-2}$ were prepared on a Fourdrinier pilot paper machine at Hercules European Research Center, Barneweld, the Netherlands. The machine speed was 12 m min^{-1}. Drying was performed using steam-heated drying cylinders at T_{max} 105°C. Six different AKD addition levels from 0.03 to 0.30 wt%, based on active size were used. The papers were made at pH 8.0. As the drying of papers was considered to be similar or even better than on full-scale paper machines, the papers were not cured at a high temperature after preparation. Instead, they were sealed with aluminium foil in plastic bags and stored at 23°C and 50% relative humidity for a week prior to analysis.

5.2.2.2 X-ray photoelectron spectrometer

The surface chemical composition of pilot papers was determined with a Kratos AXIS HS x-ray photoelectron spectrometer (Kratos Analytical, UK) at a photoelectron takeoff angle of 90°. The samples were analyzed using a monochromatic Al x-ray source for high-resolution carbon spectra and Mg x-ray source for wide and detail spectra. Analyses were made on an area <1.0 mm^2. In the analysis, wide spectra were run to detect elements present in the surface layer. The relative surface compositions and O/C were obtained from quantification of detailed spectra run for each element. High-resolution carbon spectra were also run.

The relative amounts of carbon species with different bonds to oxygen were determined from high-resolution carbon C1s spectra using a Gaussian curve-fitting program from the spectrometer manufacturer. The chemical shifts relative to C—C (C1) used in the convolution were 286.8 eV for C—O (C2), 288.2 eV for O—C—O or C=O (C3), 289.4 eV for O=C—O (C4), and 290.1 for carbonate carbon (C5).

Table 5.1 Physical properties of test liquids used.

Liquid	Surface tension (mN/m)[a] (23 ± 2°C)	Density (g/cm^3) (24 ± 2°C)	Viscosity (mPa) (25°C)
Water	72.0 ± 0.1	0.998	0.89
Ethylene glycol	48.8	1.110	16.1

[a] Measured in our lab.

5.2.2.3 Contact angle measurements

Apparent dynamic contact angles of test liquid drops were determined from side images of the drop profile, which was monitored as a function of time with a Dynamic Absorption Tester Fibro 1100 DAT (Fibro Systems AB, Sweden). With the DAT, the spreading and absorption of liquid drop can be followed with a time resolution of ∼20 ms. The instrument applies a liquid drop to the surface while a high-speed video camera captures images as the drop spreads and/or is absorbed. Liquid is automatically pumped out from a syringe and the drop is formed at a tip of Teflon-coated tube, thus avoiding liquid sticking to the tip. An electromagnetic dispenser automatically applies the drop on the substrate. During the first second, 50 images are captured and stored for later analysis. After the first second, the images are analyzed online and fewer images are captured, ∼5–10 s^{-1}. After the measurement, the saved images are evaluated by image analysis in terms of drop volume, height, base diameter/area, and contact angle. The test liquids used in the measurements comprised distilled deionized (MilliQ Plus) water and ethylene glycol (Table 5.1). The drop volume of the liquids was 4 μL. Measurements were performed at five different locations on each sample and the average values were calculated. The measurements were performed at a relative humidity of 50 ± 3% and temperature of 23 ± 1°C.

Figure 5.2 illustrates how water wets papers with different AKD coverage. The paper with a low AKD coverage (Figure 5.2a) shows a rapid absorption of the water droplet into the paper. For the paper with medium AKD coverage (Figure 5.2b) two different regimes can be distinguished. One concerns wetting, whereas the other is dominated by liquid absorption into the paper. In comparison, the paper with a high AKD coverage (Figure 5.2c) shows stable contact angles over time. The baseline and volume of the drop are in this case constant because of the hydrophobic character of the paper.

5.2.2.4 Surface roughness of paper and other tests

The surface roughness was determined by a ZYGO profilometer (New View 5010 from ZYGO, Middlefield, CT, USA). The principle is a noncontact three-dimensional metrology based on white light interferometry. The roughness value is given as Rq (root mean square) roughness. Papers were also characterized in terms of thickness and density according to standard methods. The results are summarized in Table 5.2.

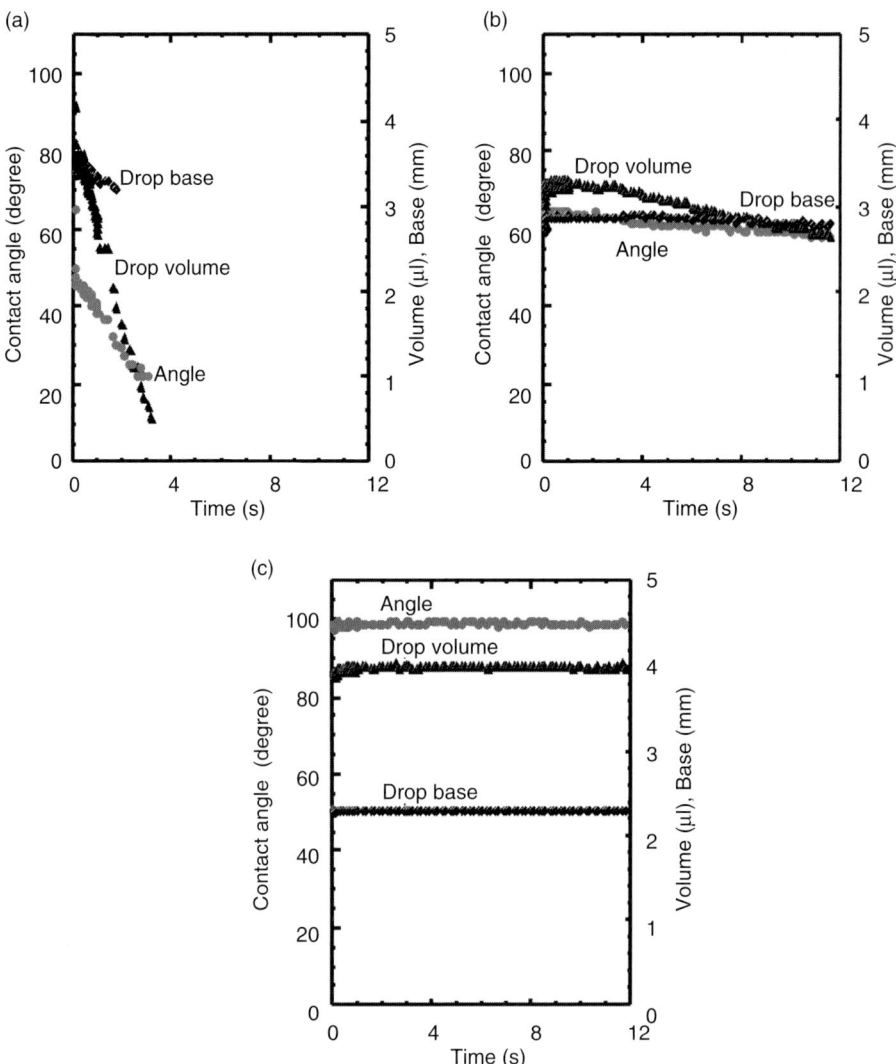

Figure 5.2 Time-dependence of the apparent dynamic contact angle, water drop volume, and baseline determined for unfilled papers with (a) low, (b) medium, and (c) high surface coverage of AKD.

5.3 Results and discussion

5.3.1 Surface chemical composition

The purpose was to study the pilot papers more in steady state with respect to sizing development. Therefore, the surface chemical composition of the papers was determined 1 week after production. The papers were stored sealed in aluminium foil and plastic bag at

Table 5.2 Thickness and density of unfilled and PCC-filled pilot papers sized with AKD.

Physical property	AKD-sized, unfilled papers		AKD-sized, PCC-filled paper 0.30%
	0.03%	0.30%	
Thickness (μm)	149	138	133
Covariance% (CoV%)	7.3	7.2	7.7
Density (kg/m^3)	583	585	597

23°C and 50% relative humidity. The time delay was chosen because of a well-known fact that AKD sizing develops slowly and may continue for several days after papermaking.

Figure 5.3 illustrates a typical XPS low-resolution detail spectra for unfilled and PCC-filled papers sized with AKD. The XPS spectra were run using a takeoff angle of 90° meaning that the information was collected from a ~10 nm thick layer. As hydrogen cannot be detected by XPS, the only atoms that can be determined in pure cellulose fibers are carbon and oxygen. In this work, the AKD-sized pilot papers without any filler contain mainly carbon (C) and oxygen (O) in its uppermost surface, the amount of nitrogen and calcium being near the noise level. Both carbon and oxygen originate from cellulose fibers, AKD, and retention aid polymer. The corresponding spectra, for PCC-filled papers sized with AKD studied, comprise carbon (C), oxygen (O), calcium (Ca), and silica (Si). C and O originate from cellulose fibers, AKD, retention system polymers, and calcium carbonate, Ca from calcium carbonate and raw water, and Si from the bentonite used in the retention aid system.

Representative highly resolved C1s XPS spectra of AKD wax and an unsized paper, as well as an unfilled and a PCC-filled paper sized by AKD, are shown in Figure 5.4. AKD used in this work to modify fibers contains aliphatic C_{18}-carbon chains in its structure, and thus gives rise to a large C1 carbon (C—C) signal, as is exemplified in Figure 5.4a. Pure cellulose fibers contain only C2 (C—O) and C3 (O—C—O) carbons. However, a small amount of extractives gives a small C1 carbon peak to the unsized paper (Figure 5.4b). Consequently, an increase of AKD on the fiber/paper surface can be detected as an increase of the aliphatic carbon, C1-carbon, originating from the C—C carbon in the size. This means that the atomic O/C ratio decreases. Figure 5.4d shows that the C1 carbon is significantly lower for the PCC-filled paper than for the unfilled one, although the amount of AKD added was the same. The C1s spectrum for the PCC-filled paper also shows a small C5 band originating from inorganic carbon, that is, carbonate carbon from PCC as expected.

It is evident from Figure 5.5 that, with increasing size addition, the C1 increases nearly linearly at low and medium addition levels and after a certain surface coverage, the C1 carbon starts to level off. The results suggest a good retention and distribution of AKD molecules on fibers at low and medium addition levels, whereas at high levels AKD retention seems to decrease. Another explanation is that AKD particles end up in multilayer batches/aggregates as seen to occur according to a time-of-flight secondary ion mass spectrometry (ToF-SIMS) analysis [27].

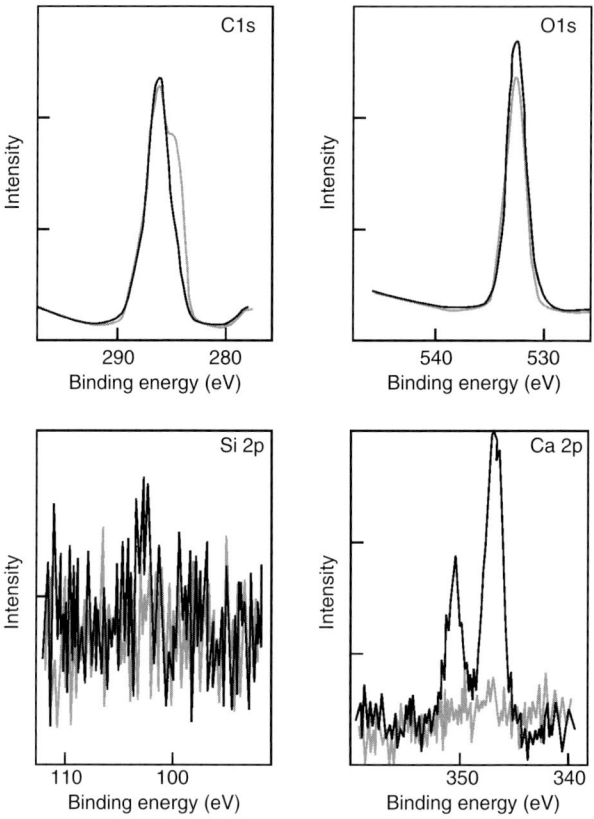

Figure 5.3 Low-resolution XPS detail spectra C1s, O1s, Si 2p, and Ca 2p for (grey line) unfilled and (black line) PCC-filled pilot papers made from ECF bleached pulp fibers and sized with AKD. The size addition in both papers was 0.12 wt%.

In a monolayer, AKD molecules are tightly packed having their polar groups against fiber surface and hydrocarbon chains upwards, while in a double-layer structure, the polar ends of the second AKD molecular layer would be upwards. This raises questions of the physical states of AKD present on the fiber surface, either as monolayer film, or in the form of multilayer structure. For AKD, the coverage has been observed both as monolayer coverage [2] and as being in the form of isolated islands with a thickness of 3 nm [3]. We have earlier reported that after initial wetting AKD can spread/migrate further over time through a monolayer film consisting of active AKD that extends from the foot of each AKD drop retained at the surface of fibers [6]. This type of spreading mechanism via surface diffusion over the air–solid interface is relatively slow, the apparent surface diffusion coefficient being on the order 10^{-11} $m^2\,s^{-1}$ at 45°C. Shchukarev et al. [7] obtained a diffusion coefficient of the same order for AKD on paper surface. However, in the current work, the papers were studied 1 week after preparation, so surface diffusion type spreading should also have occurred to a certain degree. Figure 5.6 illustrates fiber/paper surfaces covered by AKD as isolated islands, as a monolayer film, or as both.

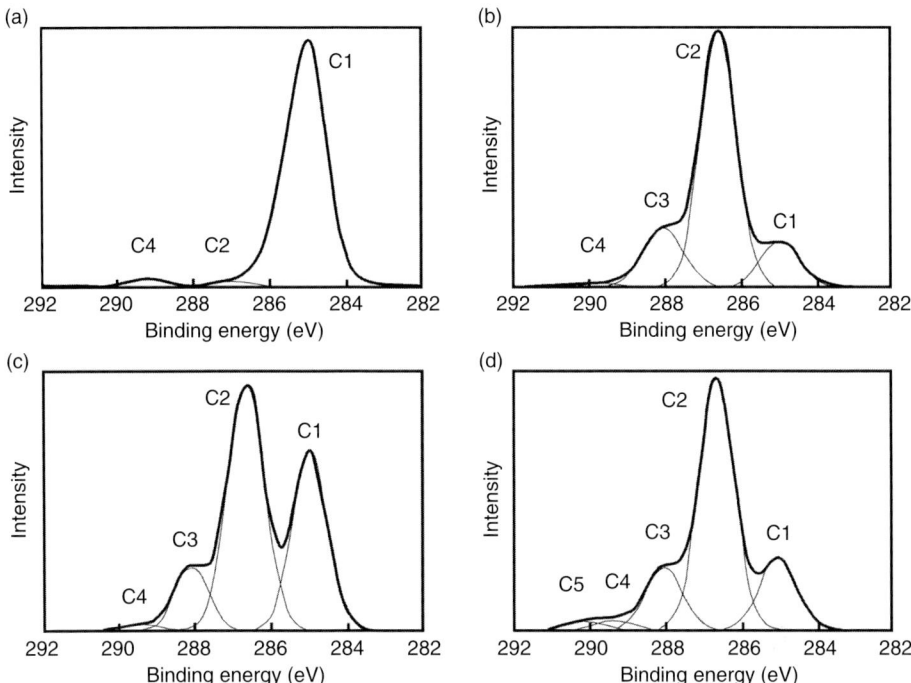

Figure 5.4 High-resolution C1s XPS spectra of (a) AKD wax, (b) unsized paper, (c) AKD-sized, unfilled paper, and (d) AKD-sized, PCC-filled pilot paper. C1: C—C from AKD and extractives, C2: C—O mainly from C—OH groups in fibers, C3: O—C—O mainly from fibers, C4: O—C=O from AKD and possible carboxyl acid/carboxylate groups in unsized paper and C5: carbonate carbon from PCC. Amount of AKD added was 0.12 wt% in (c) and (d).

Figure 5.5 also shows clearly that the increase in C1 (aliphatic carbon) with increasing addition of AKD is less significant for the PCC-filled papers, although the added amount of AKD was the same as for the unfilled papers. To enable comparison of the PCC-filled papers with the unfilled ones, the C1 carbon is given in both cases as a relative amount, that is, it has been normalized with respect to the total amount of carbon. This was done as carbon and oxygen originate from fibers as well as from carbonate. In the filled papers used in this study, the PCC content based on paper was 20 wt%. Calculated roughly on the dry specific surface areas of 1.5 m^2 g^{-1} for bleached soft and hardwood fibers [2] and of 6.0 m^2 g^{-1} for PCC, approximately a 60% higher dose of AKD is needed to reach the same size coverage in PCC-filled than in unfilled papers. However, the size concentration on the surface of the PCC-filled papers is lower than expected, especially at high addition levels. This may be due to the fact that the specific surface area of the fiber surface is higher than the assumed 1.5 m^2 g^{-1}, and also due to the different retention and spreading behavior of AKD on PCC compared to that on fiber surfaces. It should be noted that according to liquid chromatography mass spectroscopy analysis, the total amount of AKD was slightly higher in the PCC-filled papers.

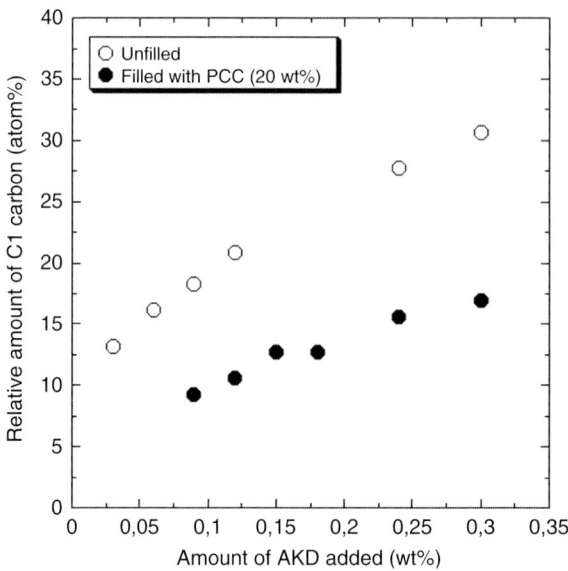

Figure 5.5 Relative amount of C1 carbon in atom% vs amount of size added for unfilled and PCC-filled pilot papers sized with AKD.

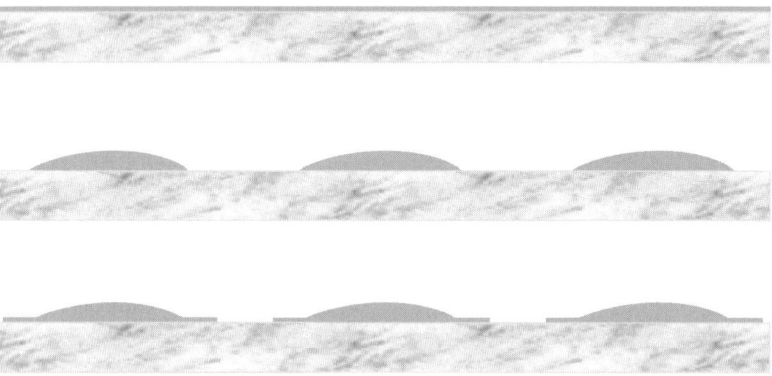

Figure 5.6 Schematic illustration of fiber/paper surfaces covered by AKD either as a monolayer, as patches or as a combination of both (see also Plate 3).

The hydrophobicity of the papers can be followed by evaluating the C1 carbon of the high-resolution C1s spectra as shown earlier or by determining the atomic ratio between oxygen and carbon obtained in the low-resolution spectra. A small round-robin investigation concluded that the correlation of high-resolution C1s data with O/C atomic ratios could also be used in testing instruments and experimental setups for pulp and paper materials [28]. Figure 5.5 shows how the hydrophobic character of the papers builds up with increasing size addition. A closer examination of the XPS spectra reveals a linear relationship between the C1 and O/C atomic ratio for both series of papers as displayed in Figure 5.7. This means

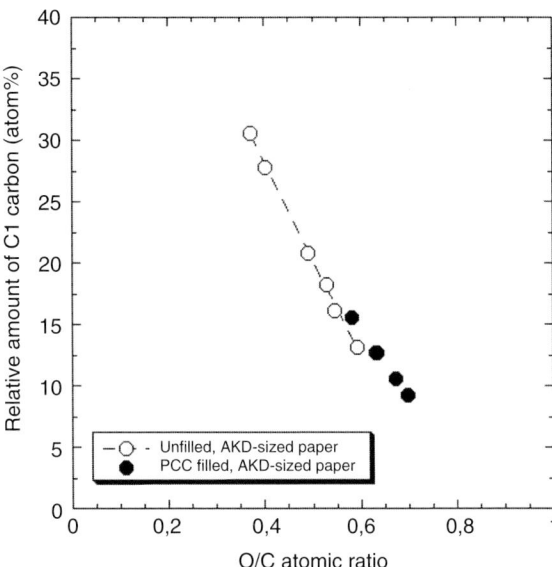

Figure 5.7 Relative amount of C1 carbon in atom% vs O/C atomic ratio for unfilled and PCC-filled pilot papers sized with AKD. The linear regression given is for the unfilled, AKD-sized papers.

that with increasing size coverage on the fibers, the signal from oxygen originating mainly from fibers decreases.

Figure 5.8 shows that, with increasing C1 carbon from AKD, the C2 and C3 carbon signals originating mainly from fibers decrease. A slight increase in the C4 carbon originating from a carbon with three bonds to oxygen, for example, —O—C=O, can also be observed. The C4 carbon could be present in the unreacted AKD or covalently bonded AKD, or both. It has been assumed that there is no significant contribution from fatty acids to the C4 carbon, as their content is low. However, it can be concluded that the level of this carbon signal is more or less constant with the increased AKD addition.

As can also be seen from Figure 5.8, the highest C1 carbon value is 42%. The values are low compared with the corresponding values for pure AKD wax, which is 96%. This result suggests that, as the C1 carbon value from AKD is reduced due to contribution from underlying fibers, the AKD is present in either a layer with thickness below the analysis depth of 10 nm or that AKD is present in patches. The most probable explanation is that there are isolated AKD patches that have spread out to give somewhat thinner layer. Further investigation of Figure 5.4 shows that the AKD wax also contained 2% C2 and 2% C4 carbons. When evaluating AKD sizing, one must also consider the hydrolysis of AKD. Under papermaking conditions, AKD will also react with water to give a ketone containing an acid group [29]. The reaction is catalyzed by ions, such as Ca^{2+}, Mg^{2+}, and HCO_3^- [30]. Thus, the C3 carbon shown in Figure 5.8 may originate from fibers (O—C—O) or hydrolyzed AKD (—C=O), or both. The former contributes more to the C3 carbon, and therefore the detection of hydrolyzed AKD is not possible. However, in a forthcoming publication, ToF-SIMS analyses revealing the existence of hydrolyzed AKD on the paper surface will be reported.

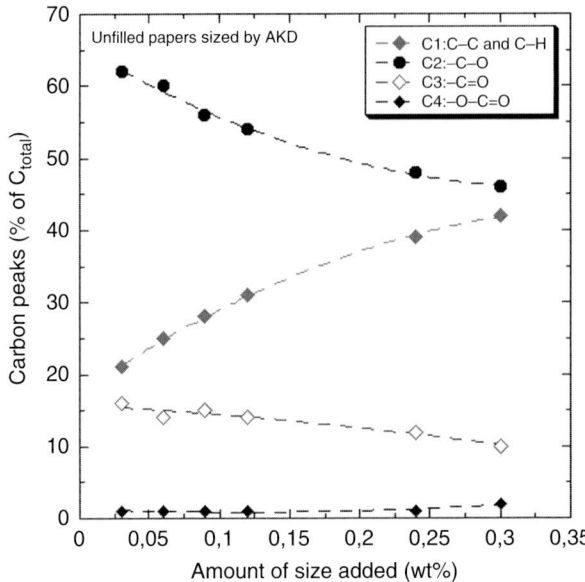

Figure 5.8 Relative amount of various carbon peaks of the high-resolution C1s spectra vs AKD addition level for unfilled pilot papers sized with AKD.

We also studied the distribution of AKD in the uppermost surface of the sized papers by the spectral background analysis of the O1s signal according to the method used by Tougaard and Ignatiev [31]. Using this approach, a thin surface layer from clusters on rough surfaces can be detected. We calculated peak-to-background D values for the O1s signal by dividing the elastic peak area by the increase in the background intensity at a distance of 40 eV as reported [32,33]. The background evaluation data is presented in Figure 5.9. The results of the unfilled papers show that at the lowest size addition of 0.03 wt% the D value is ~24 eV for AKD. In comparison, for bleached ECF kraft pulp [32] and for homogeneous polymers matrices [34], a D value of 25 eV has been shown. In our study, with increasing AKD addition the D value clearly decreases. The highest D value at low additions indicates that AKD is present mainly as patches or islands thicker than the analysis depth of ~10 nm. With increasing amounts of AKD added, the D value clearly decreases, thus indicating the formation of a thin AKD film on the surface of the papers. The oxygen atoms, mainly originating from cellulose backbone, are gradually covered with AKD that contains less oxygen. As the lowest D value of 18 is reached for the unfilled paper at the same AKD addition and retention, this shows more of a thin overlayer film for unfilled than for PCC-filled paper. The latter contains also oxygen from the filler, calcium carbonate. Also important to stress is that the decrease in the D value does not exclude the presence of patches or islands, but it does confirm surface film formation. It is an interesting finding that the D value of 18 eV obtained for the unfilled papers in this work is the same as for fibers covered by a thin film of extractives [32].

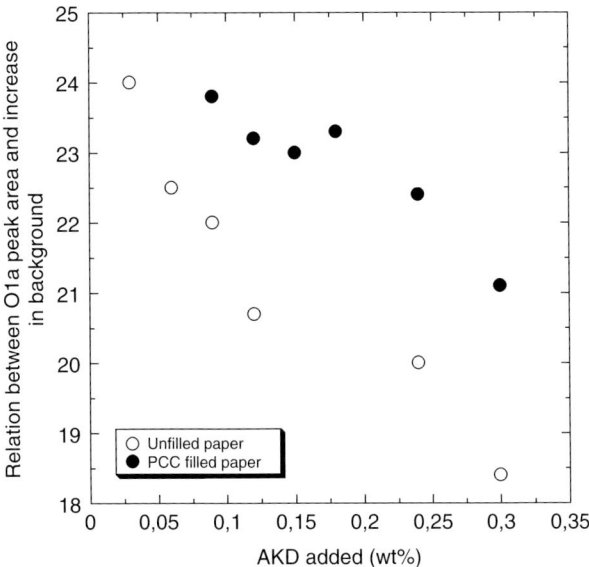

Figure 5.9 Background evaluation of the oxygen O1s region for the unfilled and PCC-filled papers sized with AKD.

5.3.2 Wettability

The chemical nature of the unfilled and PCC-filled papers was also studied using two test liquids of different surface tensions. From a thermodynamic viewpoint, the wettability of paper by a liquid is determined by the surface tensions of the paper and the liquid. The wetting dynamics may, however, depend on a number of other properties such as the liquid viscosity [35] and the heterogeneity, physical [25] or chemical [36] properties of the substrate, as well as sheet structure [18]. It is well known that the apparent contact angle on a rough surface is higher than the value on the corresponding smooth surface [18,19,25,37]. The uneven structure creates barriers against the spreading, which increases the contact angle compared with the situation of a flat surface.

In this work, the hydrophobic papers were evaluated using the $S_{apparent}$ contact angle approach. The contact angles determined were dynamic contact angles. The size of a liquid drop was 4 µL and the Rq surface roughness among the papers in this study was ~9 µm. Figure 5.10 shows a representative two-dimensional image and a cross-sectional profile of an unfilled paper surface. Roughness on different scales including varied textures with, for example, slope discontinuities can be observed. The roughness profile suggests that during contact angle determination, air can remain trapped under the drop. Consequently, the water droplet with a diameter of millimeter size rests on the composite of surface made up of both the hydrophobic paper and air, thus increasing the apparent contact angle. Indeed, the surface area measured including height variations was approximately twice of that excluding the height variations.

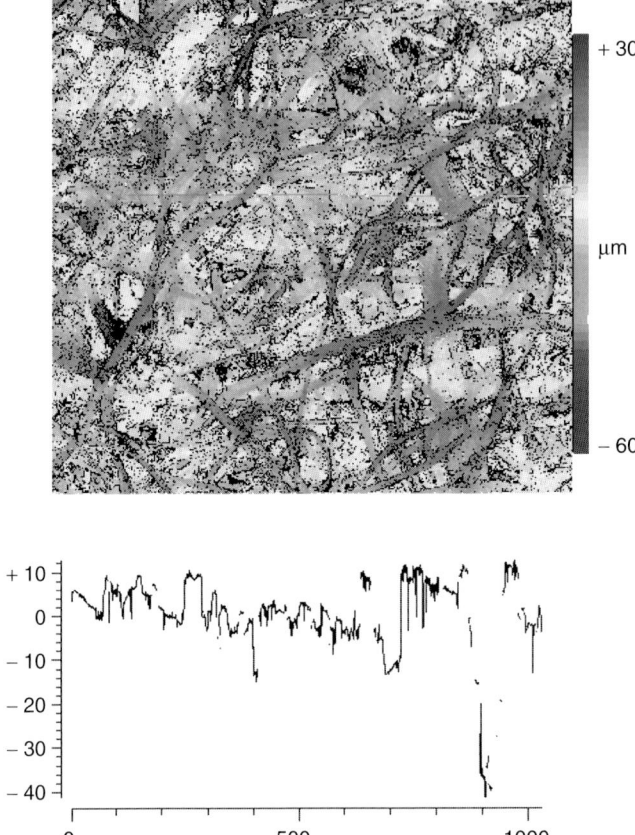

Figure 5.10 Two-dimensional (2D) topography (top; see also Plate 4) and 2D cross-sectional profile (bottom) of an unfilled pilot paper surface determined by white-light profilometry; the surface size is 1.15 mm × 1.23 mm.

To study the chemical nature of the papers as a function AKD addition, it was chosen to use initial dynamic contact angles of test liquids determined at ∼0.05 s on the AKD-sized papers. The time chosen was considered to be long enough with respect to the viscous ethylene glycol having a slower drop relaxation and to be short enough to minimize possible liquid absorption in paper. Figure 5.11 illustrates how the sizing buildup for unfilled papers increases with an increasing amount of aliphatic carbon, C1 originating from AKD. The symbols shown represent the six AKD addition levels from 0.03 to 0.30 wt% resulting in varying sizing degrees as discussed earlier. The paper containing the lowest AKD addition of 0.03 wt% shows a contact angle of ∼95° against water, whereas the contact angle for the 0.06 wt% is close to 120°. The contact angles of papers with the highest AKD additions are ∼125°. In comparison, a water equilibrium contact angle of 109° on a smooth surface of pure AKD wax has been reported [38]. The difference in the contact angles is most probably due to the roughness of the papers studied in this work.

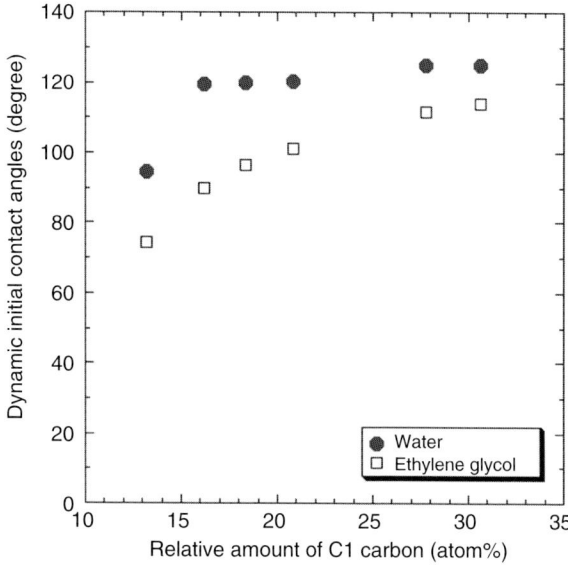

Figure 5.11 Initial dynamic contact angles at 0.05 s of water and ethylene glycol for unfilled pilot papers internally sized with AKD. The contact angles are average values of five different measurements.

It is clear from Figure 5.11 that with increased size coverage the contact angles of water and ethylene glycol increase, indicating that the total surface energy of the papers decreases. In the case of water with a high surface tension (72 mN/m), the contact angles clearly reach a plateau level when the relative amount of C1 carbon mainly originating from AKD is below 20 atom% of the total carbon. This agrees well with the water resistance results obtained by Ström et al. [3] and indicates that, with increasing size addition, the surface polarity of the papers decreases. In a previous paper [39] we showed how a small AKD addition reduced clearly the polar component in the total surface energy: $\gamma^{tot} = \gamma^{LW} + \gamma^{AB}$, where γ^{LW} is nonpolar (dispersive) and γ^{AB} polar (acid–base) component. Moreover, the LW contribution correlated well with the total surface energy. It has to be kept in mind that the hydrophobic paper surface is a heterogeneous composite of cellulose fibers, AKD and, most probably, entrapped air voids. Thus, the physical meaning of the measured surface energies is under such circumstances questionable. The results can, however, be used to predict, for example, polymer adhesion on paper.

In many applications such as in water-based printing, optimization of liquid penetration is necessary. Having a same sizing degree in a paper, liquid wetting and penetration time can be affected by choosing a liquid with a lower surface tension than water. In liquid carton, the opposite is required. The influence of the surface tension is illustrated in Figure 5.11 where the above-mentioned papers were studied also by ethylene glycol with a lower surface tension (48.0 mN/m) and a higher viscosity than water (Table 5.1). The liquids also differ in terms of basic character [40]. The contact angles increase continuously with increasing amount of C1 (mainly originating from AKD) on the paper surface. The results display clearly and surprisingly well differences also among papers with high sizing degrees, that is, high amounts of C1 carbon. The results also indicate that at the highest C1 amount of ~31 atom% (corresponding to the highest added AKD amount) the paper surface is nearly totally covered by AKD. It agrees well with the XPS results discussed in the previous section.

The contact angles shown above were extracted from the wetting study where the time-dependence wetting of each paper by water and ethylene glycol was followed as illustrated

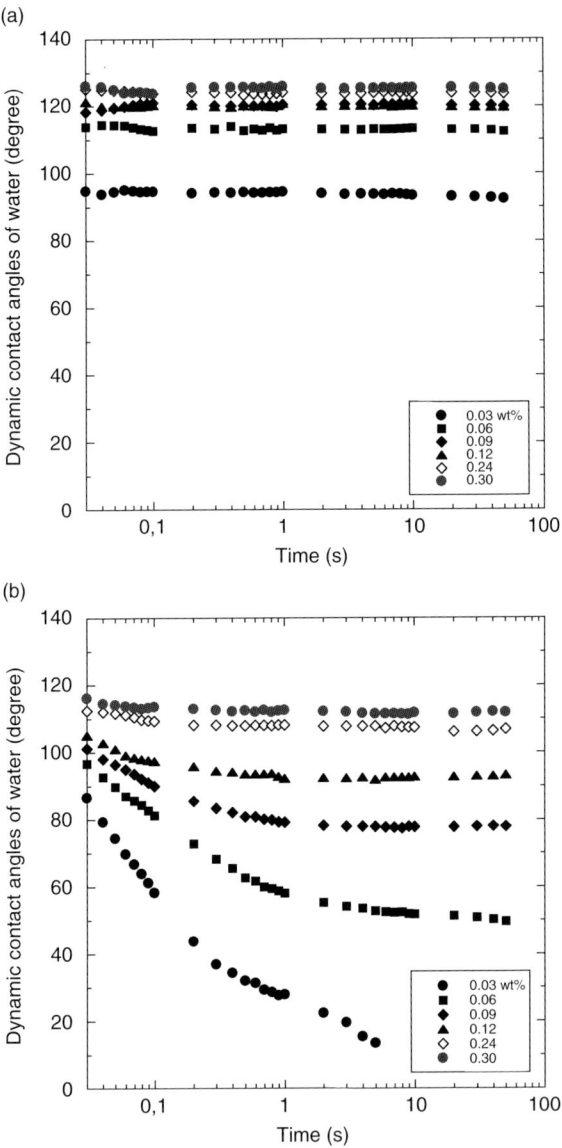

Figure 5.12 Time dependence of dynamic contact angles of (a) water and (b) ethylene glycol for unfilled pilot papers sized with AKD.

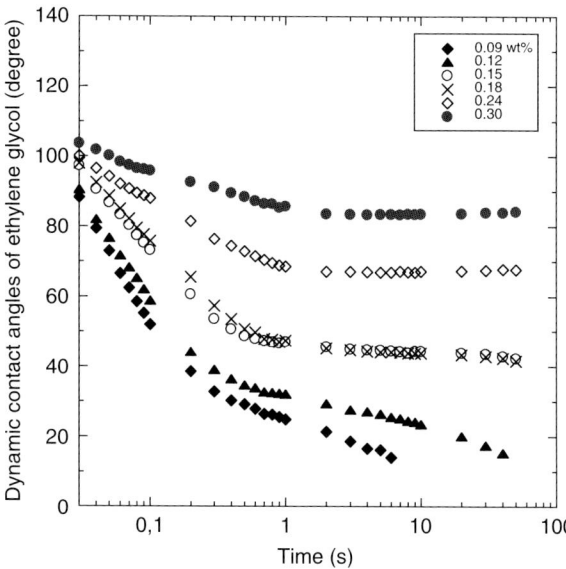

Figure 5.13 Time dependence of dynamic contact angles of ethylene glycol for PCC-filled pilot papers sized with AKD.

in Figures 5.12 and 5.13. It is obvious that water wets the surfaces of the unfilled papers only partly over the time of 60 s, whereas ethylene glycol with lower surface tension clearly shows different wetting behavior on the papers of different sizing degree. Moreover, the results suggest that the dynamics of wetting is dependent on the sizing degree of the papers. It is most clearly seen as differences in initial wetting behavior of ethylene glycol (Figures 5.12b and 5.13). Three different wetting regimes can be seen, most clearly for the papers with low sizing degree and tested by ethylene glycol. The initial wetting up to 0.1 s occurs most rapidly following one or two slower wetting regimes. For the papers with high AKD coverage, equilibrium contact angles are reached, whereas for those with low AKD coverage, the contact angles decrease after ~10 s because of the absorption of the liquid drop into paper structure. The long initial wetting by ethylene glycol is most likely due to its higher viscosity compared to water. The viscosity retards subsurface absorption more than surface spreading [35]. It has to be noted that there is most likely a difference in the z-structure of the filled paper due to the PCC particles compared to the unfilled one, a fact that may affect how a liquid absorbs into paper [41].

It has been concluded that the wetting dynamics is independent of the chemical heterogeneity of the paper surface [19]. Our results do not support such a conclusion.

The XPS and contact angle results in this work suggest that AKD does not cover totally fiber surfaces at low and medium addition levels of AKD. Consequently, it can lead to wetting hysteresis due to chemical heterogeneity [36]. For example, water vapor starts to condense on hydrophilic spots on the paper surface/pore walls at pressures below the liquid vapor pressure. As a result, the contact angle is gradually shifted from advancing to receding and the sustained pressure head is reduced, enabling water uptake.

Conclusions

The XPS results showed that AKD covered the paper surface to a higher degree in unfilled than in PCC-filled papers. The results suggest that AKD covers fiber surfaces mainly in the form of patches or islands. In addition, the background analysis of the O signal confirms the formation of a thin surface film. As expected, with an increased amount of AKD in a paper, the initial contact angles of water and ethylene glycol increased. However, the contact angles of ethylene glycol displayed a more clear increase with an increase in the sizing degree of the papers than those of water. Furthermore, by using ethylene glycol, differences in the sizing degree among the hard-sized papers could also be observed. At high AKD addition, the contact angles of water were higher than those measured on smooth pure AKD wax. This was concluded to be due to the roughness of paper.

The XPS and contact angle results indicate that only the surface of the unfilled paper with the highest AKD addition is totally covered by AKD.

Acknowledgments

The author would like to thank Hercules, Kemira Oyj, M-real, Stora Enso Oyj, Tetra Pak, and UPM for financial support. The author is very grateful to Hercules, who supplied pilot papers. Karin Hallstensson and Mikael Sundin are thanked for help with experimental work. Marie Ernstsson is acknowledged for the help with the background analysis of XPS.

References

1. Roberts, J.C. (1996) Neutral and alkaline sizing. In: Robert, J.C. (ed) *Paper Chemistry*, Blackie Academic & Professional, London, p. 140.
2. Lindström, T. and Söderberg, G. (1986) On the mechanism of sizing with alkyl ketene dimers, Part 1. Studies on the amount of alkyl ketene dimer required for sizing different pulps, *Nordic Pulp Pap. Res. J.*, 1(1), 26.
3. Ström, G., Carlsson, G., and Kiaer, M. (1992) Alkyl ketene dimer distribution in handsheets as determined by ESCA, *Wochenblatt für Papierfabrikation*, 15, 606.
4. Garnier, G., Wright, J., Godbout, L., and Yu, L. (1998) Wetting mechanism of alkyl ketene dimers on cellulose films, *Colloids Surf. A: Physicochem. Eng. Aspects*, 145(1–3), 153.
5. Garnier, G. and Godbout, L. (2000) Wetting behaviour of alkylketene dimer on cellulose and model surfaces, *J. Pulp Pap. Sci.*, 26(5), 194.
6. Seppänen, R., Tiberg, F., and Valignant, M.-P. (2000) Mechanism of internal sizing by alkyl ketene dimers (AKD). The role of the spreading monolayer precursor and autophobicity, *Nordic Pulp Pap. Res.*, 15, 452.
7. Shchukarev, A.V., Mattsson, R., and Ödberg, L. (2003) XPS imaging of surface diffusion of alkylketene dimer on paper surfaces, *Colloids Surf. A: Physicochem. Eng. Aspects*, 219, 35.
8. Dorris, G.M. and Gray, D.G. (1978) The surface analysis of paper and wood fibers by ESCA (Electron Spectroscopy for Chemical Analysis). I. Application to cellulose and lignin. *Cellul. Chem. Technol.*, 12(1), 9.
9. Laine, J. and Stenius, P. (1994) Surface characterization of unbleached kraft pulps by means of ESCA, *Cellulose*, 1, 145.

10. Fardim, P., Hulten, A.H., Boisvert, J.-P. et al. (2006) Critical comparison of methods for surface coverage by extractives and lignin in pulps by x-ray photoelectron spectroscopy (XPS), *Holzforschung*, 60(2), 149.
11. Koljonen, K., Österberg, M., Johansson, L.-S., and Stenius, P. (2003) Surface chemistry and morphology of different mechanical pulps determined by ESCA and AFM, *Colloids Surf. A: Physicochem. Eng. Aspects*, 228, 143.
12. Gellerstedt, F. and Gatenholm, P. (1999) Surface properties of lignocellulosic fibers bearing carboxylic groups, *Cellulose*, 6, 103.
13. Freire, C.S.R., Silvestre, A.J.D., Pascoal Neto, C., Gandini, A., Fardim, P., and Holmbom, B. (2006) Surface characterization by XPS, contact angle measurements and ToF-SIMS of cellulose fibers partially esterified with fatty acids, *J. Colloid Interface Sci.*, 301(1), 205.
14. Östenson, M., Järund, H., Toriz, G., and Gatenholm, P. (2006) Determination of surface functional groups in lignocellulosic materials by chemical derivatization and ESCA analysis, *Cellulose*, 13, 157.
15. Carlsson, G. and Ström, G. (1995) Water sorption and surface composition of untreated or oxygen plasma-treated chemical pulps, *Nordic Pulp Pap. Res. J.*, 1, 17.
16. Toriz, G., Ramos, J., and Young, R.A. (2004) Lignin-polypropylene composites Part II. Plasma modification of kraft lignin and particulate polypropylene, *J. Appl. Polym. Sci.*, 91, 1920.
17. Shen, W., Filonanko, Y., Truong, Y. et al. (2000) Contact angle measurement and surface energetics of sized and unsided paper, *Colloids Surf. A: Physicochem. Eng. Aspects*, 173, 117.
18. Wågberg, L. and Westerlind, C. (2000) Spreading of drops of different liquids on specially structured papers, *Nordic Pulp Pap. Res. J.*, 15(5), 598.
19. Modaressi, H. and Garnier, G. (2002) Mechanism of wetting and absorption of water drops on sized paper: Effects of chemical and physical heterogeneity, *Langmuir*, 18(3), 642.
20. von Bahr, M., Seppänen, R., Tiberg, F., and Zhmud, B. (2004) Dynamic wetting of AKD-sized papers, *J. Pulp Pap. Sci.*, 30(3), 74.
21. Zeno, E., Carré, B., and Mauret, E. (2005) Influence of surface active substances on AKD sizing, *Nordic Pulp Pap. Res. J.*, 20(2), 253.
22. Werner, O., Wågberg, L., and Lindström, T. (2005) Wetting of structured hydrophobic surfaces by water droplets, *Langmuir*, 21, 12,235.
23. Meiron, T.S., Marmur, A., and Saguy, I.S. (2004), Contact angle measurement on rough surfaces, *J. Colloid Interface Sci.*, 274, 637.
24. Bico, J., Thiele, U., and Quere, D. (2002) Wetting of textured surfaces, *Colloids Surf. A: Physicochem. Eng. Aspects*, 206, 41.
25. Wentzel, R.N. (1936) Resistance of solid surfaces to wetting by water, *Ind. Eng. Chem.*, 28, 988.
26. Cassie, A.B.D. and Baxter, S. (1944) Wettability of porous surfaces, *Trans. Faraday Soc.*, 40, 546.
27. Kleen, M. (2000) ToF-SIMS as a new analytical tool in pulp and paper research, ToF-SIMS-Problem Solving through Molecular Spectroscopy and Imaging, Proceedings, Physical electronics and YKI Seminar, Stockholm, Sweden.
28. Johansson, L.-S., Campbell, J.M., Fardim, P., Hulten, A.H., Boisvert, J.-P., and Ernstsson, M. (2005) An XPS round robin investigation on analysis of wood pulp fibres and filter paper, *Surf. Sci.*, 584(1), 126.
29. Marton J, (1990) Practical aspects of alkaline sizing. On kinetics of alkyl ketene dimer reactions: Hydrolysis of alkyl ketene dimer. *Tappi J.*, 73(11), 139.
30. Lindström, T. (2006) The catalytical mechanisms of AKD-hydrolysis, Proceedings, Intertech Pira Sizing 2006 Conference, November 2006, Stockholm, Sweden.
31. Tougaard, S. and Ignatiev, A. (1983) Concentration depth profiles by XPS: A new approach, *Surf. Sci.*, 129, 355.
32. Johansson, L.-S., Campbell, J.M., Koljonen, K., and Kleen, M. (2004) On surface distributions in natural cellulosic fibers, *Surf. Interface Anal.*, 36, 706.

33. Österberg, M., Koljonen, K., and Johansson, L.-S. (2005) Detecting wood extractives on pulp fiber surfaces using AFM and ESCA, *Appita J.*, 2, 69.
34. Zeggane, S. and Delamar, M. (1987) Composition depth information in organic materials by measurement of XPS background signal, *Appl. Surf. Sci.*, 29(4), 411.
35. Ranabothu, S.R., Cassandra, K., and Dai, L.L. (2005) Dynamic wetting: Hydrodynamic or molecular-kinetic? *J. Colloid Interface Sci.*, 288, 213.
36. Extrand, C.W. (2004) Contact angles and their hysteresis as a measure of liquid–solid adhesion, *Langmuir*, 20, 4017.
37. Wolansky, G. and Marmur, A. (1998) The actual contact angle on a heterogeneous rough surface in three dimensions, *Langmuir*, 14, 5292.
38. Shibuichi, S., Onda, T., Satoh, N., and Tsujii, K. (1996) Super water-repellent surfaces resulting from fractal structure, *J. Phys. Chem.*, 100, 19,512.
39. Seppänen, R., von Bahr M., Tiberg, F., and Zhmud, B. (2004) Surface energy characterization of AKD-sized papers, *J. Pulp Paper Sci.*, 30(3), 70.
40. Della Volpe, C. and Siboni, S. (2000) Acid–base surface free energies of solids and the definition of scales in the Good-van Oss-Chaundry theory, *J. Adhes. Sci. Technol.*, 14(2), 235.
41. Roberts, R.J., Senden, T.J., Knackstedt, M.A., and Lyne, M.B. (2003) Spreading of aqueous liquids in unsized papers by film flow, *J. Pulp Pap. Sci.*, 29(4), 123.

Chapter 6
Chemical Microscopy of Extractives on Fiber and Paper Surfaces

Pedro Fardim and Bjarne Holmbom

Abstract

Chemical microscopy methods are nanoscale measurements that couple spatial resolution with chemical specificity. Among the techniques available, time-of-flight secondary ion mass spectrometry (ToF-SIMS) has unique capabilities of imaging combined with surface mass spectrometry, while x-ray photoelectron spectroscopy (XPS) is useful for estimation of surface components and their different oxidation states. In this chapter we describe our previous investigations on the distribution and composition of extractives in pulp fibers and newsprint papers using a combination of XPS and ToF-SIMS analyses. Our results suggest that extractives are usually evenly distributed on fiber and paper surfaces, and that the surface coverage by extractives is affected by the pulping and papermaking processes rather than by the wood itself. We suggest that extractives play an important role in the nanostructure and surface energy of fiber and paper surfaces and that the role of extractives should be carefully investigated when innovative fiber and paper products are designed.

6.1 Introduction

"Wood extractives" is a generic term for a large number of compounds present in wood which are soluble in organic solvents. They are usually low-molecular-mass compounds and are nonstructural wood constituents [1]. The composition of extractives is dependent on the wood species, and the amount is generally much lower than cellulose, hemicelluloses, and lignin. However, high amounts of extractives can be found in heartwood and knots in softwoods [2] and in certain hardwood and softwood cells [3].

Wood resin is a term frequently used in the pulp and paper industry for the wood extractives that are lipophilic and soluble in nonpolar organic solvents such as acetone, dichloromethane, or petroleum ether [1]. Wood resin consists mainly of fatty acids and their esters, resin acids, sterols and other alcohols and their ester, and hydrocarbons. The resin is mainly located in parenchyma cells both in hardwoods and softwoods, as well as in resin ducts in softwoods. The process used for pulp manufacture significantly affects the amount and the distribution of extractives that affect important interactions in the different bleaching and papermaking processes. Generally, the fiber surface coverage of extractives is much higher than the amount present inside the fiber [4]. Extractives can deposit onto fiber surfaces during chemical pulping [5], bleaching [6], and papermaking [7], and can also cover the fiber surfaces of mechanical and chemimechanical pulps [8]. Surface extractives are believed to affect both pulp [9,10] and paper [11] strength, refining [12] and wetting properties [13], and is an important parameter that influences product performance.

Ideally, chemical microscopy methods should give information both on the morphology and on the chemical composition of extractives. A great number of techniques can be used for this purpose [14], although techniques with the best spatial resolution usually give limited information on chemical composition, and vice versa. Analyses of surface extractives can be carried out by x-ray photoelectron spectroscopy (XPS), also known as electron spectroscopy for chemical analysis (ESCA), and more recently also with ToF-SIMS. XPS has been used for the estimation of surface coverage of extractives [5,15–19] while ToF-SIMS has been used for assessing the detailed surface composition in addition to surface distribution [17,20]. ToF-SIMS is a technique that is most close to an ideal chemical microscopy, once the detailed chemical information in the form of surface mass spectra at depths lower than 1 nm can be obtained simultaneously with the surface distribution of secondary ions at a lateral resolution of ~200 nm.

Quantitative analysis using ToF-SIMS has many limitations caused by the effects of the electronic state of the surface on the secondary ion yield [21]. Imaging of fiber surfaces by ToF-SIMS has the advantage of chemical specificity, but has limitations regarding spatial resolution and surface damage at small raster size. XPS quantification is also limited due to surface heterogeneity, charging, and overlapping in C1s curve fitting [22–24]. The strengths and weaknesses of XPS and ToF-SIMS techniques make them complementary. In this chapter, we review our previous investigations on the composition and distribution of extractives in mechanical, chemical, and recycled pulps and in newsprint paper by a combination of XPS and ToF-SIMS analyses.

6.2 Surface analysis of extractives by XPS

XPS has been extensively used to assess the fiber surface composition and to estimate the surface coverage by lignin and extractives [5,15,17,19,25–29]. It has also been used to obtain information on chemical bonding and different oxidation states [21]. Estimation of surface coverage by extractives of pulp samples has been made based on the O/C ratios [15] and C1s peak-fitting analysis [6,29] before and after solvent extraction. The most commonly used solvent is acetone; however, dichloromethane [30] and acetone–phosphate [31] have also been used.

The chemical and morphological heterogeneity of pulp fibers presents a serious challenge in this context. Moreover, the analysis is further complicated by contamination, a problem that is common to all surface analysis. The most abundant surface contaminants are carbon and oxygen. Furthermore, there are sources of errors inherent to XPS, such as x-ray-induced irradiation damage, adsorption/desorption of volatile species in ultra-high vacuum, and the reliability of charge compensation. Recently, topics such as resistance to solvent extraction [17], contamination and degradation problems [18,32,33], and differences in the results derived from the methods for lignin surface coverage [19,29,34] have been described. Recent advances in XPS instrumentation are promising for quantitative XPS on paper-like surfaces, and recommendations on the experimental XPS setups have been proposed for paper-like materials [32]. A critical comparison of the available methods for determining surface coverage by extractives has also been reported recently [16]. These methods are based on two different modes of spectral acquisitions, that is, low- and high-resolution. Low-resolution XPS (Figure 6.1a) is used to determine the elemental surface composition and then to calculate O/C ratios. The O/C ratios are used to calculate the surface coverage by extractives, θ_{ext}, according to Equation (6.1). High-resolution XPS is used to record C1s spectra (Figure 6.1b). It is assumed that the C1s peak comprises a contribution from four different carbon functionalities: C1 (C—C, C—H, C=C), C2 (C—O or C—O—C), C3 (C=O or O—C—O), and C4 (O—C=O). The curve-fitting of C1s peak is usually performed using a Shirley or linear background, Gauss or Gauss–Lorentzian character and a full width at half-maximum (FWHM) of 0.9–1.6 eV, depending on the instrument used. Binding energy (BE) of all spectra is related to C1 at 285.0 eV. The following BE, relative to

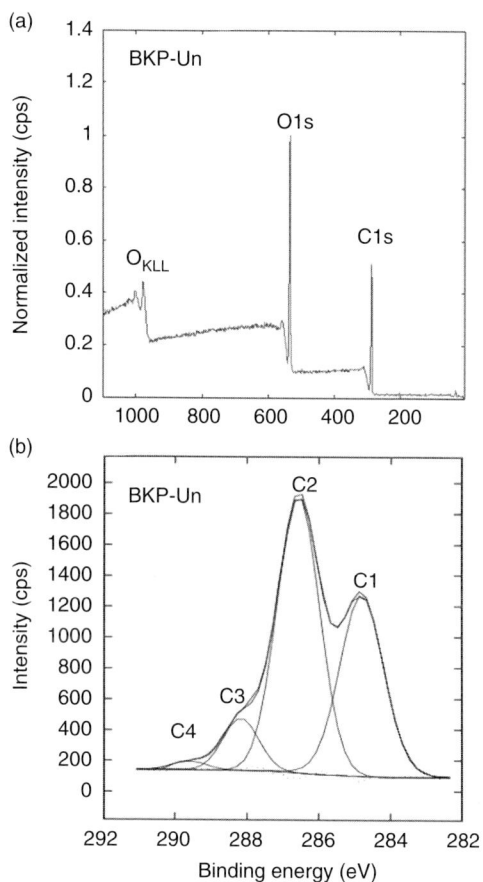

Figure 6.1 An example of XPS spectra of a birch bleached kraft pulp (BKP) before extraction (Un). Acquisition modes: low-resolution (a), high-resolution and curve fitting of C1s peak (b). Data recorded at Top Analytica, Turku, Finland.

the C1 position, are employed for the respective groups, 1.7 ± 0.2 eV for C2, 3.1 ± 0.3 eV for C3, and 4.6 ± 0.3 eV for C4. In the literature [15,26] the same assumption was made to account for the chemical and morphological heterogeneity of pulp fibers. The C1 values are used to calculate θ_{ext} according to Equations (6.2) and (6.3):

$$\phi_{extractives} = \frac{O/C_{after\ extraction} - O/C_{before\ extraction}}{O/C_{after\ extraction} - O/C_{extractives}} \quad (6.1)$$

$$\phi_{extractives} = C1_{before\ extraction} - C1_{after\ extraction} \quad (6.2)$$

$$\phi_{extractives} = \frac{C1_{before\ extraction} - C1_{after\ extraction}}{C1_{extractives} - C1_{after\ extraction}} \quad (6.3)$$

where $O/C_{extractives}$ and $C1_{extractives}$ are the theoretical O/C ratio (0.11) and C1-carbon content (94%) of oleic acid, respectively.

The O/C method based on Equation (6.1) and one of the C1 methods [29] based on Equation (6.3) work on the following assumptions:

1. Extractives are distributed as patches on the top of lignin and cellulose regions (model proposed by Ström and Carlsson 1992 [16]);
2. The thickness of the extractive layer exceeds the depth of analysis (~10 nm);
3. Complete removal of surface extractives can be achieved by solvent extraction.

The other C1 method based on Equation (6.2) [6] makes no specific assumption about the distribution of extractives, and only the difference in the C1 components (C—C, C—H, C=C) of unextracted and extracted pulp samples has to be taken into consideration. However, assumptions 2 and 3 also have to be considered.

6.3 Surface analysis of extractives by ToF-SIMS

Information on the chemical composition of the outermost surface layer can be obtained by ToF-SIMS spectrometry using either a positive or negative operation mode. The absolute counts cannot be taken as quantitative information in this technique as the secondary ion yield is influenced by the electronic and chemical states of the surface. Usually the positive mode is taken for peak identification in pulp samples due to the lower intensity of characteristic peaks in the negative mode as shown in Figure 6.2. Positive secondary ions from carbohydrates [35] and lignin [36] are originated by fragmentation processes, while extractives are desorbed from the surface, giving quasimolecular ions of the type $[M+H]^+$ and $[M+2H]^+$. Fatty acid salts of sodium and calcium can also give quasimolecular ions of the type $[M+Na]^+$, $[M+H+Na]^+$, $[M+2Na]^+$, $[M+Ca]^+$, $[M+H+Ca]^+$, and $[2M+Ca]^+$. Extractive quasimolecular ions and fragments from carbohydrates can lose small molecules such as H_2O and CO. In the case of fatty acid quasimolecular ions, losses of CH_2, C_2H_4, HCOOH, and HCOH have also been observed [37]. Secondary ion peaks due to Na, Al, Si, Ca, and other metals can also be detected at low-mass regions of the spectra. Minor peaks due to siloxane are often observed at 73, 147, 207, 221, and 281 Da, as well as

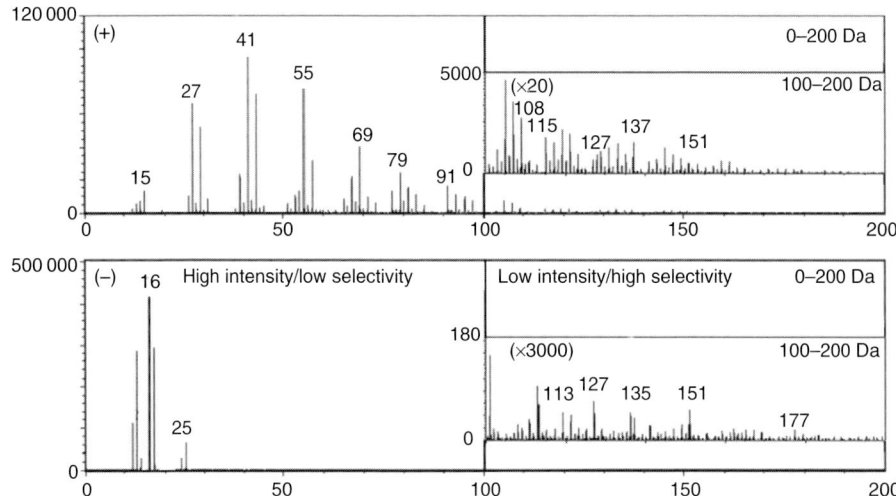

Figure 6.2 Comparison of ToF-SIMS spectra obtained in positive (+) and negative (−) modes for the same softwood bleached kraft pulp sample. The negative mode causes an extensive fragmentation of surface components, resulting in very low intensity peaks in the spectral regions where characteristic peaks of different pulp components are detected.

from phthalates at 149, 167, 279, and 391 Da. The origin of these peaks is usually difficult to determine. It must be made clear that ToF-SIMS has a very high surface sensitivity and enough mass resolution (here, up to 3000) to distinguish peaks from contaminants and pulp components. Another reasonable explanation is that traces of siloxane and phthalates are usually present in most industrial samples, and thus are part of the surface composition.

ToF-SIMS imaging is another feature of the ToF-SIMS technique where characteristic secondary ions of extractives can be mapped with a lateral resolution similar to optical microscopy, that is, ~200 nm using $^{69}Ga^+$ as the primary ion beam. Recent development of new liquid metal ion sources (LIMS) has improved the lateral resolution to a level of 40–100 nm [38]. ToF-SIMS imaging is sensitive to artifacts due to the effect of the impact angle of the primary ion gun on the sputtering yield, which are clearly observed in the bright areas of the images, and any interpretation regarding these areas should be done carefully [39]. This artifact is intrinsic in the ToF-SIMS experiments, in addition to differences in ionization probability of different molecules and clusters, and the effects of topography. Topographical effects are supposed to be eliminated in the experiments due to the large angle of collection of the instruments.

6.4 Extractives on fiber and paper surfaces as investigated by XPS and ToF-SIMS

A combination of spectrometric and imaging methods is the best available strategy to investigate the location, morphology, and distribution of extractives on the fiber and

Table 6.1 Low- and high-resolution XPS spectral data for mechanical, chemical, recycled fibers and filter paper (as reference), and theoretical values for cellulose, lignin (spruce), and oleic acid.

Material	O/C	C1 (%)	C2 (%)	C3 (%)	C4 (%)
Cellulose	0.83	0	83	17	0
Lignin (spruce)	0.33	49	49	2	0
Oleic acid	0.11	94	0	0	6
Filter paper	0.84 (0.02)	4.9 (0.3)	76.2 (1.1)	17.8 (1.3)	1.1 (0.2)
Mechanical pulp (spruce)	0.55 (0.01)	29.4 (1.5)	53.6 (0.8)	16.0 (0.9)	1.1 (0.2)
After acetone extraction	0.66 (0.01)	19.1 (1.3)	64 (1.2)	15.4 (0.9)	1.6 (0.2)
U-chemical pulp (birch)	0.45 (0.01)	38.4 (0.5)	46.2 (2.3)	13.8 (1.9)	1.6 (0.2)
After acetone extraction	0.78 (0.01)	11.5 (0.5)	70.4 (2.3)	16.0 (0.9)	2.1 (0.2)
B-chemical pulp (birch)	0.67 (0.02)	18.5 (3.0)	64.3 (2.3)	15.3 (0.3)	1.6 (0.2)
After acetone extraction	0.82 (0.04)	5.3 (1.0)	73.8 (0.2)	18.9 (1.2)	1.9 (0.3)
Recycled pulp	0.77 (0.04)	33.9 (2.3)	53.6 (1.2)	10.5 (1.0)	2.0 (0.3)
After acetone extraction	1.02 (0.08)	17.9 (3.0)	64.3 (2.0)	14.7 (0.8)	3.1 (1.0)

Note: U = unbleached and B = bleached.

paper surfaces. The content of surface extractives can be assessed by XPS and the chemical identification and localization determined by ToF-SIMS.

6.4.1 Surface coverage by extractives on mechanical, chemical, and recycled fibers

The low-resolution spectra of unextracted mechanical and chemical pulps generally have only C and O peaks, while spectra of recycled fibers have additional Al, Si, and Ca peaks originating from mineral particles used as paper fillers and coatings, such as clay, silicates, and $CaCO_3$. After solvent extraction, the O/C ratio of the pulps increases depending on the pulp type (Table 6.1). Fully bleached chemical pulps have their O/C increased to values close to 0.83, which is the typical ratio for cellulose or hemicelluloses [27]. Mechanical and unbleached chemical pulps have lower O/C ratios because of contribution from lignin. Higher O/C ratios are observed in recycled pulps than in chemical and mechanical pulps. After extraction, the O/C ratio of recycled pulps is higher than 0.83 because of the presence of oxides. The acetone extraction also affects the high-resolution spectrum. A significant reduction of C1 area accompanied by an increase in C2 and C3 due to solvent extraction is observed for all pulp samples. However, relative areas of 5–19% of C1 are still observed after extraction for both fully bleached and lignin-containing samples. This may have been due to a residual layer of lignin, extractives, or contaminants originating from contact with the environment or in the instrument vacuum chamber [32]. The effect of contaminants can be easily seen in the C1 value of filter paper; this value can be used as a reference when C1 is used for the calculation of surface coverage by lignin [16]. However, the origin of the contamination is practically impossible to determine.

Table 6.2 Surface coverage by extractives (θ_{ext}) for different pulp samples calculated using Equation (6.1) (O/C method) except for recycled pulp where the Equation (6.2) (C1 method) was used.

Pulp	θ_{ext} (area-%)	References
U-PGW spruce	15	[8]
B-PGW spruce	7	
U-TMP spruce	16	
B-TMP spruce	15	
U-CTMP spruce	15	
B-CTMP spruce	8	
U-Kraft pine (kappa 26)	8	[30]
B-ECF pine	3–7	
B-TCF pine	2–5	
U-Kraft birch (kappa 18)	37	[40]
B-ECF birch	20	[31]
U-Kraft eucalypt (kappa 15)	5.2	[17]
B-ECF eucalypt	5.0	[41]
B-ECF acacia	24	[41]
Recycled pulp (office waste)	21	[31]

Note: U = unbleached, B = bleached, PGW = pressurized groundwood pulp.

Estimations of surface coverage by extractives, θ_{ext}, for a number of pulps show that unbleached mechanical and chemical pulps have higher θ_{ext} values than the bleached ones (Table 6.2). Among these pulp samples, birch kraft pulps have the highest θ_{ext} values and are probably the main component of the recycled pulp studied. The removal of surface extractives during bleaching may be ascribed to pulp washing rather than chemical degradation. The fatty acid soaps used in de-inking may also be a source of surface extractives for the recycled pulp. This pulp contains oxides and the application of the O/C method is not adequate. In this case the C1 method is the best alternative available. It should be mentioned that the θ_{ext} values presented here reflect the contribution of the fiber portion in the pulps. It is expected that different wood cells will also differ in their θ_{ext}; for example, parenchyma and ray cells have higher contents of extractives than fibers. Therefore, the θ_{ext} for mechanical pulps need a clear definition of what types of cells have been measured. In case of fully bleached chemical pulps the relocation of extractives during the various bleaching stages is much more intensive, and a lower difference in θ_{ext} between fibers and fines might be expected.

6.4.2 Surface coverage by extractives on newsprint papers

Newsprint papers can be manufactured by using only mechanical pulp or by using mixtures of recycled, mechanical, and chemical pulps as raw material. The surface coverage of extractives in newsprint papers can be analyzed using the C1 method [Equation (6.2)],

Table 6.3 Surface coverage by extractives (θ_{ext}) estimated using the C1 method [Equation (6.2)] for two unextracted newsprint papers, and the O/C ratios and high-resolution C1s data for both the unextracted and the extracted samples.

Paper	θ_{ext}	O/C	C1	C2	C3	C4
TMP	34.3	0.35	60.6	33.2	4.6	1.6
After acetone extraction		0.82	26.3	58.6	12.0	3.0
Recycled pulp (RP)	33.6	0.46	52.3	37.0	10.0	0.6
After acetone extraction		0.81	18.7	68.7	11.5	1.2

Note: The TMP paper was made from 100% spruce TMP while the RP paper was made from 56% recycled fiber, 41% TMP and 3% chemical pulp.

but not the O/C method, because of possible interferences of oxygen-containing fillers or papermaking chemicals with the O/C method. The θ_{ext} of two newsprint papers is presented in Table 6.3. The θ_{ext} is higher than that of the similar pulps listed in Table 6.2. The reason for the higher surface coverage in newsprint papers may be the migration of extractives to paper surfaces during drying in the paper machine or the contribution of acetone-extractable components such as fatty alcohols and surfactants used as defoamers and emulsifiers in papermaking chemicals, respectively. Another possible explanation might be the deposition of colloidal dissolved substances present in papermaking water systems where mechanical pulps are used as raw materials.

The addition of recycled fibers to the newsprint furnishes does not seem to affect the θ_{ext} value when compared with newsprint paper containing only mechanical pulp (Table 6.3). This is a clear indication that the fatty acids usually employed in de-inking are not extensively attached to the fiber surfaces, being largely removed in flotation and pulp washing. This assumption is also in agreement with the values presented in Table 6.2, where bleached birch pulp and recycled pulp had very similar θ_{ext} values. Thus, it seems that the papermaking process rather than the pulping process affects the surface extractives of newsprint papers.

6.4.3 Surface composition and distribution of extractives on fiber and paper surfaces

The surface composition of unbleached and unextracted spruce TMP can be assessed in detail by ToF-SIMS spectrometry. Characteristic peaks of cellulose (127 and 145 Da) and softwood lignin (137 and 151 Da) can be distinguished from extractives that have characteristic peaks in the spectral region of 200–650 Da in positive mode (Figure 6.3). Resin acids, sterols, steryl esters, and triolein are the main components of extractives from spruce TMP. Resin acids can be detected as molecular ions of the type M^+ (302 Da) and of that from the loss of CH_3 (287 Da). Quasimolecular ions of the type $[M + H]^+$ and $[M - H]^+$ are also observed at 303 and 301 Da, respectively. Sterols have characteristic ions at 414 Da $[M^{+\bullet}]$, 415 Da $[M + H]^+$, 413 $[M - H]^+$ and 397 Da $[M + H-H_2O]^+$. Oxo-sitosterol gives the ions 429 Da $[M + H]^+$ and 411 Da $[M + H-H_2O]^+$. Steryl esters and trioleins

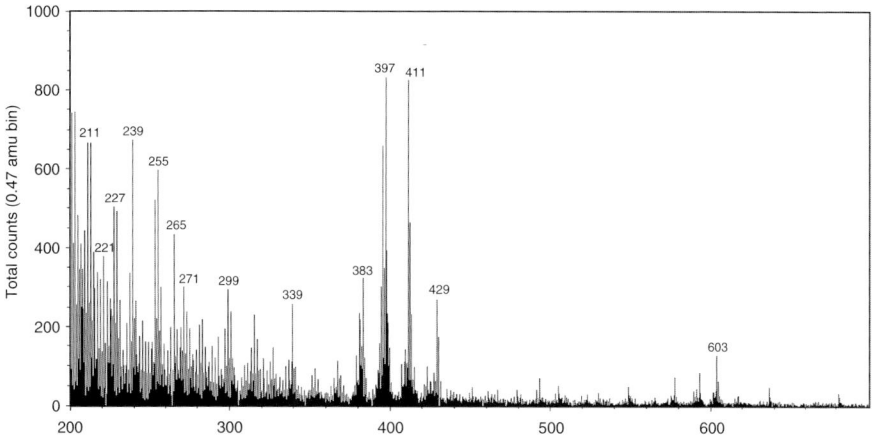

Figure 6.3 ToF-SIMS spectrum of a spruce TMP (unbleached, unextracted). Positive mode, $^{69}Ga^+$ ion gun operated at 15 kV accelerating voltage, raster size 200 μm × 200 μm with charge compensation using an electron flood gun.

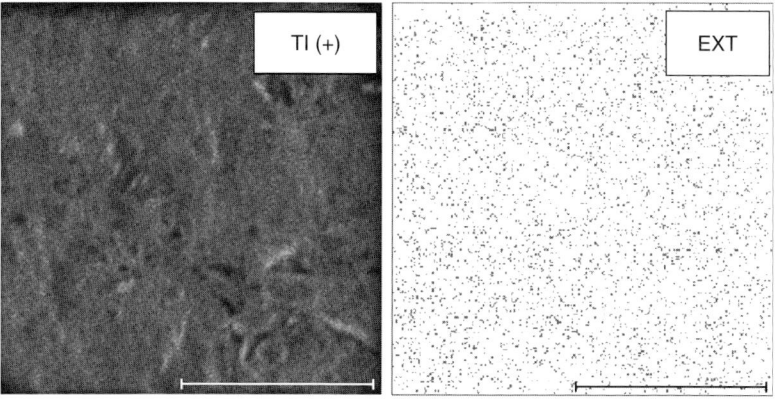

Figure 6.4 ToF-SIMS imaging of a pulp handsheet made from an unbleached TMP showing total ion (TI) and extractives (EXT) ion images. Characteristic secondary ions of resin acids, sterols, steryl esters and trioleins were mapped on handsheet surfaces.

are cleaved in the ToF-SIMS experiments in characteristics patterns where fragments of fatty acid units are generated. Steryl esters have a peak at 397 Da due to fragmentations of the type $[M + H-RCOO]^+$, while trioleins have peaks at 603, 339, and 265 Da due to fragmentations of the types $[M - RCOO]^+$, $[M + H-RCOO-RCO]^+$, and $[RCO]^+$ respectively.

The lateral distribution of the unbleached TMP extractives assessed by ToF-SIMS imaging is presented in Figure 6.4. The images were obtained in pulp handsheets and the region analyzed in this case was enriched with fibrils and fines, as can be seen in the total ion image (TI). The extractive ion image (EXT) shows that the extractives are evenly distributed

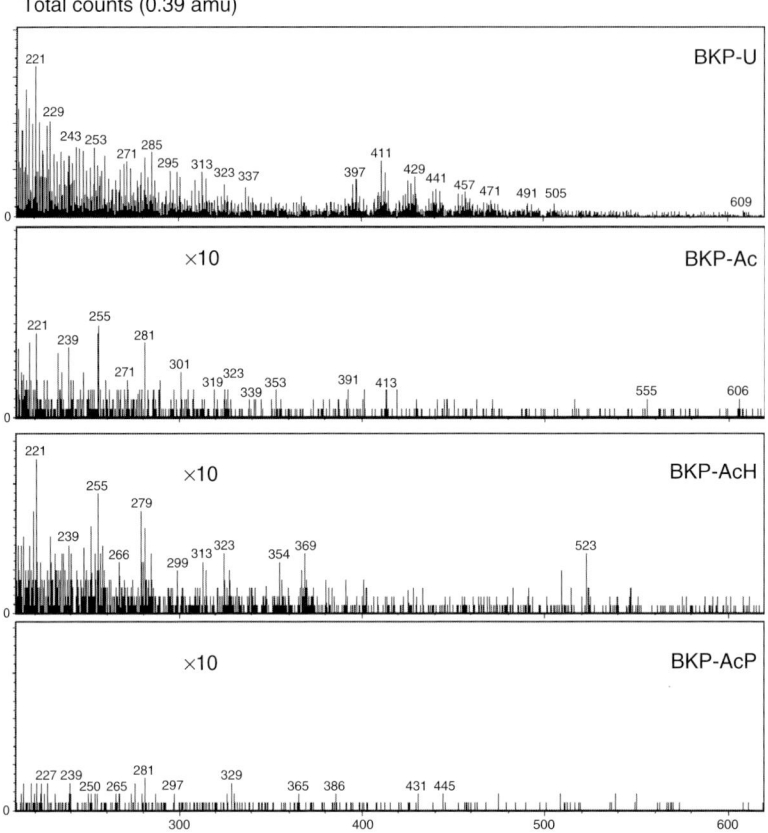

Figure 6.5 ToF-SIMS spectra of bleached birch kraft pulps unextracted (BKP-U), acetone-extracted (BKP-Ac), acetone–acetic acid extracted (BKP-AcH), and acetone–phosphate extracted (BKP-AcP), respectively. From Fardim et al. [31], copyright (2005) with permission from Elsevier.

at this field of view (200 μm × 200 μm) with no particular regions where patches or continuous layer is observed.

The composition of surface extractives by ToF-SIMS spectrometry for an unextracted bleached chemical pulp BKP-U (Figure 6.5) showed ions due to methyl betulinate (457 Da), betulinol (443, 427, 411 Da), sitosterol (415, 397 Da), sitostanol (417, 399 Da), palmitic acid (257, 239, 221, 207 Da), heptadecanoic acid (271, 253 Da), stearic acid (285, 267 Da), oleic acid (283, 265 Da), linoleic acid (281, 263 Da), arachidic acid (313, 299 Da), behenic acid (341, 327 Da), and lignoceric acid (369, 355 Da). Calcium salts of palmitic acid (551, 295 Da) and stearic acid (606, 325 Da) were also observed. Sodium salt of lignoceric acid was also present (413, 391 Da). After extraction with acetone (BKP-Ac), peaks of fatty acid calcium salts (palmitic and stearic), fatty acids (palmitic, linoleic, stearic) and sterols were still observed in the ToF-SIMS spectrum. Extraction with acetone–acetic acid (BKP-AcH) removed fatty acid calcium salts, but peaks due to fatty acid were still present, especially linoleic and lignoceric acids. After extraction using acetone–phosphate (BKP-AcP)

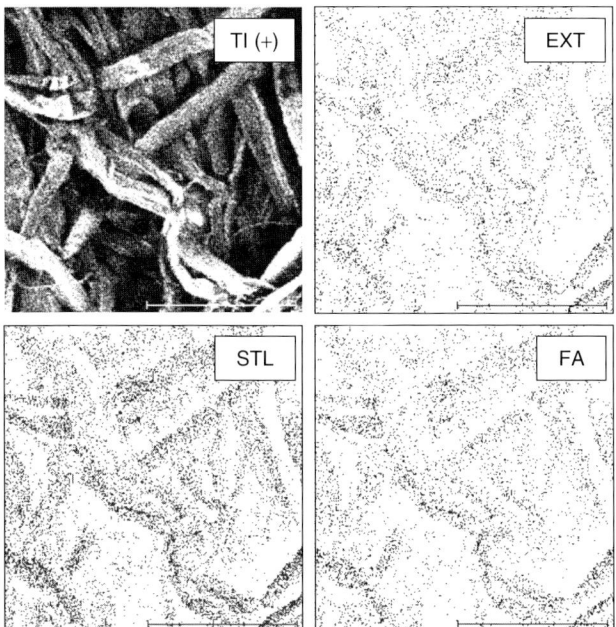

Figure 6.6 ToF-SIMS imaging of birch BKP-U; images of TIs, EXTs, sterols (STL), and fatty acids (FA) were taken using raster size of 200 μm × 200 μm and positive mode. From Fardim et al. [31], copyright (2005) with permission from Elsevier.

the ToF-SIMS spectrum showed extractive secondary ion peaks of very low intensity, close to the background level (Figure 6.5).

ToF-SIMS imaging was employed to assess the distribution of extractives in the unextracted birch BKP. The distribution of extractives (EXT), fatty acids (FA), and sterols (STL) for the BKP are presented in Figure 6.6. TI was also taken. Artifacts due to the effects of the impact angle of the primary ion gun in the sputtering yield were clearly observed in the bright areas of the TI image, and any interpretation regarding these areas should be done carefully [39]. This artifact is intrinsic in the ToF-SIMS experiments in addition to differences in ionization probability of different molecules and clusters, and the effects of topography. Topography effects are supposed to be removed in the experiments as a result of the large angle of collection of the instrument used. The distributions of EXT, STL, and FA were even for the BKP at the raster size investigated. Some agglomeration of spots was observed in the EXT and FA images.

The ToF-SIMS spectrum of the unbleached de-inked pulp (DIP-U) (Figure 6.7) is much more complex than the spectra of BKP or TMP. Inorganic components such as $CaCO_3$, clay and talc were identified in the low-mass region. Peaks of Na, Mg, Al, Si, K, and Ca were present in the high-intensity region. The spectral region of extractives showed secondary ions from sitosterol (415, 397 Da), sitostanol (417, 399 Da), palmitic acid (257, 239, 221, 207 Da), heptadecanoic acid (271, 253 Da), stearic acid (285, 267 Da), linoleic acid (281, 253 Da), arachidic acid (313, 299, 295 Da), behenic acid (341, 327, 323 Da), lignoceric acid (369, 355 Da), and resin acids (303, 302 Da). Calcium salts of palmitic acid (551, 295 Da) and

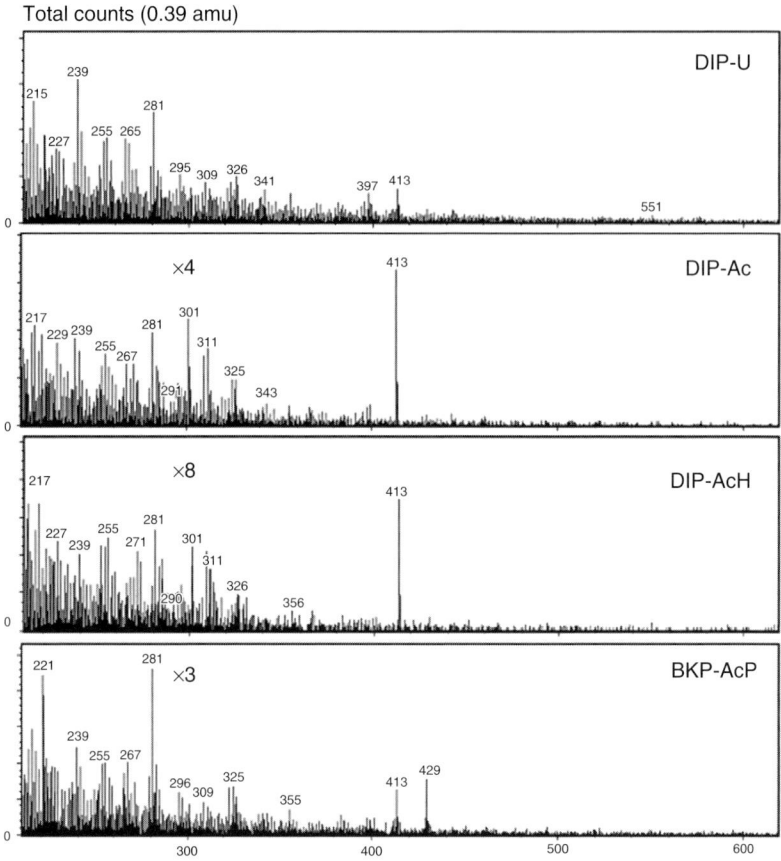

Figure 6.7 ToF-SIMS spectra of deinked pulps unextracted (DIP-U), acetone-extracted (DIP-Ac), acetone–acetic acid extracted (DIP-AcH), and acetone–phosphate extracted (DIP-AcP), respectively. From Fardim et al. [31], copyright (2005) with permission from Elsevier.

stearic acid (606, 325 Da) and sodium salt of lignoceric acid (413, 391 Da) were identified. Even just after extraction with acetone (DIP-Ac), the spectrum was different. Peaks due to calcium salts of palmitic and stearic acids as well as sodium salts of lignoceric acid were dominant, but linoleic acid peaks were also present. Extraction with acetone–acetic acid (DIP-AcH) gave a spectrum similar to that as observed for DIP-Ac, but with lower peak intensities. After extraction with acetone–phosphate (DIP-AcP), peaks due to fatty acid calcium salts, particularly from palmitic and stearic acids and sodium lignocerate, were still detected. A peak of sodium–potassium lignocerate was also observed at 429 Da. The latter was a cationization product as a consequence of addition of K^+ ions in the AcP method (Figure 6.7). Furthermore, typical peaks of siloxane (281, 221, and 207 Da) and phthalates (161, 149, 91 and 77 Da) were observed in the ToF-SIMS spectra for BKP and DIP, before and after extraction. These contaminants are commonly present in industrial samples [17,20,31]. Their peak intensities were significantly reduced after different extractions for

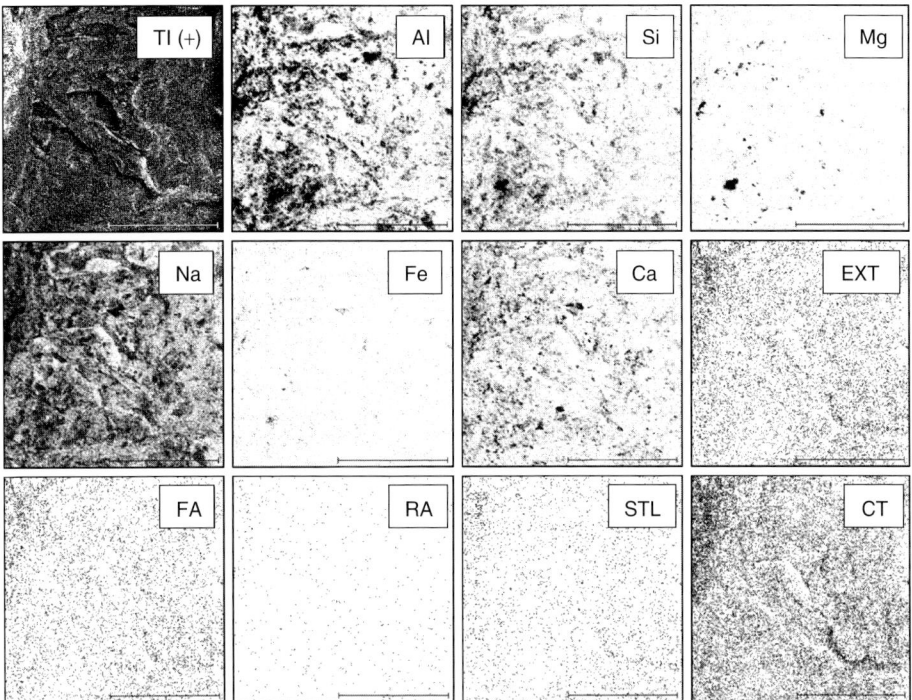

Figure 6.8 ToF-SIMS imaging of DIP-U; images of TIs, metals (Al, Si, Mg, Na, Fe, Ca), extractive ions (EXT), fatty acids (FA), resin acids (RA), sterols (STL), and contaminants (CT) were taken using raster size of 200 μm × 200 μm and positive mode. From Fardim *et al.* [31], copyright (2005) with permission from Elsevier.

the BKP, but only slightly for the DIP, where they probably originated from components of printing residuals or from the contact with plastics in the recycling process.

The distribution of metals, extractives, and contaminants was also assessed by ToF-SIMS imaging (Figure 6.8). Similar artifacts to those observed for the BKP could be observed in the TI image. The distributions of Al, Si, and Mg showed that clay and talc were present and confirmed the previous observation by XPS. The Ca image was due to calcium carbonate, while the origin of Na is difficult to determine because it is present in many paper chemicals as salts or counterions. The Fe image indicated that traces of toner pigments were probably present. EXT, FA, resin acids (RA), and STL were evenly distributed. Contaminants (CT) such as siloxane and phthalates were also observed, and they showed an even distribution with agglomeration in some spots. Similar agglomeration was observed in the FA and EXT images, but not in RA or STL. Extractives and contaminants were distributed over both fibers and minerals, with no particular dominance.

The ToF-SIMS spectrum of a newsprint paper produced using 100% of mechanical pulp shows that the composition of surface extractives is predominantly steryl esters and trioleins (Figure 6.9). In contrast to mechanical pulp, the sterols and resin acids had much lower secondary ion peak intensities. This might indicate a preferential migration or deposition of steryl esters and trioleins on the paper surfaces.

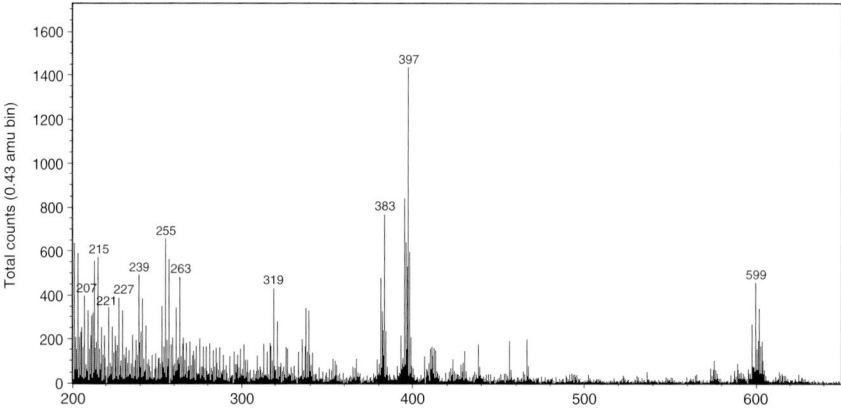

Figure 6.9 ToF-SIMS spectrum of a 100% mechanical pulp newsprint paper surface.

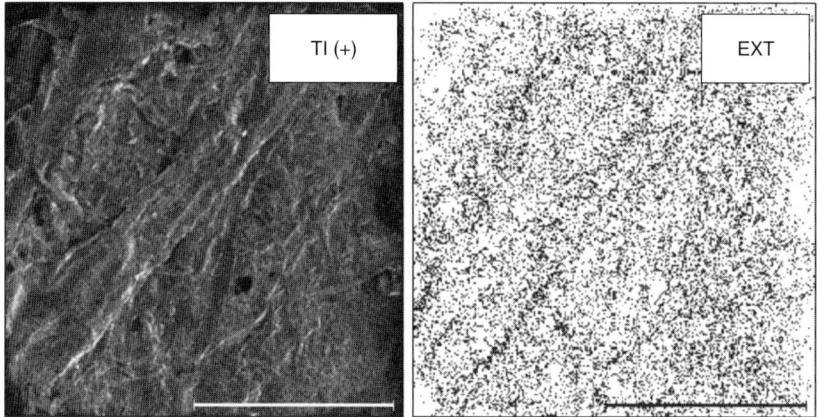

Figure 6.10 ToF-SIMS imaging of a newsprint paper; images of TIs and EXTs were taken using raster size of 200 μm × 200 μm and positive mode.

The imaging of extractives on the newsprint paper shows an even distribution in the whole field of view (Figure 6.10). However, some large agglomerates can be seen as dark spots. The dimensions of these agglomerates measured by image analysis are in the range of 0.8–1.2 μm. Probably the spots are caused by an agglomeration of colloidal particles containing triolein and steryl esters in papermaking waters and subsequent deposition on paper surfaces during the sheet formation in the wet end. The surface extractives may contribute to reducing the surface energy of paper and affecting interactions with water and printing inks.

6.5 Role of extractives on fiber and paper surfaces

Extractives with low molar mass are present in specific cells or tissues in wood. The different processes of fiber liberation affect their distribution in a way such that they are spread to other pulp components. It has been believed that extractives have a patchy distribution pattern on fiber surfaces [15]. In fact, terms such as "patchy" and "even" are empirical and should be carefully considered because they depend on the scale observed. A good illustration is the EXT images obtained by ToF-SIMS for BKP-U (Figure 6.6) and DIP-U (Figure 6.8): At the 200 μm × 200 μm raster size, that is, at micrometer scale, the EXT spots appeared evenly distributed, even though agglomeration of spots could also be noted. These agglomerations observed at nanometer scale as achieved by other techniques such as atomic force microscopy (AFM) can be described as patchy. We have reported recently that different extractives form different aggregates on fiber surfaces [31]. Our results suggested that saturated fatty acids and their salts had a tendency for patchy distribution or a formation of aggregates, while oleic acid formed a uniform layer as indicated by AFM. Considering this, one may expect that a patchy or even tendency will depend on the composition and proportion of different extractive components. This suggestion should be further investigated.

The fact that extractives are small molecules compared with cellulose, lignin, and hemicelluloses also needs to be considered. It is assumed that carbohydrates are organized in a porous and high-surface-area fiber wall [42]. The surface energy of this system is high [43] and the adhesion of compounds that lower the surface energy such as extractives and contaminants are thermodynamically favored at the fiber–air interface [44]. Extractives and contaminants can move freely through different regions in the fiber wall according to the interface involved. After drying, and in air, it is reasonable to formulate that a fiber surface is promptly covered by these low-molar-mass compounds that migrate from different fiber regions. After extraction, however, a residual layer of extractives has a favored thermodynamic tendency to remain attached to the surface or be replaced by a contamination layer from the air. This can explain why C1 peaks were observed for XPS spectra of pure cellulose films in other investigations [45]. It should also be stressed that ToF-SIMS and XPS operate at the fiber–vacuum interface and it is difficult to determine if any artifact is introduced by this feature.

There are still critical issues regarding application of chemical microscopy method such as XPS and ToF-SIMS to investigate extractives on fiber and paper surfaces. It should be mentioned that estimations of surface extractives by XPS are not absolute and only relative comparisons are possible because of incomplete removal of extractives by solvents, the contamination in the vacuum chamber and the model used for calculation. The presence of different nano-assemblies of surface extractives poses a challenge to the models available for fiber surfaces and requires investigations with imaging and spectrometry method combined. This approach demand high spatial resolution techniques to resolve both morphological and chemical composition differences. ToF-SIMS can meet these requirements to a certain extent, although improvements in instrumentation to reduce the fragmentation of high-m/z secondary ions and the raster size of analysis are needed. Other demanding capability for this technique is for improving analytical identification of several peaks present in heterogeneous samples. The possibility of controlled fragmentation of selected ions to improve identification should be considered by the instrument manufacturers, along with new ion sources to enhance secondary ion yield and imaging capability. Meanwhile, development of chemical labeling strategies is a good alternative to enhance analytical and imaging capabilities, and it will be the focus of our future work.

Finally, from a papermaking point of view, the effects of composition, distribution, and nanostructure of surface extractives on the mechanical and wetting properties of pulp fibers and papers should be further investigated. In this case, we suggest a combination of different techniques such as AFM, XPS, and ToF-SIMS as a reliable strategy for a detailed characterization of fiber and paper surfaces.

Acknowledgment

Top Analytica Oy is acknowledged for providing ToF-SIMS and XPS instrumental facilities for our research activities.

References

1. Ekman, R., Holmbom, B. (2000) The chemistry of wood resin. In: *Pitch Control, Wood Resin and Deresination* (eds E.L. Back and L.H. Allen), pp. 37–76. Tappi Press, Atlanta, GA, USA.
2. Willför, S., Hemming, J., Reunanen M., Holmbom, B. (2003) Phenolic and lipophilic extractives in Scots pine knots and stemwood. *Holzforschung* 57(4), 359–372.
3. Rydholm, S.A. (1965) *Pulping Processes*. Interscience Publishing, New York.
4. Ohno, N., Sawatari, A., Yoshimoto, I. (1992) Studies on distribution of extractives in pulp fibers by means of ESCA. *Japan Tappi J.* 46(10), 1295–1312.
5. Laine, J., Stenius, P., Carlsson, G., Ström, G. (1994) Surface characterization of unbleached kraft pulps by means of ESCA. *Cellulose* 1(2), 145–160.
6. Kleen, M., Sjöberg, J., Dahlman, O., Johansson, L-S., Koljonen, K., Stenius, P. (2002) The effect of ECF and TCF bleaching on the chemical composition of soda-anthraquinone and kraft pulp surfaces. *Nordic Pulp Pap. Res. J.* 17(3), 357–363.
7. Holmbom, B., Åman, A., Ekman, R. (1995) Sorption of glucomannans and extractives in TMP waters onto pulp fibers. In: *Proc. 8th Intern. Symp. Wood Pulping Chem.* Vol. I, pp. 597–604. KCL, Espoo, Finland.
8. Mustranta, A., Koljonen, K., Lappalainen, A. et al. (2000) Characterization of mechanical pulp fibres with enzymatic, chemical and immunochemical methods. In: *Proc. 6th European Workshop on Lignocellulosics and Pulp*, pp. 15–18. University of Bordeaux, Bordeaux, France.
9. Fuhrmann, A., Kleen, M., Koljonen, K., Österberg, M., Stenius, P. (2002) Precipitation of lignin and extractives from bleaching filtrates – effects on pulp properties. *2002 Intern. Pulp Bleaching Conf.*, Tappi Press, Atlanta, GA, USA.
10. Fardim, P., Durán, N. (2005) Influences of surface chemical composition on the mechanical properties of pulp as investigated by SEM, XPS and multivariate data analysis. *J. Braz. Chem. Soc.* 16(2), 163–170.
11. Kokkonen, P., Korpela, A., Sundberg, A., Holmbom, B. (2002) Effects of different types of lipophilic extractives on paper properties. *Nordic Pulp Pap. Res. J.* 17(4), 382–386.
12. Widsten, P., Laine, J.E., Qvintus-Leino, P., Tuominen, S. (2001) Effect of high-temperature fiberization on the chemical structure of softwood. *J. Wood Chem. Technol.* 21(3), 227–245.
13. Shen, W., Parker, I.H., Sheng, Y.J. (1998) The effects of surface extractives and lignin on the surface energy of eucalypt kraft pulp fibres. *J. Adhes. Sci. Technol.* 12(2), 161–174.
14. Cooke, P.M. (2000) Chemical microscopy. *Anal. Chem.* 72(12), 169R–188R.
15. Ström, G., Carlsson, G. (1992) Wettability of kraft pulps – effect of surface composition and oxygen plasma treatment. *J. Adhes. Sci. Technol.* 6(6), 745–761.

16. Fardim, P., Heijnesson-Hultén, A., Boisvert, J-P. et al. (2006) Critical comparison of methods for surface coverage by extractives and lignin in pulps by X-ray photoelectron spectroscopy (XPS). *Holzforschung* 60(2), 149–155.
17. Fardim, P., Durán, N. (2002) Surface chemistry of eucalyptus wood pulp fibres: effects of chemical pulping. *Holzforschung* 56(6), 615–622.
18. Johansson, L.-S. (2002) Monitoring fibre surfaces with XPS in papermaking processes. *Mikrochimica Acta* 138(3–4), 217–223.
19. Heijnesson-Hultén, A., Paulsson, M. (2003) Surface characterization of unbleached and oxygen delignified kraft pulp fibres. *J. Wood Chem. Technol.* 23(1), 31–46.
20. Fardim, P., Durán, N. (2003) Modification of fibre surfaces during pulping and refining as analysed by SEM, XPS and ToF-SIMS. *Colloids Surf. A* 233(1–3), 263–276.
21. Briggs, D. (1998) *Surface Analysis of Polymers by XPS and Static SIMS*, Cambridge University Press, Cambridge.
22. Gray, D. (1978) The surface analysis of paper and wood fibres by ESCA. III. Interpretation of carbon (1s) peak shape. *Cellul. Chem. Technol.* 12(6), 735–743.
23. Leclerc, G., Pireaux, J.J. (1995) The use of least square for XPS peak parameters estimation. Part 1. Myths and realities. *J. Electron Spectrosc. Relat. Phenom.* 71(2), 141–164.
24. Cimino, A., Gazzoli, M., Valigi, M. (1999) XPS quantitative analysis and models of supported oxide catalysts. *J. Electron Spectrosc. Relat. Phenom.* 104(1–3), 1–29.
25. Johansson, L.S., Campbell, J.M., Koljonen, K., Stenius, P. (1999) Evaluation of surface lignin on cellulose fibres with XPS. *Appl. Surf. Sci.* 92–95, 144–145.
26. Dorris, G.A., Gray, D.G. (1978) The surface analysis of paper and wood fibres by ESCA. I. Application to cellulose and lignin. *Cellul. Chem. Technol.* 12(1), 9–23.
27. Dorris, G.A., Gray, D.G. (1978) The surface analysis of paper and wood fibres by ESCA. II. Surface composition of mechanical pulps. *Cellul. Chem. Technol.* 12(6), 721–734.
28. Li, K., Reeve, D.W. (2004) Determination of surface lignin of wood pulp fibres by x-ray photoelectron spectroscopy. *Cellul. Chem. Technol.* 38(3–4), 197–210.
29. Heijnesson Hultén, A., Basta, J., Larsson, L., Ernstsson, M. (2006) Comparison of different XPS methods for fiber surface analysis. *Holzforschung* 60(1), 14–19.
30. Laine, J., Stenius, P., Carlsson, G., Ström, G. (1996) The effect of ECF and TCF bleaching on the surface chemical composition of kraft pulp as determined by ESCA. *Nordic Pulp Pap. Res. J.* 11(3), 201–210.
31. Fardim, P., Gustafsson, J., von Schoultz, S., Peltonen, J., Holmbom, B. (2005) Extractives on fibre surfaces investigated by XPS, ToF-SIMS and AFM. *Colloids Surf. A* 255(1–3), 91–103.
32. Johansson, L.-S., Campbell, J.M., Fardim, P., Heijnesson Hultén, A., Ernstsson, M., Boisvert, J.-P. (2005) An XPS round robin investigation in analysis of pulps and filter paper. *Surf. Sci.* 584, 126–132.
33. Li, K., Reeve, D.W. (2004) Sample contamination in analysis of wood pulp fibres with x-ray photoelectron spectroscopy. *J. Wood Chem. Technol.* 24(3), 183–200.
34. Risén, J. Heijnesson-Hultén, A., Paulsson, M. (2004) Surface characterization of softwood and hardwood kraft pulp fibers from different stages in a bleaching sequence. *J. Wood Chem. Technol.* 24(4), 307–321.
35. Dell, A. (1987) FAB-mass spectrometry of carbohydrates. *Adv. Carbohydr. Chem. Biochem.* 45, 19–72.
36. Saito, K., Kato, T., Tsuji, Y., Fukushima, K. (2005) Identifying the characteristic secondary ions of lignin polymer using ToF-SIMS. *Biomacromolecules* 6(2), 678–683.
37. Varmuza, K., Werther, W., Krueger, F.R., Kissel, J., Schmid, E.R. (1999) Organic substances in cometary grains: comparison of secondary ion mass spectral data and californium-252 plasma desorption data from reference compounds. *Int. J. Mass Spectrom.* 189(1), 79–92.

38. Hagenhoff, B., Breitenstein, D. (2006) ToF-SIMS characterisation of biological materials using cluster sources and imaging. *Nano-Molecular Analysis for Emerging Technologies II*, National Physics Laboratory, Teddington, UK, 17–18 October 2006.
39. Tyler, B.J. (2001) ToF-SIMS image analysis. In: *ToF-SIMS – Surface Analysis by Mass Spectrometry* (eds J.C. Vickerman and D. Briggs), pp. 475–493. IMP Publications, Chichester.
40. Buchert, J., Carlsson, G., Viikari, L., Ström, G. (1996) Surface characterizing of unbleached kraft pulp by enzymatic peeling and ESCA. *Holzforschung* 50(1), 69–74.
41. Pascoal Neto, C., Silvestre, A.J.D., Evtuguin, D. et al. (2004) Bulk and surface composition of ECF bleached hardwood kraft pulps. *Nordic Pulp Pap. Res. J.* 19(4), 513–520.
42. Herrington, T.M., Midmore, B.R. (1984) Adsorption of ions at the cellulose/aqueous electrolyte interface. Part 2. Determination of the surface area of cellulose fibers. *J. Chem. Soc., Faraday Trans.* 80, 1539–1552.
43. Berg, J.C. (1993) The importance of acid–base interactions in wetting, coating, adhesion and related phenomena. *Nordic Pulp Pap. Res. J.* 8(1), 75–85.
44. Somorjai, G.A. (1981) *Chemistry in Two Dimensions: Surfaces*. Cornell University Press, London, 1981.
45. Kontturi, E., Thuene, P.C., Niemantsverdriet, J.W. (2003) Cellulose model surfaces – simplified preparation by spin coating and characterization by X-ray photoelectron spectroscopy, infrared spectroscopy, and atomic force microscopy. *Langmuir* 19(14), 5735–5741.

Part II
Characterization of Cellulose, Lignin, and Modified Cellulose Fibers

Chapter 7
Studies of Deformation Processes in Cellulosics Using Raman Microscopy

Wadood Y. Hamad

Abstract

A Raman microscope, or microprobe, incorporates a dispersive spectrometer with photon multiplier detectors in conjunction with a light microscope for the alignment of samples and collection of spectra from very small domains (∼1–2 μm) to lessen structural heterogeneity. The Raman effect, or the inelastically scattered light by an atom or molecule, carries useful information on the molecular vibrations in a material, and the technique has successfully been applied to characterizing deformation in nonmetallic materials by monitoring the shift of the Raman-sensitive bands upon the application of external stress or strain.

This chapter focuses on demonstrating the application of Raman microscopy to studying deformation processes in native and regenerated cellulosics. The 1095 cm^{-1} Raman-sensitive band has been shown to characteristically shift upon the application of external stress to various groups of cellulose fibers, and to the extent that the Raman technique could be calibrated, it provides a unique "microscopic strain gauge" by which internal microstrains/microstresses could be monitored and predicted. The chapter also discusses how the Raman technique is used to quantify deformation in cellulose fibers by relating the fiber macrodeformation to that of the cellulose chains through the concepts of *strain* and *stress sensitivities*.

7.1 Introduction

Vibrational spectra are measured by two very different techniques, infrared (IR) and Raman spectroscopy. The energy of light[i] in the region 100–5000 cm^{-1}, 1.2–60 kJ mol^{-1} (or 0.3–15 kcal mol^{-1}) is sufficient to excite vibrations of the molecules that absorb it [2]. Rotational energies of molecules are even smaller than vibrational energies, and therefore light energetic enough to excite vibrations would simultaneously excite rotations as well. The *excitation* of rotation or vibration, in this context, means that the molecule is promoted to a state of higher energy in which its rotational frequency or vibrational amplitude is increased.[ii]

In IR spectroscopy, light of all different frequencies is passed through a sample and the intensity of the transmitted light is measured at each frequency. In Raman spectroscopy, however, light scattered by the sample, not transmitted light, is observed – from any convenient direction with respect to the incident light. Raman experiments require light of a single frequency, v_0, or monochromatic light, and the *Raman effect*, or *Raman scattering*, refers to the *inelastic scattering of a photon*.[iii] When light is scattered by an atom or molecule, most photons are elastically scattered (the Rayleigh scattering), that is, scattered photons have the same energy and, therefore, the same wavelength as the incident photons. However, a small fraction of the scattered light (~1 in 1000 photons) is scattered from excitations with optical frequencies different from, and usually lower than, the frequency of incident photons. It is these inelastically scattered photons that carry information about molecular vibrations in the material, and thus are useful for the characterization of nonmetallic materials.

The Raman effect corresponds to the absorption and subsequent emission of a photon via an intermediate electron state having a virtual energy level. If no energy exchange between the incident photons and molecules occurs, then there is no Raman effect. If, however, energy exchanges do occur, then the energy differences are equal to the vibrational and rotational energy levels of the molecule. In crystals, only specific photons are allowed by the period structure, and Raman scattering can thus appear at certain frequencies. For amorphous materials such as glass, more photons are allowed and the discrete spectral lines become broad. Raman-scattered radiation comprises two components, *Stokes* and *anti-Stokes* (Figure 7.1). When a molecule absorbs energy, the resulting photon of lower energy generates a Stokes line on the red side of the incident spectrum. If the molecule loses energy, however, the incident photons are shifted to the blue side of the spectrum, thus generating an anti-Stokes line. Since the distortion of the molecule is determined by its polarizability, Raman transition from one state to another, or a Raman shift, occurs when there is a polarizability change – during vibration or rotation.

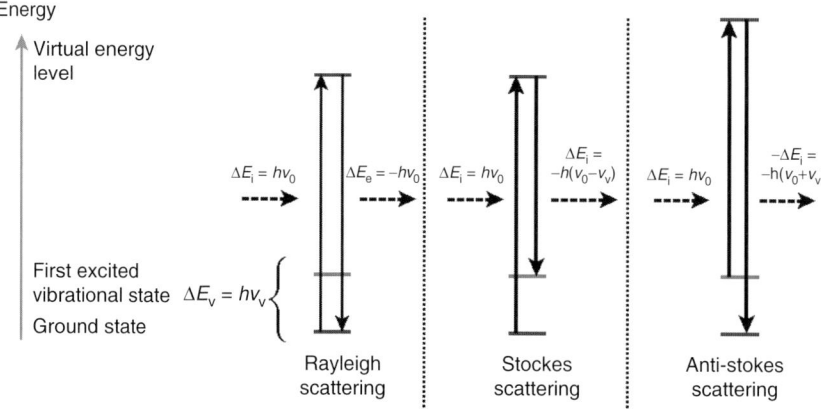

Figure 7.1 The different possibilities of visual light scattering: Rayleigh (no Raman effect), Stokes (molecules absorb energy), and anti-Stokes scattering (molecules lose energy). The symbols in the diagram are: E = the energy carried by each photon, h = Planck's constant, v = frequency. As $\lambda v = c$, where c is the speed of light, and the wave number defined as, $\bar{v} \equiv 1/\lambda$, the energy carried by the photon, E, becomes $E = hv = hc/\lambda = hc\bar{v}$. (Diagram assembled from information in References 1 and 2).

If the spectroscopist's task is to interpret – and perhaps predict – the interaction of light with matter, then the objective of the present chapter is to present a materials scientist's approach to utilizing spectroscopy to study deformation in materials. In particular, we shall discuss the principles of Raman microscopy and its application to characterize deformation processes in polymers, focusing on cellulosics. To facilitate the discussion on Raman deformation processes, a concise treatment of the structure and solid-state characteristics of cellulosics will be given.

7.2 Principles of Raman microscopy

Raman spectroscopy has enjoyed a remarkable growth in applications since the significant development of (air-cooled rather than water-cooled) lasers as exciting sources in 1960s, and the development of high-throughput monochromators and highly sensitive photon detectors such as charge-coupled device (CCD) cameras. Moreover, the innovation of the *Raman microscope* or *Raman microprobe* system has proved a particularly powerful characterization tool. The Raman microprobe is basically the coupling of a conventional Raman spectrometer with a specially designed optical microscope [3–5]. The microscope performs two key functions. It focuses the exciting light on the sample down to a diameter of ~1–2 μm, then gathers the scattered light and transmits it to the slit entrance of the spectrometer. Because the system acquires spectra from such small domains, the structural heterogeneity of the domains is greatly reduced relative to the domains examined in conventional Raman spectroscopy. It, therefore, becomes possible to identify morphological features and to use the collected spectra to relate orientation, composition, and structure to morphology.

Raman microscopy has also proved to be particularly useful in characterizing and quantifying the deformation micromechanics of polymer fibers and composites. The technique monitors the shift of the Raman-sensitive band as the sample is subjected to external stress or strain (see Figure 7.2). Data thus gathered on the *internal*, or molecular, state of deformation may be related to the macroscopic properties of the material under investigation.

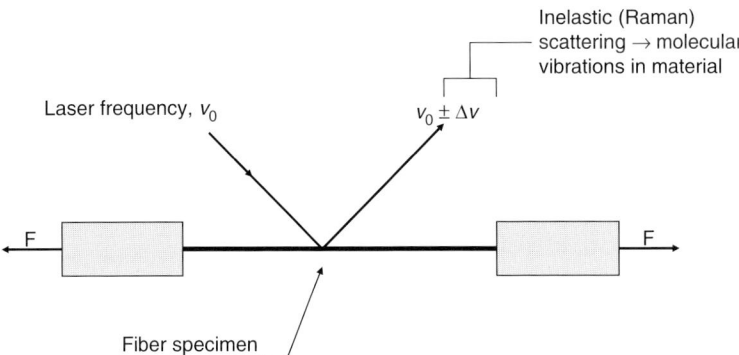

Figure 7.2 Schematic representation of the Raman effect for a fiber deformation experiment. The technique monitors the shift of the Raman-sensitive band (the inelastically scattered light), which carries quantifiable information about molecular vibrations in the material. Deformation micromechanics studies are concerned with relating this information to the material's macroscopic properties.

Research in Raman microscopy has seen an unyielding upsurge over the past three decades owing, primarily, to the important development of several new polymers that were particularly amenable to characterization by Raman spectroscopy. For instance, highly-crystalline fibers obtained from liquid crystalline polymer solutions – for example, Kevlar – have been shown to have particularly well-defined Raman spectra because of a combination of their highly oriented molecular structure and the aromatic nature of the molecular backbone. Another contemporary development is that of conjugated polymers with interesting and unusual optical and electronic properties, such as substituted polydiacetylenes, whose molecular structures also give well-defined Raman spectra.

7.2.1 Instrumentation for Raman microscopy

A Raman spectrum is obtained by the exposure of a sample to a monochromatic source of exciting photons and the measurement of the light intensity at the frequencies of the scattered light. Because the intensity of the scattered Raman component is much lower than that of the Rayleigh-scattered component, filters and diffraction gratings are used to suppress the latter component.

In conventional Raman spectroscopy, where visible laser radiation is used, the exciting photons are typically of much higher energies than those of the fundamental vibrations of most chemical bonds or systems of bonds, usually by a factor ranging from 6 for O—H and C—H bonds to ~200 for bonds between very heavy atoms [4,6]. The 514.5 and 488 nm lines from an argon-ion laser, or 632.8 nm red line of a helium–neon laser with power <50 mW power, are typically used as exciting frequencies. However, because most of fluorescent compounds have no electric absorption bands near the IR, Fourier transform (FT) Raman spectroscopy is utilized with a 1064 nm light source. The FT-Raman system is potentially efficient and has high calibration accuracy. However, an important difference between conventional and near-IR FT approaches is the difference in the scattering efficiencies at the absolute Raman frequencies. Raman scattering intensity is directly proportional to the fourth power of the excitation frequency, v_0^4, and as such the intensity of the Raman scattering is reduced by an order of magnitude by the use of IR radiation rather than excitation in the visible region. The choice between the use of dispersive or FT systems is primarily dictated by the scattering characteristics of the material under investigation and the type of information being sought. A laser with 785 nm can, for instance, combine fluorescence prevention with a high spatial resolution of dispersive Raman spectroscopy in comparison with the FT-Raman approach.

Most of the work in the literature concerning Raman spectroscopy of synthetic or biopolymers has been performed using dispersive spectrometers with photon multiplier detectors. Many of these systems incorporate a microscope for the alignment of the sample and collection of spectra from very small domains (~1–2 μm) to lessen their structural heterogeneity, as alluded to earlier. Almost all modern Raman microscopic systems incorporate array detectors to enable a particular region of a spectrum to be acquired simultaneously rather than stepping through single points, which drastically reduces the time required to obtain the spectra. Further, a modern Raman microscope incorporates a CCD camera typically with quantum efficiencies between 20% and 40% in the region 450–900 nm and a dark noise $<10^{-3}$ electrons pixel^{-1} [7].

7.2.2 Raman deformation studies of polymers

Some of the most common applications utilizing Raman spectroscopy have investigated the deformation micromechanics of high-performance fibers, where it has been found that the frequencies or wave numbers of the Raman-sensitive bands shift upon the application of stress or strain to the fibers. The phenomenon was first reported for the deformation of polydiacetylene single crystal fibers [8], and subsequently confirmed by others [9–12]. Batchelder and Bloor [9] explained that the behavior was due to the macroscopic deformation being transformed directly into stressing the covalent bonds along the polymer backbone and changes in the bond angles as discussed, almost half a century ago, by Terloar [13]. The shift in the Raman band wave numbers for polydiacetylenes are particularly large (-20 cm^{-1}/% strain) since their unique conjugated backbone structure produces strong and well-defined resonance Raman spectra [9].

Following the original work on polydiacetylene single crystals, it became evident that stress- or strain-induced shifts in the position of the Raman-sensitive bands occur in several high-performance fibers. This phenomenon was first observed in aromatic polyamides, such as Kevlar [14,15], then in rigid-rod polymer fibers such as poly(p-phenylene benzobisthiazole) (PBT) [16], poly(p-phenylene benzoxazole) (PBO) [17], and poly(2,5(6)-benzoxazole) (ABPBO) [18]. All these fibers have well-defined Raman spectra due to the high degree of molecular orientation and the possibility of resonance enhancement. As they also have high Young's moduli, it was purported that the macroscopic deformation was translated into the direct stressing of the backbone bonds in the molecules. This, in effect, materializes as shifts in the Raman-sensitive bands [14–18]. Similar behavior has been observed in gel-spun polyethylene fibers [19,20], as well as nonpolymeric fibers, such as carbon [21], silicon carbide [22], and alumina [23].

The strain sensitivities, or the rate of shift of the Raman-sensitive band with respect to percentage strain, of high-performance fibers range from -4.4 cm^{-1}/% strain for aramids [14,15] to -12.1 cm^{-1}/% strain for rigid-rod PBT fibers [16]. However, the levels of the frequency shifts measured during deformation of conventional polymers, such as oriented polyesters [24] or polypropylene [25], are found to be much smaller: typically ≤ 1.0 cm^{-1}/% strain, up to yield.

The phenomenon of stress- or strain-induced Raman-sensitive band shifts during fiber deformation was subsequently used by several researchers to study different aspects of the structure–property relations in materials (see e.g., Reference 26), and to characterize interfaces in polymers and polymer–fiber-reinforced composites (see, amongst others, [27]).

7.3 Deformation processes in cellulosics

Raman and IR spectral studies of cellulosics have typically concentrated on discerning polymorphic [6] and structural changes [28]. However a decade ago, Hamad and Eichhorn [29] were first to utilize Raman microscopy for the study of deformation processes in regenerated cellulose fibers, where it was reported that the cellulose Raman-sensitive bands shift upon the application of external stress to the fibers. Thereafter, a number of publications appeared that confirmed those findings for regenerated as well as native cellulose – see, for instance, References 30–33.

7.3.1 The structure and solid-state properties of cellulose fibers

It is well established from x-ray diffractometric studies that the crytallinity in cellulose is polymorphic: celluloses, native, regenerated, and mercerized represent two crystallographic polymorphs [34].

VanderHart and Atalla [35] concluded, based on high-resolution solid-state NMR spectroscopy, that most native celluloses crystallized as a mixture of two allomorphic forms: celluloses Iα and Iβ. Cellulose Iα was found to predominate in alga and bacterial celluloses, whereas the Iβ form was predominant in the higher plant celluloses, such as ramie or cotton. This critical finding was subsequently confirmed through careful electron diffraction studies by Sugiyama et al. [36]. By recording electron diffraction patterns from small areas along the microfibrils of the alga *Microdictyon tenuius*, these researchers were able to demonstrate that the Iα and Iβ forms coexisted in the same microfibril. The two forms were separated along the long direction of the microfibril changing sharply, but seamlessly, from one form to another. Many crystalline "blocks" of the two forms were found to alternate along the microfibril. Within the domain of each allomorph the entire width of the microfibril crystallized only in that form, with no mixing of allomorphs in the lateral direction of the microfibril. More critically, the Sugiyama et al. [36] study was successful in characterizing the unit cell of the Iα crystal form. Whereas Iβ still possessed the familiar two-chain, monoclinic unit cell identical to that proposed in earlier studies of native celluloses, the Iα form was radically different. It crystallized in a one-chain, triclinic unit cell [36]. The shape of this cell was identical to the two-chain, triclinic cell proposed by Sarko and Muggli [37] for *Valonia*, but in a crucial difference, its size was such as to accommodate only one cellulose chain. This result furnished the necessary proof for the parallel-chain structure of native celluloses. While the Iα form cannot, by virtue of its unit cell, be anything but parallel-chain, the seamless merging and alteration of the Iα and Iβ forms along the microfibril rules out anything but the parallel structure for the Iβ cellulose. The differences in chain packing of the two allomorphs are schematically shown in Figure 7.3.

The controversy concerning chain polarity in cellulose – how a parallel-chain structure could easily be converted to an anti-parallel-chain crystal structure, via mercerization without any apparent disruption to the fiber – was resolved in 1994 through a powerful method used in polymer analysis, viz., crystallography of model compounds [38]: the crystal structure of β-D-cellotetraose, a known model for cellulose II, was resolved in view of the similarities of the representative diffraction diagrams and spectroscopic data. The characteristics of the crystal structure of β-D-cellotetraose definitively predicted that cellulose II should be an anti-parallel-chain structure. With reference to Figure 7.4, the unit cell of cellotetraose, which can nearly be superimposed on the cellulose II unit cell, contains two molecules whose conformations are virtually identical with that of cellulose II chains. The molecules are packed anti-parallel, occupying the exact locations of the corresponding chains in the cellulose II cell [38]. During the mercerization process, and other solid-phase transformations, a number of crystal-to-crystal phase transforamations take place sequentially [39]. The conversion from a parallel to an anti-parallel chain crystal

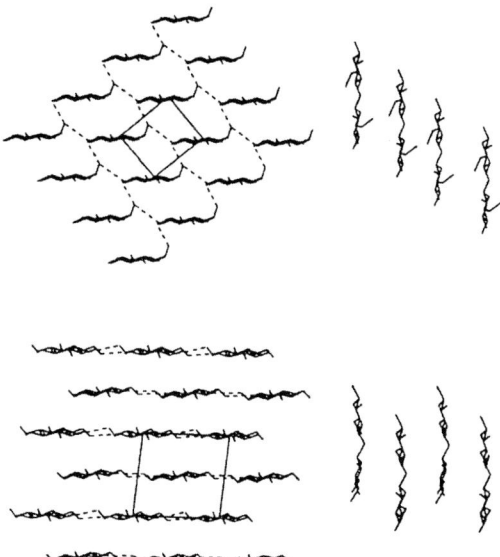

Figure 7.3 Projections of the crystal structures of cellulose Iα (top) and Iβ (bottom) on a plane perpendicular to the fiber axis. Illustrated to the right are the respective chain staggers along the fiber axis. After Sugiyama *et al.* [36], Gardner and Blackwell [52], and Hardy and Sarko [54].

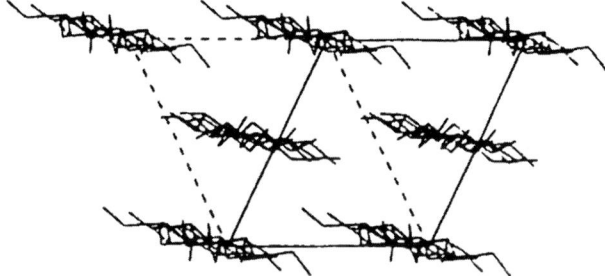

Figure 7.4 Projection of the crystal structure of β-D-cellotetraose on the ab plane. The unit cell outline of cellulose II is shown in a dashed line. After Geßler *et al.* [38], Stipanovic and Sarko [55], and Kolpak and Blackwell [56].

structure therefore occurs during the first of these transformations – change of cellulose I to Na-cellulose I [40]. Careful analysis of the crystallite sizes further showed that conversion initially began in the amorphous phase of the fiber, then quickly progressed to the smallest, least stable crystallites [41].

7.3.2 Molecular deformation processes in celluloses

Spectral features associated with the molecular vibrations that are most sensitive to polymorphic change particularly occur in the regions below 1500 cm^{-1}. Figure 7.5 depicts a typical Raman spectrum of a bleached, softwood, kraft pulp fiber from our studies. The relatively high intensity and good resolution of the skeletal modes reflect the basis for activity of molecular vibrations in Raman spectroscopy. That is, activity in the Raman spectrum requires finite transition moments involving the polarizability of the bonds.[iv] The two peaks at 895 and 1095 cm^{-1} are prominent in the Raman spectrum of cellulose, and the molecular deformation of the cellulose fiber can be ascertained by following these peaks as a cellulose fiber is subjected to external strain [29].

Wiley and Atalla [42] reported that the potential energy distributions in the region between 950 and 1180 cm^{-1} were dominated by C—C and C—O stretching motions, and that the motions were highly coupled – often involving coupling between the D-glucose rings. The high Raman intensity of the bands evident in Figure 7.5 is consistent with the large band density and the dominance of the C—C and C—O stretching motions predicted by normal-coordinate calculations [6,42,43]. As the 1095 cm^{-1} Raman band is most intense when the electric vector of the incident light is parallel to the fiber axis, it must result from C—C and C—O stretching motions parallel to the chain axis [42].

The Raman band at ~895 cm^{-1} is fairly intense in the spectrum shown in Figure 7.5. Two decades ago, Wiley and Atalla [42] suggested, in their landmark study of the Raman band assignments in celluloses, that the intensity of this band is related to the lateral size of the cellulose crystallites. It had been found earlier that the intensity of the 895 cm^{-1} correlates with the intensity of the broad, upfield shoulders for the C4 and C6 atoms in the solid-state ^{13}C NMR spectra of native cellulose [35]. The broad shoulders arise from the cellulose chains on the crystallite surfaces and in the amorphous regions, which suggest that the intensity of the 895 cm^{-1} Raman-sensitive peak is proportional to the amount of disorder in cellulose. As the likely sites of disorder in the cellulose molecules are the glycosidic linkages, the C6 atoms and the hydroxyl groups, the 895 cm^{-1} band is likely to

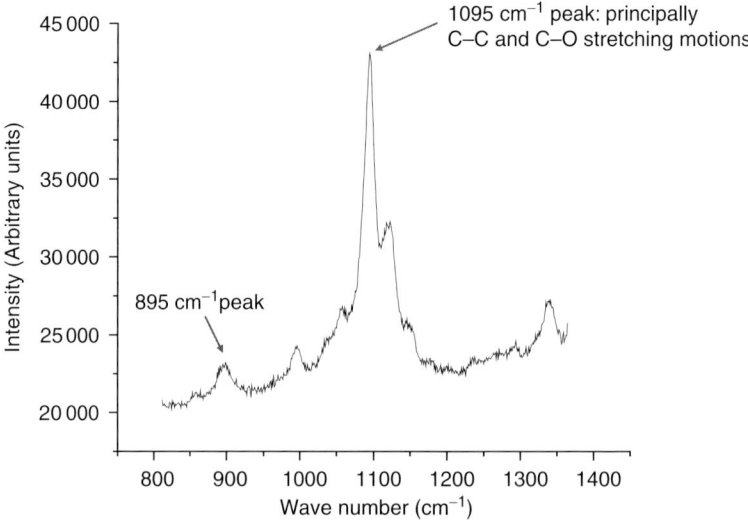

Figure 7.5 Raman spectrum of a single bleached, softwood, kraft pulp fiber in the region 800–1400 cm^{-1} using the 632.8 nm radiation from a He–Ne laser, collected over ~100 s.

involve one or more of these sites: potentially H—C—C and H—C—O bending localized at the C6 atom [42,44].

The Raman spectrum of a regenerated cellulose fiber (e.g., tencel) is similar but with two primary differences [29]. First, the intensity of the 895 cm^{-1} Raman peak is higher than that of native cellulose, indicating a higher degree of disorder. Second, there are much weaker, almost indistinguishable, peaks at ~1000 cm^{-1}. This band is dominated with C—C and C—O stretching motions, with small contributions from H—C—C, H—C—O and skeletal atom bending [42].

The effect of deformation upon the dominant Raman-sensitive bands due to skeletal modes from our studies is clearly shown in Figure 7.6. The deformation causes a shift of the bands to lower wave number accompanied by a (small) decrease in peak intensity and a (slight) broadening of the band peaks. Hamad and Eichhorn [29] were first to recognize this phenomenon in regenerated cellulose (both tencel and viscose), which was later confirmed with native cellulose ([33], see also our data in Figure 7.8) and other regenerated fibers [30,31]. It has been shown that primarily the cellulose bands identified with C—C and C—O stretching motions, the 895 and 1095 cm^{-1} Raman bands, shift upon the application of external stress/strain to the fiber. The shift is an indication of the straining of the fiber causing molecular deformation, and the broadening shows that there is a distribution of the stresses and strains over the cellulose molecules within the fiber [29].

Figure 7.7 depicts the well-correlated dependence of the peak position of the 1095 cm^{-1} Raman band upon strain for native cellulose fibers. It can be seen that the peak shifts to

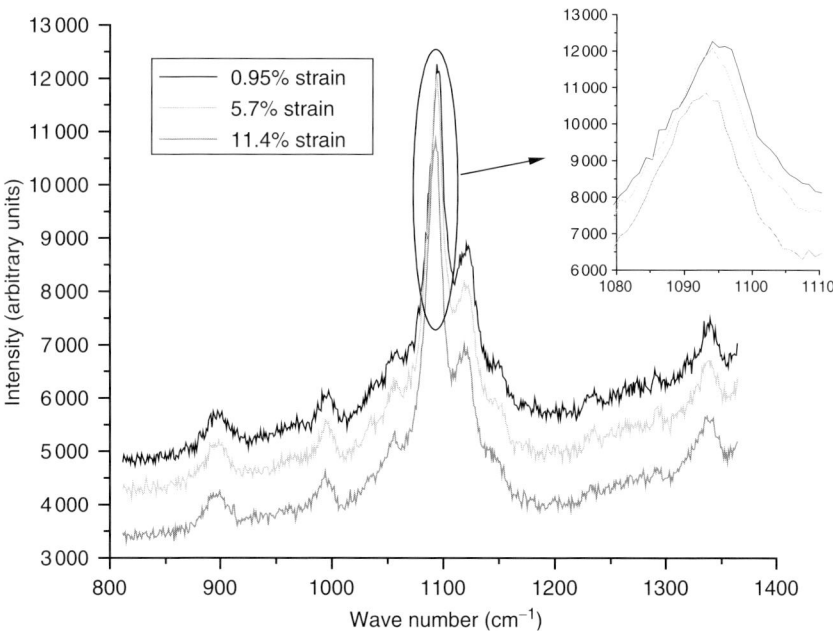

Figure 7.6 Shift in the Raman-sensitive bands, particularly the 1095 cm^{-1}, for the native cellulose shown in Figure 7.5 at three strain levels.

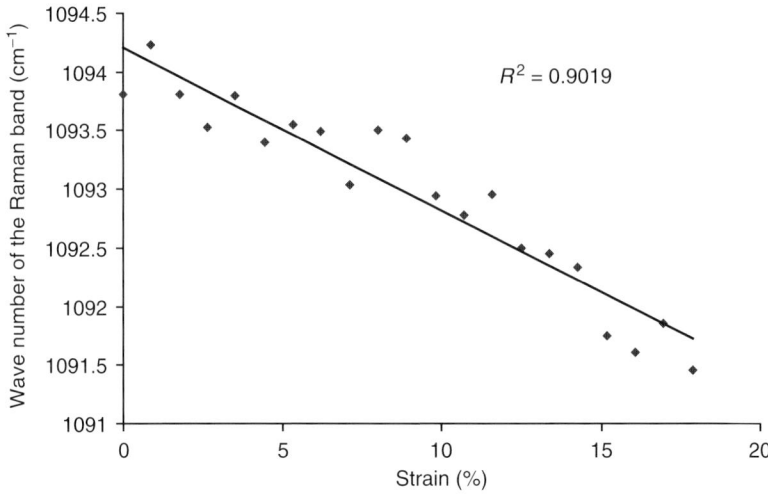

Figure 7.7 Linear dependence ($R^2 > 0.9$) of the wave number of the 1095 cm^{-1} Raman band peak of our single bleached softwood kraft pulp fiber (whose original Raman spectrum is shown in Figure 7.5) on strain.

lower wave numbers and that the dependence upon strain is clearly linear ($R^2 > 0.9$). It is worth pointing out that both the 895 and 1095 cm^{-1} have been shown to shift to lower wave numbers upon the application of external stress to regenerated cellulose fibers. However, the correlation associated with the Raman-shift vs strain response for regenerated cellulose is lower than that for the native cellulose shown in Figure 7.7. There are two main reasons. First, the intensity and resolution of the Raman-sensitive peaks for native cellulose are shown to be higher than those associated with regenerated celluloses [29–31], and there is indication of a higher degree of order (leading to a higher degree of crystallinity) associated with native cellulose. Recalling from earlier discussions, as the 895 cm^{-1} is indicative of the degree of disorder in cellulose, it is expected to be prominent in the molecular deformation of solvent-spun regenerated celluloses that are known to possess a high degree of disorder, or amorphicity – see Reference 45 for a detailed investigation of the structure of regenerated cellulose fibers. Second, scatter in the Raman measurements of regenerated celluloses stems from the intrinsic material behavior resulting from the solvent-spinning technique and the chemical/structural changes thus caused [29]. These changes seemingly influence the monoclinic character of the "unit cells"[v] for the cellulose chains and the bent conformations resulting from the glucose units' rotation around the glucose linkage [47]. The scatter, therefore, is not due to instrument-induced error.

The data scatter particularly associated with the 895 cm^{-1} Raman-shift vs strain response presented by Hamad and Eichhorn [29] for regenerated cellulose may further be explained by examining the macroscopic Young's modulus of the fiber. Figure 7.8 clearly illustrates that strain sensitivity, or the rate or Raman-sensitive band shift with respect to strain, is proportional to the material's Young's modulus: the lower the modulus, the lower the strain sensitivity. It therefore follows that the 1095 cm^{-1} band represents a more reproducible, accurate Raman shift with respect to applied strains for cellulose – native or

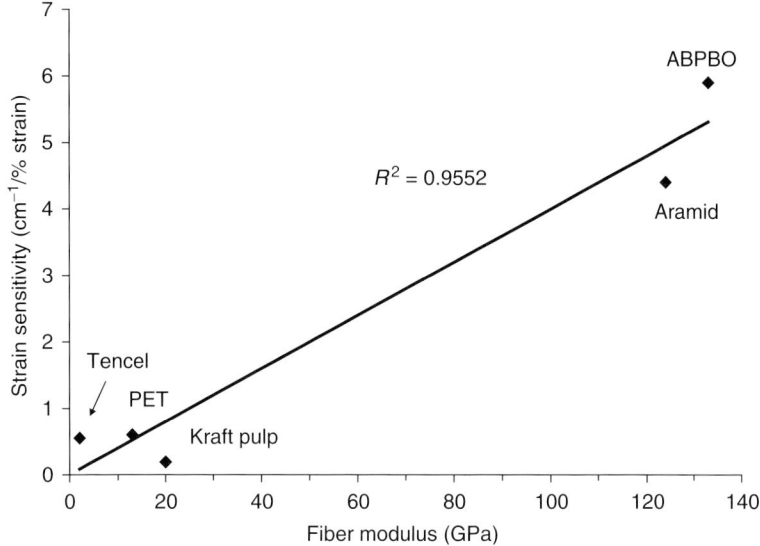

Figure 7.8 Variation of the strain sensitivity of the position of 1610 cm^{-1} Raman band in aramid fibers [51], the 1555 cm^{-1} for rigid-rod ABPBO (poly(2,5(6)-benzoxazole) fibers [18], the 1616 cm^{-1} in PET fibers [50], and the 1095 cm^{-1} for tencel [29] and our bleached softwood kraft pulp fibers as a function of the fiber's Young's modulus.

regenerated – which may subsequently be used for characterizing the interfacial phenomena in cellulosic networks or composites.

The strain sensitivity for celluloses – native and regenerated – falls within the range for conventional polymers: ≤1.0 cm^{-1}/% strain for polyethylene terephthalate (PET) films [24] and polypropylene fibers [25]. In their examination of regenerated cellulose fibers, Hamad and Eichhorn [29] showed that the macroscopic deformation of tencel is translated directly into molecular deformation. From the foregoing discussions and in reference to Figure 7.5, it is recognized that the deformation-induced skeletal modes most receptive to the Raman effect are those due to C—C and C—O stretching motions together with ring stretching. The magnitude of the 1095 cm^{-1} Raman-sensitive band shift, Δv, for a typical regenerated cellulose fiber is −3.26 cm^{-1} and for a native cellulose (kraft pulp) fiber −1.97 cm^{-1}, and the strain sensitivity, $d(\Delta v)/d\varepsilon$, is −0.55 and −0.19 cm^{-1}/% strain, respectively.

Moreover, the changes noted in Figures 7.5 and 7.6 indicate that skeletal bending and torsional modes would be altered to a greater degree than the skeletal stretching modes by any rotation about the bonds in the glycosidic linkage.

The linear correlation shown in Figure 7.7, to the extent that the Raman effect is calibrated, serves to provide a "microscopic strain gauge" by which internal strains/microstresses could be monitored [29]. If we consider the linear portion of the stress–strain relationship for celluloses,[vi] then the correspondence between the Raman vibrational frequency and stress would also be linear. Therefore, the *strain sensitivity*, $d(\Delta v)/d\varepsilon$, will be proportional to the fiber modulus, and the *stress sensitivity*, $d(\Delta v)/d\sigma$, the same for (all) cellulose fibers/materials. The analysis illustrates that the Raman technique is directly monitoring the molecular stress of the fibers rather than strain [29]. We shall return to this point through a comparison to conventional polymers.

Figure 7.8 depicts comparative results for cellulose, conventional and rigid-rod polymer, as well as high-performance fibers obtained from liquid crystalline polymer solutions. It is reasonable to assume from the figure that the basic mechanism of deformation for these (*different*) fibers is similar. It then follows that once a polymeric material is calibrated for the Raman technique, the modulus of the sample could be determined indirectly from $d(\Delta \nu)/d\varepsilon$. It should be cautioned, however, that the figure, and ensuing relations, merely give an indication that the deformation processes in these fibers are similar, but details pertaining to molecular backbone configurations and inherent structure–property relation are far from being similar. Indeed, for regenerated cellulose [29] and PET fibers [50], the Raman technique monitors molecular stress, whereas for aramid fibers it gives a direct measure of chain stretching [51]. A closer examination of the deformation mechanisms for PET and regenerated cellulose fibers seems warranted.

It has been stated earlier that the strain sensitivities for regenerated cellulose and PET fibers fall within the expected region of conventional polymers: -0.55 and -0.6 cm^{-1}/% strain, respectively. Table 7.1 reveals that the *stress sensitivities* are higher for fibers with higher values of tensile modulus. This indicates that there is necessarily more molecular stressing in the higher-modulus fibers. Consequently, the Raman technique gives a direct measure of molecular stressing in these types of fibers. It is worth noting that, for both regenerated cellulose fibers indicated in Table 7.1, the stress sensitivities, $d(\Delta \nu)/d\sigma$, while within the same range, are higher for the higher-modulus viscose fibers.

The rate of Raman-sensitive band shift with respect to stress, $d(\Delta \nu)/d\sigma$, for the regenerated celluloses are on the same order of magnitude as $d(\Delta \nu)/d\varepsilon$. When comparing these results to those of a conventional polymer, such as PET yarn, there exists a markedly higher rate of change of band shift with stress (Table 7.1). This indicates that a much lower level of molecular stressing takes place in regenerated cellulose fibers than in PET – the latter known for the presence of amorphous material and a poor level of molecular orientation. The stresses and strains in a (typical) regenerated cellulose fiber are more uniformly distributed owing to the "good" packing if unit cells forming the cellulose chains – a phenomenon accounted for in the spatial configuration by the presence of intermolecular and intramolecular bonding [52,53]. Structurally, the cellulose units are composed of two β-D-glucose residues, which are linked by an oxygen bridge to adjacent residues and are also rotated with respect to one another about a screw-axis so that they form continuous chain

Table 7.1 Comparative results for stress sensitivity variations for tencel and viscose [29] vs PET fibers [50].

Fiber	E (GPa)	Raman band (cm^{-1})	$d(\Delta \nu)/d\sigma$ (cm^{-1} GPa^{-1})
Tencel	2.1	1095	-0.36
Viscose	3.0	1095	-0.43
PET	13.0	1616	-5

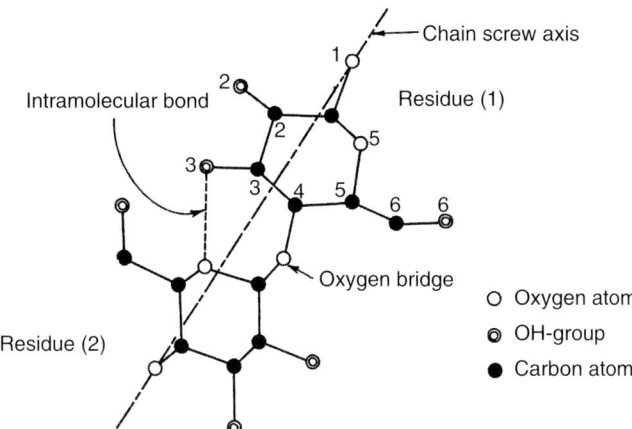

Figure 7.9 A structural model of the repeating unit cells forming cellulose chains. After Jacobson et al. [46].

segments [46]. Besides, there is an intramolecular bond between the hydroxyl group of one glucose residue and the oxygen ring of the next residue – see Figure 7.9.

Concluding remarks and future applications

It is hoped that this chapter has provided ample evidence that, beyond the interpretation of the interaction of light with matter (which is the typical purview of spectroscopic studies), Raman spectroscopy in conjunction with an optical microscope, the *Raman microprobe* or *microscope*, is a powerful tool for quantifying deformation processes in celluloses by following the molecular/chain stretching upon application of external stress or strain to the material. It has been identified that the 1095 cm^{-1} Raman-sensitive band for celluloses – native and regenerated – reproducibly provides the characteristic band for monitoring microstructural changes in these materials, or the interface of celluloses and other materials – as in composites, where experimental quantification of the interfacial phenomena are critical for understanding/predicting the mechanical response of the composite structure. The 1095 cm^{-1} Raman band is more representative in this regard than the 895 cm^{-1} band, as the latter is proportional to the amount of disorder in cellulose and typically results in more data scatter. The 1095 cm^{-1} band corresponds, in the structure of cellulose, to a combination of C—C and C—O stretching motions (and some contribution from angle bending coordinates involving heavy atoms).

Significant shifts of -1.97 and -3.26 cm^{-1} have been recorded for native (softwood kraft pulp) and regenerated (tencel and viscose) cellulose fibers when monitoring the 1095 cm^{-1} band as strain is applied to the fibers. The shifts result in strain sensitivities of -0.19 and -0.55 cm^{-1}/% strain, respectively – typical of conventional polymers. The Raman technique has been shown to give a direct measure of molecular stress, and it could be calibrated to provide a "microscopic strain gauge" by which internal strains/microstresses could be monitored and predicted. As such, the Raman technique may be used to monitor time-dependent phenomena or unravel structure–property relations in celluloses subjected to thermal, mechanical, or a combination of treatments.

Dedication

The completion of this work could not have been possible without the understanding and patience by my wife, Dima, and toddler son, Riam. Throughout this exercise and other research-related activities, they lovingly endured my noticeable absence and seeming absent-mindedness over many a weekend and an evening. It is to both of them that I dedicate this humble contribution.

Notes

i. Light is electromagnetic radiation with a wavelength that is visible to the eye (visible light), or strictly speaking, it is electromagnetic radiation of any wavelength. The elementary particle that defines light is the *photon*, and the three basic dimensions of electromagnetic radiation are: (1) intensity, or alternatively amplitude, which is related to the perception of brightness; (2) frequency, or alternatively wavelength, perceived by humans as color of light; and (3) polarization, or angle of vibration, which is only weakly perceptible by humans ordinarily [1].
ii. The interested reader is referred to Harris and Bertolucci's classic textbook, *Symmetry and Spectroscopy*, for further details and elaboration on spectroscopic techniques and principles.
iii. In 1922, the Indian physicist C.V. Raman published, with his collaborators, a series of investigations on the molecular diffraction of light. Their experimental work culminated in the discovery of the namesake radiation effect, which was published in *Nature* (Vol. 121, No. 3048, p. 501) as "A new type of secondary radiation" on March 31, 1928, for which he was awarded the Nobel Prize in 1930.
iv. It should be noted, however, that the activity of molecular vibrations in obtaining IR spectra requires finite transition moments involving the permanent dipoles of bonds undergoing displacement, hence the problem of interference.
v. Cellulose crystallizes partially or completely within a fiber in such a manner that "unit cells are formed." These cells are repeated in form of a chain, and a structural model of repeating units may thus be visualized as in Figure 7.8 [46].
vi. The relation is almost entirely linear for Kraft pulp fibers, that is, there is no yielding [48], or there may exist a noticeable yield point in the case of thermally or mechanically treated fibers [49]. For solvent-spun regenerated celluloses the yield point may vary depending upon treatment and spinning variables [45]. In the case of tencel, for instance, the stress–strain response is almost entirely linear [29].

References

1. Bohm, D. (1979) *Quantum Theory*, Dover Publications, New York.
2. Harris, D.C. and Bertolucci, M.D. (1978) *Symmetry and Spectroscopy – An Introduction to Vibrational and Electronic Spectroscopy*, Dover Publications, New York.
3. Dhamelincourt, P., Wallart, F., Leclerco, M., N'Gyuen, A.T. and Landon, D.O. (1979) Laser Raman molecular microprobe (MOLE). *Analytical Chemistry*, 51(3), 414A–421A.

4. Rosasco, G.J. (1980) Raman microprobe spectroscopy. In: *Advances in Infrared and Raman Spectroscopy*, Vol. 3 (eds. R.J.H. Clark and R.E. Hester), pp. 223–282. Heyden, London.
5. Andersen, M.E. and Muggli, R.Z. (1981) Microscopical techniques with the molecular optics laser examiner Raman microprobe. *Analytical Chemistry*, 53, 1772–1777.
6. Atalla, R.H. (1976) Raman spectral studies of polymorphy in cellulose. Part I: Cellulose I and II. *Applied Polymer Symposium*, 28, 659–669.
7. Batchelder, D.N. (1988) Multichannel Raman spectroscopy with a cooled CCD imaging detector. *European Spectroscopy News*, 80, 28.
8. Mitra, V.K., Risen, W.M. and Baughman, R.H. (1977) A laser Raman study of the stress dependence of vibrational frequencies of a monocrystalline polydiacetylene. *Journal of Chemical Physics*, 66(6), 2731–2736.
9. Batchelder, D.N. and Bloor, D. (1979) Strain dependence of the vibrational modes of a diacetylene crystal. *Journal of Polymer Science: Polymer Physics Edition*, 17, 569–581.
10. Galiotis, C., Young, R.J. and Batchelder, D.N. (1983) The solid-state polymerization and physical properties of bis(ethyl urethane) of 2,4-hexadiyne-1,6-diol. II. Resonance Raman spectroscopy. *Journal of Polymer Science: Polymer Physics Edition*, 21, 2483–2494.
11. Galiotis, C. and Batchelder, D.N. (1988) Strain dependences of first- and second-order Raman spectra of carbon fibres. *Journal of Materials Science Letters*, 7, 545–547.
12. Wu, G., Tashiro, K. and Kobayashi, M. (1989) Vibrational spectroscopic study on molecular deformation of polydiacetylene single crystals: Stress and temperature dependences of Young's modulus. *Macromolecules*, 22, 188–196.
13. Terloar, L.R.G. (1960) Calculations of elastic moduli of polymer crystals: I. Polyethylene and nylon 66. *Polymer*, 1, 95–103.
14. Galiotis, C., Robinson, I.M., Young, R.J., Smith, B.E.J. and Batchelder, D.N. (1985) Strain dependence of the Raman frequencies of a Kevlar 49 fibre. *Polymer Communication*, 26, 354–355.
15. Young, R.J., Lu, D. and Day, R.J. (1991) Raman spectroscopy of Kevlar fibres during deformation – Caveat Emptor. *Polymer International*, 24, 71–76.
16. Day, R.J., Robinson, I.M., Zakikhani, M. and Young, R.J. (1987) Raman spectroscopy of stressed high modulus poly(*p*-phenylene benzobisthiazole) fibres. *Polymer*, 28, 1883–1840.
17. Young, R.J., Day, R.J. and Zakikhani, M. (1990) The structure and deformation behaviour of poly(*p*-phenylene benzobisoxazole) fibres. *Journal of Materials Science*, 25, 127–136.
18. Young, R.J. and Ang, P.P. (1992) Relationship between structure and mechanical properties in high-modulus poly(2,5(6)-benzoxazole) (ABPBO) fibres. *Polymer*, 33, 975–982.
19. Prasad, K. and Grubb, D.T. (1989) Direct observation of Taut Tie molecules in high-strength polyethylene fibers by Raman spectroscopy. *Journal of Polymer Science: Part B: Polymer Physics*, 27, 381–403.
20. Kip, B.J., van Eijk, M.C.P. and Meier, R.J. (1991) Molecular deformation of high-modulus polyethylene fibers studied by micro-Raman spectroscopy. *Journal of Polymer Science: Part B: Polymer Physics*, 29, 99–108.
21. Robinson, I.M., Zakikhani, M., Day, R.J., Young, R.J. and Galiotis, C. (1987) Strain dependence of the Raman frequencies for different types of carbon fibres. *Journal of Materials Science Letters*, 6, 1212–1214.
22. Day, R.J., Piddock, V., Taylor, R., Young, R.J. and Zakikhani, M. (1989) The distribution of graphitic microcrystals and the sensitivity of their Raman bands to strain in SiC fibres. *Journal of Materials Science*, 24, 2898–2902.
23. Yang, X., Hu, X., Day, R.J. and Young, R.J. (1992) Structure and deformation of high-modulus alumina-zirconia fibres. *Journal of Materials Science*, 27, 1409–1416.
24. Fina, L.J., Bower, D.I. and Ward, I.M. (1988) Raman spectroscopy of stressed samples of oriented poly(ethylene terephthalate). *Polymer*, 29, 2146–2151.

25. Evans, R.A. and Hallam, H.E. (1976) Laser-Raman spectroscopic studies of mechanically-loaded polymers. *Polymer*, 17, 838–839.
26. Sengonul, A. and Wilding, M.A. (1995) Modelling of time dependence in ultra-high-modulus polyethylene based on Raman microscopy. *Polymer*, 36(23), 4379–4384.
27. Young, R.J. (1996) Evaluation of composite interfaces using Raman spectroscopy. *Key Engineering Materials*, 116–117, 173–192.
28. Marchessault, R.H. and Liang, C.Y. (1960) Infrared spectra of crystalline polysaccharides. III. Mercerized cellulose. *Journal of Polymer Science*, 43, 71–84.
29. Hamad, W.Y. and Eichhorn, S. (1997) Deformation micromechanics of regenerated cellulose fibres using Raman spectroscopy. *Journal of Engineering Materials and Technology*, 119, 309–313.
30. Eichhorn, S.J., Sirichaisit, J. and Young, R.J. (2001) Deformation mechanisms in cellulose fibres, paper and wood. *Journal of Materials Science*, 36, 3129–3135.
31. Eichhorn, S.J., Young, R.J. and Yeh, W.-Y. (2001) Deformation processes in regenerated cellulose fibres. *Textile Research Journal*, 71, 121–129.
32. Röder, T. and Sixta, H. (2005) Confocal Raman spectroscopy – applications on wood, pulp, and cellulose fibres. *Macromolecules Symposium*, 223, 57–66.
33. Gierlinger, N., Schwanninger, M., Reinecke, A. and Burgert, I. (2006) Molecular changes during tensile deformation of single wood fibres followed by Raman microscopy. *Biomacromolecules*, 7(7), 2077–2081.
34. Howsmon, J.A. and Sisson, W.A. (1954) *Cellulose and Cellulose Derivatives*, 2nd edition, Interscience, New York.
35. VanderHart, D.L. and Atalla, R.H. (1984) Studies of microstructure in native celluloses using solid-state ^{13}C NMR. *Macromolecules*, 17, 1465–1472.
36. Sugiyama, J., Vuong, R. and Chanzy, H. (1991) Electron diffraction study on the two crystalline phases occurring in native cellulose from an algal cell wall. *Macromolecules*, 24, 4168–4175.
37. Sarko, A. and Muggli, R. (1974) Packing analysis of carbohydrates and polysaccharides. III. Valonia cellulose and cellulose II. *Macromolecules*, 7(4), 486–494.
38. Geßler, K., Krauß, N., Steiner, T., Betzel, C., Sandmann, C. and Saenger, W. (1994) Crystal structure of β-D-cellotetraose hemihydrate with implications for the structure of cellulose II. *Science*, 266, 1027–1029.
39. Okano, T. and Sarko, A. (1984) Mercerization of cellulose. I. X-ray diffraction evidence for intermediate structures. *Journal of Applied Polymer Science*, 29, 4175–4182.
40. Nishimura, H., Okano, T. and Sarko, A. (1991) Mercerization of cellulose. 6. Crystal and molecular structure of Na-cellulose I. *Macromolecules*, 24, 759–770.
41. Nishimura, H. and Sarko, A. (1987) Mercerization of cellulose. III. Changes in crystallite sizes. *Journal of Applied Polymer Science*, 33, 855–866.
42. Wiley, J.H. and Atalla, R.H. (1987) Band assignments in the Raman spectra of celluloses. *Carbohydrate Research*, 160, 113–129.
43. Atalla, R.H. and Dimick, B.E. (1975) Raman-spectral evidence for differences between the conformations of cellulose I and II. *Carbohydrate Research*, 39, C1–C3.
44. Atalla, R.H. and Nagel, S.C. (1974) Celluloses: Its regeneration in the native lattice. *Science*, 185, 522–523.
45. Schurz, J. and Lenz, J. (1994) Investigation of the structure of regenerated cellulose fibres. *Macromolecule Symposium*, 83, 273–289.
46. Jacobson, R.A., Wunderlich, J.A. and Lipscomb, W.N. (1961) The crystal and molecular structure of cellobiose. *Acta Crystallography*, 14, 598–607.
47. Pedersen, B. (1974) The geometry of hydrogen bonds from donor water molecules. *Acta Crystallography*, B30, 289–291.
48. Hamad, W.Y. (2002) *Cellulosic Materials – Fibers, Networks and Composites*, Kluwer Academic Publishers, Boston, MA.

49. Hamad, W.Y. (1998) On the mechanisms of cumulative damage and fracture in native cellulose fibres. *Journal of Materials Science Letters*, 17, 433–436.
50. Young, R.J. and Yeh, W.-Y. (1994) Chain stretching in a poly(ethylene terephthalate) fibre. *Polymer*, 35, 3844–3847.
51. Andrews, M.C. and Young, R.J. (1993) Analysis of the deformation of Aramid fibres and composites using Raman spectroscopy. *Journal of Raman Spectroscopy*, 24, 539–544.
52. Gardner, K.H. and Blackwell, J. (1974) The structure of native cellulose. *Biopolymers*, 13, 1975–2001.
53. Axelrad, D.R. (1979) Theory of bond failure in hydrogen-bonded solids. *Advances in Molecular Relaxation and Interaction Processes*, 15, 51–69.
54. Hardy, B.J. and Sarko, A. (1993) Conformational analysis and molecular dynamics simulation of cellobiose and larger cellooligomers. *Journal of Computational Chemistry*, 14(7), 831–847.
55. Stipanovic, A.J. and Sarko, A. (1976) Packing analysis of carbohydrates and polysaccharides. 6. Molecular and crystal structure of regenerated cellulose II. *Macomolecules*, 9(5), 851–857.
56. Kolpak, F.J. and Blackwell, J. (1976) Determination of the structure of cellulose II. *Macromolecules*, 9(2), 273–278.

Chapter 8
Lifetime Prediction of Cellulosics by Thermal and Mechanical Analysis

Tatsuko Hatakeyama and Hyoe Hatakeyama

Abstract

Cellulose is a complex biopolymer the physical properties of which depend markedly on its plant source and method of isolation and the durability (lifetime) of which is influenced by various factors. Thermogravimetry studies and mechanical measurements have been performed on cellulose fabric and cellulose powder. Selection of thermal degradation and mechanical stress as the main factors affecting the lifetime of these cellulose materials and consideration of the time and temperature conversion are shown to allow reasonable lifetime prediction of the materials.

8.1 Introduction

Lifetime of cellulosics is an important property when plant-based materials are widely used in practical fields. Long-term properties of cellulosics are affected by various factors such as temperature, humidity [1], biological degradation [2], irradiation of light, and mechanical stress. Under ordinary conditions, these factors affect the durability of cellulosics in a complex manner. Because of this, it is difficult to choose appropriate experimental conditions to accelerate damage on cellulosics in a similar manner to that in natural conditions. Prediction necessarily introduces variation due to oversimplification.

For the long-term properties of synthetic polymers, researchers have investigated physical aging [3,4]. Molecular relaxation occurs when amorphous polymeric materials are maintained at a temperature lower than their glass transition temperatures, which causes a change of mechanical properties. The mechanical properties can be determined when the annealing temperature and time of a sample are known. Accelerated weathering, which is widely carried out in polymer industry using weather meters, is another method to shorten the time interval in order to compare the results with those obtained from weathering in nature [5].

To decrease the time scale of experimental procedure, theories of lifetime prediction necessarily assume time–temperature conversion based on linear relationship between these two factors. Cellulosics is a crystalline biopolymer and the main chain motion is restricted via inter- and intramolecular hydrogen bonding [6]. Thus, methods applied to synthetic polymers are not always appropriate for cellulosics and a certain limitation needs to be placed. Of the factors affecting the durability of cellulosics, thermal stability and mechanical strength are the major ones. In this chapter, thermal and mechanical properties of cellulosics as a function of time over a long span are described.

8.2 Materials and methods

8.2.1 Samples

Cellulose powder (Avicel powder grade PH-101) was obtained from Asahi Chemicals Co. Ltd. The size of the powder was measured using a light microscope, and the size distribution histogram was obtained. To obtain a pellet-shaped sample, the powder was pressed at room temperature for 5 min under the pressure of 1.0×10^7 Pa.

Cellulose fabric was obtained from the Japanese Standardization Association [JIS L 0803, cotton plane fabric No. 3]. The warp and weft were 20 and 16 tex, respectively. Circularly shaped fabrics with a diameter of 7.0 mm obtained using a punch were stored in a dessicator prior to the thermogravimetric (TG) studies.

For the mechanical measurements, the cellulose fabric was used as a cellulose I sample. The fabrics were soaked with a surfactant and washed several times in water. Rectangular samples of 20 mm (width) ×100 mm (length) were prepared, stored in a desiccator and then used for the mechanical measurements. Samples were annealed at various temperatures from 323 to 493 K for 1 to 2×10^4 min.

8.2.2 Thermogravimetry

A Seiko Instruments' thermogravimeter-differential thermal analyzer, TG/DTA220, SSC 5200, was used. The temperature ranged from 300 to 870 K. A sample of ∼7.0 mg was tightly placed in order to attach samples at the surface of platinum crucible. The heating rate was 20 K min^{-1}, and air or N$_2$ gas was used as an atmospheric gas with a flowing rate of 100 mL min^{-1}. The decomposition temperature and mass residue were determined as previously reported [7]. In this study, the point where the extension line of mass in the solid-state crosses with that of the steep mass decrease was used as the extrapolated temperature (T_d). The mass residue was obtained at 770 K. To calculate the activation energy by the Ozawa–Flynn–Wall method [6–9], the heating rate was varied between 2, 5, 10, 20, and 50 K min^{-1}.

8.2.3 Mechanical measurements

An Instron-type mechanical tester, Orientec Tensiron RTA, was used. The stretching rate was 50 mm min^{-1}. Measurements were carried out at 298 K (25°C). The tensile strength, tensile elongation at break, and Young's modulus were calculated. Three test pieces at each annealing temperature and time were tested and the average value was used.

8.3 Prediction of durability

In this study, the dynamic data obtained by thermogravimetry is used for the prediction of thermal stability based on the Ozawa–Flynn–Wall method [8–10]. The mechanical strength was measured using pre-annealed cellulose samples and the time–temperature relationship was established. Thermal and mechanical properties are predicted in a temperature range from 300 to 470 K and in a time range from several minutes to several hundred years.

8.3.1 Prediction by TG

The Ozawa–Flynn–Wall method for analyzing reaction kinetics using TG data has been used for the studies of thermal decomposition and crystallization kinetics of various kinds of polymers [11,12]. The method is based on a simple assumption that the reaction rate, dx/dt (x is reacted mass and t is time), is correlated with the function of reacted x, $f(x)$, via the Arrhenius equation. Using the kinetic data, the lifetime of polymeric materials can be estimated.

Thermal decomposition of cellulose and related compounds has been extensively studied [13–17]. Although conflicting results are evident in the whole temperature range of decomposition of cellulose, the thermal decomposition in the initial stage is well explained. Using kinetic data from the initial stage of thermal decomposition, the lifetime of cellulose materials can be estimated. In this study, the initial stage of thermal decomposition of the cellulose fabric is investigated by thermogravimetry.

As stated earlier, the following equation can be established between the rate of reacted mass and the activation energy (ΔE).

$$-\frac{dx}{dt} = A \exp\left(-\frac{\Delta E}{RT}\right) f(x) \tag{8.1}$$

where x is the reacted mass, t the time, A the frequency factor, T the absolute temperature in degrees Kelvin, and R the gas constant. According to the definition of thermal analysis, $dT/dt = $ constant (β) corresponds to the heating rate, and the following equation can be obtained.

$$-\int_{x_0}^{x} \frac{dx}{f(x)} = A \int_{t_0}^{t} \exp\left(-\frac{\Delta E}{RT}\right) dt = \frac{A}{\beta} \int_{T_0}^{T} \exp\left(-\frac{\Delta E}{RT}\right) dT \tag{8.2}$$

where x_0 and T_0 are the values of x and T at $t = t_0$.

As the rate of reaction is usually low at low temperatures, the next approximation can be obtained.

$$\int_{T_0}^{T} \exp\left(-\frac{\Delta E}{RT}\right) dT = \int_{0}^{T} \exp\left(-\frac{\Delta E}{RT}\right) dT \tag{8.3}$$

The right-hand side of the equation can be expressed using the following p-function [18,19].

$$\frac{\Delta E}{R} p\left(\frac{\Delta E}{RT}\right) dT = \int_{0}^{T} \exp\left(-\frac{\Delta E}{RT}\right) dT \tag{8.4}$$

When $(\Delta E/RT) > 20$, $p(\Delta E/RT)$ can be approximated by the following equation.

$$\log p\left(\frac{\Delta E}{RT}\right) \cong -2.315 - 0.4567\left(\frac{\Delta E}{RT}\right) \tag{8.5}$$

At a given mass loss, the left-hand side of the Equation (8.2) is a constant and independent of β. Thus, if a constant mass fraction is attained at a temperature T_1 for β_1, T_2 for β_2, and so on, the following equation is obtained.

$$\frac{A\Delta E}{\beta_1 R}p\left(\frac{\Delta E}{RT_1}\right) = \frac{A\Delta E}{\beta_2 R}p\left(\frac{\Delta E}{RT_2}\right) = \cdots \tag{8.6}$$

Using Equation (8.5), the following linear relations can be derived.

$$-\log \beta_1 - 0.4567\left(\frac{\Delta E}{RT_1}\right) = -\log \beta_2 - 0.4567\left(\frac{\Delta E}{RT_2}\right) = \cdots \tag{8.7}$$

$$\log \beta = -0.4567\left(\frac{\Delta E}{R}\right)\left(\frac{1}{T_i}\right) + \text{const} \tag{8.8}$$

If $1/T_i$ is plotted against $\log \beta$, ΔE can be obtained from the gradient (Ozawa–Wall–Flynn method).

Figure 8.1 shows TG curves of the cellulose fabric measured in air at various heating rates (β). It is clearly seen that the thermal decomposition takes place in two stages at ~573 and 673 K at $\beta \leq 20$ (K min^{-1}), with the first stage decomposition being reached at a mass residue of $\leq$~30%. Using TG-Fourier transform infrared spectrometry (TG-FTIR), it was identified that the gas evolved during the thermal decomposition of the samples [15]. At >570 K, the evolution of CO_2 was clearly observed. To minimize the complications from the evolution of gases such as CO_2, the Ozawa–Wall–Flynn method was applied in a range where x was smaller than 0.2, that is, the mass residue was higher than 80%.

Figure 8.2 shows the linear relationship between $\log \beta$ and the reciprocal absolute temperature (T^{-1}) of the cellulose fabric measured in air. From the gradient, the activation

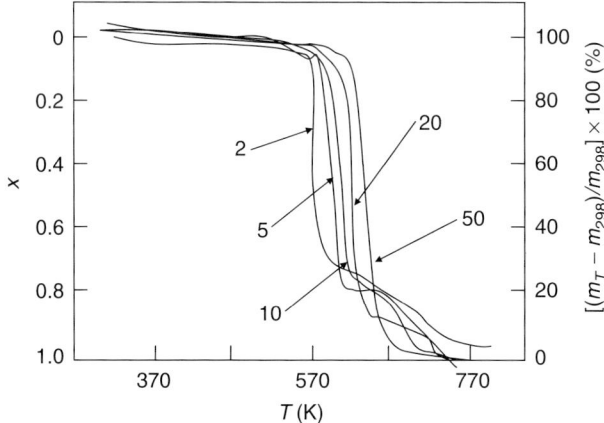

Figure 8.1 TG curves of the cellulose fabric measured in air; numerals in the figure are the heating rates, β (K min^{-1}); m_T = mass at a temperature T, and m_{298} = mass at 298 K.

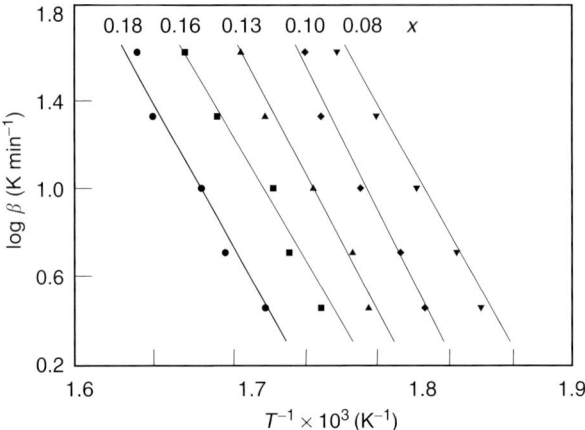

Figure 8.2 Ozawa plot of the cellulose fabric measured in air; β = the heating rate (K min^{-1}) and x = reacted mass.

Table 8.1 Activation energy of the initial stage of thermal decomposition of the cellulose fabric and the cellulose powder.

Sample	Atmosphere	ΔE (kJ mol^{-1})
Cellulose fabric	N$_2$	136
	Air	125
Cellulose powder	N$_2$	190
	Air	179

energy (ΔE) was calculated. Similar results were obtained when the flowing gas was N$_2$, although the cellulose decomposed in one stage. Calculated ΔE values for both the cellulose fabric and the cellulose powder are shown in Table 8.1.

When the temperature is assumed to maintain at a constant, T_c, Equation (8.1) can be rewritten as follows.

$$-\frac{dx}{dt} = A \exp\left(-\frac{\Delta E}{RT_c}\right) f(x) \tag{8.9}$$

Since T_c is independent of time (t) and the lifetime τ is defined as the time interval in which x changes from x_0 to x_1, Equation (8.10) can be obtained.

$$-\int_{x_0}^{x_1} \frac{dx}{f(x)} = A \int_{t_0}^{t_0+\tau} \exp\left(-\frac{\Delta E}{RT_c}\right) dt$$

$$= A \exp\left(-\frac{\Delta E}{RT_c}\right) \tau \tag{8.10}$$

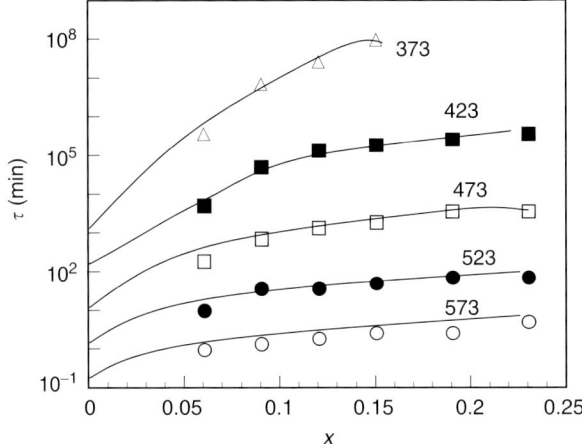

Figure 8.3 Relationship between lifetime (τ), x, and the holding temperature (T_c) of the cellulose fabric in air; numerals in the figure are the values of T_c (K).

By combining Equation (8.10) with Equation (8.2), the lifetime τ is obtained as follows.

$$\tau = \frac{\int_{T_0}^{T_1} \exp(-(\Delta E/RT))\mathrm{d}T}{\beta \exp(-(\Delta E/RT_c))} \tag{8.11}$$

Using the Equation (8.11), τ value at the chosen T_c could be calculated, as ΔE was calculated from the Equation (8.8) and β had been selected. Figure 8.3 shows the plots of τ calculated thus vs x of the cellulose fabric maintained at various T_c in air. It is clearly seen that the τ value varies from 1 to 10^8 min (~1500 years). Each plot was extended to $x = 0$ to give a value corresponding to the time interval where no mass loss would occur at the given T_c. The cellulose fabric maintained in N_2 (data not shown) showed a similar behavior. The cellulose powder also shows very similar relationships between τ, x, and T_c as those shown in Figure 8.3.

Figure 8.4 shows the relationships between T_c and τ values where $x = 0.1$ (10% mass decrease) and the interpolated τ values at $x = 0$ of the cellulose fabric maintained in air. The relationships between T_c and τ values at $x = 0.1$ and interpolated τ values at $x = 0$ of the cellulose fabric maintained in N_2 (data not shown) were similarly established. The effect of atmosphere was profound at a low-temperature region. The 10% mass decrease ($x = 0.1$) of the cellulose fabric at 370 K ($1/T_c \times 10^3 = 2.70$) is predicted to take ~72 years in air but ~150 years in N_2. At a temperature lower than 370 K, extrapolated values are not reliable because of the long extrapolation. When the Ozawa–Wall–Flynn method is applied to the initial stage of the thermal decomposition, the lifetime prediction at a constant temperature (T_c) > 370 K appears to be reliable.

8.3.2 Prediction of mechanical properties

The tensile strength of the cellulose fabric annealed at various temperatures (from 323 to 493 K) as a function of annealing time from 1 to 2×10^4 min is shown in Figure 8.5.

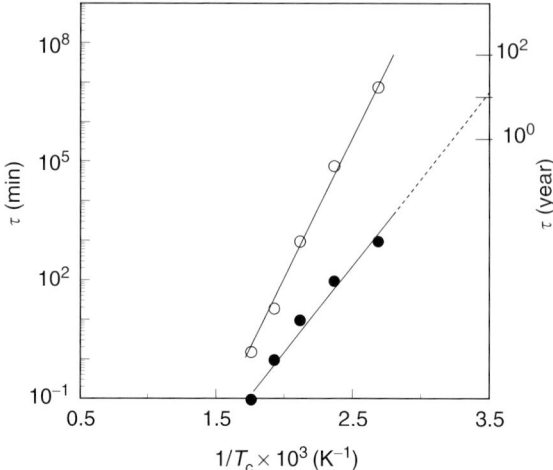

Figure 8.4 Lifetime (τ) at $x = 0$ and 0.1 (10% mass decrease) of the cellulose fabric; ●: $x = 0$, ○: $x = 0.1$.

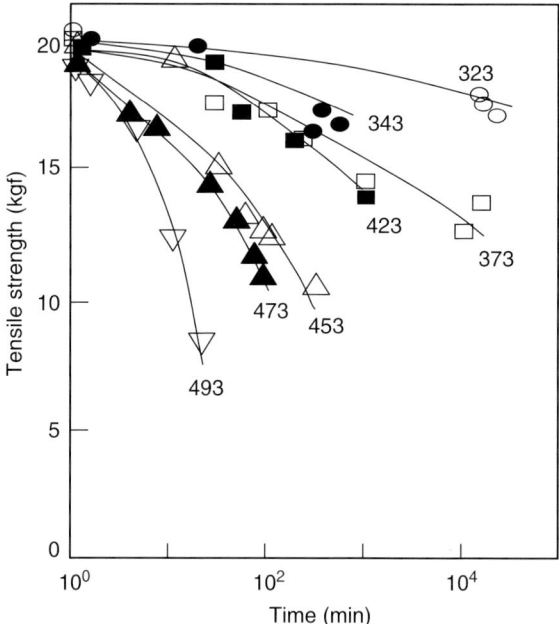

Figure 8.5 Relationships between tensile strength and annealing time; numerals in the figure are the annealing temperature, T_a (K).

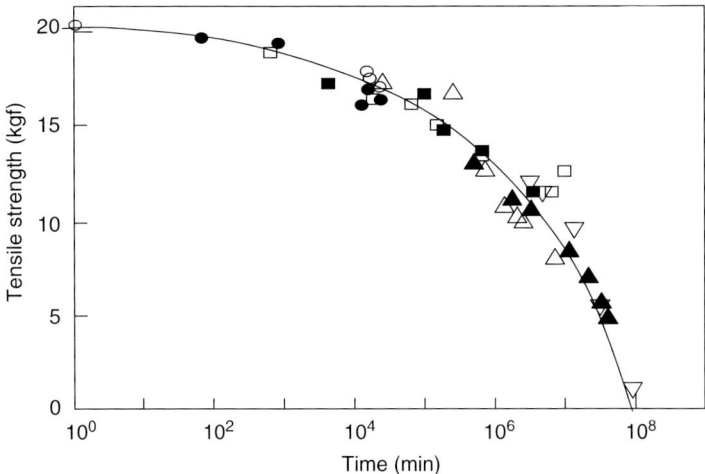

Figure 8.6 Master curve of tensile strength of the cellulose fabric; $T_R = 323$ K, ○: $T_a = 323$, ●: 343, □: 373, ■: 423, △: 453, ▲: 473, and ▽: 493 K.

As seen from the figure, the tensile strength gradually decreased at 323 K. With increasing temperature, the tensile strength decreased rapidly. Similar trends were observed for the values of elongation at break and Young's modulus.

Curves obtained at various annealing temperatures other than 323 K were horizontally shifted and a smooth master curve was obtained (Figure 8.6). The William–Landau–Ferry (WLF) equation is not applicable for crystalline polymers such as cellulose having no molten state. By assuming a simple Arrhenius-type Equation (8.12), the activation energy (ΔE) could be calculated [20].

$$\log \alpha_T = \frac{\Delta E}{2.3R} \left(\frac{1}{T} - \frac{1}{T_R} \right) \tag{8.12}$$

where α_T is the shift factor, T is the annealing temperature, T_R is the reference temperature (323 K in this case), and R is the gas constant. The calculated $\Delta E \simeq 115$ kJ mol^{-1}.

The fact that a master curve can be established indicates that the loss of the tensile strength occurs in the same mechanism regardless of the annealing temperature. To predict the mechanical properties, the time interval to reach a predetermined tensile strength by each annealing temperature (T_a) was evaluated from Figure 8.5. The predetermined tensile strength of annealed samples as a percentage was calculated based on the tensile strength of the original sample. In this study, the time intervals to reach the 5%, 10%, 12%, 15%, and 20% decreases of the tensile strength were obtained at each annealing temperature. The obtained values were plotted against the reciprocal annealing temperature. Figure 8.7 shows the representative relationship between the time interval and the reciprocal annealing temperature at 10% decrease of the tensile strength. The fitted line was extrapolated to 298 K ($T_a^{-1} \times 10^3 \simeq 3.36$ K^{-1}) and the assumed time interval at which the tensile strength of the cellulosic fabric decreases by 10% was obtained. A similar procedure was repeated to obtain

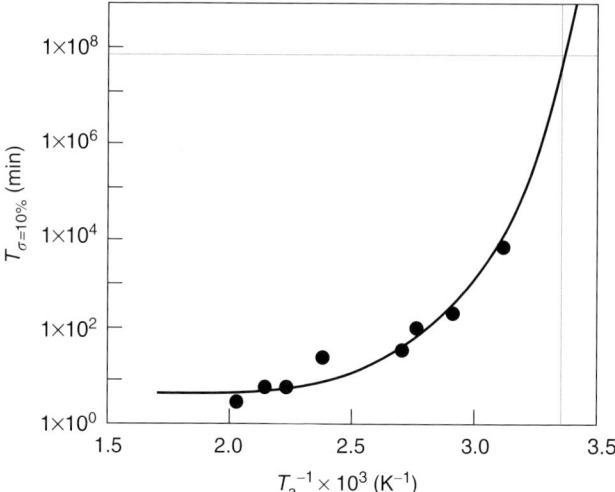

Figure 8.7 Relationship between time at 10% decrease of tensile strength ($t_\sigma = 10\%$) and reciprocal absolute annealing temperature.

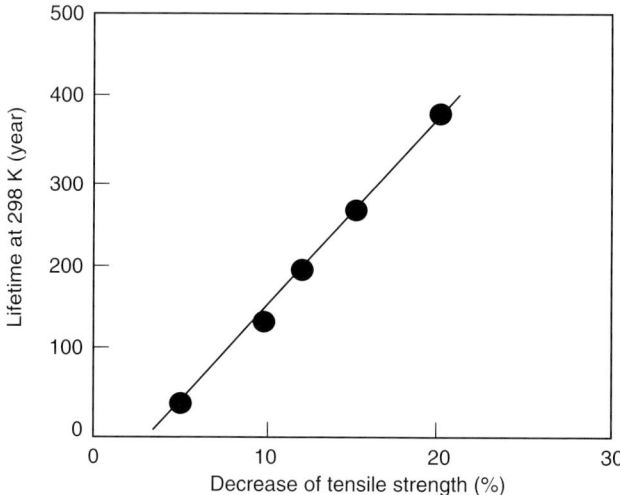

Figure 8.8 Relationship between the lifetime at 298 K and the decrease of tensile strength.

the assumed/predicted time intervals at which the tensile strength decreases by 5%, 12%, 15%, and 20% at 298 K (25°C).

Figure 8.8 shows the relationship between the decrease of tensile strength and the predicted time interval at room temperature (25°C, or 298 K). From this figure, one can predict that it would take 137 years for the tensile strength of the cellulose fabric to decrease by 10%, if no other factors are at play.

References

1. Graminski, W.L.; Parks, E.J.; Totoh, E.E., The Effect of Temperature and Moisture on the Accelerated Aging of Paper. In Eby, R.K. (ed.); *Durability of Macromolecular Materials*; ACS Symp. Ser., 95; ACS; Washington, D.C., 1979; pp. 341–354.
2. Amdradu, A.L., Assessment of Biodegradability in Organic Polymers. In Hamid, S.H. (ed.); *Handbook of Polymer Degradation*, 2nd edn; Marcel Dekker; New York, 2000; pp. 441–459.
3. Krevelen, D.W., *Properties of Polymers, their Correlation with Chemical Structure, their Numerical Estimation and Prediction from Additive Group Contributions*, 3rd edn; Elsevier; Amsterdam, 1997; pp. 397–402.
4. Struik, L.C.E., *Physical Aging in Amorphous Polymers and Other Materials*; Elsevier; Amsterdam, 1978.
5. Satoto, R.; Subowo, W.S; Yusiasih; R.; Takane, Y.; Watanabe, Y.; Hatakeyama, T., Weatherbility of High Density Polyethylene in Different Latitudes; *Polym. Degrad. Stab.* 1997, 56, 275–279.
6. Hatakeyama, T.; Hatakeyama, H., *Thermal Properties of Green Polymers and Biocomposites*; Kluwer Academics; Dordlecht, 2004.
7. Hatakeyama, T.; Quinn, F.X., *Thermal Analysis, Fundamentals and Applications to Polymer Science*, 2nd edn; John Wiley; Chichester, 1999; p. 65.
8. Ozawa, T., A New Method of Analyzing Thermogravimetric Data; *Bull. Chem. Soc. Japan* 1965, 38, 1881–1886.
9. Flynn, J.H.; Wall, L.A.; Quick, A., Direct Method for the Determination of Activation Energy from Thermogravimetric Data; *Polym. Lett.* 1966, 4, 323–328.
10. Ozawa, T., Non-isothermal Kinetics (1) Singly Elementary Process; *Netsu Sokutei (J. Soc. Calorimetry Thermal Anal., Japan)* 2004, 31, 125–132.
11. Hirose, S.; Hatakeyama, H., A Kinetic Study on Lignin Pyrolysis Using the Integral Method; *Mokuzai Gakkiashi (J. Wood Sci., Japan)* 1986, 32, 621–625.
12. Tan, J.K.; Kitano, T.; Hatakeyama, T., Crystallization of Carbon Fibre Reinforced Polypropylene; *J. Mater. Sci.* 1990, 25, 3880–3384.
13. Nguyen, T.; Zavarin, E.; Barrall II, E.M., Thermal Analysis of Lingocellulosic Materials. Part I. Unmodified Materials; *J. Macromol. Chem.* 1981, C20(1), 1–65.
14. Nguyen, T.; Zavarian, E.; Barrall II, E.M., Thermal Analysis of Lingocellulosic Materials. Part II. Modified Materials; *Macromol. Sci. Rev. Macromol. Chem.* 1981, C21, 1–60.
15. Hatakeyama T.; Zhenhai, L., *Handbook of Thermal Analysis*; John Wiley; Chichester, 1998; p. 143.
16. Soldi, V., Stability and Degradation of Polysaccharides. In Dumitriu, S. (ed.); *Polysaccharides, Structural Diversity and Functional Versatility*, 2nd edn; Marcel Dekker; New York, 2005; pp. 41–68.
17. Hirata, T., Thermal Degradation. In Cellulose Soc., (ed.); *Cellulose Handbook*; Asakura Publisher; Tokyo, 2000; p. 188.
18. Doyle, C.D., Kinetic Analysis of Thermogravimetric Data; *J. Appl. Polym. Sci.* 1961, 5, 285–292.
19. Doyle, C.D., Estimating Isothermal Life from Thermogravimetric Data; *J. Appl. Polym. Sci.* 1962, 6, 639–642.
20. Krevelen, D.W., *Properties of Polymers, Their Correlation with Chemical Structure; Their Numerical Estimation and Prediction from Additive Group Contributions*, 3rd edn; Elsevier; Amsterdam, 1997; p. 406.

Chapter 9

Recent Advances in the Isolation and Analysis of Lignins and Lignin–Carbohydrate Complexes

Mikhail Yurievich Balakshin, Ewellyn Augsten Capanema, and Hou-min Chang

Abstract

This review discusses advantages and limitations of various methods for the isolation and characterization of native and technical lignins and lignin–carbohydrate complexes (LCCs) based on our recent studies as well as achievements of other groups in the field. The use of high-resolution correlation and quantitative nuclear magnetic resonance (NMR) techniques in the analysis of lignocellulosics provides a large amount of information on a high variety of structures in relatively short experimental time. The NMR data should be complemented by appropriate wet chemistry techniques that are focused on specific substructures. In addition, coupling selective modification of lignin with high-resolution NMR analysis gives valuable information, especially in the studies of new and underinvestigated lignin and LCC preparations.

9.1 Introduction

Understanding the mechanisms of various processes for the chemical utilization of lignocellulosics requires good knowledge on the lignin structure in the starting raw material and its transformation during the processing. In addition to mechanistic studies, the structures of technical lignins are of primary importance for their utilization as chemical feedstock. Increasing interest in biomass utilization necessitates the analysis of large amounts of lignocellulosics materials (soft- and hardwoods, nonwoody plants, agricultural and municipal wastes, etc.) and the understanding of their transformations in various biorefining processes. These demands require efficient and informative methodology for characterization of lignin in lignocellulosics of different types.

Lignin is an irregular aromatic biopolymer composed by phenylpropane (or C_9) units of the *p*-hydroxyphenyl (H), guaiacyl (G), and syringyl (S) types (Figure 9.3). These monomeric units are linked in lignin macromolecule by various ether and C—C bonds [1–3]. According to current understanding, almost all lignin in softwood and softwood pulps is linked to polysaccharides, mainly hemicelluloses [4]. The occurrence of stable lignin–carbohydrate (LC) bonds is one of the main reasons preventing selective separation of the wood components in biorefining processes. Linkages between lignin and carbohydrates also create significant problems in the selective isolation of lignin preparations/samples from lignocellulosics.

In spite of extensive works on lignin and LCC chemical structures our knowledge in this field is still insufficient. Further progress requires development of new analytical methodologies. This chapter summarizes our recent studies, along with relevant achievements of other groups, on the isolation and characterization of lignin and LCC from wood and chemical pulps. On the basis of these data, optimal methodology to assess lignin and LCC structure is discussed.

9.2 Isolation of lignin and LCC preparations/samples from wood and pulps

Most methods for lignin analysis require the isolation of lignin preparations/samples from lignocellulosic materials. The main problems in lignin isolation are associated with the complex structure of the cell wall and the interaction of its components. As almost all lignin is linked to polysaccharides, it is not possible to isolate pure lignin without any chemical cleavage. An appropriate isolation procedure should produce a representative lignin preparation and minimize structural changes during isolation. Generally, two approaches are used to isolate lignin from lignocellulosics: acidolysis methods [5–10] and extraction of lignin after ball milling and/or enzymatic hydrolysis of carbohydrates [10–15]. Although the acidolytic methods are quite fast and produce lignin preparations of high purity, the acidic conditions employed trigger some changes in lignin structures [5,7–10,16]. Apparently, the stability of most LC linkages (probably of the ether type) is higher than that of certain lignin linkages. Therefore, the use of acidolysis to obtain pure lignin preparations results in not only extensive cleavage of LC bonds, but also cleavage of certain amounts of lignin–lignin ether bonds. A decrease in the acidity of dioxane solution to avoid degradation of major lignin structures results in preparations with a high amount of carbohydrates [9,17]. Isolation of lignins by enzymatic hydrolysis of polysaccharides is accompanied by minimal structural changes, but the resulting lignin preparations contain 5–15% carbohydrates [10–16,18,19] that complicate the analysis of the lignin samples. However, application of advanced analytical methods such as multidimensional heteronuclear NMR techniques overcomes complications due to the presence of carbohydrates after enzymatic hydrolysis [19–22]. In addition, LC linkages can be investigated [10,16,19,23,24].

9.2.1 Isolation of lignin from wood

Various methods for the isolation of lignin and LCC preparations/samples from wood are summarized in Figure 9.1. Enzymatic hydrolysis cannot be directly applied to wood, and a preceding efficient milling of the wood sample is required. During ball milling significant degradation of polysaccharides occurs along with a certain breakdown of the lignin macromolecule. Extraction of the milled wood with 96% dioxane produces

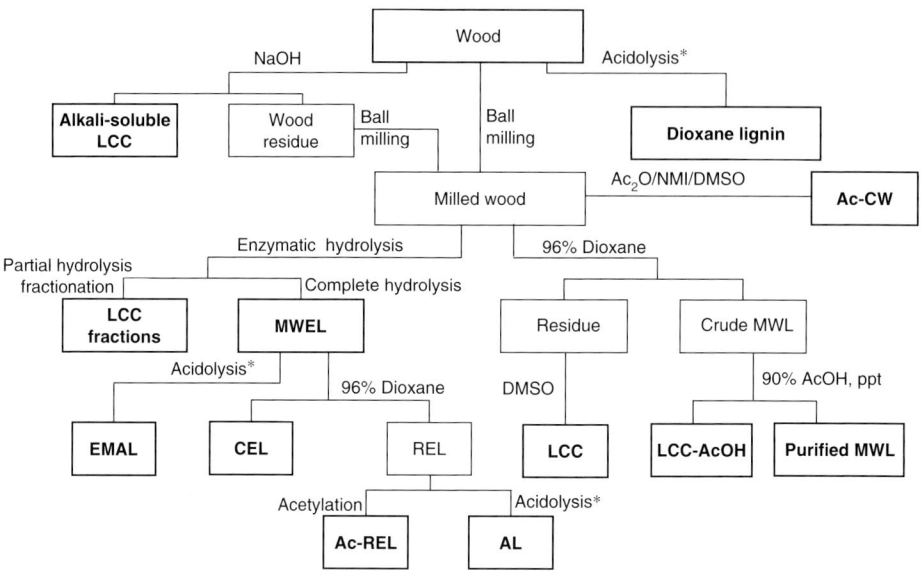

Figure 9.1 Isolation of lignin and LCC preparations/samples from wood. *Acidolysis is usually carried out in aqueous dioxane (85–90%) with addition of HCl (0.05–0.2 M) under reflux.

milled wood lignin (MWL) [11]. This is currently the most common procedure for the isolation of lignin from wood. There are two main issues/questions when transferring knowledge on the structure of MWL to lignin *in situ*: (1) how representative of the whole wood lignin MWL preparations with typical yields of 20–30% are, and (2) how significant lignin degradation is during the milling.

As the yield of MWL has been considered insufficient to reproduce the whole lignin in wood, significant efforts were made recently to increase the yield of the isolated lignins. Lu and Ralph [25] suggested a method that results in complete solubilization of wood by dissolving acetylated milled wood in N-methylimidazole/dimethyl sulfoxide (NMI/DMSO) to obtain a preparation called "acetylated cell walls" (Ac-CW). This allowed the characterization of wood components by solution-state NMR that provides much higher resolution than solid-state NMR.

Enzymatic hydrolysis of carbohydrates in the milled wood followed by dioxane extraction of the residue produces cellulolytic enzyme lignin (CEL) [12]. Very recently Hu *et al.* [26] were able to obtain pine CEL preparations with high yields, up to 86% (calculated on true lignin). The material in the dioxane insoluble residue was DMSO-soluble after acetylation (Figure 9.1). Moreover, all the lignin has been isolated as CEL preparation from birch wood [27]. Thus, the whole wood lignin was analyzed with high-resolution NMR methods. The much lower carbohydrate content made this analysis easier than the NMR analysis of the Ac-CW preparation.

Argyropoulos and coworkers [17,28] suggested a more complex three-stage isolation procedure involving ball milling, enzymatic hydrolysis followed by acidolysis of the residue to obtain enzymatic mild acidolysis lignin (EMAL) (Figure 9.1). This method provides higher yields of lignin preparations at the same milling time than the CEL protocol. However, the maximal yield obtained (<75%) [17,29] was lower than the yield of CEL obtained without acidolysis (86–100%) [26,27]. A modification of the EMAL approach was used [22]. The authors extracted lignin released after the enzymatic hydrolysis (equivalent to CEL) and used only the insoluble residue for acidolysis (Figure 9.1) to avoid the acidic degradation of the CEL fraction.

As all the methods discussed require ball milling of wood, it is important to understand how significant lignin is degraded during the milling. It is known that some changes in the lignin structure occur during ball milling, particularly the increase of the content of carbonyl and phenolic OH groups and the decrease of molecular mass [11,12]. Recent studies showed that ball milling did not cause changes in aromatic ring of lignin units, but results in some cleavage of β-O-4 structures in the *whole wood lignin* [26,30,31]. However, the amount of β-O-4 units detected by ozonolysis in the pine *MWL and CEL preparations* in the yield range of 20–65% was practically the same [26]. Moreover, NMR and FTIR (Fourier transform infrared spectroscopy) studies demonstrated that the chemical structures of MWL and CEL preparations isolated in high yields (up to 65%) were fairly similar to those isolated in the common 20–30% yields [26,32,33]. Further increase in the yield of pine CEL >65% resulted in a decrease in the amount of β-O-4 units [26], likely because of the accumulation of degradation products in the CEL preparation. Therefore, the classical MWL isolated from softwoods with the yield of 20–30% is as much representative, in terms of morphological origin, as high-yield MWL and CEL preparations [26,32] (but might contain somewhat less degradation products).

Various LCC preparations/samples can be obtained by methods similar to those used for lignin isolation. Extraction of the wood residue with DMSO after extraction of MWL produces Bjorkman's LCC [34]. Other solvents and water were also used to extract LCC from the residue obtained after MWL isolation [35,36]. Partial enzymatic hydrolysis of the milled wood followed by wet chemistry separation allows the isolation of various LCC fractions from wood and pulps [4]. Complete enzymatic hydrolysis of milled wood results in a preparation called milled wood enzymatic lignin (MWEL) [37] (Figure 9.1). (CEL is the dioxane-soluble fraction of MWEL.) A preparation enriched in LCC fragments, called LCC-AcOH, can be obtained during purification of the crude MWL with 90% AcOH [38]. Similar preparation (LCC-W) was obtained earlier [39] using a more complex protocol. Therefore, most of lignin and LCC preparations obtained give information on both lignin and LCC structure, and from this point of view can be defined as carbohydrate-rich LCC (Bjorkman's LCC and similar ones) and lignin-rich LCC (MWEL, CEL).

On the basis of our current knowledge, the advantages and limitations of various lignin and LCC preparations can be discussed. Ideally, Ac-CW protocol allows the structural characterization of all the wood components using a single preparation. However, identification and especially quantification of various structures is a challenge, even using very advanced NMR methods, as the composition of the preparation is very complex [25]. In addition, even the Ac-CW protocol cannot be considered as nondestructive as 13–27% of β-O-4 linkages in wood lignin is degraded depending on milling intensity (Y. Matsumoto, personal communication). As the Ac-CW preparation contains the whole wood lignin, it also contains the degraded lignin fragments. No important differences have been detected in the chemical structure of lignin in MWL (with yields above *c*.20%), CEL, and EMAL [17,22,26,28,40] preparations, except the degradation of some acid-labile lignin units in EMAL [22]. (Some structural differences observed [29] were very likely due to very low yields of the MWL used for comparison.) As purified MWL contains much less carbohydrates than the corresponding CEL preparation, it is more suitable for precise analysis of the lignin structure. CEL can be very useful for the analysis of LC linkages. MWEL is very representative as it contains most of the LC linkages of the original wood [37]. However, in contrast to CEL, it is not completely soluble. This limits its analysis with high-resolution spectroscopic methods. Optimal application of EMAL preparation is still to be found. Much higher carbohydrate content in the EMAL than that in purified MWL results in less accurate quantitative lignin analysis. On the other hand, as LC bonds are partially degraded during acidolysis, the use of EMAL preparation for the analysis of LCC is less suitable than that of CEL. Moreover, the structure of MWL and CEL lignin is independent of the milling intensity and apparatus used [26,27,31–33]. In contrast, the composition of EMAL is significantly affected by milling intensity [17]. This implies that intensive milling affects the subsequent acidolysis stage of the EMAL protocol (but not dioxane extraction of MWL!), triggering oxidation and condensation reactions [17]. Therefore, the optimized EMAL method requires very mild milling resulting in very long experimental time. For example, to obtain ~45% yield, a milling time of almost 1 month is required [17]. In contrast, intensive ball milling can produce ~30% yield of MWL (or 60% yield of CEL) within 1–2 h of milling [41].

The major problem in detailed analysis of most LCC preparations is the low frequency of LC bonds. The frequency of LC bonds in wood is only ~0.03 per C_9-unit [37]. From this point of view LCC-AcOH preparation is very promising as it contains large amounts of LC linkages [38]. In spite of a lower yield, as compared to Bjorkman's and Koshijima's LCCs [34–36], this preparation is not less representative as the frequencies of LC bonds is much higher. It has been shown [38] that LCC-AcOH preparation qualitatively represents both the middle lamellae and secondary wall regions of the cell wall. However, whether the amounts of various LC linkages in the LCC-AcOH are proportional to their amounts in the whole cell wall is still under investigation.

In contrast to softwood, the composition of MWL preparations isolated from *hardwood* using the common protocols is dependent on the lignin yield and usually contains a significant amount of carbohydrates (5–10%) [27,42,43]. However, MWL isolated from *Eucalyptus globulus* and *Eucalyptus grandis* after alkali pre-extraction of the wood sample under mild conditions (to remove tannins) contained very small amounts of sugars, even without purification common for MWL [16,44,45]. This finding implies that in eucalyptus a significant amount of carbohydrates are attached to lignin via alkali-labile linkages, probably of the ester type [44]. Moreover, comparison of *E. grandis* MWL preparations with the yields of 55% and 25%, respectively, showed very little differences [33]. The NaOH pre-extraction of wood resulted in the loss of a part of lignin enriched in H- and G-units and phenolic OH as compared to the bulk lignin; significant differences in the amounts of inter-unit linkages were also observed [16,44]. The alkali-soluble lignin contained ~20% of carbohydrates and therefore was referred to as LCC [16,44] (Figure 9.1). Similar results were obtained with other hardwood species [27]. Thus, it could be suggested that hardwood native lignins consist of two fractions: an "easy soluble lignin" and the bulk ones. The differences in the structure of the traditional MWL preparations as milling proceeds are therefore due to variations in the proportion of the "easy soluble" and "bulk" lignin fractions in MWL. Pre-extraction of the "easy soluble lignin" from wood with NaOH under mild conditions (Figure 9.1) allows the separation of these fractions and results in the independence of the structure of MWL isolated from the pre-extracted wood on its yield [27,33].

Our very recent results [27] have shown that the structure of CEL with yields above ~75% is somewhat different from that of the "bulk" MWL or CEL, showing a noticeable decrease in S/G ratio, in agreement with earlier work on isolation of dioxane lignin from aspen wood [5]. Therefore, it appears that the structure of hardwood native lignins is rather heterogeneous. "Bulk lignin" (equivalent to MWL isolated from NaOH pre-extracted wood) represents ~50–60% of the total hardwood lignin. The "easy extractable lignin" (equivalent to NaOH-soluble lignin) and the "residual" 25–30% of lignin are somewhat different from the MWL. Comprehensive studies in this area are in progress in our laboratory.

9.2.2 Isolation of lignin from pulps

Similar to the isolation of lignin from wood, isolation of lignin from pulps can be performed using acidolysis or enzymatic hydrolysis. Ball milling is not needed for the isolation of lignin from most chemical pulps, and mechanical degradation of lignin can be therefore avoided.

Isolation of enzymatic residual lignins from softwood kraft pulps is a well-established procedure producing lignin preparations with high yields and relatively low enzyme impurities [10,13–15,19]. In contrast, significant problems for the isolation of enzymatic residual lignins from hardwood kraft pulps [46] and semibleached pulps have been reported [15]. Low yields (25–30%) make these preparations nonrepresentative for the whole pulp lignin. Very large amounts of protein contaminants (15–35%) result in significant problems in lignin analysis, particularly with spectroscopic methods [46]. Recently we were able to dramatically decrease protein contaminations in hardwood residual lignin preparations (to 1–6% in non-purified lignins) and to increase the yields to 40–50% by applying cellulase preparations with high activity and by optimizing the enzyme charge [16,18,23]. The optimal enzyme charge was the highest for the *E. globulus* pulp and the lowest for the birch pulp with similar lignin content [16,18].

It is important to mention that the low yields of hardwood residual lignins (which decreases with decrease in lignin content in pulps) indicate that a significant portion of lignin is soluble in aqueous solution after enzymatic hydrolysis of the pulp even after its acidification [18]. Apparently, this lignin consists of low-molecular-mass fragments, which should be linked to carbohydrates. The frequency of LC bonds (per lignin unit) in this fraction should be much higher than that in the fraction of higher molecular mass residual lignin. Therefore, it is of interest to isolate the water-soluble lignin/LCC fraction and characterize it. This work is currently in progress in our laboratory.

Similar to the isolation of lignin from wood, the protocol combining enzymatic hydrolysis and acidolysis (EAL) inherits shortcomings of both methods [9]. An increase in the purity of EAL preparations was very insignificant as compared to the purity of the enzymatic lignin. At the same time the acidolytic step triggers changes in the lignin structure (concentration of HCl was higher than that for the isolation of EMAL from wood). Improvement in EAL lignin solubility after the acidolysis step was obviously due to degradation of lignin in acidic dioxane as an increase in lignin purity was very insignificant [49]. An optimized protocol for the isolation of enzymatic residual lignin produces lignin preparations well soluble in DMSO [18,19], a typical solvent for NMR analysis.

9.3 Lignin and LCC analysis

9.3.1 Methods in lignin and LCC analysis

The majority of analytical methods in lignin chemistry can be divided into wet chemistry methods and spectroscopic techniques. Most of the wet chemistry methods were developed a long time ago and can be considered as traditional methods in lignin chemistry. They can be subdivided into elemental and functional group analysis (OMe, total OH, phenolic OH, carbonyl group, carboxyl groups) and degradation methods for inter-unit linkages analysis.

In contrast to wet chemistry, dramatic progress has been achieved recently in spectroscopic and spectrometric methods. Among them, multi-dimensional and quantitative ^{13}C NMR techniques are the most informative methods in lignin analysis [48,49]. Multidimensional heteronuclear NMR methods offer much better separation of signals in complex lignin preparations than one-dimensional (1D) NMR and also provide more reliable interpretation of the signals. However, routine two-dimensional (2D) sequences are not quantitative. Some promising attempts to develop a quantitative heteronuclear single quantum correlation (HSQC) sequences have been made [50,51], but a robust quantitative HSQC technique is not available at the present time. Therefore, a combination of correlation NMR methods with quantitative ^{13}C NMR is the best approach as the former allows the identification of various lignin moieties in a given preparation and thus significantly helps in their reliable quantification by ^{13}C NMR [7,45,52]. Recently, the methods for the quantification of various lignin moieties used in lignin analysis with ^{13}C NMR were reviewed [45,52]. The most reliable methods were chosen and the limitations of other methods were explained, along with a few new suggested approaches. This allows building up a comprehensive algorithm for the quantification of a large variety of lignin units (Table 9.1) using a combination of ^{13}C NMR of acetylated and nonmodified lignin [45,52].

Recently, application of an internal standard for the quantification of lignin by ^{13}C NMR has been suggested [53], similar to ^{1}H [54] and ^{31}P NMR [17]. Although this approach can be a useful addition in some particular cases, under most circumstances it is less informative and accurate than the traditional method for the calibration of ^{13}C spectra of lignin using the resonance of the aromatic carbons, as discussed earlier [55]. Moreover, while good results were obtained with model compounds [53], application of the internal standard for the quantification of spruce MWL and EMAL preparations [28] resulted in totally abnormal data. For example, conversion of the data obtained in mmol g^{-1} to structures/C9-unit indicates that 70% of G-units have a substituent at C-2 position of the aromatic ring as compared to only trace amounts of these structures detected by other methods [1,2,48,52]. The integral assigned to the sum of β-O-4 and dibenzodioxocin structures (pinoresinol and phenyl coumarane structures must be also included in this value [48,52]) was very low, ~25/100 C9 units [28]. This indicates that a number of issues must be addressed before this approach [28] can be applied to lignin quantification.

The major advantages of high-resolution NMR spectroscopy over wet chemistry methods are:

- observation of the lignin as a whole, not only specific moieties;
- analysis of a large variety of lignin moieties in a relatively short experimental time and with relatively small amounts of sample;

Table 9.1 Amounts of various moieties in soft- and hardwood MWLs (per 100 C₉-units).

Moieties	Spruce (wet chemistry) [1,2]	Spruce (quantitative 2D NMR) [47]	Spruce (^{13}C NMR) [49]	Birch (wet chemistry) [1,2]	Eucalyptus grandis (NMR, PO) [42]
β-O-4/α-OH (**1**)	34	36	36		55
β-O-4/α-CO (**4**)		<1	2		2
Phenylcoumaran (**5**)	9–12	12	9	6	3
Pino/syringoresinol (**6**)	2–4	2.5	2	3	3
DBDO (**7**)		5	7		<1
β-1/α-OH (**9**)	7–9	2	1	7	1
Spirodienone (**8**)					
Guaiacyl		2			1
Syringyl			2		4
Secoisolaricerisinol (**10**)		1	1		nd
Ar–CH=CH–CH₂OH (**18**)	4	1	2		<1
Ar–CH₂–CH₂–CH₂OH (**17**)	4	1	2		<1
Ar–CO–CH₂–CH₂OH (**16**)		3	2		<1
Ar–CH=CH–CHO (**19**)	4	5	4		1
Ar–CHO (**13**)		3	5		3
Ar–COOH (**14**)			2		2
Ar–CH₂OH (**15**)			nd		2
β-O-4 total	45–50		45[a]	60	61
α-O-4 non-cyclic	6–8 (11)	nd	nd	6–8	nd
Alkyl-O-alkyl total[b]			39–45		–
α-O-alkyl total[b]			15–21		
γ-O-alkyl total[b]			24		23
5–5′ total[b] (**12**)	19–22	22	24–27	9	6
Etherified[b]			19		
Non-etherified[b]			5–8		
4-O-5′ (**11**)					
Guaiacyl	4–7			1	3
Syringyl				5.5	6
6(2)-condensed					
Guaiacyl	3–4			1	3
Syringyl				1	3
OMe	95		95		160
Total OH	131		138		144
Aliphatic OH	100		107		125
Primary	75		68		70
Secondary	25		39		55
Benzylic	16–25		38		54
Phenolic OH	20–31		31	20	19
Total carbonyl	20	15	21		17
Aldehyde			9		4
Ketone			10		8
α-CO			10		8
Nonconjugated			5		3
COOH	2–4		5		5
Degree of condensation	45		38		21
S/G/H				50:50	62:36:2

See Figure 9.3 for structures of **1**, **4–10**, and **13–19**.
nd: not detected.

[a] Sum of identified moieties.
[b] Amount of C₉-units involved; in a case of symmetric moieties the amount of these structures will be half of the value present.

- detailed characterization of side-chain moieties; for example, different types of β-O-4 moieties (see structures **1–4, 7, 21** in Figure 9.3) can be distinguished;
- direct estimation of the amounts of lignin structural moieties;
- generation of a comprehensive fingerprint for lignin and LCC molecules with ^{13}C NMR, even when precise calculations are not possible.

The main disadvantage of high-resolution spectroscopic and spectrometric methods is the necessity to isolate soluble-lignin preparations. In contrast, the major advantage of degradation techniques is the possibility of analyzing lignin without its isolation. The relative accuracy of degradation techniques is high. However, most of the methods require the use of correction coefficients to achieve the absolute values. Therefore, the degradation techniques allow high precision for *relative* comparison of different lignins, but quantitative ^{13}C NMR is preferable to access the *absolute* amounts of lignin subunits. In contrast to ^{13}C NMR, wet chemistry methods are focused on specific lignin moieties, and a combination of various wet chemistry techniques is needed to obtain comprehensive information.

Some of the common degradation methods are nitrobenzene oxidation (NBO), thioacidolysis (TA), ozonolysis, and permanganate oxidation (PO). Derivatization followed by reductive cleavage (DFRC) method has been recently developed [56] as an alternative to the TA method for the estimation of β-O-4 moieties. Modified DFRC and TA methods have been suggested to evaluate various types of β-O-4 units [30,57]. However, DFRC usually gives significantly lower yields of the reaction products than TA. The reason is incomplete cleavage of β-O-4 units in softwood lignin [57], although the corresponding lignin model compounds reacted completely [56]. As the consequence, DFRC-^{31}P NMR method gave a much lower amount of β-O-4 units in softwood lignins (0.25–0.30/C9-unit) [17,58] than other methods (see Table 9.1). In contrast, TA enables a complete cleavage of β-O-4 units in lignin [57,59], and therefore, it is more suitable for quantitative lignin analysis. It is noteworthy that the amount of β-O-4 units in eucalyptus lignins obtained by DFRC-^{31}P NMR method [29] was very similar to the values obtained by other methods [7] (see also Table 9.1). Therefore, the low value obtained for the softwood lignin could be due to the resistance of condensed β-O-4 lignin moieties to cleavage under the conditions of DFRC protocol. This suggestion is in good agreement with the observation that a very low amount of phenolic 5–5′ units were released after DFRC treatment of softwood lignin [60].

Taking into account the advantages and limitations of the aforementioned methods, the best strategy is to combine NMR analysis with appropriate wet chemistry techniques [7,26]. For example, structural studies using NMR spectroscopy and PO were very informative [7,45]. The former provides detailed information on the side-chain structure and the latter produces data on various substituents on the aromatic ring. Another example of the efficient combination of NMR and wet chemistry methods is a correlation between the relative values obtained by degradation methods with absolute values obtained for the same samples with NMR techniques. For example, a correlation between syringaldehyde/vanillin (SA/V) numbers obtained by NBO and S/G values obtained by ^{13}C NMR has been established for various hardwood MWLs [27]. This correlation allows calculation of S/G (absolute) ratio from SA/V numbers obtained by NBO analysis of hardwood lignin *in situ*, without isolation.

9.3.2 Coupling of wet chemistry and NMR methods

Improved analysis can be achieved by a coupling of selective wet chemistry methods with NMR spectroscopy. For example, analysis of lignin degraded by TA with ^{31}P NMR allows significant improvement in the resolution of ^{31}P NMR spectra and semiquantitative analysis of β-5, 5–5′, and 4-O-5 condensed lignin moieties [59].

The method of selective incorporation of ^{13}C isotopes into specific positions of lignin followed by ^{13}C and 2D NMR studies, developed by Terashima and coworkers, appreciably increases the capacity of NMR spectroscopy in lignin analysis, both high-resolution and solid-state methods [61,62]. Unfortunately, it is a time-consuming and expensive method, and high experimental skill is required to produce labeled lignin preparations.

Selective modification of specific lignin functional groups by reduction, acetylation, and methylation is often used in UV- and FTIR. Coupling these specific reactions with high-resolution NMR techniques was especially valuable [63,64]. In addition to the widely used acetylation of OH groups [45,48,49,52], other reactions such as $NaBH_4$ reduction (carbonyl groups), methylations with methanolic HCl (conjugated OH, COOH groups), CH_2N_2 (phenolic OH and COOH groups), and Me_2SO_4 (total OH), and oxidation of conjugated OH groups with DDQ (2,3-dichloro-5,6-dicyano-1,4-benzoquinone) were studied using comprehensive NMR methods [64]. The use of these additional reactions was very useful in the identification of unknown signals in 2D and ^{13}C NMR spectra of native and technical lignins. Moreover, selective modification of specific lignin functionalities allowed separations of signals, which were overlapped in the spectra of unmodified lignin preparations, and thus their accurate quantification by ^{13}C NMR.

9.3.3 Strategy in the analysis of lignocellulosics

Nowadays, research targets often require not only reliable and informative methodology but also an efficient one to enable comprehensive lignin analysis in a short time and sometimes using a minimal amount of sample [32]. There is not a perfect or universal method for obtaining the complete information on the lignin structure. An optimal approach for the analysis of lignin is dependent on the research objective, the nature of the lignocellulosics, the number of samples to be analyzed, the quantity of samples available, and so on. Figure 9.2 shows our proposal on a general analysis strategy for various situations.

If there are a large number of samples to be analyzed, it would be reasonable to screen the samples with *express methods* such as near infrared (NIR) spectroscopy and pyrolysis gas chromatography–mass spectrometry (Pyr-GC-MS) methods. If the samples are soluble in an NMR solvent, express semiquantitative NMR methods such as ^{1}H and ^{31}P can be applied. Express methods enable the analysis of a large number of samples in a short time, but provide only limited information. Nonetheless, the result of using such methods allows selection of samples with significant differences for more detailed analysis.

The combination of high-resolution NMR, preferably a ^{1}H–^{13}C HSQC or heteronuclear multiple quantum coherence (HMQC) 2D technique, and quantitative ^{13}C NMR [45,52] is a good example of *comprehensive methods* in lignin analysis. This approach provides a significant amount of information on various lignin moieties (functional groups, side-chain structures, G/S ratio, degree of condensation, semi-quantitative estimation of various

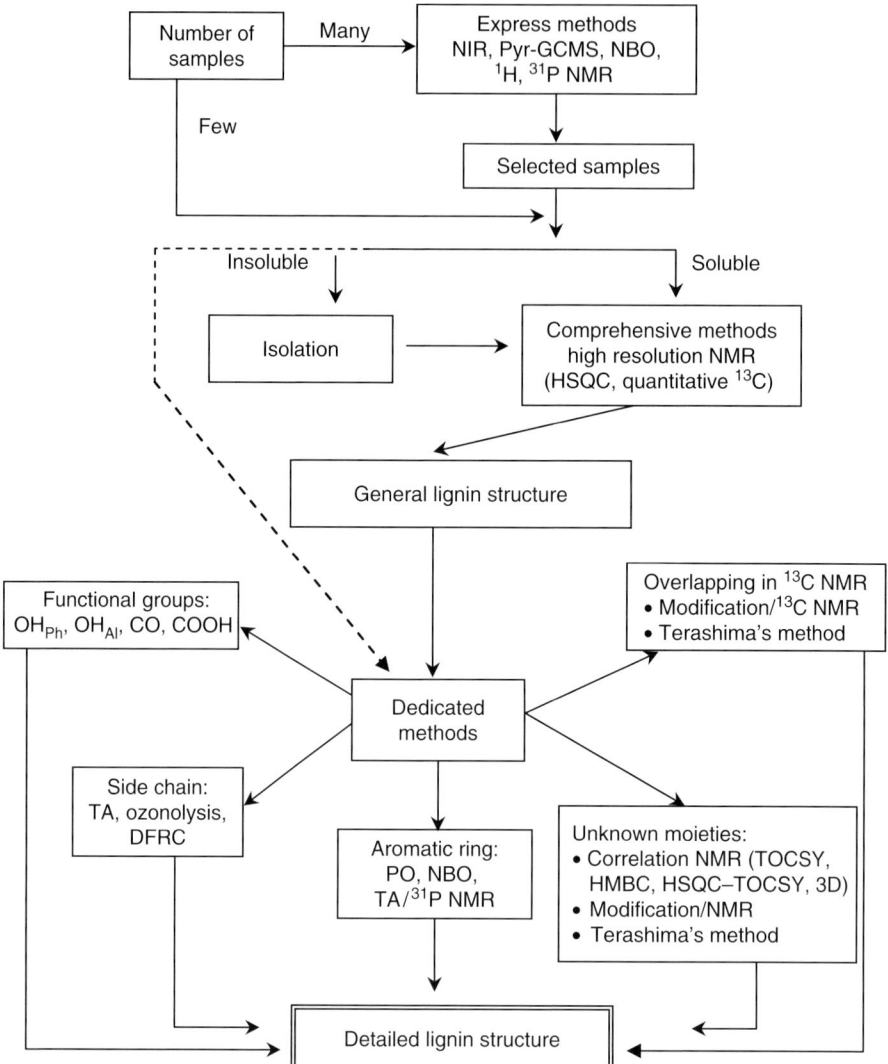

Figure 9.2 General strategy in lignin analysis.

condensed moieties, etc.) in an acceptable experimental time using a relatively small amount of samples (40–70 mg) [52]. The use of a relaxation reagent in an appropriate concentration allows a three- to fourfold decrease in experimental time for quantitative ^{13}C NMR without any damage to the spectra quality when DMSO-d_6 is used as the NMR solvent [52]. However, the use of the relaxation reagent significantly decreases the spectral resolution in acetone-d_6 and therefore is not recommended for this NMR solvent. Further decrease in experimental time and the amount of sample can be achieved by the use of a cryogenic probe [24,25]. As has been mentioned earlier, high-resolution NMR spectroscopy requires the isolation of soluble lignin preparation/sample that usually consumes an appreciable amount of time.

However, improvement of isolation methods such as ball milling, and/or the yields of lignin preparations significantly decreased the experimental time and the amount of sample (wood) required to obtain a sufficient amount of lignin. Moreover, the use of accelerated solvent extraction (ASE, Dionex) allows the extraction of MWL preparation in 15 min (instead of 4 days in classical MWL protocol) at 80°C without any change in lignin structure [41] resulting in tremendous decrease in the experimental time. Thus, combination of all the achievements in the isolation techniques and NMR spectroscopy has allowed comprehensive lignin analysis using 100–200 mg of wood (20–30 mg of lignin) within a few days (including isolation and analysis) [25,27,32,41].

Dedicated methods can be used to obtain additional information on specific features of the lignin structure and/or to verify previous information with another independent technique. A few examples are given in Figure 9.2. In addition, comparison of detailed data obtained for isolated lignin preparations with available results from the analysis of lignin without its isolation is valuable for understanding the structure of lignin *in situ*.

9.3.4 Current understanding of the structure of lignin

The structure and biosynthesis of lignin, including newly discovered moieties, were comprehensively discussed in a recent review [3]. However, the authors did not present much quantitative information, and therefore we will focus on this issue here. Table 9.1 summarizes our recent results [45,52], along with literature data, on the structure of soft- and hardwood MWLs. As spruce MWL is the most investigated lignin preparation, a few analytical approaches to its analysis are compared (Table 9.1). Birch lignin is presented as the most investigated hardwood lignin, whereas the data on *E. grandis* MWL are an example of the application of our recent approach. Recent findings in lignin chemistry are in general agreement with earlier suggested comprehensive models [1,2], however, some important exceptions have been indicated:

- The amounts of benzylic OH groups were significantly underestimated by a wet chemistry method [52]. The benzyl alcohol groups were detected by DDQ oxidation of premethylated lignin followed by UV detection of the α-carbonyl groups formed. Our studies on this reaction using the HMQC technique showed [64] that the reaction is not selective; the oxidation of α-OH groups is not complete, and some yet unidentified structures are formed in notable amounts in addition to the target α-carbonyl products. Therefore, NMR techniques give more reliable values (∼35/100 C_9-units) than the wet chemistry method (16–25/100 C_9-units).
- Modern NMR methods do not confirm the presence of non-cyclic α-O-4 moieties (structure **2**) (Figure 9.3); dibenzodioxocin (DBDO) moieties (structure **7**) (Figure 9.3) have been suggested instead [49,65]. The quantity of the dibenzodioxocin structures has not been determined very accurately yet. ^{13}C NMR apparently gives somewhat inflated values due to the incomplete resolution of DBDO signal [66]. Calculation of DBDO moieties from quantitative HSQC spectra [50] should be more accurate. However, the experimental conditions allowing the quantitative 2D analysis of lignin have not been completely established yet. The DFRC-^{31}P NMR approach [58,60] should be considered with care due to the inaccuracies discussed earlier [66].

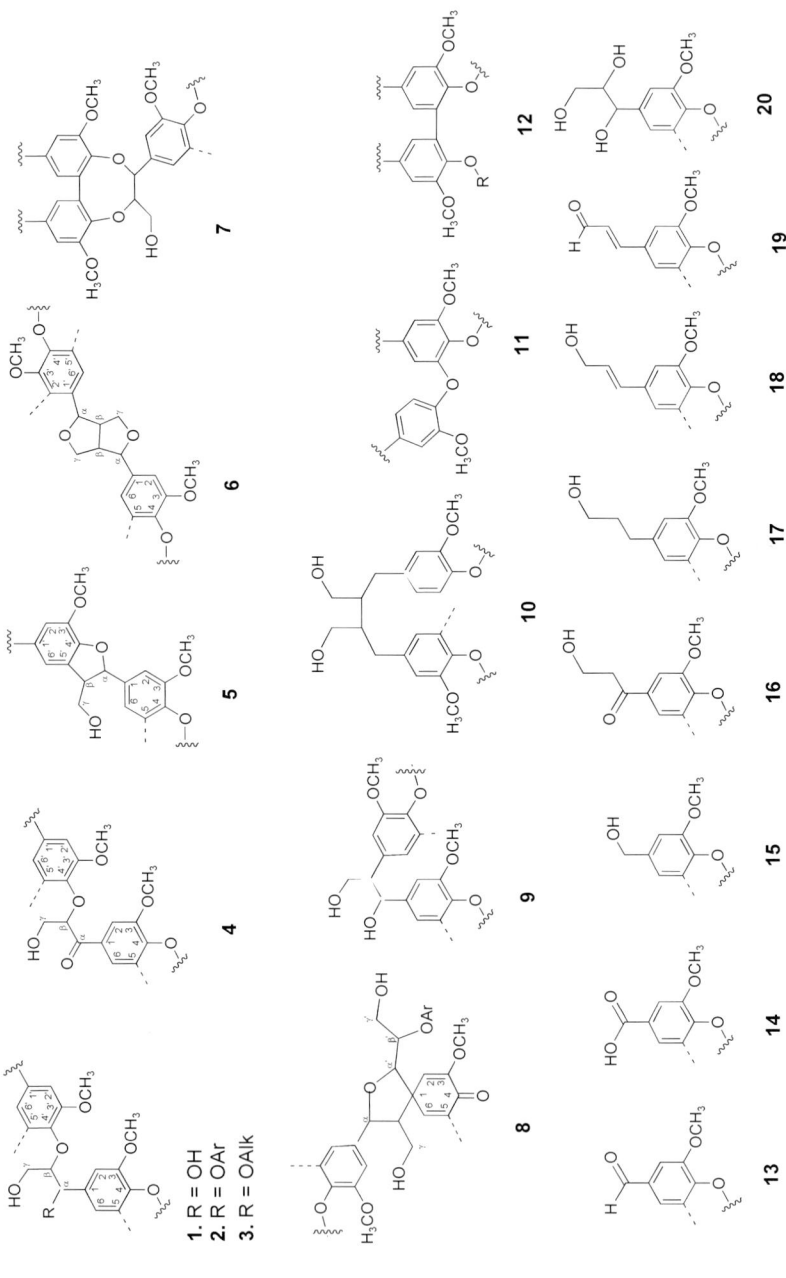

Figure 9.3 Structural moieties of MWL (**1–20**), kraft lignins (**21–30**), hydroxyphenyl (H), guaiacyl (G), and syringyl (S) units, and major LC bonds (**A–C**) as detected by high-resolution NMR techniques.

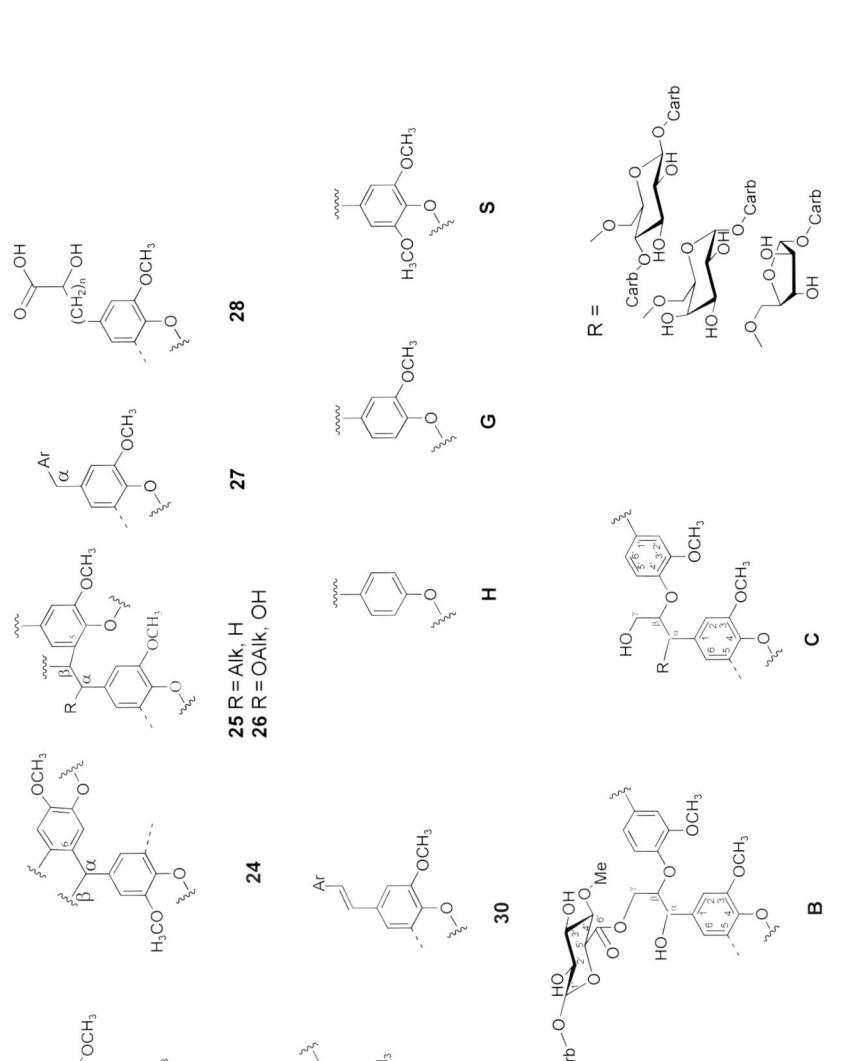

Figure 9.3 Continued.

- The amount of β-1/α-OH moieties (structure **9**) was significantly overestimated by degradation techniques; significant amount of β-1 units are of spirodienone type (structure **8**) (Figure 9.3) [52,67].
- Significant amounts of Alk-O-Alk moieties (Table 9.1) have been suggested based on the analysis of quantitative ^{13}C NMR spectra [45,52]. However, this suggestion and the exact structures of these moieties require further investigations.

Generally, the structure of lignin is very heterogeneous. Although most of the side-chain structures are β-O-4 units (45–50% for softwood and up to 65% for typical hardwoods), they are of different structural types (structures **1–4** and **7**; Figure 9.3). Various combinations of G- and S-units in these structures in hardwood lignins further increase the diversity. In addition, stereochemical isomers (E and Z, R and S) should be taken into account. A large variety of different moieties are present in lignins in small (1–5%) amounts (Figure 9.3, Table 9.1). Furthermore, it is very difficult to come to 100% balance when quantifying lignin structure in absolute values with quantitative NMR methods. All identified side-chain units make up ∼80–85%, and the exact structure of 15–20% lignin moieties is still unknown.

It is noteworthy to mention that the pictorial structures of lignin suggested [1,2] are very tentative as very little is known about the sequence of various lignin moieties in the macromolecule. The only method allowing the assessment of the sequences of lignin units is the electrospray ionization-mass spectrometry (ESI-MS) technique. First interesting results reported tentative assignments of some of the lignin oligomers [68]. However, most of the lignin fragments are still to be assigned.

9.3.5 Analysis of technical lignins

The structures of technical lignins are even more heterogeneous than those of native lignins. Degradation of lignin during pulping results in a notable decrease in the amounts of original lignin moieties, particularly β-O-4 units, and the formation of a large variety of new structures. NMR approach to assess the structure of technical lignins is different from what we use for native lignins [45,52]. Due to broad and overlapping peaks ^{13}C NMR allows quantification typically on the level of functional groups and inter-unit linkages (see Table 9.2), but not much on the amounts of specific structures [55]. Correlation 2D NMR was very useful in the identification of various moieties formed during kraft pulping, such as units **21–28** (Figures 9.3 and 9.4), along with moieties of the original native lignin [19–21,23,24,69]. However, their quantification still remains a challenge. Comparison of relative signal intensities was used to estimate the abundance of various moieties [19,20,23,24]. Development of semi- and quantitative 2D NMR methods [22,50,51] is very important for more accurate analysis of technical lignins.

Investigation of technical lignins with 2D NMR demonstrated that the classical theory of lignin condensation, developed by using lignin model compounds, is very limited in kraft pulping of wood [19,24,69]. The role of Na_2S in kraft pulping mechanism should also be revised [70]. A new radical mechanism of condensation reactions involving sulfur species has been recently suggested [71]. However, this mechanism does not explain lignin condensation in alkaline conditions in the absence of Na_2S (in soda pulping), and therefore, the mechanism of lignin condensation in alkaline pulping should be further investigated.

Table 9.2 Amount of different functionalities (per 100 Ar) in residual and technical/dissolved kraft lignins determined by ^{13}C NMR.

Moieties	EGL RL	EGR RL	BRL	EGL DL	EGR DL	BDL	PRL	PDL
Al–COOH	22	17	30	16	13	18	37	20
Ar–COOH	2	2	3	2	3	2	4	1
Total OH	162	143	123	128	119	107	127	108
Alip OH	123	78	56	51	39	27	55	34
OHpr	71	43	28	29	24	23	36	23
OHsec	52	35	28	22	15	4	19	11
OHph	39	65	67	77	80	80	72	74
S/G ratio	2.3	1.1	1.3	2.5	1.4	1.7	–	–
OMe	155	129	124	141	125	141	98	81
Ar–H	194	197	162	191	186	172	210	218
Degree of condensation	37	52	82	37	55	65	90	82
β-5	1	4	4	1	2	2	4	3
β–β	3.5	3	3	2	1.5	1.5	2.5	2.5
β-O-4	37	21	3	12	5	2	5	3
Oxygenated aliphatic	221	161	123	110	79	86	114	72
Alk-Ether (total)	112	105	81	61	52	68	69	43
Alk-O-Alk	51	70	48	38	32	48	41	17

Abbreviations: EGL RL = *Eucalyptus globulus* residual lignin; EGR RL = *Eucalyptus grandis* residual lignin; BRL = birch residual lignin; EGL DL = *E. globulus* dissolved lignin; EGR DL = *E. grandis* dissolved lignin; BDL = birch dissolved lignin; PRL = pine residual lignin; PDL = pine dissolved lignin.

Recently, we revised the quantification of technical lignin structures with ^{13}C NMR that in many aspects is different from the analysis of native lignins [55]. The results are summarized in Table 9.2. It is important to mention that the variations in the results obtained from different analytical methods and calculation algorithms are significantly higher than in the case of native lignins.

It is interesting to note that the differences in the structure of hardwood lignins after kraft pulping are much bigger than those in the corresponding native lignins [23,24,55]. Lignin in some hardwood species (birch, aspen, and poplar) is strongly degraded whereas lignin in other species (*E. globulus*, *E. grandis*, and sweetgum) undergoes much less degradation. This interesting phenomenon is worthy of further investigation.

9.3.6 Linkages between lignin and carbohydrates

The main types of LC linkages in wood are believed to be phenyl glycoside bonds (A), benzyl esters (B) and benzyl ethers (C) (Figure 9.3). Different degradation techniques are commonly used for the analysis of LCC isolated from plant tissues. This approach includes cleavage of LC bonds and identification of the resulting products by alkaline hydrolysis, acid hydrolysis, Smith degradation, methylation analysis, and DDQ oxidation [36,37,39,72,73]. However, most of these methods give information on the carbohydrate part of LCC only and do not provide direct evidences on the presence LC bonds. Spectroscopic methods

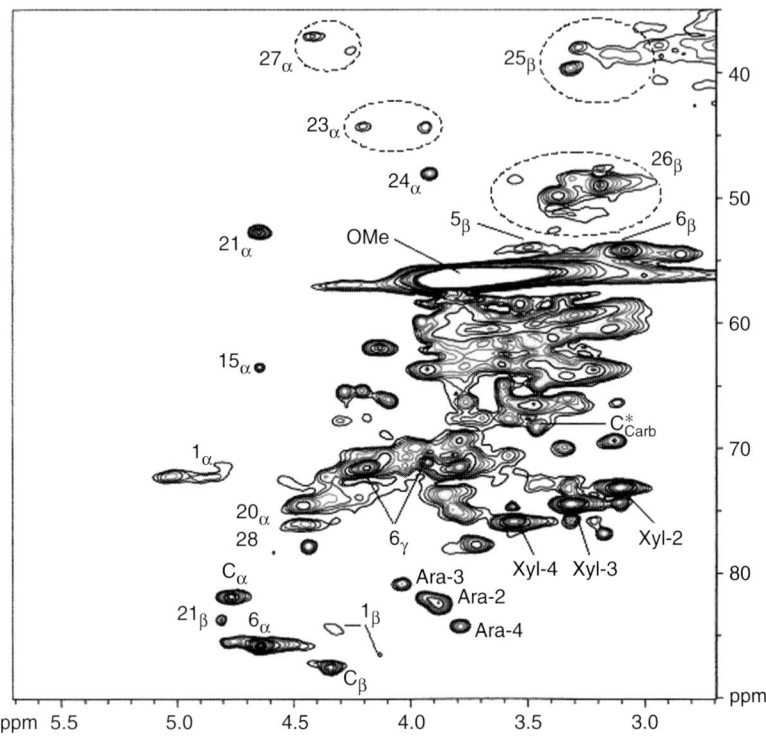

Figure 9.4 HSQC spectrum of aspen dissolved kraft lignin obtained using Bruker CryoProbe™. Numbers and letters correspond to structures shown in Figure 9.3. Ara- and Xyl-designate arabinose and xylose units, respectively. C^*_{carb} correspond to C-5 of arabinose and/or C-6 of glucose and galactose in benzyl ether lignin–carbohydrate units (**C**).

faces significant difficulties because of heavy overlapping signals originated from different functionalities of LCCs.

Recent application of multi-dimensional NMR spectroscopic techniques provides a great advantage in the LCC analysis offering direct observation of various LC bonds in a model LCC [74] as well as in preparations isolated from wood and kraft pulps [19,23,24,32,38,44]. This allows, for the first time, direct detection of phenyl glycoside bonds in LCC preparations isolated from eucalypt and pine [32,38,44] (Figure 9.5). Another important finding was the detection of γ-ester LC bonds instead of commonly believed benzyl (α-) esters (Figure 9.5). The 2D NMR showed the presence of benzyl ether linkages between lignin and various carbohydrate sites in preparations/samples of native and kraft residual and dissolved lignins [19,23,24,38] (Figure 9.5).

Problems resulted from the low frequency of LC bonds in LCC can be overcome by the use of newly developed NMR equipment of high sensitivity. The use of a cryogenic probe allows 3–4 folds increase in the sensitivity of an NMR spectrometer. Application of the cryoprobe in the analysis of residual and dissolved kraft lignins with high-resolution HSQC and heteronuclear multiple bond correlation (HMBC) techniques [24] allows the efficient detection of various LC bonds presented in these preparations in low concentrations.

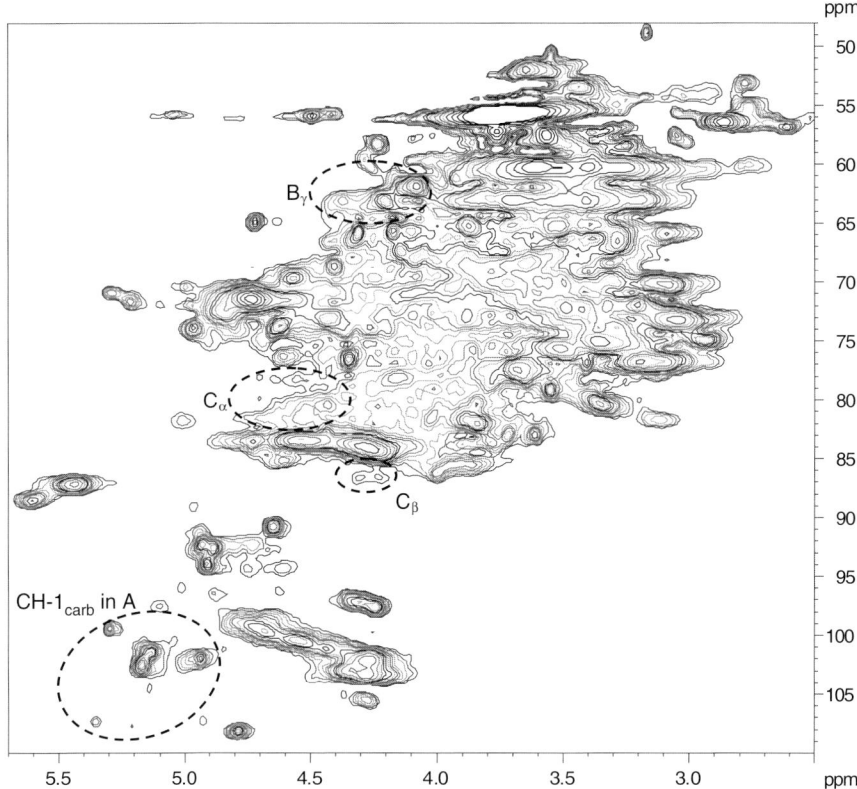

Figure 9.5 Expanded aliphatic region of the HMQC spectrum of the LCC-AcOH preparation. The letters A, B, and C designate lignin–carbohydrate bonds of phenyl glycoside, γ-ester and benzyl ether types, respectively (see Figure 9.3). The letters α, β, and γ in subscript designate positions in lignin C$_9$-unit. From Balakshin et al. [38], figure 2, copyright (2007), with permission from Holzforschung.

Another way to improve the analysis of LCCs is to isolate preparations/samples enriched in LC linkages. LCC–AcOH preparation described in the previous section has high amounts of various LC linkages (∼0.20–0.25/C$_9$-unit) and is more suitable for 2D NMR analysis [32,38] (Figure 9.5). However, some important signals were still overlapped in the HSQC spectra due to a very complex sample nature. Detailed information on LC bonds, particularly carbohydrate linkage sites, can be obtained by using 3D NMR. Another approach is selective sample pretreatments (mild alkaline and acidic hydrolysis, reduction, methylation) followed by the analysis of the pretreated preparations with correlation 2D and quantitative ^{13}C NMR methods [41].

In spite of great progress in the analysis of LCC with modern NMR techniques, accurate quantification of LC bonds using NMR spectroscopy is still a challenge. Developing quantitative 2D NMR methods can overcome this problem. Overall, a combination of multidimensional NMR techniques and quantitative ^{13}C NMR with wet chemistry methods is likely the best approach in the comprehensive analysis of LCC preparations.

Conclusions

There is not a perfect method for the isolation and analysis of lignin and LCC. Each analytical approach has specific advantages and limitations. Therefore, combination of appropriate methods and confirmation of experimental data by various independent techniques give the best results. The choice of optimal methodological strategy depends on the research objective, the nature of lignocellulosics, the number of samples to be studied and other variables. Significant progress in the analysis of lignocellulosics has been made recently with the development of new analytical methods, generating new structural information while allowing the comprehensive analysis to be done with a small amount of samples in a short experimental time.

Acknowledgment

The authors are very grateful to Prof. Y. Matsumoto (The University of Tokyo) for his criticism and valuable suggestions.

References

1. Adler, E. (1977) Lignin chemistry: past, present and future. *Wood Sci. Technol.* 11, 169–218.
2. Sakakibara, A. (1991) Chemistry of lignin. In: Hon, D.N.-S., Shiraishi, N. (eds) *Wood and Cellulose Chemistry*. Marcel Dekker Inc., New York, pp. 113–175.
3. Ralph, J., Lundquist, K., Brunow, G. et al. (2004) Lignins: natural polymers from oxidative coupling of 4-hydroxyphenylpropanoids. *Phytochem. Rev.* 3, 29–60.
4. Lawoko, M., Henriksson, G., Gellerstedt, G. (2005) Structural differences between the lignin–carbohydrate complexes in wood and in chemical pulps. *Biomacromolecules* 6, 3467–3473.
5. Pepper, J.M., Baylis, P.E.T., Adler, E. (1959) The isolation and properties of lignins obtained by the acidolysis of spruce and aspen woods in dioxane–water medium. *Can. J. Chem.* 37, 1241–1248.
6. Gellerstedt, G., Lindfors, E.-L. (1991) On the structure and reactivity of residual lignin in kraft pulp fiber. *Proc. Intern. Pulp Bleaching Conf.*, SPCI, Stockholm, Vol. 1, p. 73.
7. Evtuguin, D.V., Pascoal Neto, C., Silva, A.M.S. et al. (2001) Comprehensive study on the chemical structure of dioxane lignin from plantation *Eucalyptus globulus* lignin. *J. Agric. Food Chem.* 49, 4252–4261.
8. Mortha, G., Nikandrov, A., Robert, D., Lachenal, D., Zaroubine, M.Ya. (2001) Characteristics of lignins extracted from oak wood and kraft pulps by acetic acid/$ZnCl_2$ acidolysis: Comparison with other methods. *Proc. 11th Intern. Symp. Wood Pulping Chem.*, Nice, France, Vol. I, pp. 245–250.
9. Jääskeläinen, A.S., Sun, Y.J., Argyropoulos, D.S., Tamminen, T., Hortling, B. (2003) The effect of isolation method on the chemical structure of residual lignin. *Wood Sci. Technol.* 37, 91–102.
10. Tamminen, T.L., Hortling, B.R. (1999) Isolation and characterization of residual lignin. In: Argyropoulos, D.S. (ed.) *Advances in Lignocellulosics Characterization*. Tappi Press, Atlanta, GA, pp. 1–42.
11. Bjorkman, A. (1956) Studies on finely divided wood. Part I. Extraction of lignin with neutral solvents. *Svensk Papperstidn.* 59, 477–485.

12. Chang, H.-m., Cowling, E.B., Brown, W., Adler, E., Miksche, G. (1975) Comparative studies on cellulolytic enzyme lignin and milled lignin of sweetgum and spruce. *Holzforschung* 29, 153–159.
13. Chang, H.-m. (1992) Isolation of lignin from pulp. In: Lin, S.Y., Dence, C.W. (eds) *Methods of Lignin Chemistry*. Springer-Verlag, Heidelberg/Berlin/New York, pp. 71–74.
14. Yamasaki, T., Hosoya, S., Chen, C.-L., Gratzl, J.S., Chang, H.-m. (1981) Characterization of residual lignin in kraft pulp. *Proc. 9th Intern. Symp. Wood Pulping Chem.*, Stockholm, Sweden, Vol. II, pp. 34–42.
15. Jiang, J.-E., Chang, H.-M., Bhattacharjee, S.S., Kwoh, D.L.W. (1987) Characterization of residual lignins isolated from unbleached and semi-bleached softwood kraft pulps. *J. Wood Chem. Technol.* 7, 81–96.
16. Balakshin, M.Yu., Evtuguin, D.V., Pascoal Neto, C. (2001) Lignin–carbohydrate complexes in *Eucalyptus globulus* wood and kraft pulps. *Proc. 7th Brazilian Symp. Chem. of Lignin and Other Wood Components*, Belo Horizonte, MG, Brazil, pp. 53–60.
17. Guerra, A., Filpponen, I., Lucia, L.A., Saquing, C., Baumberger, S., Argyropoulos, D.S. (2006) Toward a better understanding of the lignin isolation process from wood. *J. Agric. Food Chem.* 54, 5939–5947.
18. Capanema, E.A., Balakshin, M.Yu., Chen, C.-L. (2004) An improved procedure for isolation of residual lignins from hardwood kraft pulps. *Holzforschung* 58, 464–472.
19. Balakshin, M.Yu., Capanema, E.A., Chen, C.-L., Gracz, H. (2003) Elucidation of the structures of residual and dissolved pine kraft lignin using an HMQC technique. *J. Agric. Food Chem.* 51, 6116–6127.
20. Capanema, E.A., Balakshin, M.Yu., Chen, C.-L., Gratzl, J.S., Gracz, H. (2001) Structural analysis of residual and technical lignins by $^1H–^{13}C$ correlation 2D NMR-spectroscopy. *Holzforschung* 54, 302–308.
21. Liitia, T.M., Maunu, S.L., Hortling, B. Toikka, M., Kilpelainen, I. (2003) Analysis of technical lignins by two- and three-dimensional NMR spectroscopy. *J. Agric. Food Chem.* 51, 2136–2143.
22. Ralph, J., Akiyama, T., Kim, H. *et al.* (2006) Effects of coumarate 3-hydroxylase down-regulation on lignin structure. *J. Biol. Chem.* 281, 8843–8853.
23. Capanema, E.A., Balakshin, M.Yu., Chen, C.-L., Gratzl, J.S., Gracz, H. (2001) Studies on kraft pulp lignins using HMQC NMR technique. *Proc. 7th Brazilian Symp. Chem. of Lignin and Other Wood Components*, Belo Horizonte, MG, Brazil, September 2–5, 2001, pp. 61–68.
24. Capanema, E.A., Balakshin, M.Yu., Chen, C.-L., Colson, K.L., Gracz, H.S. (2003) Use of cryogenic NMR probes in ^{13}C and $^1H–^{13}C$ 2D NMR techniques for structural analysis of lignin preparation. *Proc. 12th Intern. Symp. Wood Pulping Chem.*, Madison, Wisconsin, USA, June 9–12, pp. 179–182.
25. Lu, F., Ralph, J. (2003) Non-degradative dissolution and acetylation of ball-milled plant cell walls: high resolution solution-state NMR. *Plant J.* 35, 535–544.
26. Hu, Z., Yeh, T.-F., Chang, H.-m., Matsumoto, Y., Kadla, J.F. (2006) Elucidation of the structure of cellulolytic enzyme lignin. *Holzforschung* 60, 389–397.
27. Capanema, E.A., Balakshin, M.Yu., Katahira, R., Chang, H.-m., Jameel, H. (2007) Structural variations in hardwood lignins. *Proc. 14th Intern. Symp. Wood Fibre Pulping Chem.* CD-ROM, Durban, South Africa, June 25–28.
28. Wu, S., Argyropoulos, D.S. (2003) An improved method for isolating lignin in high yield and purity. *J. Pulp Pap. Sci.* 29, 235–240.
29. Guerra, A., Filpponen, I., Lucia, L.A., Argyropoulos, D.S. (2006) Comparative evaluation of three lignin isolation protocols for various wood species. *J. Agric. Food Chem.* 54, 9696–9705.
30. Ikeda T., Holtman, K., Kadla, J.F., Chang, H.-m., Jameel, H. (2002) Studies on the effect of ball milling on lignin structure using a modified DFRC method. *J. Agric. Food Chem.* 50, 129–135.
31. Fujimoto, A., Matsumoto, Y., Meshitsuka, G., Chang, H.-m. (2005) Quantitative evaluation of milling effects on lignin structure during the isolation process of milled wood lignin. *J. Wood Sci.* 51, 89–91.

32. Capanema, E., Balakshin, M.Yu., Heermann, M.L. et al. (2005) Chemical properties in CAD-deficient pine and their effect on pulping. *Proc. 13th Intern. Symp. Wood, Fibre Pulping Chem.*, Auckland, New Zealand, Vol. II, pp. 173–180.
33. Capanema, E.A., Balakshin, M.Yu., Kadla, J.F., Chang, H.-m. (2007) On isolation of milled wood lignin from eucalyptus wood. *O Papel*, (N5), 74–79.
34. Björkman, A. (1957) Studies on Finely Divided Wood. Part 3. Extraction of lignin–carbohydrate complexes with neutral solvents. *Sven. Papperstidn.* 60, 243–251.
35. Aimi, H., Matsumoto, Y., Meshitsuka, G. (2004) Structure of small lignin fragment retained in water-soluble polysaccharide extracted from sugi MWL isolation residue. *J. Wood Sci.* 50, 415–421.
36. Koshijima, T., Watanabe, T. (2003) *Association between Lignin and Carbohydrates in Wood and Other Plant Tissues.* Springer-Verlag, Berlin, Heidelberg.
37. Obst, J. (1982) Frequency and alkali resistance of lignin–carbohydrate bonds in wood. *Tappi* 65, 109–112.
38. Balakshin, M.Yu., Capanema, E.A., Chang, H.-m. (2007) A fraction of MWL with high concentration of lignin–carbohydrate linkages: Isolation and analysis with 2D NMR spectroscopic techniques. *Holzforschung* 61, 1–7.
39. Azuma, J., Takahashi, N., Koshijima, T. (1981) Isolation and characterization of lignin–carbohydrate complexes from milled-wood lignin fraction of *Pinus densiflora Carbohydr. Res.* 93, 91–104.
40. Holtman, K.M., Chang, H.-m., Kadla, J.F. (2004) Solution-state nuclear magnetic resonance study of the similarities between milled wood lignin and cellulolytic enzyme lignin. *J. Agric. Food Chem.* 52, 720–726.
41. Balakshin, M.Yu., Capanema, E.A., Chang, H.-m., Jameel, H. (2007) Structural variations in pine and birch lignin–carbohydrate complex preparations. *Proc. 14th Intern. Symp. Wood Fibre Pulping Chem.* CD-ROM. Durban, South Africa, June 25–28.
42. Lee, Z.Z., Meshitsuka, G., Cho, N.S., Nakano, J. (1981) Characteristics of milled wood lignins isolated with different milling time. *Mokuzai Gakkaishi* 27, 671–677.
43. Fujimoto, A. (2005) *Effect of milling on the structure of lignin.* Ph.D. Thesis, The University of Tokyo.
44. Balakshin, M.Yu., Evtuguin, D.V., Pascoal Neto, C., Silva, A.M.S., Domingues, P.M., Amado, F.M.L. (2001) Studies on lignin and lignin–carbohydrate complex by application of advanced spectroscopic techniques. *Proc. 11th Intern. Symp. Wood Pulping Chem.*, Nice, 2001, Vol. 1, pp. 103–110.
45. Capanema, E.A., Balakshin, M.Yu., Kadla, J.F. (2005) Quantitative characterization of a hardwood milled wood lignin by NMR spectroscopy. *J. Agric. Food Chem.* 53, 9639–9649.
46. Duarte, A.P., Robert, D., Lachenal, D. (2000) *Eucalyptus globulus* kraft residual lignins. Part 1. effect of extraction methods upon lignin structure. *Holzforschung* 54, 365–372.
47. Koda, K., Gaspar, A.R., Yu, L., Argyropoulos, D.S. (2005) Molecular weight-functional group relations in softwood residual kraft lignins. *Holzforschung* 59, 612–619.
48. Chen, C.-L., Robert, D. (1988) Characterization of lignin by 1H and 13C NMR spectroscopy. In: Wood, W.A., Kellogg, S.T. (eds) *Methods in Enzymology.* Academic Press Inc., New York, Vol. 161B, pp. 137–174.
49. Ralph, J., Marita, J., Ralph, S.A. et al. (1999) Solution-state NMR of lignins. In: Argyropoulos, D.S. (ed.) *Advances in Lignocellulosics Characterization.* Tappi Press, Atlanta, GA, pp. 55–108.
50. Zhang, L., Gellerstedt, G. (2000) Achieving quantitative assignment of lignin structure by combining ^{13}C and HSQC NMR techniques. *Proc. 6th European Workshop on Lignocellulosics and Pulp*, Bordeaux, France, September 3–6, pp. 7–10.

51. Heikkinen, S., Toikka, M.M., Karhunen, T., Kilpelainen, I.A. (2003) Quantitative 2D HSQC (Q-HSQC) via suppression of J-dependence of polarization transfer in NMR spectroscopy: Application to wood lignin. *J. Am. Chem. Soc.* 125, 4362–4367.

52. Capanema, E.A., Balakshin, M.Yu., Kadla, J.F. (2004) A comprehensive approach for quantitative lignin characterization by NMR spectroscopy. *J. Agric. Food Chem.* 52, 1850–1860.

53. Zia, P., Akim, L., Argyropoulos, D.S. (2001) Quantitative ^{13}C NMR of lignins with internal standards. *J. Agric. Food Chem.* 49, 3573–3578.

54. Lundquist, K. (1980) NMR studies of lignins. Investigation of spruce lignin by 1H NMR spectroscopy. *Acta Chem. Scand.* B34, 497–501.

55. Capanema, E.A., Balakshin, M.Yu., Chang, H.-m., Jameel, H. (2005) Isolation and characterization of residual lignins from hardwood pulps: Method improvements. *Proc. 13th Intern. Symp. Wood Fibre Pulping C*, Auckland, New Zealand, Vol. III, pp. 57–64.

56. Lu, F., Ralph, J. (1997) Derivatization followed by reductive cleavage (DFRC method), new method for lignin analysis: Protocol for analysis of DFRC monomers. *J. Agric. Food Chem.* 45, 2590–2592.

57. Holtman, K.M., Chang, H.-m., Jameel, H., Kadla, J.F. (2003) Elucidation of lignin structure through degradative methods: Comparison of modified DFRC and thioacidolysis. *J. Agric. Food Chem.* 51, 3535–3540.

58. Tohmura, S., Argyropoulos, D.S. (2001) Determination of arylglycerol-ß-aryl ethers and other linkages in native and technical lignins. *J. Agric. Food Chem.* 49, 536–542.

59. Smit, R., Suckling, I.D., Ede, R.M. (1997) A new method for the quantification of condensed and uncondensed softwood lignin structures. *Proc. 9th Intern. Symp. Wood Pulping Chem.*, Montreal, Canada, p. L4–1.

60. Argyropoulos, D.S., Jurasek, L., Kristofava, L., Xia, Z., Sun, Y., Palus, E. (2002) Abundance and reactivity of dibenzodioxocins in softwood lignin. *J. Agric. Food Chem.* 50, 658–666.

61. Terashima, N., Seguchi, Y., Robert, D. (1991) Selective ^{13}C-enrichment of side chain carbons of guaiacyl lignin in pine. *Holzforschung* 45(Suppl.), 35–39.

62. Evtuguin, D.V., Balakshin, M.Yu., Terashima, N., Pascoal Neto, C., Silva, A.M.S. (2003) New complementary information on the *E. globulus* lignin structure obtained by ^{13}C selective labelling and advanced NMR techniques. *Proc. 12th Intern. Symp. Wood Pulping Chem.*, Madison, Wisconsin, USA, pp. 177–180.

63. Evtuguin, D.V., Robert, D. (1997) The detection of muconic acid type structures in oxidized lignins by 13C NMR spectroscopy. *Wood Sci. Technol.* 31, 423–431.

64. Balakshin, M.Yu., Capanema, E.A. (2005) Selective lignin modification coupled with NMR spectroscopy: A method for lignin analysis. *Proc. 13th Intern. Symp. Wood Fibre Pulping Chem.*, Auckland, New Zealand, Vol. II, pp. 353–360.

65. Karhunen, P., Rummakko, P., Sipila, J., Brunow, G., Kilpelainen, I. (1995) Dibenzodioxocins; A novel type of linkages in softwood lignins. *Tetrahedron Lett.* 36, 169–170.

66. Balakshin, M.Yu., Capanema, E.A., Goldfarb, B., Frampton, J., Kadla, J. (2005) NMR studies on Fraser fir *Abies fraseri* (Pursh) Poir. lignins. *Holzforschung* 59, 488–496.

67. Zhang, L., Gellerstedt, G., Lu, F., Ralph, J. (2006) NMR studies on the occurrence of spirodienone structures in lignins. *J. Wood Chem. Technol.* 26, 65–79.

68. Evtuguin, D.V., Amado, F.M.L. (2003) Application of electrospray ionization-mass spectrometry to the elucidation of the primary structure of lignin. *Macromol. Biosci.* 3, 339–343.

69. Gellerstedt, G., Zhang, L. (2001) Chemistry of TCF-bleaching with oxygen and hydrogen peroxide. In: Argyropoulos, D.S. (ed.) *Oxidative Delignification Chemistry: Fundamental and Catalysis*. ACS Symp. Series 785, Washington, DC, pp. 61–72.

70. Balakshin, M.Yu., Capanema, E.A., Chen, C.-L. (2003) On the role of Na_2S in alkaline pulping. *Proc. of 12th ISWPC*, Madison, Wisconsin, USA, Vol. 2, pp. 121–124.

71. Gellerstedt, G., Majtnerova, A., Zhang, L. (2004) Towards a new concept of lignin condensation in kraft pulping. Initial results. *C. R. Biologies* 327, 817–826.
72. Fengel, D., Wegener, G. (eds) (1984) *Wood: Chemistry, Ultrastructure, Reactions.* Walter de Gruyrer, Berlin.
73. Helm, R.F. (2000) Lignin–polysaccharide interaction in woody plants. In: Glasser, W.G., Northey, R.A., Schultz, T.P. (eds) *Lignin: Historical, Biological, and Material Perspectives.* ACS Symp. Series 742, Washington, DC, pp. 161–171.
74. Evtuguin, D.V., Goodfellow, B.J., Pascoal Neto, C., Terashima, N. (2005) Characterization of lignin–carbohydrate linkages in *Eucalyptus globulus* by 2D/3D NMR spectroscopy using specific carbon-13 labelling technique. *Proc. 13th Intern. Symp. Wood Pulping Chem.*, Auckland, New Zealand, 2001, Vol. 2, pp. 439–446.

Chapter 10

Chemical Composition and Lignin Structural Features of Banana Plant Leaf Sheath and Rachis

Lúcia Oliveira, Dmitry Victorovitch Evtuguin, Nereida Cordeiro, Armando Jorge Domingues Silvestre, and Artur Manuel Soares da Silva

Abstract

Two morphological regions, leaf sheath and rachis, of banana plant "Dwarf Cavendish" grown in Portugal were subjected to chemical composition studies and found to have high ashes (19–27%), extractives (13–18%), cellulose (31–37%), and hemicelluloses (8–12%) contents, and relatively low lignin (10–13%) content. The native *in situ* lignins were characterized by nitrobenzene and permanganate oxidation (PO), while the dioxane lignins (DLs) were analyzed by PO, UV, Fourier transform-infrared spectroscopy (FT-IR), solid- and liquid-state NMR spectroscopies and size exclusion chromatography. Both the leaf sheath and the rachis lignins are essentially of *p*-hydroxyphenyl (H)–guaiacyl (G)–syringyl (S) type with H/G/S ratio of (20–35)/(51–60)/(14–20). The structural variation of lignin in these morphological parts of banana plant is significant in terms of H/G/S ratio and for the content of condensed structures. Most of the H units in DLs are terminal phenolic coumarates linked to other lignin substructures by ester bonds in contrast to ferulates that are mainly ether-linked to the bulk lignin. It is proposed that banana plant leaf sheath lignin is chemically bonded to suberin-like components of the cell tissues by ester linkages via ferulates in a greater extend than rachis lignin. β-O-4 structures (0.31–0.34/C_6), the most abundant in DLs, comprise mainly S units.

10.1 Introduction

Nonwood agricultural residues are highly abundant around the world and are becoming increasingly important as raw materials for energy and chemical feedstock production. The annual production of such biomass per area is much higher than in forests, and the payback time of nonwood plantations is much shorter than that of forest plantations. Wheat straw, rice straw, and bagasse are the most well-known examples. The banana tree, an annual herbaceous plant that produces fruit all year around, could potentially serve as an inexpensive and readily available nonwood, renewable source of biomass.

About 4 500 000 ha of cultivated area worldwide are occupied by banana plantations [1]. Nowadays, India is the largest worldwide producer of banana (16.8 million tons year^{-1}) followed by Ecuador (5.4 million tons year^{-1}) and Brazil (5.3 million tons year^{-1}) [1]. Among the banana varieties, the Cavendish is the most exploited, corresponding to $\sim\frac{1}{3}$ of the worldwide production (24.0 million tons year^{-1}). Usually, after harvesting the single bunch of bananas, a great amount of residues are produced. Despite the important function of these residues for soil fertilization, their accumulation in high amounts promotes the proliferation of insects and microorganisms, leading to serious environmental and phytosanitary problems. In terms of dry weight material, the Cavendish group produces $\sim$8 tons ha^{-1} of pseudostems, 7.7 tons ha^{-1} foliage and $\sim$0.5 tons ha^{-1} of rachis [2]. The development of new applications for these residues could serve as an income for banana producers and regional economy.

Recent preliminary studies on the utilization of pseudostems and rachis from "Dwarf Cavendish" as raw materials for papermaking [3] and composite materials [4] have shown rather promising results. Further investigations on the potential industrial utilization of "Dwarf Cavendish" banana residues require a detailed knowledge of their chemical composition. The bibliography covering fundamental aspects of banana plant chemistry is rather scarce and a more comprehensive study is needed. This knowledge is crucial in identifying possible applications and profitable processing of banana plant residues. In this chapter we present our results on the chemical and structural characterization, in particular, lignin structural analysis of the most abundant fractions, leaf sheath, and rachis from "Dwarf Cavendish" banana plants.

10.2 Materials and methods

10.2.1 Preparation of plant material

Randomly selected mature "Dwarf Cavendish" banana plants were harvested from a banana plantation in Funchal (Madeira Island, Portugal). The plants were separated into five different morphological parts: petioles/midrib, leaf blades, floral stalk, leaf sheath, and rachis. The pseudostems were separated from foliage, manually separated into leaf sheath (major part of the pseudostem) and floral stalk, and air-dried. The fractions corresponding to leaf sheath and rachis were milled in a Retsch AS200 and sieved to 40–60 mesh fractions, respectively, and used for our studies.

10.2.2 Chemical analysis

The ash content was determined by complete incineration of a milled leaf sheath or rachis sample in a Nabertherm muffle furnace at 600°C for 6 h. The analysis of mineral element was performed at the Laboratories of the Service Central d'Analyse of CNRS (Vernaison, France). The extractives were isolated by successive Soxhlet extractions of a sample with dichloromethane, ethanol/toluene (1:2 v/v), and water. The solvent and water-extracted milled leaf sheath and rachis were air-dried and used for further analyses. The lignin contents were determined using the Klason method [5]. The holocellulose and the cellulose were obtained using the peracetic acid method [6] and the Kürschner–Hoffner method [7], respectively. The holocellulose was fractionated into hemicellulose A, hemicellulose B, and α-cellulose by successive extraction, in nitrogen (N_2) atmosphere, with 5% and 24% KOH aqueous solutions containing 0.014 g L^{-1} of $NaBH_4$, respectively [8]. The pentosan and starch contents were determined using the bromide/bromate method [9] and the iodine colorimetric method [10], respectively. Crude protein contents were calculated by converting the nitrogen contents ($N \times 4.4$), determined by the Kjeldahl method in a Kjeldahl Selecta Alcodest still [11,12]. The neutral sugars in the water extractives and the isolated hemicelluloses were quantified as alditol acetate derivatives by gas chromatography [13]. All chemical analysis and fractionation experiments were carried out, at least, in duplicate, and the presented results are the average values with a standard deviation <3%.

10.2.3 Isolation and characterization of lignins

10.2.3.1 Milled wood lignin

Milled wood lignins (MWLs) were isolated from the solvent and water-extracted milled leaf sheath and rachis, respectively, using a centrifugal ball mill (Retsch S100) with sintered corundum I jar and balls, and purified according to the Björkman method [14] with minor modifications [15].

10.2.3.2 Dioxane lignin

The solvent and water-extracted milled leaf sheath and rachis were, respectively, subjected to an alkaline (0.3% aqueous NaOH) extraction and then to acidolysis using the dioxane method adapted from a previously published procedure [16]. The alkali-extracted materials were subjected to three sequential extractions (30 min each) with 200 mL of dioxane/water (9:1, v/v) solution containing 0.2 M HCl at reflux under N_2, and were then washed with dioxane/water (9:1, v/v) (without HCl) [16]. The extracts were concentrated separately to ~40 mL each; and the resulting concentrates were then combined. The DLs were precipitated by adding the combined concentrates to cold water, isolated by centrifuging, washed with water until neutral pH, and freeze-dried.

10.2.3.3 Lignin purification

The isolated DLs were purified by dissolution in dioxane/methanol (9:1, v/v) (under stirring at room temperature for 2 h) and re-precipitation in cold water, followed by centrifuging and freeze-drying. The centrifuging procedure was repeated to ensure that all the purified lignin was extracted. All the purified lignin fractions were combined. The yields of purified DLs from both the leaf sheath and the rachis were ~30% of the Klason lignin in the alkali-extracted materials. The purified DLs were dispersed three times in chloroform at room temperature for 24 h with constant stirring, followed by filtration and drying under vacuum to remove the aliphatic compounds in them [17]. The purified, chloroform-treated DLs with yields of ~17% of the Klason lignin in the alkali-extracted materials were submitted to structural characterization.

10.2.3.4 Chemical analysis

Pyrolysis–gas chromatography/mass spectrometry (Py–GC/MS) of the MWLs, nitrobenzene and/or permanganate oxidation (PO) of the *in situ* lignins and the DLs were performed according to the literature procedures [18–20]. The analyses of neutral sugars in the DLs were performed according to the literature procedure [13]. Methoxyl group analysis was performed using Zeisel procedure [21], and the elemental composition was determined using a LECO CHNS-932 instrument.

10.2.3.5 Analysis of DLs by size exclusion chromatography

The weight-averaged molecular weight (M_w) of the DLs dissolved in dimethylformamide (0.5% w/v) and 0.1 N LiCl were determined by size exclusion chromatography (SEC) using a PL-GPC 110 chromatograph (Polymer Laboratories, UK) equipped with a Plgel 10 μm precolumn and two 300 mm × 7.5 mm Plgel (5 μm) Mixed D columns (Polymer Laboratories, UK). The precolumn, column, injection system, and a refractive index detector were maintained at 70°C. The eluent (0.5% w/v LiCl in DMF) was pumped at a flow rate of 0.9 mL min^{-1}. The SEC columns were calibrated using lignin preparations previously characterized by electrospray ionization–mass spectrometry (ESI–MS) [22].

10.2.3.6 Analysis of DLs by ultraviolet, FT-IR, and ^{13}C cross-polarization magic angle spinning NMR

UV spectra of the DLs were recorded in 2-methoxyethanol on a JASCO V-560 UV–vis spectrophotometer (1.0-cm cell). Fourier transform-infrared spectroscopy (FT-IR) spectra of the DLs were recorded on a Mattson 7000 FT-IR spectrometer using KBr pellets (1/250 mg). The spectra resolution was 4 cm^{-1}, and 64 scans were averaged. ^{13}C solid-state NMR spectra of the DLs were recorded at 100.6 MHz (9.4 T) on a Bruker Avance 400 spectrometer. A 7-mm double bearing Bruker rotor was spun in air at 5.0 kHz. In all experiments the ^1H and ^{13}C 90° pulses were ~0.4 μs. The cross-polarization magic angle spinning (CP-MAS) spectra were recorded with a 5 s recycle delay and a 2 ms contact time.

10.2.3.7 Analysis of the acetylated DLs by ^1H NMR

The DLs were acetylated according to the literature procedure [21]. The ^1H NMR (300 MHz) spectra of the acetylated DLs in deuterated chloroform (CDCl$_3$) solution (2% concentration) were obtained on a Bruker Avance 300 spectrometer at room temperature. The acquisition parameters used were: 12.2 μs pulse width (90°); 3 s relaxation delay; 300 scans.

10.2.3.8 Analysis of MWLs and DLs by ^{13}C NMR

^{13}C NMR (75.5 MHz) spectra of the MWLs and DLs in DMSO-d_6 (~23% concentration) were recorded on a Bruker Avance 300 spectrometer at 318 K with SiMe$_4$ (tetramethylsilane [TMS]) as an internal reference. The inverse-gated decoupling sequence, which allows quantitative analysis and comparison of signal intensities, was used with the following parameters: 4.1 ms pulse width (90° pulse angle); 12 s relaxation delay; 16 K data points; 18 000 scans.

Table 10.1 Chemical composition of leaf sheath and rachis of "Dwarf Cavendish" (% w/w).

Components	Leaf sheath	Rachis
Ashes	19.0	26.8
Extractives[a]		
Dichloromethane	1.4	1.5
Ethanol/toluene	2.1	1.4
Water	9.1	14.7
Total	12.6	17.6
Lignin		
Insoluble[a]	12.6	9.6
Soluble	0.7	0.9
Total	13.3	10.5
Holocellulose[a]	49.7	37.9
Hemicellulose A[a]	7.2	3.9
Hemicellulose B[a]	4.2	3.6
α-Cellulose[a]	37.1	28.4
Cellulose[a]	37.3	31.0
Pentosanes	12.4	8.3
Starch	8.4	1.4
Proteins	1.9	2.0

[a] The ash content was determined and the final value was corrected using the following equation: [% component corrected] = [% component obtained] × (100 − [% ashes])/100.

10.3 Result discussion

10.3.1 Chemical composition of leaf sheath and rachis

The results of the chemical composition of the "Dwarf Cavendish" leaf sheath and rachis are presented in Table 10.1. Both the leaf sheath and the rachis contain significant amounts of ashes (19.0% and 26.8%) and extractives (12.6% and 17.6%), which are considerably high when compared with some other fast growing plants [23–25]. The lipophilic extractives (∼1–2%) are likely composed mainly of long-chain fatty acids, (including several α- and ω-hydroxyfatty acids), long-chain aliphatic alcohols, sterols, and steryl glucosides. Minor amounts of aromatic acids, mono-glycerides and fatty acids steryl esters have been detected within these lipophilic extractives [26,27]. The high contents of ashes and lipophilic extractives may constitute a negative point of the leaf sheath and rachis when used as a fiber source for pulp and paper processing. Ashes have a negative effect on kraft pulping such as in the recovery of pulping chemicals and in the pulp yield, and on paper quality. Extractives will increase the consumption of chemicals during pulping and bleaching processes and may lead to pitch deposits in mill machinery and in pulp and paper, requiring higher maintenance costs and also decreasing the final product quality. However, the identification of "Dwarf Cavendish" as an abundant source of steryl glucosides [26], known for the health benefits associated with their inclusion in human diet, namely their ability to lower blood cholesterol [28,29], could provide other potential uses of the leaf sheath and rachis.

Water-soluble extractives (9.1% in leaf sheath and 14.7% in rachis) are the predominant contributors to the total extractives content ($\sim$72–84%). Preliminary results showed that the water-soluble extractives are composed mainly of ashes (49% in leaf sheath and 64% in rachis) and carbohydrates (33.3% in leaf sheath and 6.4% in rachis). Among the water-soluble/water-extracted carbohydrates in leaf sheath, starch was the most abundant constituent.

Cellulose is the major polysaccharide in the leaf sheath and rachis ($\sim$31–37%). The leaf sheath and rachis also contain significant amounts of hemicelluloses (11.4% and 7.5%, respectively; see further characterization below) and lignin ($\sim$10–13%). The amounts of lignin are typical for a large variety of gramineaceous species and very similar to those reported for other annual plants [23–25,30]. Proteins were detected in small amounts in the leaf sheath and rachis ($\sim$2%).

10.3.2 Noncellulosic polysaccharides

Noncellulosic polysaccharides, consisting mainly of pentosanes and starch, represented 42% and 26% of the total amount of polysaccharides in the leaf sheath and rachis, respectively (Table 10.1). In the rachis, starch was detected in a very small amount (1.4%), whereas in the leaf sheath its content was rather significant (8.4%). Similarly, the leaf sheath contained more pentosanes (12.4%) than the rachis (8.3%). Fractionation of holocellulose into the acidic hemicellulose (typically the glucuronoxylans), hemicellulose A (or HA), the neutral hemicellulose (typically the glucomannans), hemicellulose B (or HB), and α-cellulose showed that there was more HA than HB, particularly in the leaf sheath (Table 10.2). Further analysis of the monosaccharide composition of the HA and HB fractions (Table 10.2) showed that the HA fractions from both the leaf sheath and the rachis were composed essentially of xylans. However, the HA fraction of leaf sheath contained also a remarkable amount of arabinans. The relatively high proportion of xylose in the HB fractions may be, at least partially, due to the incomplete xylan extraction during HA isolation. The relatively high proportions of both the xylose and the glucose in the HB fractions suggested the presence of xyloglucans in these fractions. A more complete elucidation of the structures of these hemicelluloses remains to be performed.

10.3.3 Lignin characterization

10.3.3.1 Isolation and some chemical analyses of lignin

The leaf sheath together with the rachis represent the major components of the banana tree, containing long fibers suitable for potential applications in papermaking and in biocomposites [3,4]. In this context the knowledge on the lignin structure in these two morphological regions is particularly important.

Table 10.2 Monosaccharides composition in hemicelluloses A and B (HA and HB, respectively) of "Dwarf Cavendish" from leaf sheath and rachis (% molar proportions).

Monosaccharide	Leaf sheath		Rachis	
	HA	HB	HA	HB
Rhamnose	0.5	0.4	1.5	0.4
Arabinose	27.5	4.1	6.3	1.2
Xylose	57.0	52.3	72.6	50.4
Mannose	0.2	0.9	1.0	2.1
Galactose	6.9	4.1	2.9	4.4
Glucose[a]	7.9	38.2	15.7	41.5

[a] Corrected for starch content.

The isolation of MWL from the leaf sheath and rachis was a difficult task. Owing to the relatively low content and the strong structural association of the *in situ* "Dwarf Cavendish" lignin with polysaccharides and lipophilic compounds, the yields of the purified MWLs were rather low (0.8% and 4.0% for the rachis and the leaf sheath, respectively). The analysis of solid-state ^{13}C NMR spectra of these lignin samples (not shown) revealed the presence of high proportions of aliphatic, phenolic compounds (tannins), and carbohydrates. The ratios of *p*-hydroxyphenyl (H)/guaiacyl (G)/syringyl (S) units, estimated by Py–GC–MS analysis, were 31:55:14 and 40:24:36 for the MWLs from the leaf sheath and the rachis, respectively. These ratios were significantly different from those obtained for the *in situ* lignins by nitrobenzene oxidation analysis of the leaf sheath and rachis (12:25:63 and 11:18:71, respectively). These results (low yields, high contamination and different H/G/S ratios from the *in situ* lignins) show that the MWLs are not representative of the *in situ* lignins. Thus, our detailed lignin structural analyses were carried out only on the DLs.

DLs were isolated from extractives-free materials that had also been extracted with alkali to eliminate the tannins. Their yields (∼30% based on Klason lignin in the alkali-extracted materials) were about half of those reported for other annual plants [31,32]. This can be explained by a strong structural association of the *in situ* "Dwarf Cavendish" lignin with polysaccharides and, especially with lipophilic compounds in these tissues, based on a series of spectroscopic data that were discussed elsewhere [33]. For this reason, all DLs were additionally purified by dispersion/extraction in/with chloroform at room temperature for 24 h. Notable amounts of long-chain hydroxyacids (2-hydroxytetracosanoic, 24-hydroxytetracosanoic, 26-hydroxyhexacosanoic, and 28-hydroxyoctacosanoic acids among others) and *p*-coumaric (4-hydroxycinnamic) and ferulic (4-hydroxy-3-methoxycinnamic) acids have been detected in the chloroform extract [33]. These acids may be originated from suberin-like structures, saponified during the alkaline extraction prior to DL isolation [33].

Purified and chloroform-treated DL from the leaf sheath (DL$_{LS}$) and that from the rachis (DL$_R$) were subjected to methoxyl group content and elemental analyses, leading to the formulation of the empirical formula per phenylpropane unit

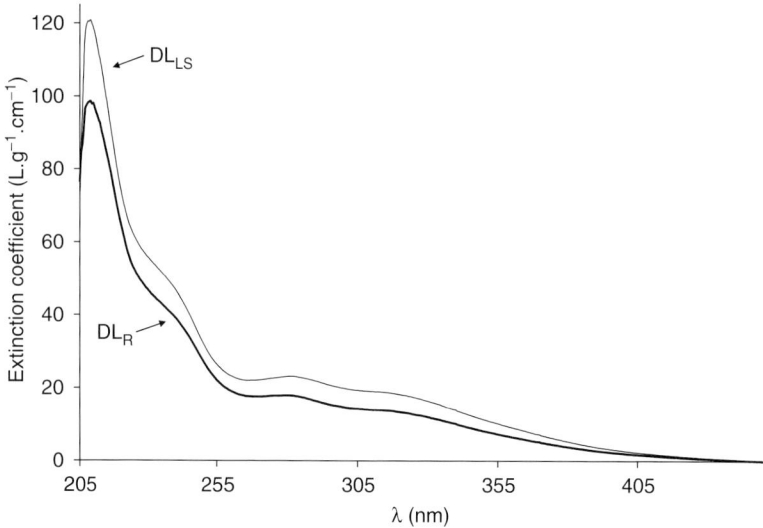

Figure 10.1 UV spectra of dioxane lignins from banana plant leaf sheath (DL_{LS}) and rachis (DL_R).

(C9 or ppu): $C_9H_{9.78}O_{2.71}(OCH_3)_{1.09}$ for DL_{LS} and $C_9H_{8.54}O_{4.18}(OCH_3)_{1.41}$ for DL_R. The frequency of methoxyl group substitution in the aromatic ring of DL_{LS} was slightly lower than that of DL_R, suggesting a higher content of S units in the later. The obtained values are similar to those reported previously for other annual plant lignins such as *Arundo donax* and *Hibiscus cannabinus* [31,32].

10.3.3.2 Analysis of DLs by UV, FT-IR, and ^{13}C CP-MAS NMR

The UV spectra of DL_{LS} and DL_R (Figure 10.1) showed absorption bands typical for DLs from other annual plants [31,32]. However, the extinction coefficient of the characteristic absorption maximum at 280 nm which corresponds to the $\pi \to \pi^*$ transition in the aromatic ring [34] for DL_{LS} was higher than that for DL_R (Table 10.3). The hypsochromic shift of this band for DL_R (Figure 10.1, Table 10.3) confirmed the higher proportion of S units in rachis than in leaf sheath lignin as suggested from the methoxyl group content analysis.

DL_{LS} also had a higher extinction coefficient at 310 nm than DL_R (Table 10.3). The absorption at 310 nm, assigned to the $n \to \pi^*$ transition in lignin units containing $C\alpha{=}O$ groups and $\pi \to \pi^*$ transitions in lignin units with $C\alpha{=}C\beta$ linkages conjugated to the aromatic ring, had previously been reported as being indicative of the presence of 4-hydroxycinnamic acid type structures [35]. The presence of 4-hydroxycinnamic acid type structures in our DLs was supported by the FT-IR spectra (Figure 10.2), which showed the characteristic bands at 1708 and 1630 (number not shown in the spectra) cm^{-1} ascribable to the C=O stretching of carboxylic acids, and to the stretching of C=C moieties conjugated to the aromatic rings, respectively [36].

In the FT-IR spectrum of DL_{LS} (Figure 10.2) the intensity of the signal at 1328 cm^{-1} that can be assigned to syringyl ring breathing with $C_{Ar}{-}OCH_3$ stretching [36] is lower than that in the spectrum of DL_R. Simultaneously, the ratio of A_{1462}/A_{1594} that is related

Table 10.3 Extinction coefficients ($L g^{-1} cm^{-1}$) of DL_{LS} and DL_R at different wavelengths.

Wavelength (nm)	DL_{LS}	DL_R
209	121	99
230	56	47
276/280[a]	23	18
310	19	14

[a] 276 nm for DL_R and 280 nm for DL_{LS}.

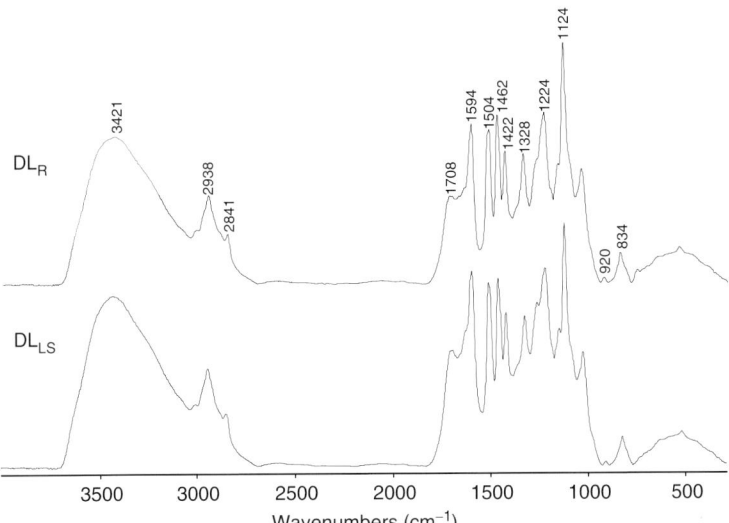

Figure 10.2 FT-IR spectra of dioxane lignins from banana plant leaf sheath (DL_{LS}) and rachis (DL_R).

to the amounts of methoxyl groups in lignins [37] is lower for DL_{LS} than for DL_R. Both of these features suggest a higher content of S units in DL_R, which is in agreement with the empirical formula and the UV analysis.

^{13}C CP-MAS NMR spectroscopy confirmed the higher content of S units in DL_R and also showed the higher content of 4-hydroxycinnamic acid type structures in DL_{LS} than in DL_R. Thus, the relative intensity of the signal at δ 152 ppm, assigned to C3/C5 resonance in S units [38], is higher for DL_R than for DL_{LS} (Figure 10.3). In contrast, the signal at $\delta \sim 160$ ppm, assigned to C4 in p-coumarates, is remarkably stronger for DL_{LS}. Additionally, the presence of aliphatic signals at δ 10–40 ppm was detected mainly in the DL_{LS} spectrum.

10.3.3.3 Lignin molecular weight

The SEC chromatograms of DL_{LS} and DL_R were significantly different from each other, and they showed bimodal and trimodal molecular weight distribution curves, respectively

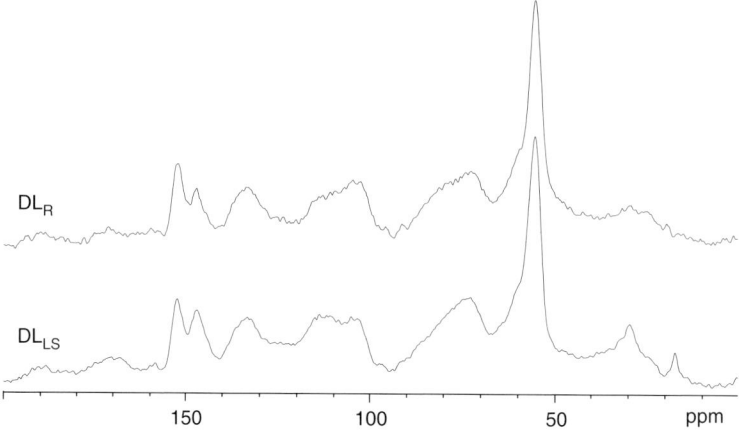

Figure 10.3 Solid-state ^{13}C CP-MAS NMR spectra of dioxane lignins from banana plant leaf sheath (DL$_{LS}$) and rachis (DL$_R$).

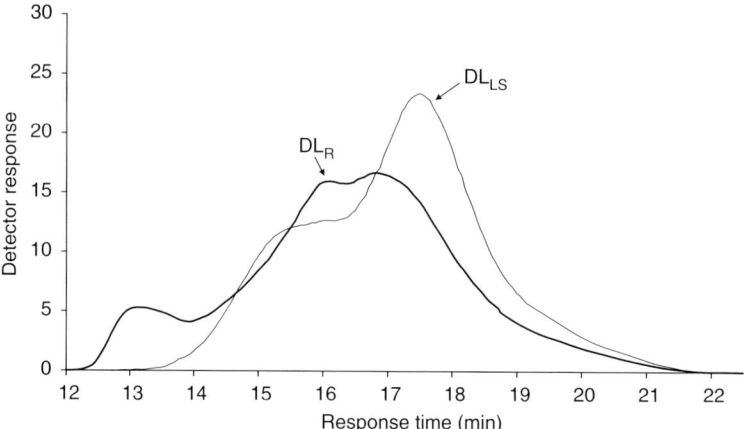

Figure 10.4 SEC molecular weight distribution of dioxane lignins from banana plant leaf sheath (DL$_{LS}$) and rachis (DL$_R$).

(Figure 10.4), providing evidence for the presence of structurally inhomogeneous lignin fractions. One of the reasons may be the branching of some of the lignin fractions with suberin-like materials that increased the hydrodynamic volume of lignin macromolecules. Formally determined M_w values of DL$_{LS}$ and DL$_R$ were 3350 and 5450 Da, respectively. However, lignin fragments possess in general rather low molecular weights, which in these cases correspond to the fractions eluted at 17–18 min (1200–1800 Da).

10.3.3.4 PO analysis of DLs and the in situ lignins

PO of the ethylated DLs and *in situ* lignins, followed by esterification according to the literature procedure [19] and GC/GC-MS analysis, allowed the identification and quantification

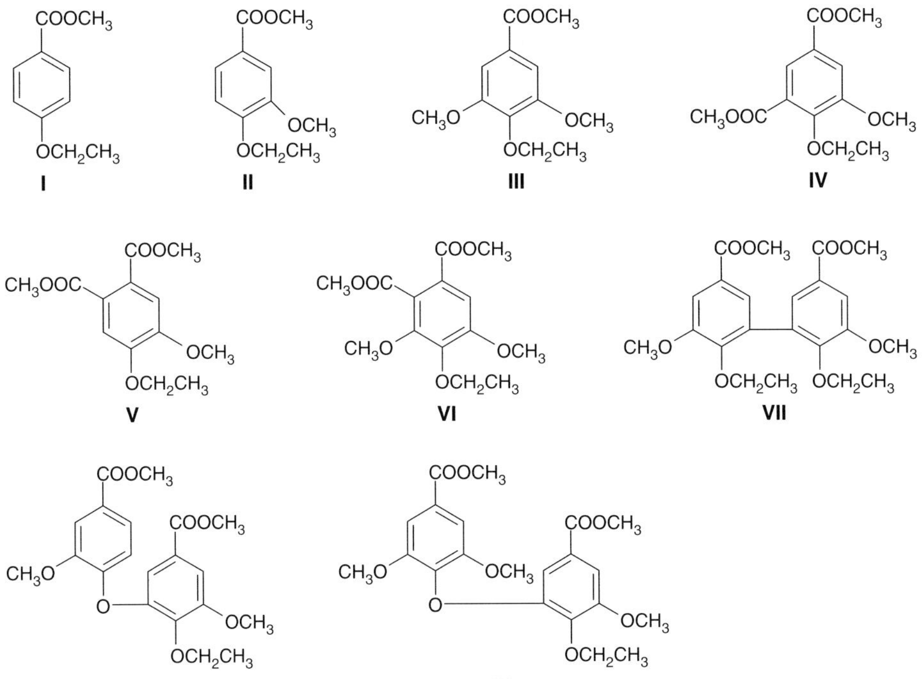

Figure 10.5 Various carboxylic acid methyl esters obtained using permanganate oxidation procedure [19].

Table 10.4 Molar proportions of carboxylic acid methyl esters (**I–IX**) (see Figure 10.5 for structures) and the corresponding H, G, and S units obtained using the permanganate oxidation procedure [19].

Samples	I (H)	II (G)	III (S)	IV (G)	V (G)	VI (S)	VII (G)	VIII (G)	IX (G-S)
Leaf sheath[a]	35	36	9	5	2	4	5	2	2
Rachis[a]	20	43	16	6	3	2	5	2	3
DL$_{LS}$	21	33	24	6	3	2	6	1	4
DL$_R$	23	28	31	4	2	2	3	1	6

[a] Alkali-extracted material.

of nine carboxylic acid methyl esters (Figure 10.5, Table 10.4), and the estimation of the molar proportions of the different lignin substructures in these samples.

Our UV and FT-IR results indicate that most of the H units are *p*-coumaric (4-hydroxycinnamic) acid type structures. The significantly higher content of H units in the *in situ* lignins (20–35 mol%, Table 10.4) than that obtained by nitrobenzene analysis (11–12 mol%) indicate that most of them are terminal phenolic units, as only such units are accessible to PO analysis. Although the proportions of condensed structures (products

IV–IX, Table 10.4) in the *in situ* lignins and in DLs are similar, the proportions of non-condensed H, G, and S structures (products **I–III**) in these samples are rather different. During the acidolytic isolation of DLs, some of the β-O-4 linkages in lignin are broken, yielding phenolic end units accessible for PO analysis. The increased proportion of product **III** and the accompanying, decreased proportion of product **I** (for DL_{LS}) or **II** (for DL_R) when compared with the corresponding *in situ* lignins (Table 10.4) suggested that the O4 units in the β-O-4 linkages involve a relatively higher S units but lower H units for the leaf sheath lignin, and relatively higher S units but lower G units for the rachis lignin. It can also be concluded based on results of PO analysis that most of the condensed structures of "Dwarf Cavendish" lignin are represented by phenylcoumarin (β-5′) and biphenyl (5-5′) type structures (products **IV** and **VII**, respectively). The noticeable amounts of product **VI** (2–4 mol%) indicate the presence of α-6′ (isotaxiresinol type) or β-6′ (phenylisochroman type) structures as suggested previously in wood lignins [39,40].

DL_{LS} had slightly more condensed units (molar ratio of products **IV–IX**/products **I–III** = 22%) than DL_R (molar ratio of products **IV–IX**/products **I–III** = 18%). The contents of condensed units in these lignins are similar to those reported for other annual plants and woods [16,31,32,41].

10.3.3.5 ^{13}C NMR analysis

The quantitative ^{13}C NMR spectra of the purified DL_{LS} and DL_R are presented in Figure 10.6, and the data on the frequency of occurrence of different structures, together with signals (δ ppm) used for the assignment of the structures, are summarized in Table 10.5. The almost complete absence of typical polysaccharide signals (δ 90–103 ppm) confirms

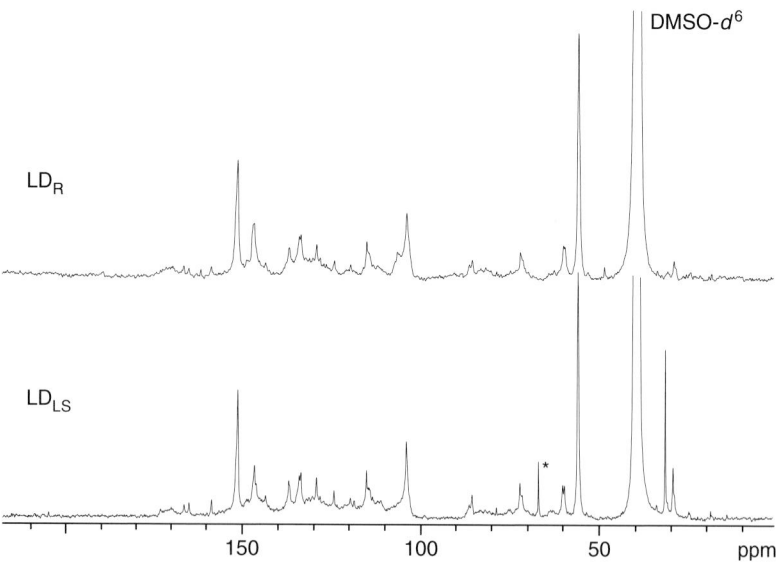

Figure 10.6 ^{13}C NMR spectra of dioxane lignins from banana plant leaf sheath (DL_{LS}) and rachis (DL_R) (∗ = solvent contamination).

Table 10.5 Frequency of various structural units in DL_{LS} and DL_R as revealed by ^{13}C NMR.

Structural units (signal used for assignment, δ)	Number/C_6	
	DL_{LS}	DL_R
OCH_3 (55.0 ppm)	1.10	1.55
β–β + β-5 structures (51.5–54.0 ppm)	0.08	0.02
β-O-4 structures without Cα=O (59.0–61.0 ppm)	0.27	0.32
β-O-4 structures with Cα=O (63.0 ppm)	0.04	0.02
Benzyl and Cγ ester structures (75.0–76.0 and 64.0–65.0 ppm)	0.09	0.05
Coniferyl alcohol structures (61.8 ppm)	0.02	0.02
p-Coumaric acid structures (Pc) (167.8 or 159.0–162.0 ppm)	0.07	0.05
Ferulic acid structures (Fe) (calculated value)	0.05	0.05
p-Hydroxybenzoic acid (163.0 ppm)	nd	0.02
Esterified Pc and Fe structures (166–167 ppm)	0.07	0.04
Etherified Pc and Fe structures (167–168 ppm)	0.05	0.06
H/G/S ratio	7:43:50	7:29:64

the low content of associated polysaccharides in both lignins as determined by analysis of neutral sugars (~1–2%). The signals at δ 103–156 and 159–162 ppm are due to aromatic (Ar) carbon and the side-chain olefinic carbon resonances [42]. We used the integral of these signals that had been corrected for the olefinic carbons (see below) to calculate the molar number of all structural units per one Ar ring (C_6); molar number of a structural unit/C_6 = integral of the characteristic carbon resonance in the structure/corrected integral at "103–156 + 159–162" ppm

The resonances at δ 167.8 and 166.4 ppm are assigned to C=O carbons in carboxyl and ester groups, respectively, in p-coumaric and ferulic acid structures [43,44]. The signals at δ 159.0–162.0 ppm also indicate the presence of p-coumaric structures and are assigned to Ar–C4. The signal at δ 163.0 ppm present only in the DL_R spectrum is assigned to Ar–C4 in p-hydroxybenzoic acid. A small resonance at δ 61.8 ppm is assigned to Cγ in cinnamyl alcohol units [42]. Using these assignments, the integral of signals at 103–156 and 159–162 ppm (abbreviated as $I_{103-162}$) was corrected for the presence of olefinic structures $[I_{103-162}(corrected) = I_{103-162} - 2I_{166.0-168.0} - 2I_{61.5-62.0}]$ and used for the calculation of the molar number of a structural unit/C_6.

Characteristic tertiary carbon signals from S units at δ 103.0–110.0 ppm (C2,6), G units at δ 110.0–125.0 ppm (C2,5,6), and quaternary carbon resonances from H units at δ 159.0–162.0 ppm (C4) were used to assess the H/G/S ratio in both lignins [16]. In DL_R the resonance at 163 ppm, due to p-hydroxybenzoic acid, was also included as H resonance. The H/G/S ratio was 7:43:50 and 7:29:64 for DL_{LS} and DL_R, respectively. However, the proportion of S units may be seriously overestimated as these structures are less condensed than G units and contribute more strongly to the number of tertiary carbon signals. The amounts of H type structures other than p-coumarates were insignificant (<0.02/C_6) as revealed from integrals of the signals at δ 156–158 ppm (C4 in corresponding structures) and were not included in our calculations. The amount of p-coumaric acid structures was assessed based on signals at δ 159.0–162.0 ppm and the amount of ferulic acid structures was calculated from the difference of integrals at 166.0–168.0 ppm (sum of p-coumarates and ferulates) and those at δ 159.0–162.0 ppm (only p-coumarates).

p-Coumaric acid structures in both lignins are mainly terminal phenolic units, because the signal at 159.8 ppm (C4 in phenolic p-coumarates) is prevalent over that at δ 160.8 ppm (C4 in nonphenolic p-coumarates). Therefore, a predominant part of p-coumaric acid structures is ester-linked to other lignin substructures. In the ^{13}C NMR spectra, the characteristic resonances at δ 75.0–76.0 ppm (Cα in benzyl acylated β-O-4 structures) and 64.0–65.0 ppm (Cγ in γ-acylated β-O-4 structures) [44] were observed. The total amount of benzyl and γ-ester structures, calculated from the corresponding integrals, was almost 30% more intense than the integral of signals at δ 166.0–167.0 ppm (Cγ=O in esterified p-coumarates and ferulates), indicating that p-coumaric acid structures are not the only ones ester-linked to the bulk lignin. These could be the suberin-like compounds, although the latter are normally linked to lignin by ester linkages via ferulates [45]. The intense resonances at δ 29.0 and 31.2 ppm in DL_{LS} (Figure 10.6) show that, despite the removal of the major proportion of aliphatic compounds during extraction with chloroform, part of them still remained in the purified DL_{LS} sample. A group of nonresolved signals at δ 171–173 ppm for DL_{LS} suggested the presence of ester-linked fatty/hydroxyl acids. Non-esterified p-coumarates and ferulates (integral at δ 167.0–168.0 ppm) should be linked to other lignin substructures by ether or carbon–carbon bonds.

Since most p-coumaric acid structures are terminal phenolic units (resonance at δ 159.8 ppm) and their content (0.07/C_6) is similar to that of the sum of esterified p-coumaric acid and ferulic acid structures (resonance at δ 166.4 ppm), almost all ferulic acid units should be ether-linked. The lignin structures involving β-O-4 linkages without Cα=O group were calculated based on the integral of signals at δ 59.0–61.0 ppm assigned to the Cγ resonance [42]. These are the most frequent structures in both lignins, representing ~0.27–0.32/C_6. The β-O-4 structures with Cα=O group (0.02–0.04/C_6) were assessed based on the small signal at δ 63.0 ppm, assigned to Cγ in the corresponding structures [42]. The amounts of these structures were confirmed by the integral at δ 196.0–198.0 ppm assigned to Cα=O groups of β-O-4 structures [46]. DL_R contains fewer β–β and β–5 structures than DL_{LS} (Table 10.5), as calculated from integral of signals at δ 51.5–54.0 ppm assigned to Cβ in the corresponding structures [42].

Comparing the intensities of the signals at δ 152.1 ppm (C3,5 in etherified β-O-4 linked S units) and 149.0–149.5 ppm (C3 in etherified β-O-4 linked G units), the ratio of which is about 8:1 for both lignins, it can be concluded that S units are mainly ether-linked contributing to linear molecular fragments, whereas G units are mainly non-etherified and linked by carbon–carbon bonds representing terminal phenolic lignin units. This explains the much higher content of G than S units in PO analysis (Table 10.4).

10.3.3.6 ^{1}H NMR analysis

^{1}H NMR spectroscopy was used to confirm the quantitative estimation of some structural units obtained from the analysis of ^{13}C NMR and to obtain additional structural information on the DLs. All calculations were made per C9 unit using the signal of methoxyl protons at 3.60–4.00 ppm as an internal standard. The ^{1}H NMR spectra of the acetylated DL_{LS} and DL_R are presented in Figure 10.7. The amounts of β-O-4 structures without

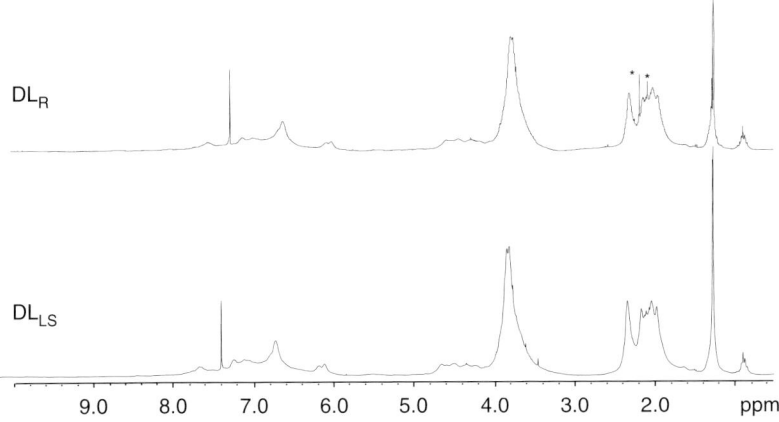

Figure 10.7 ^{1}H NMR spectra of the acetylated dioxane lignins from banana plant leaf sheath (DL$_{LS}$) and rachis (DL$_R$) (∗ = solvent contamination).

Table 10.6 Frequency of major structural units in DL$_{LS}$ and DL$_R$ as revealed by ^{1}H NMR.

Structural elements	DL$_{LS}$	DL$_R$
—CH$_2$— and —CH$_3$	1.36	1.69
Hβ in β–β structures	0.04	nd
Hα in β-5 structures (phenylcoumaran) and Hα in noncyclic benzyl aryl ethers (α-O-4 structures)	0.08	0.03
Hα in β-O-4 structures without Cα=O and Hα in β-1 structures	0.26	0.31

Note: nd – not detected.

Cα=O (Table 10.6), estimated based on the Hα signal at δ 5.8–6.2 ppm [47], are practically the same as those determined by ^{13}C NMR spectroscopy per C6 unit (Table 10.5). From the amounts of β–β structures estimated based on the Hβ signal at δ 3.0–3.15 ppm (0.04/C9 for DL$_{LS}$ and 0.0/C9 for DL$_R$; Table 10.6) and those of β-5 and β–β structures (Table 10.5), the amounts of β-5 structures can be calculated to be 0.04/C6 and 0.02/C6 for DL$_{LS}$ and DL$_R$, respectively.

The signals between 0.8 and 1.6 ppm, assigned to CH$_3$ and CH$_2$ in saturated aliphatic chains, respectively, suggest once more the presence of aliphatic compounds linked to lignin. The existence of suberin-like aliphatic chains covalently bound to the lignin polymer with similar signals has already been reported for *Quercus suber* L [17]. However, the presence of proton signals from suberin-like aliphatic structures (including hydroxy acids) at δ 1.5–2.2 ppm did not allow the quantification of aromatic and aliphatic hydroxyl groups, the acetate derivatives of which showed resonances in the same region [48]. The broad signal δ ∼7.4–7.6 ppm was assigned to H2/H6 aromatic protons and Hα in *p*-coumaric acid structures [46], confirming the presence of significant amounts of these structures in both lignins.

Conclusions

The chemical compositions of leaf sheath and rachis, the major components of the banana plant "Dwarf Cavendish" (*Musa acuminata* Colla, var. *cavendish*) biomass, have been determined to estimate their potential applications. A remarkable variability in the amounts and structures of the main constituents in these two morphological parts has been detected. Despite their rather high ash and extractive contents, the leaf sheath and rachis are rich in cellulose (31–37%) and non-(hemi)cellulosic polysaccharides (10–21%) and contain rather low amounts of lignin (10–13%). The noncellulosic polysaccharides, the structure of which has yet to be determined, may be of interest in the chemical processing of hemicelluloses (xylan, arabinan, and xyloglucan).

The similarity of lignin structures in the leaf sheath and rachis suggests that they can be processed together in pulp production. The previously reported successful pulping of the leaf sheath and rachis [3] can be explained by several lignin structural features revealed in this study: (1) high content of S units and relatively low degree of lignin condensation, and (2) relatively high content of β-O-4 structures and alkali-labile ester-linked structural units connecting the lignin with suberin-like component of the cell wall. Results of this work also suggest different mechanisms involved in the bonding of *p*-coumarates and ferulates to the main lignin backbone during biosynthesis.

References

1. FAO. FAO Statistical Databases. 2006. Available online at: http://faostat.fao.org/site/340/default.aspx (accessed on 24.04.2006).
2. Soffner, M.L.A.P., Produção de Polpa Celulósica a partir de Engaço de Bananeira, Master Thesis, Escola Superior de Agricultura "Luiz de Queiroz", Universidade de São Paulo, 2001.
3. Cordeiro, N.; Belgacem, M.N.; Chaussy, D.; and Moura, J.C.V.P., Pulp and Paper Properties from Dwarf Cavendish Pseudo-stems; *Cellul. Chem. Technol.* 2005, 39, 517–529.
4. Faria, H.; Cordeiro, N.; Belgacem, M.N.; and Dufresne, A., Dwarf Cavendish as a Source of Natural Fibers in Polypropylene-based Composites; *Macromol. Mater. Eng.* 2006, 291, 16–26.
5. Acid-insoluble Lignin in Wood and Pulp, TAPPI Test Method T222 om-02, Tappi, Atlanta, USA, 2002.
6. Obolenskaya, A.V.; Elnitskaya, E.P.; and Leonovitch, A.A., *Laboratorni raboti po Khimia Previcine I Cellulosi*, Moscow Ecologia, 1991.
7. Browning, B.L., The Isolation and Determination of Cellulose. In: *Methods of Wood Chemistry*, Vol. II; Interscience Publishers; New York, USA, 1967; p. 406.
8. Browning, B.L., The Isolation and Separation of Hemicelluloses. In: *Methods of Wood Chemistry*, Vol. II; Interscience Publishers; New York, USA, 1967; pp. 567–569.
9. Browning, B.L., Determination of Sugars. In: *Methods of Wood Chemistry*, Vol. II; Interscience Publishers; New York, USA, 1967; pp. 617–620.
10. Humphreys, F.R. and Kelly, J., A Method for the Determination of Starch in Wood; *Anal. Chim. Acta.* 1961, 24, 66–70.
11. Yeoh, H.H. and Wee, Y.C., Leaf Protein Contents and Nitrogen-to-Protein Conversion Factors for 90 Plant-Species; *Food Chem.* 1994, 49(3), 245–250.
12. Conklin-Brittain, N.L.; Dierenfeld, E.S.; Wrangham, R.H.; Norconk, M.; and Silver, S.C., Chemical Protein Analysis: A Comparison of Kjeldahl Crude Protein and Total Ninhydrin Protein from Wild, Tropical Vegetation; *J. Chem. Ecol.* 1999, 24(12), 2601–2622.

13. Blakeney, A.B.; Harris, P.J.; Henry, R.J.; and Stone, B.A., A Simple and Rapid Preparation of Alditol Acetates for Monosaccharide Analysis; *Carbohydr. Res.* 1983, 113, 291–299.
14. Björkman, A., Studies on Finely Divided Wood. Part I. Extraction of Lignin with Neutral Solvents; *Svensk Papperstidn.* 1956, 59, 477–485.
15. Obst, J.R. and Kirk, T.K., Isolation of Lignin. In: Wood, W.A. and Kellogg, S.T. (eds); *Methods in Enzymology, Biomass*; Part B, Vol. 161; Academic Press; New York, 1988; pp. 87–100.
16. Evtuguin, D.V.; Pascoal Neto, C.; Silva, A.M.S.; Amado, F.M.L.; Robert, D.; and Faix, O., Comprehensive Study on the Chemical Structure of Dioxane Lignin from Plantation *Eucalyptus globulus* Wood; *J. Agric. Food Chem.* 2001, 49, 4252–4261.
17. Pascoal Neto, C.; Cordeiro, N.; Seca, A.; Domingues, F.; Gandini, A.; and Robert, D., Isolation and Characterization of a Lignin-like Polymer of the Cork of *Quercus suber* L; *Holzforschung* 1996, 50, 563–568.
18. Meier, D. and Faix, O., Pyrolysis-Gas Chromatography-Mass Spectrometry. In: Lin, S.Y. and Dence, C.W. (eds); *Methods in Lignin Chemistry*; Springer-Verlag; Berlin, 1992; pp. 177–199.
19. Gellerstedt, G., Chemical Degradation Methods: Permanganate Oxidation. In: Lin, S.Y. and Dence, C.W. (eds); *Methods in Lignin Chemistry*; Springer-Verlag; Berlin, 1992; pp. 322–333.
20. Chen, C.-L., Nitrobenzene and Cupric Oxide Oxidations. In: Lin, S.Y. and Dence, C.W. (eds); *Methods in Lignin Chemistry*; Springer-Verlag; Berlin, 1992; pp. 301–321.
21. Zakis, G.F., *Functional Analysis of Lignins and their Derivatives*; Tappi Press; Atlanta, GA, 1994; pp. 31–33.
22. Evtuguin, D.V.; Domingues, P.; Amado, F.L.; Pascoal Neto, C.; and Ferrer Correia, A., Electrospray Ionization Mass Spectrometry as a Tool for Lignins Molecular Weight and Structural Characterisation; *Holzforschung* 1999, 53, 525–528.
23. Belgacem, M.N.; Zid, M.; Nicolsi, S.N.; and Obolenskaya, A.V., Study of the Chemical Composition of *Alpha tunisia*. In: *Chim. Technol. Drev., Mej. Sbor. Trud.*, LTA; Leningrad, 1986; pp. 111–114.
24. Pascoal Neto, C.; Seca, A.; Fradinho, D.; *et al.*, Chemical Composition and Structural Features of the Macromolecular Components of *Hibiscus cannabinus* Grown in Portugal; *Ind. Crops Prod.* 1996, 5, 51–58.
25. Pascoal Neto, C.; Seca, A.; Nunes, A.M.; *et al.*, Variations in Chemical Composition and Structure of Macromolecular Components in Different Morphological Regions and Maturity Stages of *Arundo donax*; *Ind. Crops Prod.* 1997, 6, 51–58.
26. Oliveira, L.; Freire, C.S.R.; Silvestre, A.J.D.; Cordeiro, N.; Torres, I.C.; and Evtuguin, D., Steryl Glucosides from Banana Plant *Musa Acuminata* Colla Var *Cavendish*; *Ind. Crops Prod.* 2005, 22, 187–192.
27. Oliveira, L.; Freire, C.S.R.; Silvestre, A.J.D.; Cordeiro, N.; Torres, I.C.; and Evtuguin, D., Lipophilic Extractives from Different Morphological Parts of Banana Plant 'Dwarf Cavendish'; *Ind. Crops Prod.* 2006, 23, 201–211.
28. Moreau, R.; Whitaker, B.D. and Hicks, K.B., Phytosterols, Phytostanols and their Conjugates in Foods: Structural Diversity, Quantitative Analysis and Health Promoting Uses; *Prog. Lipid Res.* 2002, 14, 457–500.
29. Quílez, J.; Garcia-Lorda, P. and Salas-Salvadó, J., Potential Uses and Benefits of Phytosteros in Diet: Present Situation and Future Directions; *Clin. Nutr.* 2003, 22, 343–351.
30. Atchison, J.E., Data on Non-wood Plant Fibers. In: Hamilton, F. and Leopold, B. (eds); *Pulp and Paper Manufacture*; TAPPI Press; Atlanta, GA, 1993; pp. 157–163.
31. Seca, A.M.L.; Cavaleiro, J.A.S.; Domingues, F.M.J.; Silvestre, A.J.D; Evtuguin, D.; and Pascoal Neto, C., Structural Characterization of the Bark and Core Lignins from Kenaf (*Hibiscus cannabinus*); *J. Agric. Food Chem.* 1998, 46, 3100–3108.
32. Seca, A.M.L.; Cavaleiro, J.A.S.; Domingues, F.M.J.; Silvestre, A.J.D; Evtuguin, D.; and Pascoal Neto, C., Structural Characterization of the Lignin from the Nodes and Internodes of *Arundo donax* Reed; *J. Agric. Food Chem.* 2000, 48, 817–824.

33. Oliveira, L.; Evtuguin, D.; Cordeiro, N.; Silvestre, A.J.D.; Silva, A.M.S.; and Torres, I.C., Structural Characterization of Lignin from Leaf Sheath of Banana Plant 'Dwarf Cavendish'; *J. Agric. Food Chem.* 2006, 54, 2598–2605.
34. Lin, S.Y., Ultraviolet Spectrophotometry. In: Lin, S.Y. and Dence, C.W. (eds); *Methods in Lignin Chemistry*; Springer-Verlag; Berlin, 1992; pp. 217–232.
35. Sun, J.-X.; Sun, X.-F.; Sun, R.-C.; Fowler, P.; and Baird, M.S., Inhomogeneities in the Chemical Structure of Sugarcane Bagasse Lignin; *J. Agric. Food Chem.* 2003, 51, 6719–6725.
36. Herbert, H.L., Infrared Spectra. In: Sarkanen, K.V. and Ludwig, C.H. (eds); Lignins Occurrence, Formation, Structure and Reactions; Wiley-Interscience; New York, 1971; pp. 267–297.
37. Sarkanen, K.V.; Chang, H.-M.; and Allan, G.G., Species Variation in Lignins. 2. Conifer Lignins; *Tappi J.* 1967, 50, 583–586.
38. Hawkes, G.E.; Smith, C.Z.; Utley, J.H.P.; Vargas, R.R.; and Viertler, H., Comparison of Solution and Solid-state ^{13}C NMR Spectra of Lignins and Lignin Model Compounds; *Holzforschung* 1993, 47, 302–312.
39. Nimz, H.H., Beech Lignins Proposal of a Constitutional Scheme; *Angew. Chem. Int. Ed. Engl.* 1974, 13, 313–321.
40. Sakakibara, A., Chemistry of Lignin. In: Hon, D.N.-S. and Shiraishi, N. (eds); *Wood and Cellulosic Chemistry*; Marcel Dekker; New York, 1991; pp. 113–175.
41. Pinto, P.C.; Evtuguin, D.V. and Pascoal Neto, C., Chemical Composition and Structural Features of the Macromolecular Components of Plantation *Acacia mangium* Wood; *J. Agric. Food Chem.* 2005, 53, 7856–7862.
42. Robert, D., Carbon-13 Nuclear Magnetic Resonance Spectrometry; In: Lin, S.Y. and Dence, C.W. (eds); *Methods in Lignin Chemistry*; Springer-Verlag; Berlin, 1992; pp. 250–273.
43. Lapierre, C., Application of New Methods for the Investigation of Lignin Structure; In: Jung, G.H.; Buxton, D.R.; Hatfield, R.D. and Ralph, J. (eds); *Forage Cell Wall Structure and Digestibility*; ASA-CSSA-SSSA; Madison, WI, 1993; pp. 133–166.
44. Ralph, J.; Marita, J.M.; Ralph, S.A.; *et al.*, Solution-State NMR of Lignins; In: Argyropoulos, D.S. (ed.); *Advances in Lignocellulosics Characterization*; Tappi Press; Atlanta, GA, 1999; pp. 55–108.
45. Bernards, M.A., Demystifying Suberin; *Can. J. Bot.* 2002, 80, 227–240.
46. Ralph, S.A.; Ralph, J. and Landucci, L.L., NMR Database of Lignin and Cell Wall Model Compounds. 1996. Available online at: http://www.dfrc.wisc.edu/software.html (accessed on 17.01.2004).
47. Chen, C.-L. and Robert, D., Characterization of lignin by ^{1}H and ^{13}C NMR spectroscopy. In: Wood, W.A. and Kellogg, S.T. (eds); *Methods in Enzymology*; Academic Press; New York, 1988; pp. 137–158.
48. Lundquist, K., Proton (1H) NMR Spectroscopy; In: Lin, S.Y. and Dence, C.W. (eds); *Methods in Lignin Chemistry*; Springer-Verlag; Berlin, 1992; pp. 242–249.

Chapter 11
Recent Advances in the Characterization of Lignosulfonates

Stuart E. Lebo, Svein Magne Bråten, Guro Elise Fredheim, Bjart Frode Lutnaes, Rolf Andreas Lauten, Bernt O. Myrvold, and Timothy John McNally

Abstract

Lignosulfonates are complex polymers isolated from spent sulfite pulping liquors and used mainly as dispersants and binders in several industrial fields. While years of research have led to the partial elucidation of the structure of these polymers, much is still not known. In this chapter, we review our results on the characterization of several industrial lignosulfonates using size exclusion chromatography and multi-angle laser light-scattering, two-dimensional (2D) nuclear magnetic resonance (NMR) spectroscopy, hydrophobic interaction chromatography (HIC), and x-ray (SEC-MALLS) fluorescence spectroscopy.

11.1 Molecular weight determination of lignosulfonates by size exclusion chromatography and multi-angle laser light scattering

11.1.1 Background

Size exclusion chromatography (SEC) is a well-established method for characterizing polymers in solution. It separates polymer samples according to differences in hydrodynamic volume. After calibration with standard polymers of known molar mass, the result can be converted into molar mass distribution. However, if the molecular configuration of the unknown is not the same as that of the calibration standards, very large errors can occur. In fact, this is the case with lignosulfonates (i.e., the determination of absolute molecular weight by SEC has been restricted by the lack of commercial absolute molecular weight standards). Calibrations performed with polymers such as polystyrenesulfonates [1,2], pullulans (polysaccharides consisting of maltotriose units) [3], or proteins [4] have been attempted with mixed results. Better results have been obtained by Lewis [5,6] and Buchholz [7], who introduced lignosulfonate fractions with molecular weights determined independently by analytical ultracentrifuge for calibration.

A major improvement in the field of molecular weight determination is the online combination of SEC and multi-angle laser light scattering (MALLS). In light scattering the observed zero-angle ($\theta = 0$) Rayleigh factor (R_θ) is related to the weight-average molecular weight through the standard equations:

$$Kc/R_\theta = 1/M_w + 2A_2c, \tag{11.1}$$

$$K = 4\pi^2 n_0^2 (dn/dc)^2 N_A^{-1} \lambda_0^{-4} \tag{11.2}$$

where (dn/dc) is the refractive index increment (which has to be known from independent measurements), n_0 is the refractive index of the solvent, N_A is Avogadro's number, λ_0 is the wavelength of the incident light (*in vacuo*), and A_2 is the second viral coefficient. For each elution slice (i) in SEC the concentration (c_i) is obtained from a concentration-sensitive detector, usually a refractive index or UV detector, and R_θ is obtained from the MALLS (after extrapolation to zero-angle). $M_{w,i}$ is then calculated according to Equation (11.1). A_2 must be known from independent (batch) light-scattering measurements, unless the sample concentration is kept so low that $A_2 c \ll 1/M_w$.

The molecular weights of lignosulfonate molecules are found in the literature to vary from 1000 to 150 000 g mol^{-1} [5,7–9], but values greater than 1.0×10^6 g mol^{-1} have been published [10,11]. Most of the molecular weight measurements on lignosulfonates have been made on fractions prepared from softwoods. Based on a limited number of results, lignosulfonates from hardwoods appear to have lower molecular weight than those from softwoods [12–14]. The lower molecular weight can be explained by the presence of sinapyl units in hardwoods, which have lesser tendency to polymerize than the coniferyl units in softwoods [13].

11.1.2 Experimental and materials

The instrument setup consisted of a SEC-column (Jordi Glucose-DVB) combined with a DAWN-F MALLS detector (Wyatt Technology Corporation) together with a Waters refractive index-detector. The mobile phase consisted of water, sodium hydrogenphosphate (Na$_2$HPO$_4$), dimethyl sulfoxide (DMSO), and sodium dodecylsulfate. This complex mobile phase was chosen to prevent formation of aggregates and absorption to the column material. Sulfonated lignins are fluorescent, and the MALLS detector was therefore equipped with suitable optical filters. A more detailed description of the setup is found in Fredheim [15].

Eight industrial lignosulfonate samples (designated as LS1–LS8) produced by Borregaard LignoTech, Norway, were analyzed. The wood type and the process of the samples are presented in Table 11.1. Six lignosulfonate fractions (F-40–F-70) prepared from LS1 were also included in this study. The fractionation was performed by eluting the lignosulfonate, LS1, from a cellulose column with aqueous ethanol at decreasing ethanol concentrations [16].

11.1.3 Results and discussion

Figure 11.1 shows elution profiles and calibration curves (plots of log M vs elution volume) of the six lignosulfonate fractions of LS1. A progressive shift towards higher elution volumes reflects the decrease in weight-average molecular weight (M_w). More than 90% calculated

Table 11.1 Wood type, counterion, and process for eight commercial lignosulfonates.

Sample	Wood type	Counterion	Process
LS1	Spruce	Na	Ion exchanged and ultrafiltered
LS2	Spruce	Na	Ion exchanged and filtered
LS3	Spruce	Na	LS1 pH adjusted to 12.5 and heat treated
LS4	Spruce	Na	LS1 desulfonated and oxidized
LS5	80 : 20 (Spruce : Birch)	Na	Ion exchanged and centrifuged
LS6	Aspen	Ca	No treatment
LS7	*Eucalyptus globulus*	Na	Ion exchanged and pH-adjusted to 9–10
LS8	*Eucalyptus grandis*	Ca	Heat treated

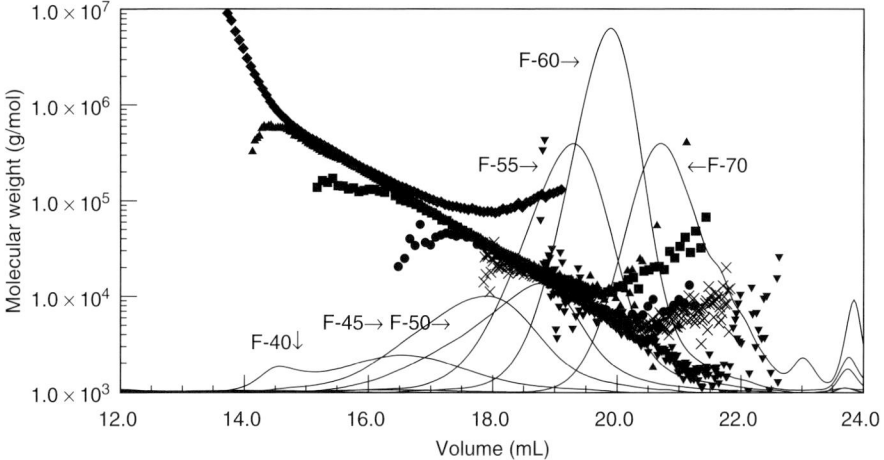

Figure 11.1 Calibration plots and elution curves for fraction F-70 (▼), F-60 (×), F-55 (●), F-50 (■), F-45 (▲), and F-40 (♦) [15].

sample recovery indicates that little material adsorbed to the column and that a negligible amount of low-molecular-weight components eluted in the salt peak [15].

The calibration curves in Figure 11.1 are basically the same for all the fractions. This indicates that the column covers the entire molecular weight range, and that peak broadening does not influence the result. It also demonstrates that all fractions belong to a family of polymer chain with the same basic shape, only differing in the molecular weight. The calculated M_w and number-average molecular weight (M_n) and polydispersities (M_w/M_n) are summarized in Table 11.2. The results demonstrate that commercial lignosulfonates are extremely polydisperse as fractions with molecular weights ranging from 4600 (F-70) to 400 000 g mol^{-1} (F-40) were obtained. Accordingly the polydispersity of the unfractionated sample was high.

The elution profiles and calibration curves obtained for the five softwood lignosulfonates, LS1–LS5 and for the three hardwood lignosulfonates, LS6–LS8, are shown in Figure 11.2a and b, respectively [17]. Only minor differences can be observed from the elution profiles

Table 11.2 dn/dc values, molecular weight averages, degrees of sulfonation and intrinsic viscosities (η) obtained for the unfractionated and the six fractions of LS1.

Fraction	dn/dc (mL g^{-1})	M_n (g mol^{-1})	M_w (g mol^{-1})	M_w/M_n	SO$_3$/C$_{9.95}$	η (mL g^{-1})
LS1	0.195	7 200	64 000	8.8		
F-70	0.186	3 200	4 600	1.5	0.64	1.8
F-60	0.196	6 100	8 000	1.3	0.53	3.0
F-55	0.200	9 900	15 000	1.5	0.49	3.8
F-50	0.204	18 000	34 000	1.9	0.44	5.2
F-45	0.203	31 000	68 000	2.3	0.41	6.6
F-40	0.205	120 000	400 000	3.3	0.39	12.1

of softwood, LS1–LS3 and LS5. They are all broad, reflecting considerable polydispersity. Sample LS4 has a narrower and symmetrical elution profile, which is shifted towards higher elution volumes, indicating both lower molecular weight and lower polydispersity than the other samples. This sample has been exposed to harsh processing conditions, and a reduction in molecular weight is therefore expected. The elution profiles for the hardwood samples are all clearly shifted towards higher elution volumes as compared to those for the softwood samples, suggesting that the average molecular weight is lower. In these cases distinct low-molecular-weight molecules appear as individual peaks between 20.5 and 23.0 mL.

The same basic calibration curve was obtained for all the samples except LS4 (Figure 11.2). Provided that non-SEC interactions do not influence the elution behavior, these data suggest that both softwood and hardwood lignosulfonates belong to a single polymer family with the same basic conformation, differing primarily in molecular weight.

The weight-averaged molecular weights of the lignosulfonates range from ~5000 to 60 000 g mol^{-1}, and the polydispersities are high (Table 11.3). Softwood lignosulfonates (LS1–LS3 and LS5) have relatively high M_w, ranging from 36 000 to 61 000 g mol^{-1}. For LS4 the low M_w (4400 g mol^{-1}) may be attributed to the harsh processing conditions. Hardwood lignosulfonates from aspen and *Eucalyptus* (LS6–LS8) have generally much lower M_w, ranging from 5700 to 12 000 g mol^{-1}. The lower-molecular-weight lignosulfonates from hardwood is attributed to the intrinsic chemical structure of hardwood lignin, and not the pulping process. Overall the SEC-MALLS method developed is fast and molecular weights can be determined over a broad range (5000–400 000 g mol^{-1}) in a single experiment.

11.2 Structure of lignosulfonates by two-dimensional nuclear magnetic resonance spectroscopy

11.2.1 Introduction

Nuclear magnetic resonance (NMR) spectroscopy is extensively used for structural determination of natural products and polymers [18,19]. Initially, the use of NMR spectroscopy to study the structure of both "native" and technical lignins was limited to one-dimensional (1D) ^{1}H and ^{13}C NMR spectroscopy [20,21]. However, the complex nature of lignins,

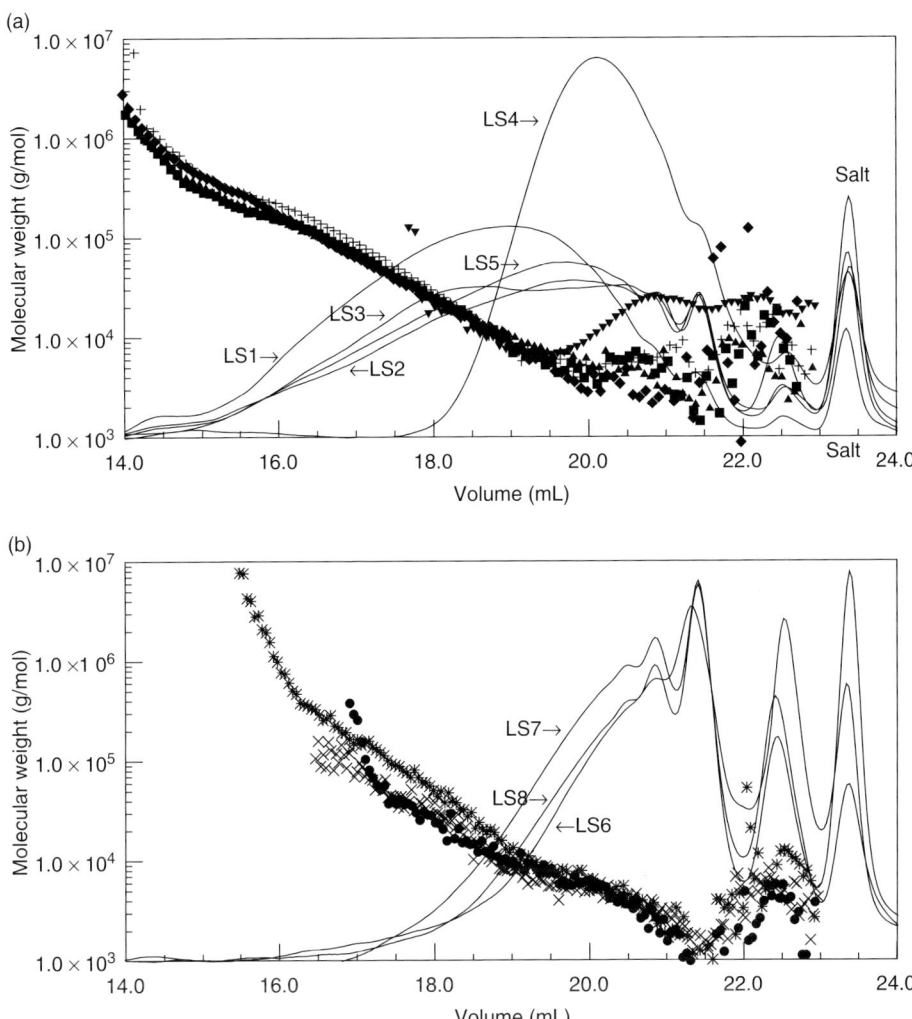

Figure 11.2 (a) Calibration plots and elution curves for softwood lignosulfonates: LS1 (♦), LS2 (■), LS3 (+), LS4 (▼) and LS5 (▲). (b) Calibration plots and elution curves for hardwood lignosulfonates: LS6 (✳), LS7 (●) and LS8 (×). From Fredheim et al. [17], Figure 1, copyright (2003) with permission from Taylor & Francis.

including lignosulfonates, often resulted in overlapping peaks in 1D spectra. Therefore, 2D ^1H–^{13}C heteronuclear correlation spectra are now being used more frequently to obtain structural information on lignins [22].

11.2.2 NMR of lignosulfonates

Although "non-polar" or "non-sulfonated" technical lignins have been studied extensively using NMR spectroscopy, lignosulfonates have not. The main reason for this is that

Table 11.3 Molecular weight averages, polydispersities and degrees of sulfonation obtained for eight lignosulfonate samples (LS1–LS8).

Sample	M_n (g mol^{-1})	M_w (g mol^{-1})	M_w/M_n	$SO_3/C_{9.95/10.5}$
LS1	5 000	61 000	12	0.52
LS2	3 200	36 000	11	0.59
LS3	5 100	41 000	8.2	0.53
LS4		4 400		0.20
LS5	4 300	36 000	8.8	0.52
LS6	2 200	12 000	5.3	0.54
LS7	2 200	6 300	3.0	0.54
LS8	1 900	5 700	3.0	0.46

lignosulfonates are insoluble in almost all organic solvents. Lignosulfonates are usually obtained as metal salts, and the resulting solutions obtained when they are dissolved in water are problematic in NMR, due to currents in the conducting solution and sample heating. Lignosulfonates also contain some paramagnetic ions, typically iron, manganese, and nickel, the presence of which causes signal broadening due to an increased relaxation rate.

11.2.3 Sample preparation

The literature contains a number of methods for derivatizing nonsulfonated lignins to make them soluble in organic solvents [20,21]. These methods do not extend well to lignosulfonates. However, lignosulfonates can be rendered soluble in methanol by ion exchange to the acid form [23]. This method was adapted to obtain lignosulfonate in the D^+ form dissolved in methanol-d_4, using Amberlite or Amberlyst ion exchange resin. In this reaction, most of the metal ions, including the paramagnetic impurities, are removed from the solution, thereby resolving the problems associated with conducting solutions and increased relaxation rates. Also, by running the analyses in methanol-d_4, the association problems observed in D_2O were reduced and spectra with better resolution could thus be obtained.

11.2.4 Equipment

All NMR spectra have been recorded on a Bruker Avance 600 NMR Spectrometer operating at 600 MHz for ^1H and 150 MHz for ^{13}C. The instrument was equipped with a TCI-cryogenic probe with cold ^1H and ^{13}C channels.

11.2.5 Results and discussion

Three lignosulfonates, LS2, LS3, and LS4 (see Table 11.1 for their descriptions), were ion-exchanged to the D^+ forms and analyzed in methanol-d_4. Figure 11.3 shows the ^1H–^{13}C HSQC spectrum of a low-molecular-weight fraction of lignosulfonate, LS2. The signal from

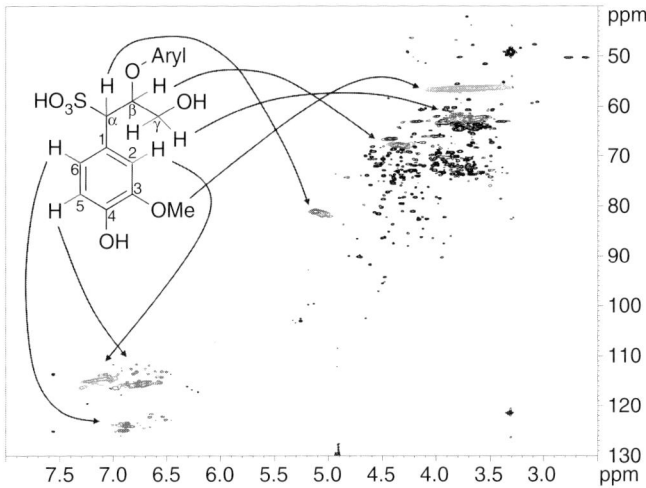

Figure 11.3 600 MHz ^{1}H–^{13}C HSQC NMR spectrum of a low-molecular-weight fraction of LS2; the structure of the α-sulfonic acid β-O-4-moiety, including the numbering of the carbon atoms, is shown together with the assignment of the various C–H signals by color and arrows from the structure (see also Plate 5).

methoxy groups in this sample can be seen in green. Signals from the three aromatic protons and carbons indicated in orange, magenta, and blue at the aromatic 2, 5, and 6 positions, respectively, can also be seen. Model compound studies indicate that the signals from the most important inter-unit linkage in lignosulfonate, the α-sulfonic acid β-O-4-moiety, can be seen in brown. The signals in the 4.5–5.5/90–105 region of the spectrum are ascribable to the protons/carbons present at the anomeric positions in glycosides, indicating that residual sugars are present in LS2.

Figure 11.4 shows the multiplicity-edited ^{1}H–^{13}C HSQC spectrum of a mid-range (i.e., moderately hydrophobic and dispersing) lignosulfonate, LS3. Signals from methylene groups in this sample is shown in red. Signals from methine and methyl groups is shown in black. Despite partial desulfonation, signals for the α-sulfonic acid β-O-4-moiety is still present, as seen in brown. The spectrum also shows that LS3 is less heterogeneous than LS2. In addition, the intensity of the aromatic signal at ~6.7/123 ppm associated with protons/carbons at the C6 position on the aromatic moieties (shown in blue in Figure 11.3), is somewhat lower than that for LS2. This indicates that LS3 contains a higher amount of 6-substituted aromatic moieties. The HSQC spectrum of LS3 contains no signals associated with residual sugars.

The ^{1}H–^{13}C HSQC spectrum of a high-end (i.e., highly hydrophobic and dispersing) ultrafiltered and oxidized lignosulfonate, LS4, shows very few signals in the aliphatic region (Figure 11.5). No signals associated with the α-sulphonic acid β-O-4-moiety, which is prominent in the HSQC spectra of LS2 and LS3, can be seen, indicating that the β-O-4 linkage between the aromatic units is broken in the oxidation process. Another interesting feature is the presence of signal at ~5.3/104 ppm that may be ascribed to non-aromatic double bonds. Also, the relative intensity of the aromatic signal associated with the proton/carbon

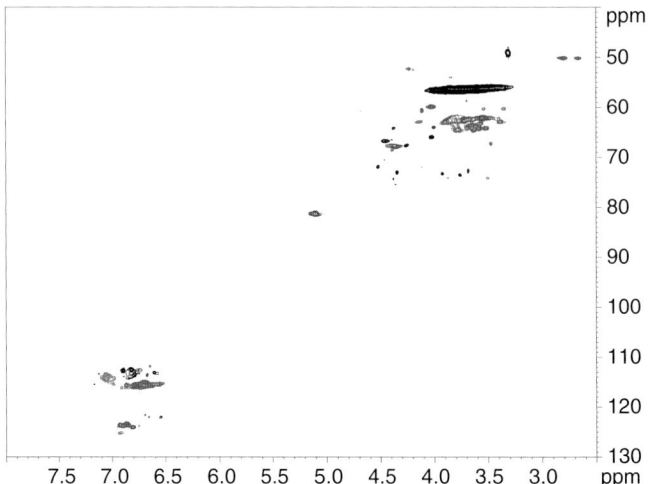

Figure 11.4 600 MHz ^{1}H–^{13}C multiplicity-edited HSQC spectrum of LS3; signals ascribable to the methylene groups, the methine or methyl groups, and the α-sulfonic acid β-O-4-moiety are shown in red, black, and brown, respectively (see also Plate 6).

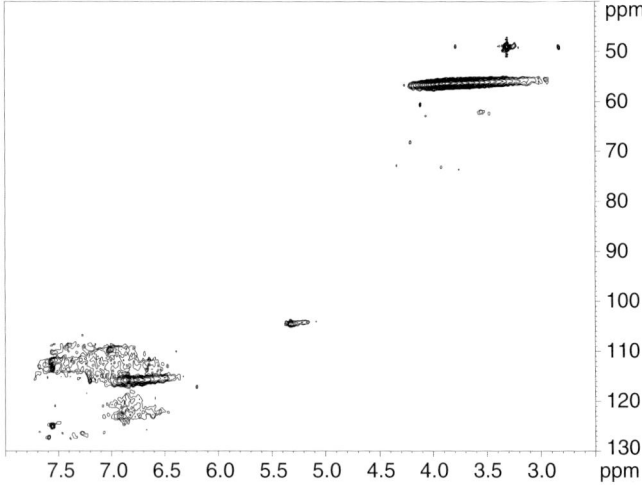

Figure 11.5 600 MHz ^{1}H–^{13}C HSQC spectrum of a high-end lignosulfonate, LS4.

at the C6 position on the aromatic moiety is even lower for LS4 than for LS3, showing that a substitution may have occurred at this position during the oxidation.

The aromatic region in the spectrum of LS4 is also much more complex than that in the spectra of LS3 and LS2. Integration of the signals in the aromatic region and the signals from the methoxyl groups shows that LS4 contains significantly fewer aromatic protons relative to methoxyl protons than LS2 or LS3.

The results discussed herein indicate that, while the study of lignosulfonates by NMR spectroscopy is still relatively new, further work is warranted because the technique produces a wealth of useful structural information.

11.3 Hydrophobicity of lignosulfonates by HIC

11.3.1 Background

HIC separates substances according to hydrophobicity. HIC is typically used for purification and analysis of peptides and protein mixtures [24–28]. A HIC column consists of a hydrophilic matrix with weak hydrophobic ligands where hydrophobic surface sites of the sample can adsorb. Adsorption is promoted by using mobile phases with high salt concentrations. Solute molecules with larger hydrophobic surface areas will interact stronger with a HIC column than solute molecules with larger hydrophilic surface areas. A gradient in the salt content in the mobile phase often facilitates separation of the solute molecules according to hydrophobicity. Initially, a high salt content favors adsorption of the hydrophobic material while the hydrophilic passes through the column almost unhindered. Reduction of the salt content in the mobile phase promotes the elution of the hydrophobic solutes adsorbed on the column. The result is a chromatogram where the hydrophilic fractions of the sample elute first, followed by increasingly hydrophobic fractions. HIC is similar to reverse-phase chromatography but the interaction with the column material is weaker in HIC. The theoretical basis of HIC is explained by solvophobic theory [29] or by preferential interaction theory [30–33].

11.3.2 Methods

Recently, we have developed a method for HIC analysis of lignosulfonates [34]. In this method, the lignosulfonates are eluted as seven peaks in the chromatogram, numbered from 1 to 7. Peaks 1–3 represent the hydrophilic portion while peaks 5–7 represent the hydrophobic portion. Hence, a comparison of the area of the first three peaks relative to the area of the last three peaks allows the classification of lignosulfonates as either more or less hydrophobic.

11.3.3 Results and discussion

Of significant academic and industrial interest is the question of which factors affect the hydrophobic character of lignosulfonates. Lignosulfonates have an abundance of charged groups such as carboxylic acids, sulfonic acids, and phenolic hydroxyl groups. Lignosulfonate fragments containing a large amount of these hydrophilic groups elute early while fragments with smaller amounts of hydrophilic groups elute later in the chromatogram [31]. In most applications, the molecular weight of the solute does not affect the order of retention in HIC. Although some unusual observations have been made [35,36], it is discussed later and demonstrated elsewhere that on a general basis the molecular weight of the sample is less important than the number of hydrophobic groups [37]. Figure 11.6

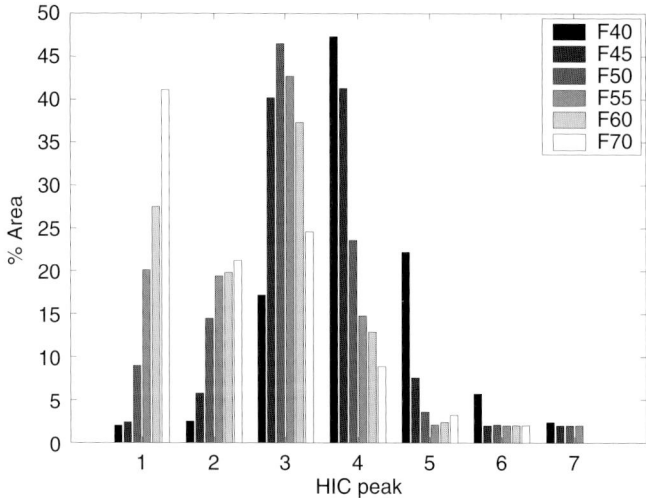

Figure 11.6 HIC characterization of six fractions from LS1.

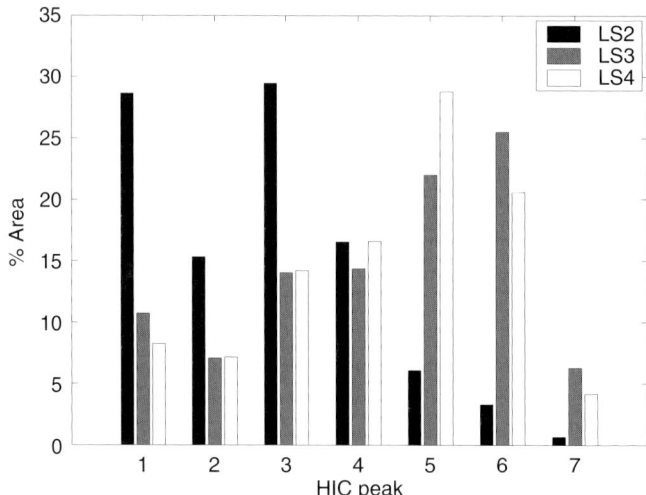

Figure 11.7 HIC characterization of LS2, LS3, and LS4.

displays the percentage areas of the seven HIC peaks for six fractions with different molecular weights from LS1 [16]. It can be seen that the high-molecular-weight fraction, which has a lower content of sulfonate groups (see Table 11.2), is more hydrophobic than the low-molecular-weight fraction. It appears that the hydrophobicity of the lignosulfonates increases with molecular weight due to the smaller amount of sulfonate groups.

Figure 11.7 shows the partition of three of the lignosulfonates listed in Table 11.1 in the HIC chromatogram. Some analytical data for these three lignosulfonates are listed

Table 11.4 HIC, molecular weight and total polar group content data for three commercial lignosulfonates.

Parameter	LS2	LS3	LS4
HIC Area peaks 1–3	73.4%	31.8%	29.6%
HIC Area peaks 5–7	10.2%	53.8%	53.5%
Σ Polar groups	18.0%	15.6%	19.1%

The polydispersity of industrial lignosulfonates is large and the variation in M_W reported for LS2–LS4 in Table 11.3 and here reflect the natural variation between batches.

in Table 11.4. The sum of polar groups, Σ Polar groups, refers to the sum of sulfonic and carboxylic acids and phenolic OH groups and was determined according to literature methods [38]. LS2 has the highest hydrophilic character for the HIC analysis but the amount of hydrophilic groups is not deviating significantly from LS3 and LS4. LS3 and LS4 have similar HIC profiles but the amount of charged groups is somewhat lower in LS3. The important difference between LS3 and LS4 appears to be the molecular weight. It is apparent from this comparison that HIC provides qualitatively new information in lignosulfonate analysis. Classical analysis (Table 11.4) do not suggest significant difference between LS2, LS3 and LS4. In light of their use as dispersants, LS2 is a lower end dispersant while LS3 and LS4 are mid-range and high-end dispersants respectively, possibly due to the more pronounced hydrophobic character.

It is difficult to draw conclusions on the behavior of LS4 with a limited understanding of its structure. However, it can be elucidating to compare LS4 with typical anionic surfactants. A typical anionic surfactant comprises one hydrophilic and one hydrophobic parts, and this clear division enables such a surfactant to form highly organized mesoscopic structure. A possible scenario can be that LS4 is a lignosulfonate that most closely resembles typical anionic surfactants. The low molecular weight (a distinct hydrophobic character) combined with a high apparent charge density gives LS4 characteristics more similar to those of synthetic surfactants than other purified lignosulfonates.

11.4 Elemental analysis of lignosulfonates by x-ray fluorescence spectroscopy

11.4.1 Background

Various elements are monitored and controlled in the manufacture of commercial lignosulfonates. These elements include calcium, sulfur, and those naturally present in trace quantities such as chloride, sodium, iron, manganese, nickel, and copper. Calcium is monitored in calcium lignosulfonates both prior to and post base exchange to other cations. Sulfur is monitored in products where the degree of sulfonation is increased or decreased depending on the desired result in the final product. To produce high-quality lignosulfonates for ceramics and refractories, sodium is controlled because it can, if present in sufficient quantity, affect the temperature at which vitrification occurs. These elements are regularly monitored as part of quality control programs to verify that they are present within set limits.

Typically, metal content is determined on an individual basis by atomic absorption spectroscopy or simultaneously with inductively coupled plasma (ICP) spectroscopy. Sulfur is analyzed by a sulfur analyzer or by ion chromatography.

X-ray fluorescence (XRF) spectroscopy was investigated as a method that could analyze the aforementioned elements quickly, simultaneously, and without destructive wet chemistry preparation. XRF can identify the elements over a wide concentration range from parts per million, ppm, to percentage levels. Each element is identified by its characteristic x-ray emission spectrum and quantified by the intensity of its characteristic line. When an atom is excited by removal of an electron from an inner shell, it usually returns to its normal state by transferring an electron from some outer shell to the inner shell and emitting energy as x-rays. In XRF spectroscopy, excitation is accomplished by irradiation with x-rays of shorter wavelength. XRF is a nondestructive analytical technique used to identify and determine the concentrations of elements in solids and liquids.

Moseley first demonstrated the relationship between atomic number, Z, and the reciprocal of the wavelength, $1/\lambda$, for each spectral line corresponding to a particular series of emission lines for each element in the periodic table. The relationship is expressed as

$$c/\lambda = a(Z - \sigma)^2 \tag{11.3}$$

where c is the velocity of light, a is a proportionality constant, and σ is a constant whose value depends on the particular series. In energy dispersive spectrometers the pulse height of the signal is proportional to the x-ray photon energy. The energy is related to the wavelength by the Duane–Hunt equation [39]:

$$\lambda_0 = hc/eV = 12.393/V \tag{11.4}$$

where λ_0 is the wavelength in angstroms (Å), V is the voltage in keV, e is the charge on the electron, h is Plank's constant, and c is the velocity of light. These wavelengths and energies are well documented and the measured spectral intensities allow an element and its concentration to be identified (Table 11.5). Dean states that chemical form (sample matrix) may shift emission lines as much as 10–20 eV [40].

Table 11.5 X-ray emission wavelengths and energies for element emission line, $K\alpha_1$.

Element	Wavelengths (Å)	Energy (keV)
Sodium	11.909	1.041
Sulfur	5.3720	2.308
Calcium	3.3583	3.691

Source: From Dean [40; Table 8-1].

Table 11.6 Conditions for XRF spectroscopic analysis of lignosulfonates.

Element	Voltage (kV)	Current (μA)	Medium	Tube filter	Counting time per sample (s)	Sample preparation
Calcium	8.00	325	Helium	Al Thin	60 and 180	Pressed powder
Sodium	5.00	150	Helium	None	180	Pressed powder
Sulfur	4.00	500	Helium	None	180	30% Solution

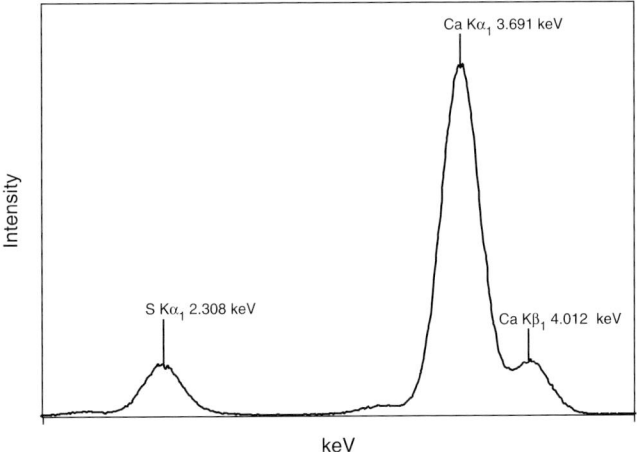

Figure 11.8 X-ray fluorescence spectrum of a calcium lignosulfonate standard analyzed with a rhodium x-ray tube showing sulfur and calcium.

11.4.2 Experimental and materials

The measurements were performed using a PANalytical MiniPal 2 benchtop, energy dispersive XRF (EDXRF) spectrometer equipped with a 30 kV rhodium tube, five filters, a helium purge, and a high resolution solid-state detector. An algorithm was used to correct for matrix effects and resolve elements. All analyses were performed in a helium environment to maximize the sensitivity of lower energy x-rays. The measurement conditions are given in Table 11.6. Powder samples were pressed in 32 mm aluminum pellet cups. To improve sample homogeneity, sulfur measurements were performed using 30% solutions.

11.4.3 Results and discussion

The contents of calcium and sodium, and the content of sulfur of a series of commercial lignosulfonates were analyzed with atomic absorption or ICP spectroscopy and with a sulfur analyzer, respectively, and used to provide the calibration curves. An example of a XRF spectrum of a calcium lignosulfonate with sulfur and calcium peaks present is shown in Figure 11.8. The calibration curve for calcium is shown in Figure 11.9.

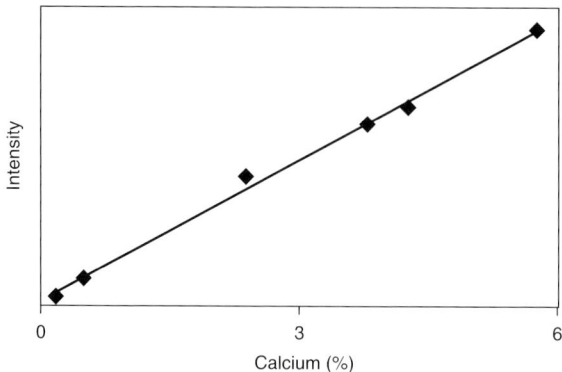

Figure 11.9 Calibration plot for calcium.

Table 11.7 Calibration results for calcium, sodium, and sulfur.

Element	Concentration range (wt%)	RMS (wt%)	Correlation coefficient	Lower limit of detection (wt%)
Calcium	0.18–5.77	0.102	0.996	0.001–0.002
Sodium	0.61–15.38	1.348	0.977	0.1–0.2
Sulfur	0.66–3.90	0.076	0.998	0.002–0.003

The calibration results for the three elements are summarized in Table 11.7. The root mean square (RMS) error listed is an indication of the magnitude of difference between the XRF measured concentration and the chemical standard. The calibration results indicate a good correlation between the known element concentration and the measured XRF intensity. The lower limits of detection for calcium and sulfur are satisfactory, while for sodium, the detection limit is at about the sodium specification. The comparatively poor sodium detection limit likely reflects the reduced detector sensitivity to the low energy x-rays the sodium atom emits.

The reproducibility is another important aspect of quantitative analysis. A sample was measured repeatedly to provide data. The average concentration, the standard deviation and the relative standard deviation are presented in Table 11.8. The reproducibility is within 2% relative for calcium, sodium, and sample **1** for sulfur. Apparently, there may be matrix effects influencing the results for sample **2**.

The work presented above represents a screening evaluation of the efficacy of XRF spectroscopy to serve as a quality control tool for the manufacture of commercial lignosulfonates. The technique has satisfactory sensitivity and correlation for calcium and sulfur. However, in the case of sodium, it is evident some technique refinement is needed to improve the correlation coefficient and the lower limit of detection. Also warranted is the investigation of the possible matrix effects in the case of sulfur. In the lignosulfonate sample matrix sulfur can be present as the sulfate ion, the sulfite ion, or the sulfonate group. It is postulated

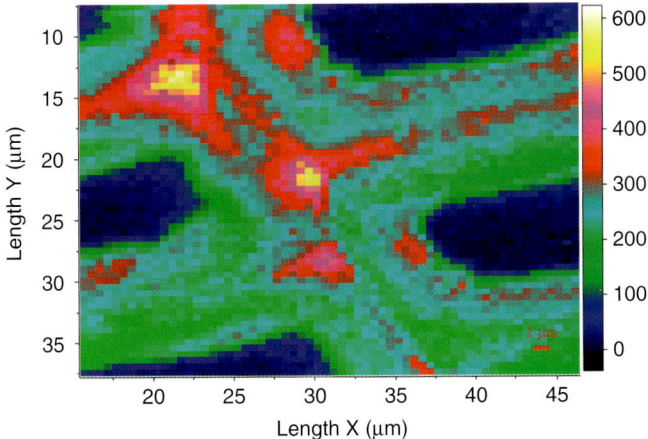

Plate 1 Raman image (false color) of lignin spatial distribution in selected cell wall area in (a) in two-dimensional representation. Intensity scale appears on the right. Bright white/yellow locations indicate high concentration of lignin; dark blue/black regions indicate very low concentration, for example, lumen area; from Reference 9.

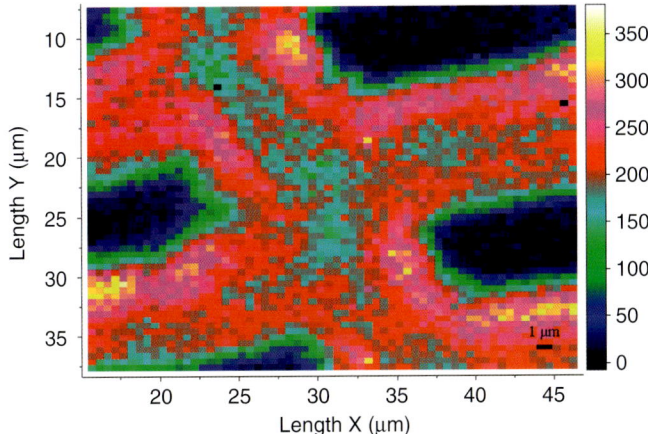

Plate 2 Raman images (false color) of cellulose spatial distribution in cell wall area selected in (a) in two-dimensional representation. Bright white/yellow locations indicate high cellulose concentration; dark blue/black regions indicate very low concentration, for example, lumen area; from Reference 9. See Plate 2 for the color image. (From Agarwal [9].)

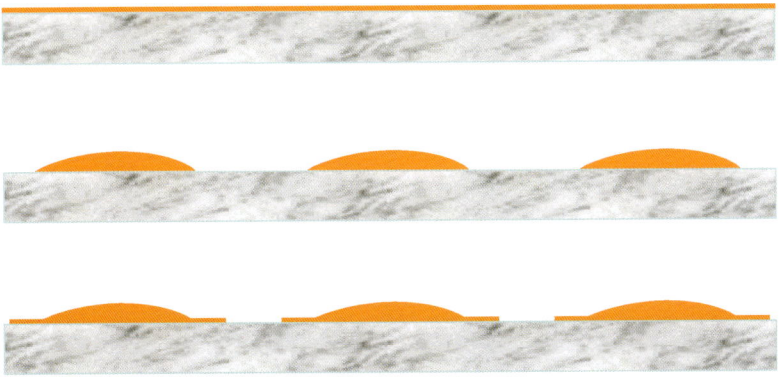

Plate 3 Schematic illustration of fiber/paper surfaces covered by AKD either as a monolayer, as patches or as a combination of both.

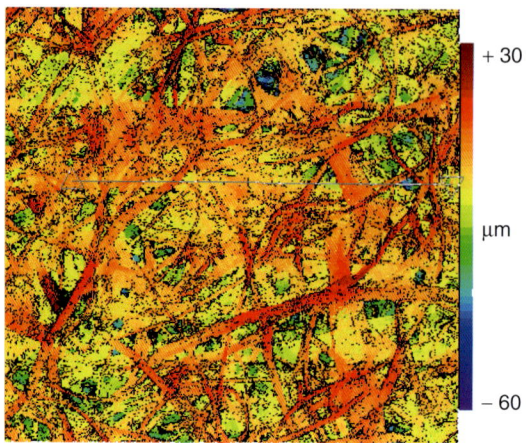

Plate 4 Two-dimensional (2D) topography of an unfilled pilot paper surface determined by white-light profilometry; the surface size is 1.15 mm × 1.23 mm.

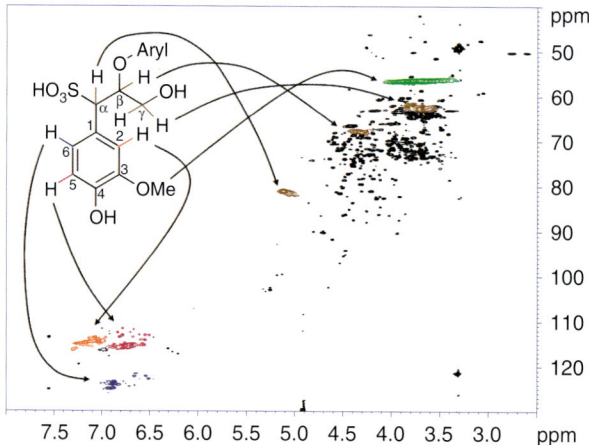

Plate 5 600 MHz ^{1}H–^{13}C HSQC NMR spectrum of a low-molecular weight fraction of LS2; the structure of the α-sulfonic acid β-O-4-moiety, including the numbering of the carbon atoms, is shown together with the assignment of the various C–H signals by color and arrows from the structure.

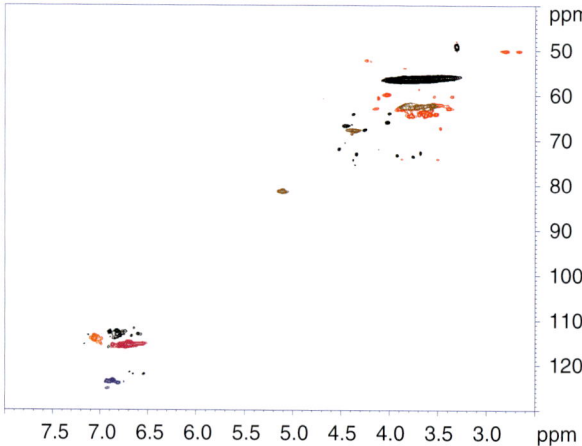

Plate 6 600 MHz ^{1}H–^{13}C multiplicity-edited HSQC spectrum of LS3; signals ascribable to the methylene groups, the methine or methyl groups, and the α-sulfonic acid β-O-4-moiety are shown in red, black, and brown, respectively.

Table 11.8 Average concentration, standard deviation, and relative standard deviation.

Element	Average concentration (wt%)	Standard deviation (wt%)	Relative standard deviation (%)	Number of measurement
Calcium	6.96	0.035	0.5	5
Sodium	8.2	0.17	2.0	3
Sulfur:				
Sample 1	2.11	0.038	1.8	7
Sample 2	2.83	0.166	5.8	9

that, as the energy spectra are affected by chemical form, these different forms of sulfur have different characteristic energy spectra. They therefore distort the correlation between energy intensity and sulfur concentration depending on their relative abundance.

References

1. Pellinen, J.; Salkinoja-Salonen, M., Aqueous Size Exclusion Chromatography of Industrial Lignins; *J. Chromatogr.* 1985, 17, 129–138.
2. Hage van der, E.R.E.; Loon van der, W.M.G.M.; Boon, J.J.; Lingeman, H.; Brinkman, U.A.T., Combined High-performance Aqueous Size-exclusion Chromatographic and Pyrolysis-gas Chromatographic-mass Spectrometric Study of Lignosulphonates in Pulp Mill Effluents; *J. Chromatogr.* 1993, 634, 263–271.
3. Chen, F.; Li, J., Aqueous Gel Permeation Chromatographic Methods for Technical Lignins; *J. Wood Chem. Technol.* 2000, 20, 265–276.
4. Budin, D.; Susa, L., Bestimmung des Molekulargewichts und der Molekularwichtsverteilung von Ligninsulfonaten aus Ca-Bisulfiteablauge; *Holzforschung* 1982, 36, 17–22.
5. Lewis, N.G.; Yean, W.Q., High-performance Size-exclusion Chromatography of Lignosulphonates; *J. Chromatogr.* 1985, 331, 419–424.
6. Lewis, N.G.; Goring, D.A.I.; Wong, A., Fractionation of Lignosulphonates Released during the Early Stages of Delignification; *Can. J. Chem.* 1983, 61, 416–420.
7. Buchholz, R.F.; Neal, J.A.; McCarthy, J.L., Some Properties of Paucidisperse Gymnosperm Lignin Sulfonates of Different Molecular Weights; *J. Wood Chem. Technol.* 1992, 12, 447–469.
8. Moacanin J.; Felicetta, N.F.; Haller, W.; McCarthy, J.L., Molecular Weights of Lignin Sulfonates by Light Scattering; *J. Am. Chem. Soc.* 1955, 77, 3470–3475.
9. Forss, K.; Stenlund, B., Molecular Weights of Lignosulfonates Fractionated by Gel Chromatography; *Papper och Trä*, 1969, 1, 93–105.
10. Rezanowich, A.; Goring, D.A.I., Polyelectrolyte Expansion of Lignin Sulfonate Microgel; *J. Colloid Interface Sci.*, 1960, 15, 452.
11. Hüttermann, A., Gelchromatographie vaon Na-Ligninsulfonaten an Sepharose CL 6B; *Holzforschung*, 1977, 31, 45–50.
12. Nokihara, E.; Tuttle, M.J.; Felicetta, V.F.; McCarthy, J.L., Molecular Weights of Lignin Sulfonates during Delignification by Bisulfite-sulfurous Acid Solutions; *J. Am. Chem. Soc.* 1957, 79, 4495–4499.
13. Sjöström, E.; Haglund, P.; Janson, J., Changes in Cooking Liquor Composition during Sulphite Pulping; *Svensk Papperstidn.* 1962, 65, 855–869.

14. Forss, K.; Stenlund, B.; Sågfors, P.E., Determination of the Molecular-weight Distribution of Lignosulfonates and Kraft Lignin; *J. Appl. Polym. Sci.* 1976, 28, 1185–1194.
15. Fredheim, G.; Braaten, S.M.; Christensen, B.E., Molecular Weight Determination of Lignosulfonates by Size Exclusion Chromatography and Multi-angle Laser Light Scattering; *J. Chromatogr. A* 2002, 942, 191–199.
16. Felicetta, V.F.; Ahola, A.; McCarthy, J.L., Distribution in Molecular Weight of Certain Lignin Sulfonates; *Can. J. Chem.* 1956, 78, 1899–1904.
17. Fredheim, G.; Braaten, S.M.; Christensen, B.E., Comparison of Molecular Weight and Molecular Weight Distributions of Softwood and Hardwood Lignosulfonates; *J. Wood Chem. Technol.* 2003, 23(2), 197–215.
18. Morris, G.A., *Structure Elucidation of Natural Products by NMR*; Wiley; Chichester, 1992.
19. Mirau, P.A., *A Practical Guide to Understanding the NMR of Polymers*; Wiley-Interscience; Hoboken, 2005.
20. Lundquist, K., Proton (1H) NMR spectroscopy. In Lin, S.Y. and Dence, C.W. (eds); *Methods in Lignin Chemistry*; Springer-Verlag; Berlin, 1992; pp. 242–249.
21. Robert D., Carbon-13 Nuclear Magnetic Resonance Spectrometry. In Lin, S.Y. and Dence, C.W. (eds); *Methods in Lignin Chemistry*; Springer-Verlag; Berlin, 1992; pp. 252–273.
22. Ralph, J.; Marita, J.M.; Ralph, S.A., Solution-state NMR of lignins. In Argyropoulos, D.S. (ed.); *Advances in Lignocellulosic Characterization*; Tappi Press; Atlanta, 1999; pp. 55–108.
23. Bialski, A.M.; Luthe, C.E.; Fong, J.L., Sulphite-promoted Delignification of Wood: Identification of Paucidisperse Lignosulfonates; *Can. J. Chem.* 1986, 64, 1336–1344.
24. Bramanti, E.; Ferri, F.; Sortino, C.; Onor, M.; Raspi, G.; Venturini, M., Characterisation of Denatured Proteins by Hydrophobic Interaction Chromatography: A Preliminary Study; *Biopolymers* 2003, 69, 293–300.
25. Porath, J.; Sundberg, L.; Fornstedt, N.; Olsson, I., Salting-out in Amphiphilic Gels as a New Approach to Hydrophobic Adsorption; *Nature* 1973, 245, 465–466.
26. Diogo, M.M.; Queiroz, J.A.; Monteiro; G.A.; Martins, S.A.M.; Ferreira, G.N.M.; Prazeres, D.M.F., Purification of a Cystic Fibrosis Plasmid Vector for Gene Therapy Using Hydrophobic Interaction Chromatography; *Biotechnol. Bioeng.* 2000, 68, 576–583.
27. Pomazal, K.; Prohaska, C.; Steffan, I., Hydrophobic Interaction Chromatographic Separation of Proteins in Human Blood Fractions Hyphenated to Atomic Spectrometry as Detector of Essential Elements; *J. Chromatogr. A* 2002, 960, 143–150.
28. Azuma, J.-I.; Koshijima, T.; Hydrophobic Interaction Chromatography of Pine Björkman-LCC; *Mokuzai Gakkaishi*, 1985, 31, 383–387.
29. Takahashi, N.; Azuma, J.-I.; Koshijima, T., Fractionation of Lignin–Carbohydrate Complexes by Hydrophobic-Interaction Chromatography; *Carbohydr. Res.* 1982, 107, 161–168.
30. Melander, W.R.; Corradini, D.; Horváth, C.S., Salt-Mediated Retention of Proteins in Hydrophobic Interaction Chromatography; *J. Chromatogr.* 1984, 317, 67–85.
31. Fausnaugh, J.L.; Regnier, F.E., Solute and Mobile Phase Contributions to Retention in Hydrophobic Interaction Chromatography of Proteins; *J. Chromatogr. A* 1986, 359, 131–146.
32. Xia, F.; Nagrath, D.; Garde, S.; Cramer, S.M., Evaluation of Selectivity Changes in HIC Systems Using a Preferential Interaction Based Analysis; *Biotechnol. Bioeng.* 2004, 87, 354–363.
33. Perkins, T.W.; Mak, D.S.; Root, T.W.; Lightfoot, E.N., Protein Retention in Hydrophobic Interaction Chromatography: Modelling Variation with Buffer Ionic Strength and Column Hydrophobicity; *J. Chromatogr. A* 1997, 766, 1–14.
34. Ekeberg, D.; Sandersen Gretland, K.; Gustafsson, J.; Braten, S.M.; Fredheim, G.E., Characterisation of Lignosulphonates and Kraft Lignin by Hydrophobic Interaction Chromatography; *Anal. Chim. Acta* 2006, 565, 121–128.
35. Gerle, M.; Fischer, K.; Roos, S.; *et al.*, Main Chain Conformation and Anomalous Elution Behavior of Cylindrical Brushes as Revealed by GPC/MALLS, Light Scattering, and SFM; *Macromolecules* 1999, 32, 2629–2637.

36. Janzen, R.; Unger, K.K.; Giesche, H.; Kinkel, J.N.; Hearn, M.T.W., Evaluation of Advanced Silica Packings for the Separation of Biopolymers by High-performance Liquid Chromatography; *J. Chromatogr.* 1987, 397, 91–97.
37. Renard, D.; Lavenant-Gourgeon, L.; Ralet, M.-C.; Sanchez, C., *Acacia Senegal* Gum: Continuum of Molecular Species Differing by their Protein to Sugar Ratio, Molecular Weight and Charges; *Biomacromolecules* 2006, 7, 2637–2649.
38. Lin, S.Y.; Dence, C.W. (eds), *Methods in Lignin Chemistry*; Springer-Verlag; Berlin, 1992, Chapter 7 Functional Group Analysis, pp. 407–484.
39. Duane, W.; Hunt, F.L., X-ray Wave-lengths; *Phys. Rev.* 1915, 66, 166–171.
40. Dean, J.A. (ed.), *Lange's Handbook of Chemistry*, 13th Edition; McGraw-Hill; New York, 1985, pp. 8–3.

Chapter 12
Integrated Size-Exclusion Chromatography (SEC) Analysis of Cellulose and its Derivatives

Akira Isogai and Masahiro Yanagisawa

Abstract

Use of size-exclusion chromatography (SEC) systems equipped with multi-angle light-scattering (MALS), quasi-elastic light-scattering, and photodiode array detectors provides significant and valuable information such as absolute molecular mass (MM), MM distribution, root mean square radius and molecular conformations of cellulose, pulp samples and cellulose derivatives in solutions, and their interactions with other coexisting components such as hemicelluloses or residual lignin. The LiCl/N,N-dimethylacetamide (DMAc)and LiCl/1,3-dimethyl-2-imidazolidinone (DMI) solvent systems are suitable for direct dissolution of cellulose, pulp samples, and cellulose derivatives for SEC-MALS analysis, although some skillful handling in the preparation of the cellulose solutions and careful interpretations of the SEC results are required.

12.1 Introduction

Polymer chains take various forms or conformations in solution- and solid-states depending on their primary structures and surroundings. Some specific properties and functionalities of polymers are influenced by their molecular mass (MM) values, MM distributions, conformations and other factors as macromolecules. This also holds true with cellulose and cellulose derivatives which have been extensively utilized in various forms such as fibers, films, gels, thickeners, retarding materials, membranes, paper and others with or without chemical and morphological modifications. Cellulosic materials generally have various MM and MM distributions, depending on their biological origins and treatment histories during isolation, purification and modification processes. Size-exclusion chromatography (SEC) is practically the sole method for MM and MM distribution measurement of cellulose and cellulose derivatives.

Recently, SEC analysis of unmodified cellulose using LiCl/N,N-dimethylacetamide (DMAc) as a mobile phase has been extensively studied [1–8]. Many constructive results on the molecular characteristics of cellulose and other related materials (e.g., hemicelluloses and residual lignin) were thus obtained [9–14]. These techniques were comprehensively reviewed by Strlič and Kolar [15], Conner [16], and Sjöholm [17]. However, some problems associated with LiCl/DMAc were also reported; for example, incomplete dissolution of several cellulose materials such as softwood kraft pulps and detrimental degradation of cellulose during dissolution at heating in this solvent system [6]. LiCl/1,3-dimethyl-2-imidazolidinone (DMI) is a transparent and colorless solvent system applicable to unmodified celluloses, and has some advantages over LiCl/DMAc in terms of solubility of some cellulose materials such as tunicate cellulose and softwood kraft pulps [18–20].

SEC systems equipped with multi-angle light scattering (MALS) (also referred to as multi-angle laser light scattering, MALLS), quasi-elastic light scattering (QELS) and photodiode array (PDA) detectors have emerged as potential tools for studying molecular structures and conformations of polymers in solution states, and the MM and MM distributions of cellulose, hemicelluloses and residual lignin as well as their interactions [17,21]. Especially, absolute MM, root mean square radius (RMS radius, R_g or $\langle S^2 \rangle^{1/2}$) and hydrodynamic radius (R_h) of each elution slice can be successively and simultaneously obtained by the SEC-MALS-QELS system. This sequential technique needs no multiple standard samples with known MM and MM distributions, and samples with various MM distributions can be utilized in the same manner in principle. Distributions of small amounts of carbonyl and carboxyl groups present in pulp samples can also be determined by selective fluorescent-labeling techniques combined with SEC-MALS analysis [5,22,23]. The LiCl/DMAc system has been used as a solvent and a mobile phase in the SEC-MALS analyses of cellulose by many researchers, while a few studies using LiCl/DMI have also been reported. In this chapter, some results of SEC-MALS-PDA analysis of cellulose and cellulose derivatives recently obtained in our laboratory and the related techniques are discussed.

12.2 Background of MALS analysis

The $R(\theta)$ values obtained from Rayleigh light scattering (LS) intensities detected with photodiodes in a MALS apparatus are expressed by the Equation (12.1). K^*, n_0 (refractive index of the solvent) and λ_0 (wavelength of the incident light) are constant, depending on the solvent and polymer used in SEC analysis, and the concentration c, weight-average MM, M_w, mean-square radius $\langle S^2 \rangle_z$ and the angle θ are variables [21].

$$\frac{K^*c}{R(\theta)} \approx \frac{1}{M_w}\left[1 + \left(\frac{16\pi^2 n_0^2}{3\lambda_0^2}\right)\langle S^2 \rangle_z \sin^2\left(\frac{\theta}{2}\right)\right] \quad (12.1)$$

The constant K^* is expressed by the Equation (12.2), and dn is the solution refractive index increment with respect to a concentration change dc of the solute molecules. The dn/dc values must be measured separately using an interferometric refractometer with laser light having the same wavelength as that of the MALS apparatus used.

$$K^* = 4\pi^2 (dn/dc)^2 n_0^2 / (n_a \lambda_0^4) \quad (12.2)$$

Figure 12.1 shows a representative relationship between $\sin^2(\theta/2)$ and $K^*c/R(\theta)$. At each slice of SEC elution patterns, the concentration c calculated from refractive index (RI) intensity is constant. Thus, M_w and RMS radius ($\langle S^2 \rangle_z^{1/2}$) of each slice can be obtained from the extrapolated value at $\theta \to 0$ and the slope, respectively, in Figure 12.1, assuming that c is

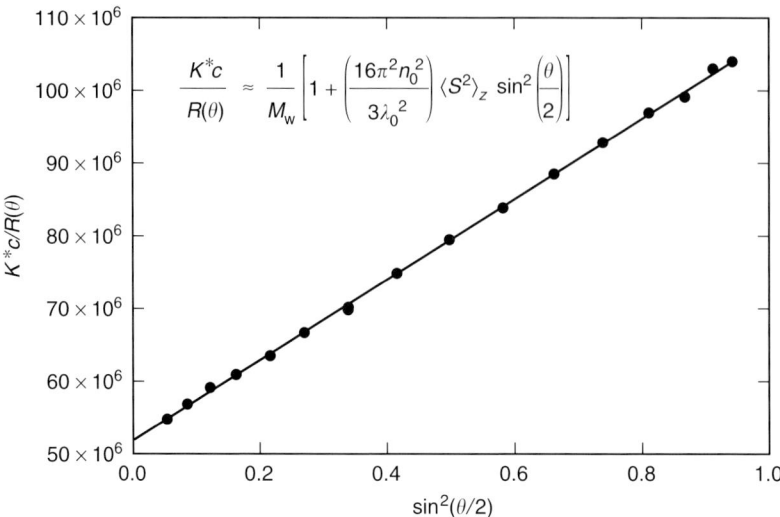

Figure 12.1 Relationship between $\sin^2(\theta/2)$ and $K^*c/R(\theta)$ of a polymer obtained by MALS.

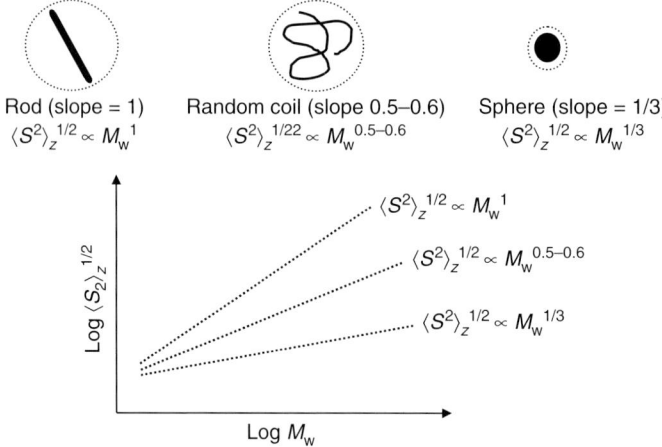

Figure 12.2 Typical conformation plots (double-logarithmic plots of M_w vs $\langle S_2 \rangle_z^{1/2}$) of polymers obtained by SEC-MALS analysis.

sufficiently small. The commercially available MALS apparatus and the attached software provide the absolute values of M_w and RMS radius simultaneously and successively at each slice of SEC elution patterns without using any monodisperse standard polymers.

The obtained M_w and RMS radius values corresponding to each slice of SEC elution patterns also impart significant information concerning conformations of the molecules in the solvent. From the slope of double-logarithmic plots of M_w vs RMS radius of polymers (Figure 12.2), their conformations can be categorized to, for example, rod-like, random-coil and sphere ones. As described later, the possibility for cellulose molecules to have some branch structures with hemicelluloses, or the intra-molecular electric repulsions of cellulose derivatives having ionic groups in water can be studied from their conformation plots obtained by the MALS analysis. The MALS detector is generally sensitive to polymer molecules with RMS radii greater than 10 nm. Thus, when an intrinsic viscosity detector and a QELS detector are equipped to the MALS system, more detailed and complementary information concerning polymer conformations in solvents can be obtained.

12.3 Dissolution of cellulose in solvents for SEC analysis

There are two ways to dissolve cellulose in a solvent for determination of MM and MM distribution by SEC analysis, derivatizations (mostly tricarbanilation) of cellulose followed by dissolution in an organic solvent such as tetrahydrofuran (THF), and direct dissolutions of cellulose in LiCl/amide solvents. Depolymerization of cellulose must be avoided during derivatization and dissolution processes, and the solvent should be intrinsically colorless for SEC analysis. SEC-MALS analysis of cellulose triphenylcarbamates (CTCs) and cellulose triethylcarbamates has been studied extensively, where chemical wood pulps and cotton celluloses are used as the starting materials [14,17]. As for direct dissolution of cellulose in a solvent for SEC analysis, 8–10% LiCl/DMAc, and 8–10% LiCl/DMI (see Figure 12.3 for structures of DMAc and DMI) are commonly used. ^{13}C-NMR spectra of low-MM cellulose dissolved in 8% LiCl/DMI and 8% LiCl/DMAc suggested that cellulose molecules are dissolved in these solvents without forming any derivatives [19]. However, some complex formation at each hydroxyl group of cellulose with the solvent component(s) may be possible.

One of the following two activations or pretreatments of cellulose is required to obtain cellulose solutions in the LiCl/amide systems: (1) solvent exchange from water to the amide via acetone and (2) heating in the amide at ~150°C for 30 min. Some active species are formed as by-products in cellulose/LiCl/amide systems during the heating process, resulting in partial depolymerization of cellulose molecules [6]. The solvent exchange method is thus more suitable. In this method, a cellulose sample is first dispersed in water and then solvent-exchanged from water to either DMAc or DMI through acetone. In each solvent, the cellulose slurry is stirred at room temperature overnight, and filtration of the cellulose is carried out using either filter paper on a Buchner funnel or grass filter.

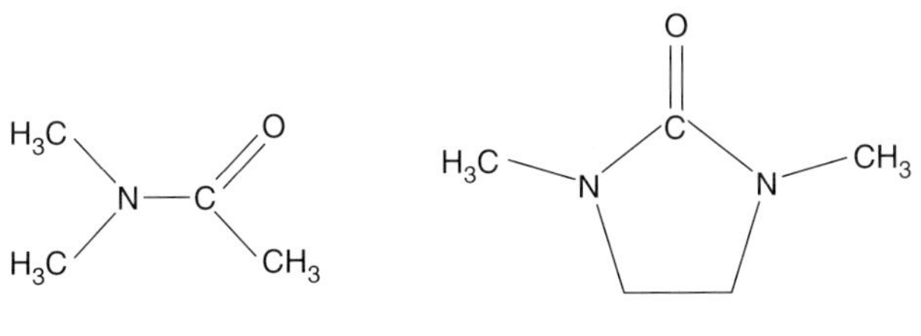

DMAc (N,N-dimethylacetamide)   DMI (1,3-dimethyl-2-imidazolidinone)

Figure 12.3 Chemical structures of DMAc and DMI.

It has been difficult to know the precise cellulose concentrations in the LiCl/amide solutions prepared, because a small amount of cellulose may be lost during the aforementioned multiple solvent exchange and filtration process. This may give inaccurate dn/dc values, resulting in inaccurate MM data by SEC-MALS analysis. Matsumoto *et al.* [18] developed the following procedures to overcome this problem. The wet cellulose containing a small amount of DMAc or DMI obtained from the final filtration of the cellulose slurry is vacuum-dried at 60°C for 40 h to completely remove the residual DMAc or DMI. The solvent-exchanged and then vacuum-dried cellulose can be stored in a sample bottle in a desiccator before use. LiCl dried overnight at 120°C is added to DMAc or DMI dried for over 40 h over molecular-sieve 4A, and the mixture is stirred until LiCl is completely dissolved in the solvent. Because LiCl, DMAc, and DMI are all hygroscopic, the preparation of 8% LiCl/DMAc or LiCl/DMI must be carried out quickly so as not to absorb moisture as far as possible. Absorption of moisture by the solution causes incomplete dissolution of cellulose or formation of cellulose aggregates when the water content exceeds a certain level [24,25]. The LiCl/amide solution thus prepared is stored in the dark under dry conditions before use. To prepare the cellulose solutions, a known amount of 8% LiCl/DMAc or LiCl/DMI is added to a suitable amount of the solvent-exchanged, vacuum-dried cellulose in a sample bottle equipped with a magnetic stirrer bar. The slurry is stirred at room temperature or below 4°C until the cellulose is completely dissolved, the time required for which depends on the cellulose samples and is generally 1–30 days.

Cellulose solutions in 8% LiCl/amide system with accurate cellulose concentrations can be, thus, obtained by the above activation and dissolution procedures. In general, cellulose concentrations are adjusted to 0.5–10 mg mL^{-1}. The LiCl/DMAc is suitable for the dissolution of regenerated celluloses, while the LiCl/DMI is suitable for tunicate cellulose, softwood kraft pulps, and holocelluloses [19,20]. The cellulose solutions in LiCl/DMI are stable without depolymerization for more than 6 months at room temperature. The cellulose solutions in 8% LiCl/DMAc or LiCl/DMI thus prepared are diluted with pure DMAc or DMI to adjust the LiCl concentration to that in the eluent for SEC analysis. Generally, the LiCl concentration is decreased to 0.5–1%, and cellulose concentrations are adjusted to 0.5–5 mg mL^{-1}.

12.4 SEC-MALS analysis of celluloses

A few steps that are useful to follow when subjecting cellulose solutions in LiCl/DMAc or LiCl/DMI to SEC analysis are (1) in some cases, it is better for the column temperature to be increased up to about 60°C to decrease the viscosity of the eluent, (2) when some insoluble residues are present in the sample solutions, they must be removed by, for example, centrifugation, (3) the sample solutions and eluents must be filtered through 0.45 or 0.10 μm PTFE membrane before use, and (4) because LiCl possibly rusts metallic parts in the MALS apparatus, the eluent used for and remained in the MALS system including the injector and the column after the measurement must be displaced with fresh DMAc or DMI to remove LiCl completely.

Details on the SEC system used in our laboratory are described in a previous paper [26]. A schematic diagram and a photograph of our SEC-PDA-MALS-QELS-RI system are shown in Figure 12.4. Data acquisition and processing were carried out using the ASTRA IV

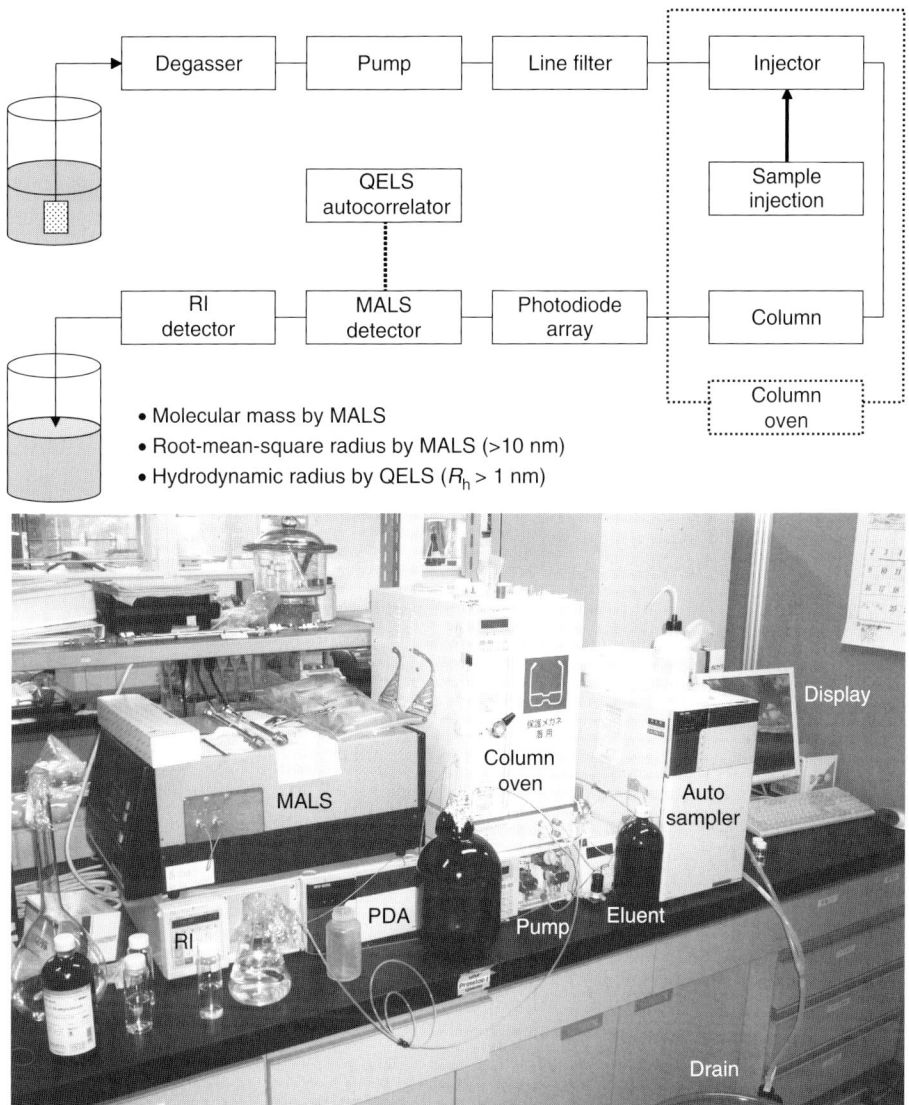

Figure 12.4 Schematic representation and photograph of our SEC-PDA-MALS-QELS system.

software (Wyatt technologies) and the LC solution software (Shimadzu). SEC conditions were: sample concentration = 0.05–0.3% (w/v), injection volume = 100 μL, flow rate = 0.5 mL min^{-1} and the column temperature = 60°C. The detector cells of PDA and RI were also kept at 60°C. Before injection, the sample solutions were filtered through a 0.45 μm PTFE disposable membrane (Omnipore; Millipore, USA). The eluents (1% LiCl/DMI and 8% LiCl/DMAc) were filtered through a 0.10 μm PTFE membrane (Millex-LG; Millipore, USA) before use.

12.5 Determination of d*n*/d*c* of cellulose in LiCl/amide solvents

As described in the previous section, an accurate measurement of dn/dc (specific refractive index increment) values of polymers in solutions is needed for obtaining correct values of M_w and RMS radius by SEC-MALS analysis. For cellulose derivatives such as CTC, triethylcarbamate or acetate, and various water-soluble cellulose ethers, the dn/dc values determined are accurate for SEC-MALS analysis. In contrast, however, determination of dn/dc values of polymers in complexing solvents such as LiCl/amide systems is difficult [8,27]. The M_w values obtained using the dn/dc values of cellulose in the LiCl/amide solvents may not be those of celluloses themselves; cellulose molecules may form transient complexes with the solvent components. This may lead to some increases in the apparent M_w values of celluloses. Thus, the MM data reported so far using the SEC method based on dn/dc values obtained by the conventional means may not be correct when LiCl/DMAc is used as the cellulose solvent.

To solve the above problem, Saalwächter et al. [28] employed an indirect method using CTC that does not form complex with the solvent. They measured MM values of CTC samples in THF and determined the correct degrees of polymerization (DP), which were then used as the DP of the original cellulose samples. This method is applicable to the dn/dc measurement on the assumption that no degradation occurs on cellulose molecules during the carbanilation process [29]. We also adopted this indirect method using CTC to determine dn/dc values of cellulose in 8% LiCl/DMAc and 1% LiCl/DMI. The calculated dn/dc values obtained by this indirect method were clearly different from the actual measured values obtained using an interferometric refractometer (Table 12.1), indicating that cellulose molecules form some complexes with the solvent component(s). It may be possible to evaluate the MM of each repeating unit of cellulose including LiCl and/or amide molecules in the solutions by comparing DP values obtained using the calculated and measured dn/dc values.

12.6 Determination of MM values of celluloses and pulps by SEC-MALS

A SEC elution pattern of a hardwood bleached kraft pulp (HBKP, α-cellulose content = 90%), detected by RI, and the corresponding M_w and RMS radius plots are shown in

Table 12.1 Specific refractive index increment (dn/dc) of cellulose in LiCl/amide solvents [26].

Solvent	Measured dn/dc (mL g^{-1})	Calculated dn/dc (mL g^{-1})
8% LiCl/DMAc	0.078	0.091
1% LiCl/DMI	0.062	0.087

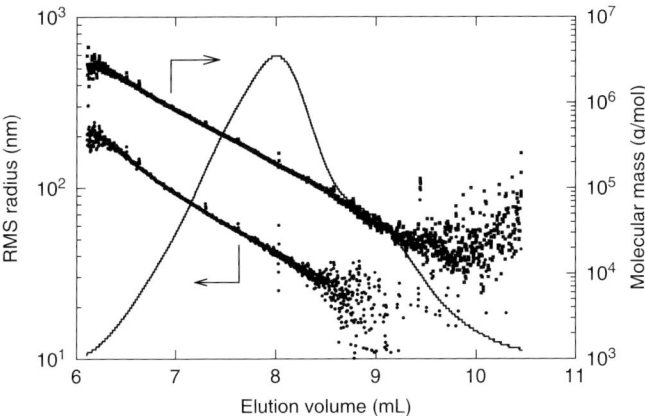

Figure 12.5 SEC elution pattern and the corresponding RMS radius and MM plots of a hardwood bleached kraft pulp dissolved in 1% LiCl/DMI. From Yanagisawa et al. [19], Figure 4, copyright (2004) with permission from Springer.

Figure 12.5, where 1% LiCl/DMAc was used as both the HBKP solvent and the SEC eluent [19]. Both these plots decrease linearly with the elution volume, clearly showing that cellulose molecules are separated properly in the reverse order of their M_w or RMS radius by the SEC system. The shoulder peak around 9 mL elution volume is due to the hemicellulose fraction present in the HBKP [1]. Because MALS can detect polymer molecules with RMS radius greater than ∼10–20 nm, the plots of RMS radius at the elution volume higher than 8.4 mL, and the MM plots at the elution volume higher than 9 mL are remarkably scattered, and no meaningful data can be obtained in these regions.

Because LS detection is very sensitive to aggregation, it can be used to pick up aggregates that may be present in the system. Chromatograms of LS at 90° and RI of the HBKP are illustrated in Figure 12.6. The RI detector gives a typical bimodal SEC elution pattern. The shapes of two elution patterns detected by LS and RI in the high-MM region are correlated with each other, and no irregular peak denoting aggregation is observed in the high-MM region of the LS chromatogram. Therefore, practically no aggregates are present in this system under the given conditions. The applicability of this HBKP solution to SEC analysis is validated from these results.

The SEC elution patterns of various cellulose and pulp samples obtained in our laboratory are shown in Figure 12.7. Eight percent LiCl/DMAc was used both as the solvent for the samples and as the eluent in SEC-MALS analysis, and the dn/dc value of 0.091 mL g^{-1} [26] in Table 12.1 was adopted. The elution volumes at the peak positions of these cellulose samples corresponded well to their viscosity average MM values obtained by the 0.5 M copper ethylenediamine method. Most of the pure celluloses such as bacterial, cotton and microcrystalline celluloses had SEC elution patterns close to normal distributions, while bleached chemical wood pulps such as the HBKP and a softwood bleached sulfite pulp (SBSP) had typical bimodal SEC elution patterns due to the presence of hemicellulose fractions. The M_w, M_n, and M_w/M_n values and the corresponding DP$_w$ and DP$_n$ values for the cellulose and pulp samples obtained by the MALS method (Figure 12.7) are summarized in Table 12.2.

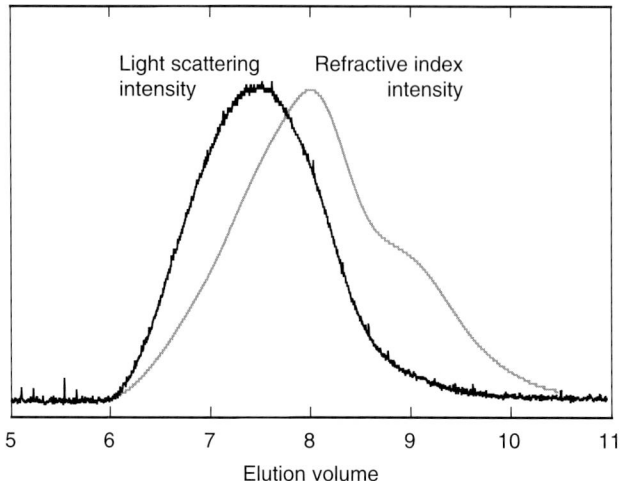

Figure 12.6 SEC elution patterns of hardwood bleached kraft pulp detected by refracrive index (RI) and laser light scattering (LS) at 90°. From Yanagisawa *et al.* [19], Figure 5, copyright (2004) with permission from Springer.

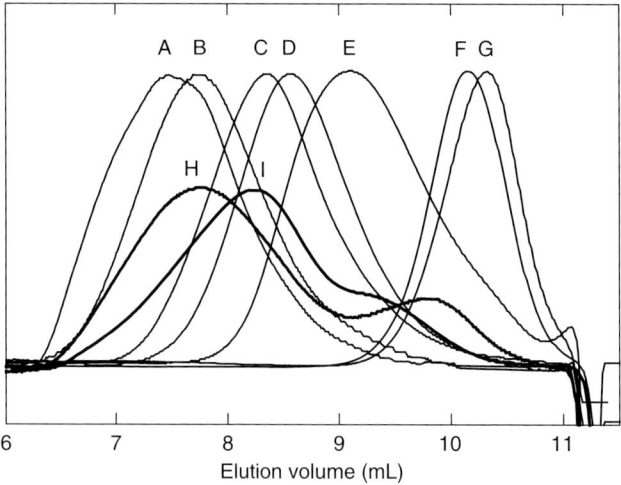

Figure 12.7 SEC elution patterns of various cellulose and pulp samples A–I (see Table 2 for sample descriptions) detected by RI with 8% LiCl/DMAc as the eluent.

12.7 Conformation analysis of cellulose molecules in LiCl/DMI

Figure 12.8 shows the conformation plots of an acid-hydrolyzed tunicate cellulose, the cotton lint and microcrystalline cellulosed, the HBKP and SBSP dissolved in 1% LiCl/DMI, where the *x*-axis shows the weight-average contour length, L_w ($L_w = 0.515 \times DP_w$ nm) of the cellulose chains. These conformation plots are on a common line in general, which is

Table 12.2 Weight- and number-average MM values (M_w and M_n) and the corresponding degrees of polymerization (DP_w and DP_n) of various cellulose and pulp samples determined by SEC-MALS using 8% LiCl/DMAc as the eluent[a].

Cellulose samples in Figure 12.7	M_w	(DP_w)	M_n	(DP_n)	M_w/M_n
A Bacterial cellulose	899 000	(5500)	667 000	(4100)	1.35
B Cotton lint	341 000	(2100)	203 000	(1300)	1.68
C Cotton linters	91 200	(600)	56 900	(350)	1.60
D Regenerated cellulose fiber[b]	81 200	(500)	39 500	(240)	2.06
E Microcrystalline cellulose	26 600	(190)	20 200	(120)	1.32
F Hydrolyzate of sample D	4 970	(31)	4 590	(28)	1.08
G Low MM cellulose[c]	3 820	(23)	3 780	(23)	1.01
H Softwood bleached sulfite pulp	307 000	(1800)	54 400	(340)	5.64
I Hardwood bleached kraft pulp	272 000	(1700)	67 100	(410)	4.05

[a] The dn/dc value of 0.091 (mL/g) was used [26].
[b] Bemlise® prepared form cotton linters using copper ammonium hydroxide solution.
[c] Prepared from microcrystalline cellulose by homogeneous hydrolysis using 85% H_3PO_4 [30].

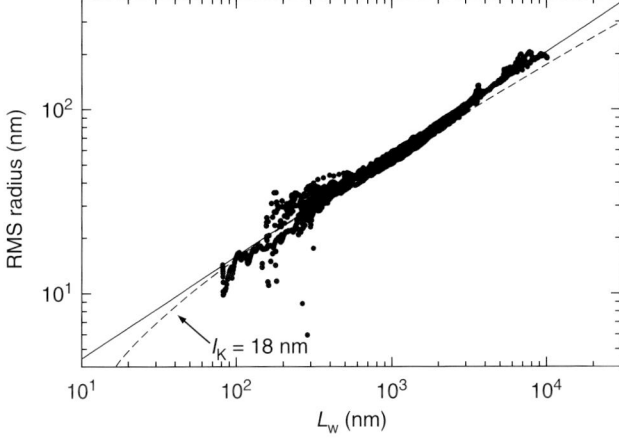

Figure 12.8 Double-logarithmic plots of RMS radius vs weight-average contour length L_w of various cellulose and pulp samples, (- - -) Benoit–Doty theoretical curve at the Kuhn segment length of 18 nm. From Yanagisawa and Isogai [26], Figure 6, copyright (2005) with permission from American Chemical Society.

to be expected for similar polymer molecules. The slope value of the linear part of the plots is ∼0.57, showing that the high-L_w parts of these cellulose and pulp samples are similar to each other in terms of the molecular shape which has random-coil conformations. It is reasonable that even semiflexible polymers such as cellulose would have random coil conformations as a whole when they have sufficiently high L_w. A slight but reproducible curvature appears in the low-L_w region; the conformation plots deviate from the linear ones in the high-L_w region. These show that the conformation of cellulose molecules varies, depending on their L_w values; the cellulose molecules in the low-L_w region have no longer random coil conformations but have more extended and semi-rod-like shapes.

Such an L_w-dependent conformation transition between random coils and rigid rods may be explained by the Benoit–Doty theory [31] for the Kratky–Porod chains [32] or the wormlike polymer chains, which are the most typical models proposed for semi-flexible polymers. According to the theory, RMS radius $\langle S^2 \rangle^{1/2}$ (or R_g) of unperturbed polymer chains are given by the Equation (12.3).

$$\langle S^2 \rangle_z = \frac{l_K L}{6} - \frac{l_K^2}{4} + \frac{l_K^3}{4L}\left[1 - \frac{l_K}{2L}(1 - e^{-2L/l_K})\right] \quad (12.3)$$

where l_K is the Kuhn segment length, which is used as a measure of chain stiffness of the polymer. With this parameter, the contour length L or $0.515 \times DP_w$ is described by the Equation (12.4)

$$L = N_K l_K \quad (12.4)$$

where N_K is the number of the Kuhn segments.

In Figure 12.8, the theoretical curve at $l_K = 18$ nm (dashed line) is depicted. This curve is almost linear with the slope of ~0.5 in the sufficiently high-L_w region, showing random-coil conformations. On the other hand, a remarkable bending is observed in the low-L_w part of the curve; this reflects the coil/rod conformation transition of cellulose. The theoretical curve imposed in Figure 12.8 seems to agree with the actual conformation plots especially in the low-L_w region ($L_w < 1000$ nm), where $N_K < 50$ and the excluded volume effect is practically negligible [33]. On the contrary, in the high-L_w region, the difference between the actual data and the theoretical curve becomes greater, and this is probably due to the expression of the excluded volume effect. Installation of a QELS autocorrelator into the MALS system enables additional determination of the hydrodynamic radius (R_h) of polymers, and the SEC-MALS-QELS data obtained for CTC in THF also show that cellulose molecules having low L_w have semi-rod-like shapes [26].

12.8 SEC-MALS analysis of cellulose in softwood kraft pulps using LiCl/DMI

Although 8% LiCl/DMAc dissolves most of cellulose and pulp samples, it does not completely dissolve softwood kraft pulps. Some insoluble residues or gels are always formed and they must be removed by centrifugation or filtration before the dissolved kraft pulps are subjected to SEC analysis. However, softwood kraft pulps are completely soluble in 8% LiCl/DMI, and thus with the LiCl/DMI as the solvent and eluent SEC-MALS analysis can be used to determine the MM and MM distribution of softwood kraft pulps as well as other cellulose and pulp samples.

The SEC elution patterns of various cellulose and pulp samples including a softwood bleached kraft pulp (SBKP) and their MM plots [20] are depicted in Figure 12.9a. Even though the cellulose and pulp samples had various SEC elution patterns, depending on their MM values and MM distributions, all the cellulose and pulp samples except the SBKP

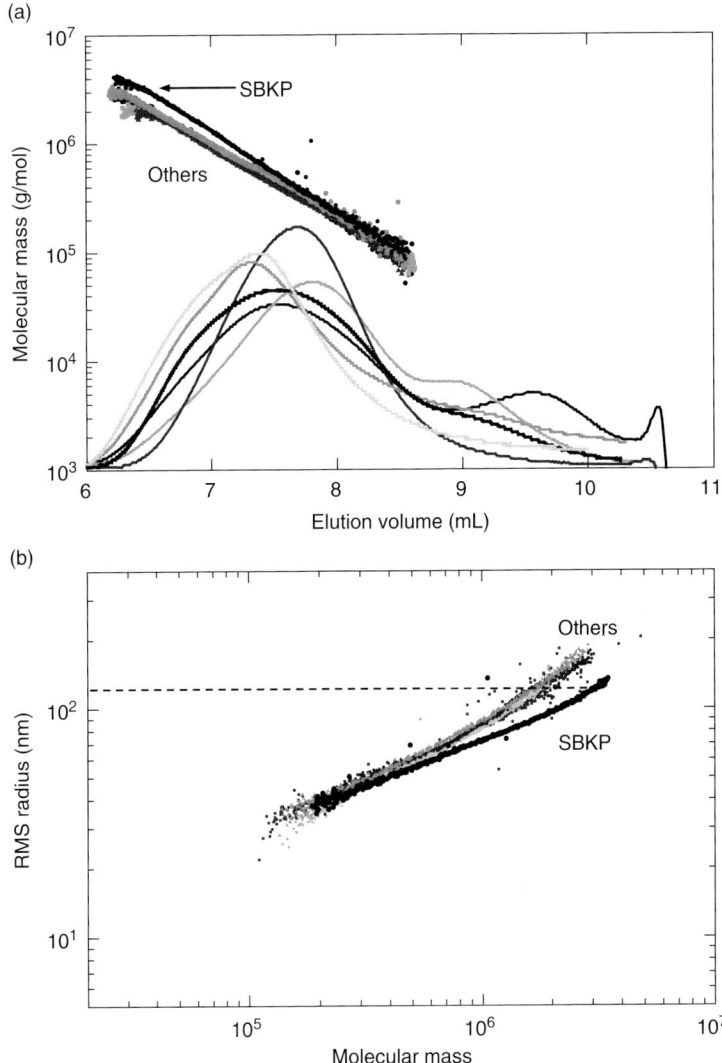

Figure 12.9 SEC elution patterns and MM plots (a) and the corresponding double-logarithmic (conformation) plots (b) of various cellulose and pulp samples including SBKP in 1% LiCl/DMI. From Yanagisawa et al. [20], Figures 6 and 7, copyright (2005) with permission from Springer.

gave a similar line of MM vs elution volume plots; these cellulose molecules intrinsically have similar conformations and structures in the LiCl/DMI solvent. On the other hand, anomalous MM plots were obtained for the SBKP. As the arrow in Figure 12.9 indicates, the position of the MM plots of the SBKP deviates from the others in the high-MM region; the MM of the cellulose in the SBKP is about twice as much as those of the other cellulose and pulp samples at the same elution volume in this region.

The particular behavior of the cellulose molecules in the SBKP becomes clear when its relationship between M_w and the RMS radius is depicted together with those of the other cellulose and pulp samples (Figure 12.9b). Because the slopes of the conformation plots for all the cellulose and pulp samples except the SBKP are in the range of 0.57–0.59, they behave as random coils in 1% LiCl/DMI. On the other hand, because the slope value for the SBKP is ~0.41, cellulose molecules of the SBKP are regarded to have some branch or aggregate structures in the LiCl/DMI solvent. Similar results of anomalous cellulose molecules were observed also for softwood unbleached kraft pulps (SUKPs) and softwood holocelluloses by SEC-MALS analysis, but not for SBSP or acid-hydrolyzed SBKP. Thus, some branch structures of glucomannan may be partially present in high-MM cellulose molecules of softwood, and these branch structures between glucomannan and cellulose may be susceptible to acid treatments and removable by pulping under acid conditions [20].

12.9 SEC-PDA-MALS analysis of unbleached chemical pulps

Because softwood and hardwood unbleached chemical pulps are soluble in 8% or 1% LiCl/DMI, SEC-MALS analysis of residual lignin in unbleached chemical pulps is possible when a ultraviolet (UV) or a PDA detector is attached to the SEC-MALS system. Relationships in the distribution between residual lignin detected by UV at 280 nm and cellulose and hemicellulose fractions detected by RI in unbleached chemical pulps can therefore be studied for unbleached chemical pulps [3,9–14]. UV-vis absorption spectrum of each slice of SEC elution patterns can also be obtained simultaneously, when the PDA is attached to the SEC system. Structural analysis of residual lignin between pulp samples or between different MM regions of the same pulp is possible by the SEC-PDA-MALS system. Typical SEC elution patterns of softwood and hardwood unbleached kraft pulps (HUKPs) dissolved in 1% LiCl/DMI, which were detected by RI and UV at 280 nm, are shown in Figure 12.10. For the SUKP, a significant amount of residual lignin coexists with high-MM fraction (mostly cellulose) of the pulp. For the HUKP, residual lignin is predominantly present in the low-MM fraction (mostly hemicelluloses), although some residual lignin is clearly present also in the high-MM fraction. Detailed studies of residual lignin in unbleached chemical wood pulps will be presented elsewhere.

12.10 SEC-MALS analysis of cellouronic acid and other water-soluble cellulose derivatives

When regenerated or mercerized cellulose suspended in water at pH 10–11 was oxidized with sodium hypochlorite and a catalytic amount of 2,2,6,6-tetramethylpiperidine-1-oxy radical (TEMPO) at room temperature, it became water-soluble within 30 min as the oxidation proceeded [34,35]. NMR analysis of the obtained products revealed that almost all C6 primary hydroxyl groups of the cellulose were regioselectively converted to carboxyl ones, giving water-soluble β-1,4-linked polyglucuronic acid sodium salt having a homogeneous chemical structure, that is, cellouronic acid sodium salt (CUA-Na) [34] (Figure 12.11). The MM and MM distribution of the CUA-Na filtered with 0.02 μM membrane was studied by SEC-MALS using 0.1 M NaCl as an eluent [35]. Carboxymethyl cellulose (CMC) sodium

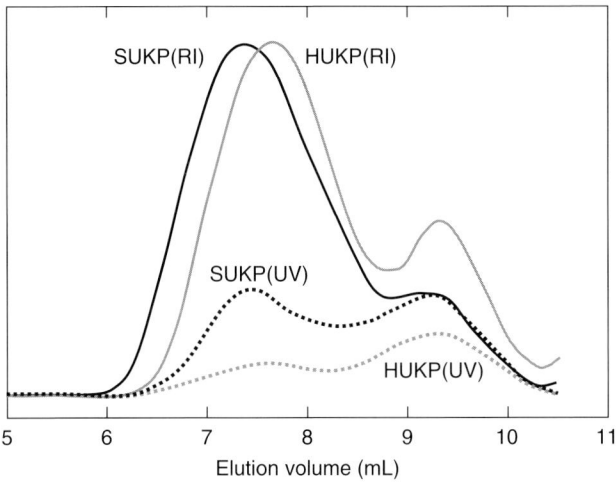

Figure 12.10 Typical SEC elution patterns of softwood and hardwood unbleached kraft pulp, detected by RI and UV at 280 nm.

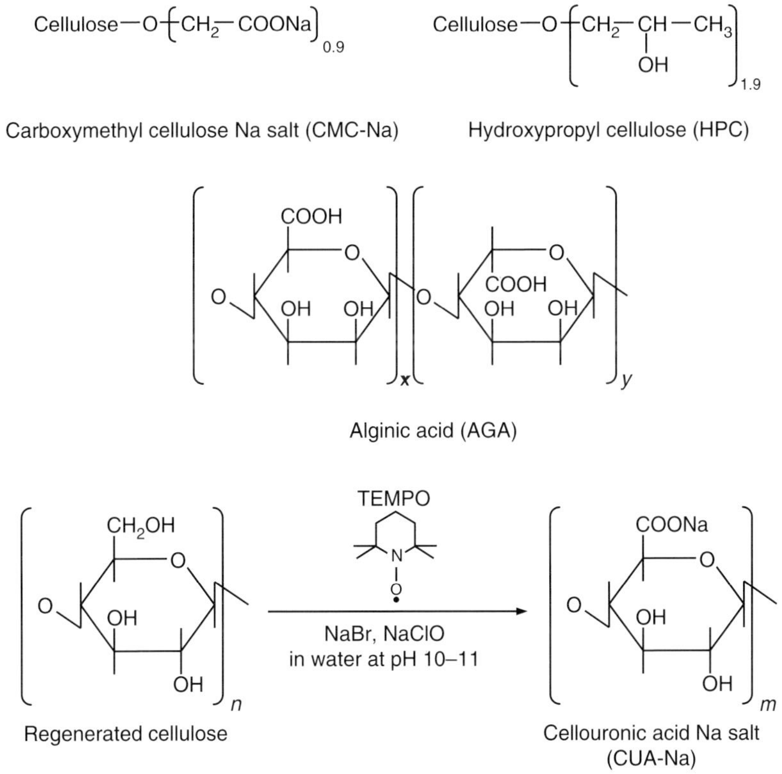

Figure 12.11 Chemical structures of carboxymethyl cellulose sodium salt, hydroxypropyl cellulose, and alginic acid, and the oxidation of regenerated cellulose to cellouronic acid sodium salt.

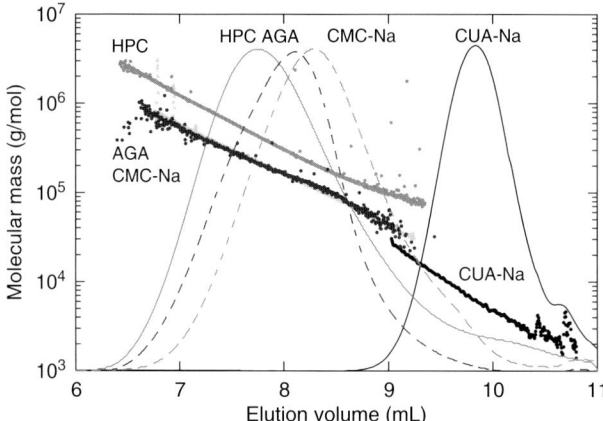

Figure 12.12 SEC elution patterns and the corresponding MM plots of water-soluble cellulose derivatives and alginic acid. From Shibata *et al*. [35], Figure 3, copyright (2006) with permission from Springer.

Table 12.3 Weight- and number-average molecular mass values (M_w and M_n) and the corresponding degrees of polymerization (DP_w and DP_n) of cellouronic acid filtered with a 0.02 µm membrane, cellulose derivatives and alginic acid, determined by the SEC-MALS system using 0.1 M NaCl as the eluent [35].

Sample	M_w	(DP_w)	M_n	(DP_n)	M_w/M_n
Cellouronic acid Na salt	7 050	(36)	4 350	(22)	1.6
Carboxymethyl cellulose Na salt	136 000	(580)	67 300	(290)	2.0
Hydroxypropyl cellulose	435 000	(1600)	139 000	(500)	3.1
Alginic acid	203 000	(1000)	120 000	(600)	1.7

salt (CMC-Na), hydroxypropyl cellulose (HPC), and alginic acid (AGA) (Figure 12.11) were also analyzed for comparison.

All these cellulose derivatives had SEC elution patterns close to normal distributions (Figure 12.12). The MM plots of these materials revealed that the MM values linearly decreased with an increase of the elution volume in the whole range, resulting from the proper size-exclusion mechanism by the SEC column. The MM plots of AGA and CMC-Na mostly overlapped, indicating that these two ionic polysaccharides have similar molecular expansions in 0.1 M NaCl. Although HPC and CMC-Na have the same cellulose backbone, CMC-Na molecules are more expanded in the solution probably by intra-molecular electric repulsions. CUA-Na filtered with 0.02 µm membrane had a SEC peak position far from those of the other samples; its MM is far lower than that of AGA, CMC-Na, or HPC. The MM plot of CUA-Na is slightly shifted to lower MM value at the same elution volume; the CUA-Na molecules are expanded in the solution more than those of AGA or CMC-Na.

The M_w and M_n values and the corresponding DP_w and DP_n, respectively, of CUA-Na, CMC-Na, HPC, and AGA are calculated from the SEC-MALS data, and summarized in Table 12.3. Because the DP_v value of the original viscose rayon used for the preparation

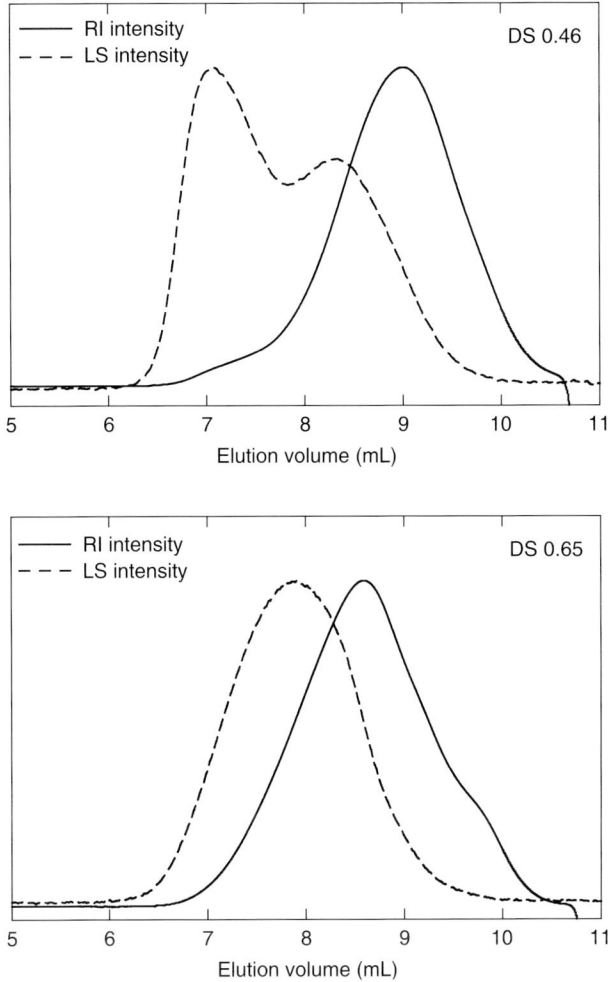

Figure 12.13 SEC elution patterns of carboxymethyl celluloses with different degree of substitution (DS), detected by RI and LS at 90° [38].

of cellouronic acid is 380, the low DP_w value for cellouronic acid prepared shows that significant depolymerization occurs on the cellulose chains during the TEMPO-mediated oxidation. The DP_w values ∼36 were always observed for cellouronic acids prepared from various regenerated cellulose samples under various TEMPO-mediated oxidation conditions. This constant DP_w of cellouronic acid is close to that of the leveling-off DP of regenerated celluloses (∼DP 40) obtained by the dilute and heterogeneous acid hydrolysis [36]. Thus, depolymerization of cellulose chains during the TEMPO-mediated oxidation may preferably occur in disordered regions of the cellulose, resulting in the DP_w values of cellouronic acids close to the leveling-off DP in a manner similar to that of the dilute acid hydrolysis.

12.11 SEC-MALS analysis of CMC used as wet-end additive in papermaking

CMC has anionic charges as well as pulp fibers that contain small amounts of carboxyl groups, depending on the wood species used and the pulping and bleaching conditions. However, it was recently found that more than 80% CMC added to the pulp slurries were irreversibly adsorbed onto pulp fibers, and that the degree of CMC adsorption on pulp fibers was influenced by the degree of substitution (DS) of CMC and the electrical conductivity of pulp slurries [37,38]. Because CMC adsorption introduces new anionic sites onto pulp fiber surfaces, efficiency of wet end additives and the resultant paper properties can be improved. To understand the CMC adsorption behavior in terms of CMC molecular conformations in aqueous solutions at various electrical conductivities, SEC-MALS analysis was applied to CMC samples with different DS values.

Figure 12.13 shows SEC elution patterns of CMC with DS 0.46 and 0.65, detected by RI and laser LS at 90° [38]. As described previously, LS is quite sensitive to aggregates present in the apparently clear solutions. In fact, the aggregate formation is clearly detected for the CMC with DS 0.46, although the absolute amount of the aggregates are not so large, judging from the SEC elution pattern detected by RI. On the other hand, no such aggregates are present in the solution of CMC with DS 0.65. It has been found that CMC molecules added to pulp slurries are preferably adsorbed on pulp fibers and become effective in improving paper qualities, only when some CMC aggregates are present in the solutions. CMC molecules tend to have these aggregates with decreasing DS and increasing electrical conductivity. Thus, the optimum effect of CMC addition to pulp slurries in papermaking

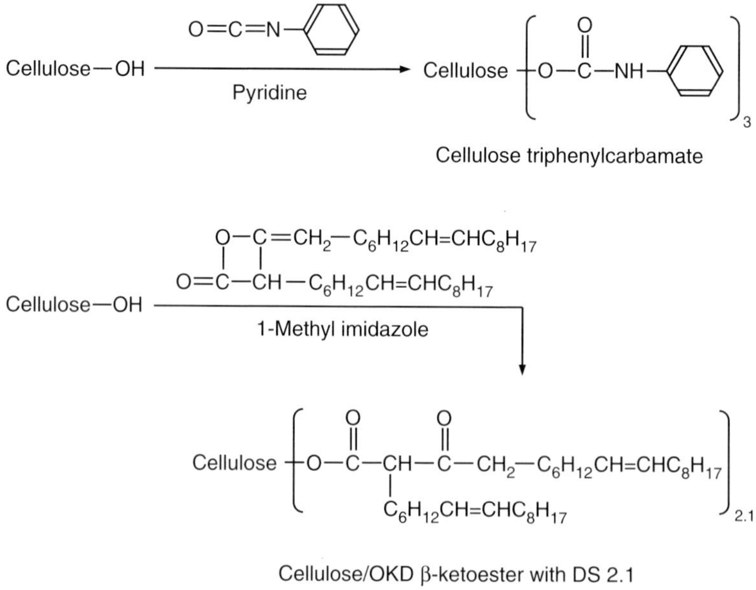

Figure 12.14 Reactions of cellulose to form cellulose triphenylcarbamate and cellulose/OKD β-ketoester.

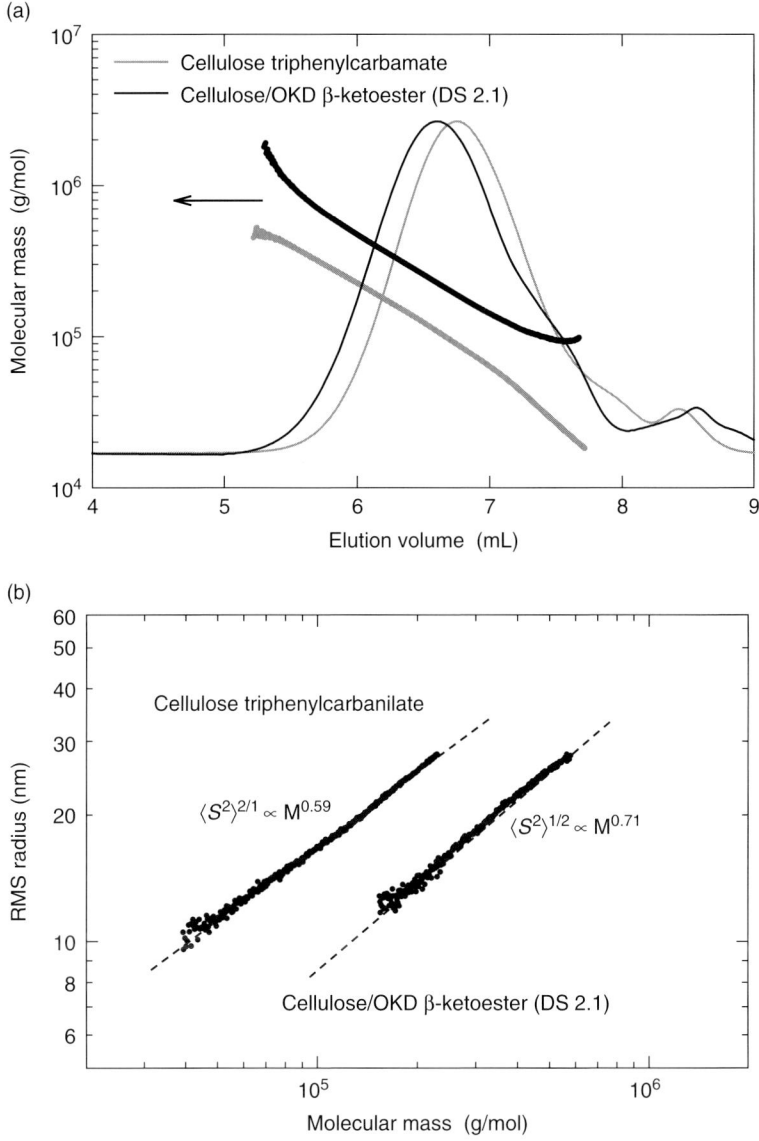

Figure 12.15 SEC elution patterns and MM plots (a) and the corresponding conformation plots (b) of CTC and cellulose/OKD β-ketoester with DS 2.1. From Yoshida *et al.* [39], Figures 13 and 14, copyright (2005) with permission from Elsevier.

can be obtained, when CMC molecules have high affinity to each other to form aggregates. Although the presence of such aggregates in solutions should be avoided or removed for general measurements of accurate MM, the molecular aggregates detected by SEC-MALS analysis often provides valuable information concerning functionalities of molecules in solution state.

12.12 SEC-MALS analysis of polymer-brush-type cellulose β-ketoesters

Cellulose β-ketoesters with long and branched alkenyl chains were prepared using *cis*-9-octadecenyl ketene dimer (OKD) and LiCl/DMI as the esterifying reagent and cellulose solvent, respectively [39]. The DS values of the cellulose/OKD β-ketoesters were controllable up to 2.1 by selecting the reaction conditions. Thus, new polymer-brush-type cellulose β-ketoesters can be prepared by the homogeneous reaction between cellulose and OKD [39]. Solution- and solid-state ^{13}C-NMR analyses revealed that cellulose backbones of the cellulose/OKD β-ketoesters with DS 2.1 behave like a solid in chloroform owing to strong restriction on the movement of cellulose chains by the long and branched alkenyl substituents introduced. SEC-MALS analysis was applied to the cellulose/OKD β-ketoester dissolved in THF for conformation analysis in comparison with CTC. Reactions of cellulose to form these two cellulose derivatives are shown in Figure 12.14.

Figure 12.15a depicts SEC elution patterns and the corresponding MM plots of CTC and cellulose/OKD β-ketoester. Both the MM plots decrease with an increase of the elution volume, showing that each molecule is properly separated according to its hydrodynamic radius in THF by the SEC system. The peak position in the SEC elution pattern of cellulose/OKD β-ketoester is higher than that of CTC, and the MM value of the former is two to three times as much as that of the latter at the same elution volume. Because average mass values of one glucose residue in CTC and cellulose/OKD β-ketoester with DS 2.1 are 519 and 1270.8, respectively, the difference in the MM plots between the two cellulose derivatives is quite reasonable. Figure 12.15b shows the double-logarithmic plots of MM vs the RMS radius for the two cellulose derivatives. The slope of the plots for CTC is 0.59, indicating that its molecules have random-coil conformation in THF. On the other hand, the cellulose/OKD β-ketoester with DS 2.1 has the slope value of 0.71, which corresponds to semi rigid-rod conformation. The long and branched alkenyl substituents introduced into the cellulose hydroxyls may have brought about such a typical molecular conformation to the cellulose/OKD β-ketoester in THF.

References

1. Westermark, U.; Gustafsson, K., Molecular Size Distribution of Wood Polymers in Birch Kraft Pulps; *Holzforschung* 1994, 48(Suppl.), 146–150.
2. Striegel, A.M.; Timpa, J.D., Molecular Characterization of Polysaccharides Dissolved in Me$_2$NAc-LiCl by Gel-Permeation Chromatography; *Carbohydr. Res.* 1995, 267, 271–290.
3. Sjöholm, E.; Gustafsson, K.; Pettersson, B.; Colmsjö, A., Characterization of the Cellulosic Residues from Lithium Chloride/*N*,*N*-Dimethylacetamide Dissolution of Softwood Kraft Pulp; *Carbohydr. Polym.* 1997, 32, 57–63.
4. Schult, T.; Hjerde, T.; Optun, O.I.; Kleppe, P.J.; Moe, S., Characterization of Cellulose by SEC-MALLS; *Cellulose*, 2002, 9, 149–158.
5. Röhring, J.; Potthast, A.; Rosenau, T. *et al.*, A Novel Method for the Determination of Carbonyl Groups in Cellulosics by Fluorescence Labeling. 1. Method Development; *Biomacromolecules*, 2002, 3, 959–968.

6. Potthast, A.; Rosenau, T.; Sartori, J.; Sixtra, H.; Kosma, P., Hydrolytic Processes and Condensation Reactions in the Cellulose Solvent System N, N-Dimethylacetamide/Lithium Chloride. Part 2: Degradation of Cellulose; *Polymer* 2003, 44, 7–17.
7. Dupont, A.L., Cellulose in Lithium Chloride/N, N-Dimethylacetamide, Optimization of a Dissolution Method Using Paper Substrates and Stability of the Solutions; *Polymer* 2003, 44, 4117–4126.
8. Berggren, R.; Berthold, F.; Sjöholm, E.; Lindström, M., Improved Methods for Evaluating the Molar Mass Distributions of Cellulose in Kraft Pulp; *J. Appl. Polym. Sci.* 2003, 88, 1170–1179.
9. Karlsson, O.; Westermark, U., Evidence for Chemical Bonds between Lignin and Cellulose in Kraft Pulps; *J. Pulp Pap. Sci.* 1996, 22, J397–J401.
10. Karlsson, O.; Westermark, U., The Significance of Glucomannan for the Condensation of Cellulose and Lignin under Kraft Pulping Conditions; *Nordic Pulp Pap. Res. J.* 1997, 12, 203–206.
11. Karlsson, O.; Pettersson, B.; Westermark, U., The Use of Cellulases and Hemicellulases to Study Lignin–Cellulose as well as Lignin–Hemicellulose Bonds in Kraft Pulps; *J. Pulp Pap. Sci.* 2001, 27, 196–201.
12. Karlsson, O.; Pettersson, B.; Westermark, U., Linkages between Residual Lignin and Carbohydrates in Bisulphite (Magnefite) Pulps; *J. Pulp Pap. Sci.* 2001, 27, 310–316.
13. Sjöholm, E.; Gustafsson, K.; Berthold, F.; Colmsjö, A., Influence of the Carbohydrate Composition on the Molecular Weight Distribution of Kraft Pulps; *Carbohydr. Polym.* 2000, 41, 1–7.
14. Berthold, F.; Gustafsson, K.; Berggrem, R.; Sjöholm, E.; Lindström, M., Dissolution of Softwood Kraft Pulps by Direct Derivatization in Lithium Chloride/N, N-dimethylacetamide; *J. Appl. Polym. Sci.* 2004, 94, 424–431.
15. Strlič, M.; Kolar, J., Size Exclusion Chromatography of Cellulose in LiCl/N, N-Dimethylacetamide; *J. Biochem. Biophys. Methods* 2003, 56, 265–279.
16. Conner, A.H., Size Exclusion Chromatography of Cellulose and Cellulose Derivatives, In: Wu, C.-S. (ed.); *Handbook of Size Exclusion Chromatography*; Chromatographic Science Ser.; Marcel Dekker; New York, 1995; 69, pp. 331–352.
17. Sjöholm, E.; Size Exclusion Chromatography of Cellulose and Cellulose Derivatives. In: Wu, C.-S. (ed.); *Handbook of Size Exclusion Chromatography and Related Techniques*; Chromatographic Science Ser.; Marcel Dekker; New York, 2004; 91, pp. 311–354.
18. Matsumoto, T.; Tatsumi, D.; Tamai, N.; Takai, T., Solution Properties of Celluloses from Different Biological Origins in LiCl/DMAc; *Cellulose* 2001, 9, 275–282.
19. Yanagisawa, M.; Shibata, I.; Isogai, A., SEC-MALLS Analysis of Cellulose using LiCl/1,3-Dimethyl-2-imidazolidinone as an Eluent; *Cellulose* 2004, 11, 169–176.
20. Yanagisawa, M.; Shibata, I.; Isogai, A., SEC-MALLS Analysis of Softwood Kraft Pulp Using LiCl/1,3-Dimethyl-2-imidazolidinone as an Eluent; *Cellulose* 2005, 12, 151–158.
21. Wyatt, P.J.; Light Scattering and the Solution Properties of Macromolecules. In: Wu, C.-S. (ed.); *Handbook of Size Exclusion Chromatography and Related Techniques*, Chromatographic Science Ser.; Marcel Dekker; New York, 2004; 91, pp. 623–655.
22. Potthast, A.; Röhring, J.; Rosenau, T.; Borgards, A.; Sixta, H.; Kosma, P., A Novel Method for the Determination of Carbonyl Groups in Cellulosics by Fluorescence Labeling. 3. Monitoring Oxidative Processes; *Biomacromolecules* 2003, 4, 743–749.
23. Bohm, R.; Potthast, A.; Schiehser, S.; Rosenau, T.; Sixta, H.; Kosma, P., The FDAM Method: Determination of Carboxyl Profiles in Cellulosic Materials by Combining Group-Selective Fluorescence Labeling with GPC; *Biomacromolecules* 2006, 7, 1743–1750.
24. Röder, T.; Potthast, A.; Rosenau, T. et al., The Effect of Water on Cellulose Solutions in DMAc/LiCl; *Macromol. Symp.* 2002, 190, 151–159.
25. Chrapava, S.; Tourand, D.; Rosenau, T.; Potthast, A.; Kunz, W., The Investigation of the Influence of Water and Temperature on the LiCl/DMAc/Cellulose System; *Phys. Chem. Chem. Phys.* 2003, 5, 1842–1847.

26. Yanagisawa, M.; Isogai, A., SEC-MALS-QELS Study on the Molecular Conformation of Celllose in LiCl/Amide Solutions; *Biomacromolecules* 2005, 6, 1258–1265.
27. Dupont, A.L.; Harrison, G., Conformation and dn/dc Determination of Cellulose in N,N-Dimethylacetamide Containing Lithium Chloride; *Carbohydr. Polym.* 2004, 58, 233–243.
28. Saalwächter, K.; Burchard, W.; Klüfer, P. *et al.*, Cellulose Solutions in Water Containing Metal Complexes; *Macromolecules* 2000, 33, 4094–4107.
29. Burchard, W., A Comparative Structure Analysis of Cellulose and Amylose Tricarbanilates in Solution; *Makromol. Chem.* 1961, 44–46, 358–387.
30. Isogai, A.; Usuda, M., Preparation of Low-Molecular Weight Celluloses Using Phosphoric Acid; *Mokuzai Gakkaishi* 1991, 37, 339–344.
31. Benoit, H.; Doty, P., Light Scattering from Non-Gaussian Chains; *J. Phys. Chem.* 1953, 57, 958–963.
32. Kratky, O.; Porod, G., X-ray Investigation of Dissolved Chain Molecules; *Rec. Trav. Chim.* 1949, 68, 1106–1122.
33. Fujita, H., Some Unsolved Problems on Dilute Polymer Solutions; *Macromolecules* 1988, 21, 179–185.
34. Isogai, A.; Kato, Y., Preparation of Polyuronic Acid from Cellulose by TEMPO-Mediated Oxidation; *Cellulose* 1998, 5, 153–164.
35. Shibata, I.; Yanagisawa, M.; Saito, T.; Isogai, A., SEC-MALS Analysis of Cellouronic Acid Prepared from Regenerated Cellulose by TEMPO-Mediated Oxidation; *Cellulose* 2006, 13, 73–80.
36. Watanabe, S.; Hayashi, J., Molecular Chain Conformation and Crystallite Structure of Cellulose I. 1. Fine-Structure of Rayon Fibers; *J. Polym. Sci. Polym. Chem. Ed.* 1974, 12, 1065–1087.
37. Watanabe, M.; Gondo, T.; Kitao, O., Advanced Wet-End System with Carboxymethyl Cellulose; *TAPPI J.* 2004, 3(5), 15–19.
38. Gondo, T.; Watanabe, M.; Soma, H.; Kitao, O.; Yanagisawa, M., Isogai, A., SEC-MALS Study on Carboxymethyl Cellulose (CMC): Relationship between Conformation of CMC Molecules and Their Adsorption Behavior onto Pulp Fibers; *Nordic Pulp Pap. Res. J.* 2006, 21(5), 591–597.
39. Yoshida, Y.; Yanagisawa, M.; Isogai, A.; Suguri, N.; Sumikawa, N., Preparation of Polymer Brush-Type Cellulose β-Ketoesters Using LiCl/1,3-Dimethyl-2-imidazoldinone as a Solvent; *Polymer* 2005, 46, 2548–2557.

Chapter 13
^{13}C CPMAS NMR Studies of Wood, Cellulose Fibers, and Derivatives

Sirkka Liisa Maunu

Abstract

The research area in lignocellulosic materials is an extremely large field, and it can be approached in a variety of ways. The diversity of nuclear magnetic resonance (NMR) spectroscopy, together with the increased knowledge and the development of the hardware and software possibilities during recent years, open up many possibilities for characterizing lignocellulosic materials. Solid-state NMR methods provide a way to do chemical structure analyses in a native state. Even if the resolution is lower compared with liquid-state measurements, it is a remarkable choice for samples with restricted solubility such as residual lignins, or when a physical structure such as cellulose morphology is studied. This chapter demonstrates how advanced ^{13}C cross-polarization magic angle spinning (CPMAS) NMR methods can be applied to obtain morphological as well as chemical information of such complicated substances – woods, cellulose fibers, and cellulose derivatives.

13.1 Introduction

Solid-state nuclear magnetic resonance (NMR) is a remarkable choice for samples with limited solubility or when a physical structure such as material morphology is studied, despite its much lower resolution compared to liquid-state NMR [1]. ^{13}C cross-polarization magic angle spinning (CPMAS) NMR methods provide a way to conduct chemical structure analyses in a native state without any chemical modifications or fractionation pretreatment. A relatively high amount of sample is needed; however, good information of a bulk sample is obtained.

The combination of cross-polarization (CP) and magic angle spinning (MAS) together with proton dipolar decoupling (DD) initiated the utilization of this method also in the area of wood, fiber, and cellulose research. Accumulation of knowledge together with the development of the hardware and software provides a way to reach increasingly more detailed information about the structure of heterogeneous materials.

For pure cellulose samples the crystallinity index (CrI) obtained from ordinary ^{13}C CPMAS spectra has been shown to correlate well with the corresponding crystallinities obtained by x-ray diffraction. Hemicelluloses and lignin resonate in the same region as amorphous cellulose, interfering thus the determination of cellulose crystallinity in wood and pulp samples. To obtain quantitative cellulose spectra, these amorphous noncellulosic components must be removed prior to CrI determination. This can be done chemically or spectroscopically using special spectral edition that is based on proton spin relaxation differences [2,3]. The same procedure is utilized when the heterogeneity of cellulose derivatives is studied [4].

The information in the basic ^{13}C CPMAS NMR spectra of lignin samples remains limited because of the low resolution. To obtain more detailed structural information, the use of special techniques is required. In the dipolar dephasing (DD) technique, dipolar interactions between protons and carbons cause the fast decay of the signals from carbons attached to protons (i.e., tertiary aromatic carbons in lignins) in the CPMAS spectra. This technique allows more precise analysis of the quaternary carbons, and it is utilized for example in the studies of the degree of condensation in lignins [5,6].

Despite the known disadvantages of the solid-state NMR, that is, long measurement time and limited resolution, valuable information of solid materials, which is not accessible with other methods, has been obtained during the past few decades. With this method the physical information of the samples is obtained on the grounds of the molecular-level structure via chemical shifts.

This chapter reviews our recent studies dealing with solid-state NMR research of wood, isolated wood components, cellulose fibers, and cellulose derivatives.

13.2 Experimental

All the measurements were conducted with a Varian ^{UNITY}INOVA 300 NMR spectrometer operating at 75.5 MHz for carbons. The 7-mm sample rotor was rotated at a spinning speed of 5000 Hz in most cases. Contact times were 1–2 ms, delay between pulses 2–4 s depending on the sample, and acquisition time 20 ms. Time of accumulations was long enough to obtain a good signal-to-noise ratio. Before NMR measurements the wood and pulp samples were moistened with deionized water ($\sim$50 wt% H_2O). The chemical shifts in CPMAS spectra were referenced using the cellulose C1 signal as an internal standard (105 ppm) or methoxyl signal (56 ppm) when lignin samples were studied. Detailed description of the samples and parametres used during the measurements in various cases studied are available in the primary publications referenced.

13.3 Special characteristics of ^{13}C CPMAS NMR

High-resolution solid-state ^{13}C NMR has been successfully exploited in the structural and morphological studies of lignocellulosics ever since the early work done in the 1980s and the 1990s [7,8]. The potential of the combination of three special techniques, CP, MAS, and DD was demonstrated also in the studies of cellulose morphology. It became possible to evaluate supramolecular arrangement in the solid state from the line shapes or from the splitting of resonance signals.

Solid-state NMR is a bulk technique in which particle size effects have a minor impact on the intensity of the measured signal. There is no need for a single crystal of sufficient quality as in the case of x-ray crystallography. A further advantage is that samples can be studied without fractionation or isolation of components and thus all changes that might occur due to chemical treatments are avoided. In wood samples, cellulose, hemicelluloses, lignin, and to some extent also extractives give characteristic signals in the solid-state NMR spectrum. Relatively sharp signals are assigned to ordered cellulose and hemicelluloses while broader signals are assigned to disordered hemicelluloses and lignin.

For samples with diverse chemical structures, such as wood or heterogeneous cellulose derivatives, the CP kinetics differs between aromatic, carboxylic carbons, and aliphatic carbons. The individual carbon intensities in the CPMAS spectra depend upon the contact time and the relaxation rates. It follows that the comparison of intensities requires the recording of a contact time series together with relaxation evaluations [9].

It is generally accepted that the ^{13}C CPMAS NMR spectrum of cellulose is quantitative. The intensities for each of the six carbons in the anhydroglucose ring are expected to be equal because of the rather rigid hydrogen-bonded structure and because all carbons have directly bonded protons. The CP rate is equal for these carbons, which opens the possibility to study cellulose morphology.

During the measurement, a solid sample is rotated rapidly about an axis in a so-called magic angle relative to the external field. In practice, the rotation rate is lower than is needed to remove the chemical shift anisotropies of aromatic and carbonyl resonances, and hence for these, a narrow center band with various side bands is detected. The appearance of spinning side bands (SSB) increases the complexity of quantitative analysis because their intensities have to be included to the main signal intensity. Further, the overlapping of signals has to be avoided.

An ordinary ^{13}C CPMAS NMR spectrum of wood contains frequent overlaps of various signals from different wood components (see Figure 13.1). In chemical pulp samples the overlapping is still considerable, though less important due to the lower amount of hemicelluloses and residual lignin. In the proton spin relaxation edition (PSRE) method, the differences in the proton spin relaxation times ($T_{1\rho H}$) of different spatial domains, crystalline cellulose, and amorphous matrix are utilized to separate the phases into subspectra of their own. In addition to the ordinary CP, PSRE pulse sequence has a spin-lock delay between proton preparation pulse and the contact time. In this way, the interference of lignin and hemicellulose signals can be removed by purely spectroscopic means and all the possible structural changes caused by chemical treatments are avoided [2,3]. The crystallinity indices of cellulose can hence be obtained for wood samples using solid-state NMR [2,3,10–13]. An important feature, discovered with the subspectra separation for pulp samples, is that the intimate association (interaction/bonding) between hemicelluloses and cellulose leaves hemicellulose residuals in the crystalline phase [14].

Solid-state ^{13}C CPMAS spectroscopy is a much utilized characterization method for lignin studies. The major drawback is, however, its low resolution compared with the solution-state measurements. This can and needs to be compensated by special techniques. In the DD technique [5,6] the high-power decoupler is turned off for a short while (d2) after CP and then turned on again for data acquisition. During the delay, the dipolar interactions between protons and carbons induce a fast dephasing of the signals, which is most rapid for the tertiary aromatic carbons. The extent of lignin condensation in softwoods and softwood pulps has been estimated by this technique [15–19]. Recently chemical modifications in

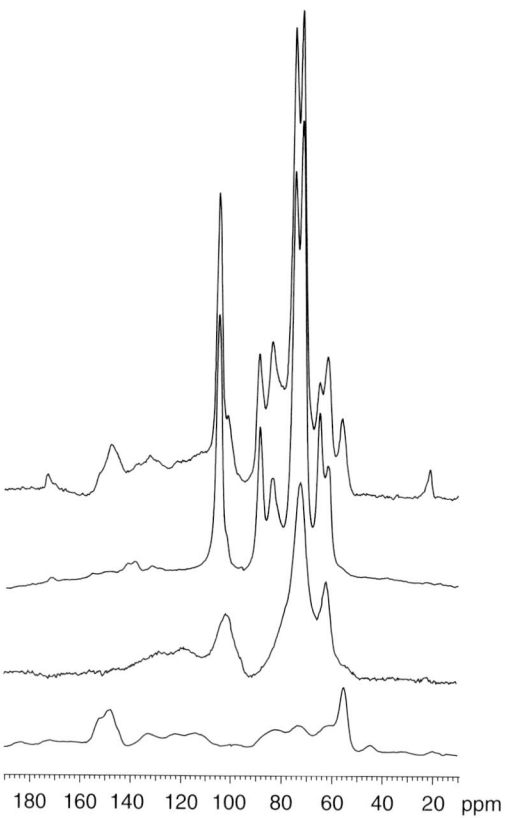

Figure 13.1 The 75 MHz ^{13}C CPMAS spectra of a wood sample, a chemical pulp sample, a mixture of hemicelluloses, and milled wood lignin (MWL), respectively, from top to bottom. From Maunu [1], Figures 5 and 9, copyright 2002 with permission from Elsevier.

wood lignin structures occurring at high temperatures were also evaluated on the basis of DD measurements [10,11,13,20,21].

13.4 Applications

13.4.1 Wood

Solid-state ^{13}C CPMAS NMR spectroscopy has been extensively applied to the structural studies of natural wood. In native wood samples the main structural components – cellulose, hemicelluloses, lignin and, to some extent, extractives – give their characteristic shifts (Table 13.1) to the spectrum shown in Figure 13.1. Proportional comparisons can be done directly on the basis of the conventional CPMAS measurements, but comparisons between results from different studies require special accuracy focused on the measurement

Table 13.1 Signal assignments for ^{13}C CPMAS spectra of wood.

Chemical shift (ppm)	Assignment
173	COOH in acetyl
153	S3e, S5e, G4e
148	S3f, S5f, G3
146	G4f
136	S1e, S4e, G1e
133	S1f, S4f, G1f
120	G6
116	G5
112	G2
105	C1 of cellulose
102	C1 of hemicelluloses
89	C4 of crystalline cellulose
84	C4 of amorphous cellulose
72–75	C2/C3/C5 of cellulose
65	C6 of crystalline cellulose
62	C6 of amorphous cellulose
56	Methoxyl groups in lignin
21	CH_3 in acetyl groups

S = syringyl, G = guaiacyl, e = etherified C4, f = free phenolic C4, S3e refers to aromatic carbon-3 of the S unit with etherified C4, and so on.

conditions, that is, sample humidity, spectrometer hardware and pulse sequence, values of parametres, as well as data handling [1].

13.4.1.1 Cellulose morphology in wood

For wood samples characterized by solid-state NMR, a wide range of percentual values of cellulose crystallinity are reported in literature. The main reason for this is the way how the CPMAS spectrum is divided into subspectra or deconvoluted to individual signals, in other words, how successfully lignin and hemicelluloses have been eliminated either chemically or spectroscopically [1].

Digital resolution enhancement of the ^{13}C CPMAS spectrum was the first method used to distinguish between the different crystalline cellulose forms in wood samples. Although the method remains questionable for solid-state spectra, comparisons of signal strengths at the crystalline C4 region in the wood samples studied show a higher relative amount of I_α for the softwoods and higher relative amount of I_β forms for the hardwoods [22–24]. Normal and tension woods from *Populus maximowiczii* have been proved to be rich in the cellulose I_β form [1,25].

The conventional ^{13}C CPMAS spectra of wood contain many signal overlaps. The interfering signals of lignin and hemicelluloses must be removed before CrI of cellulose can be

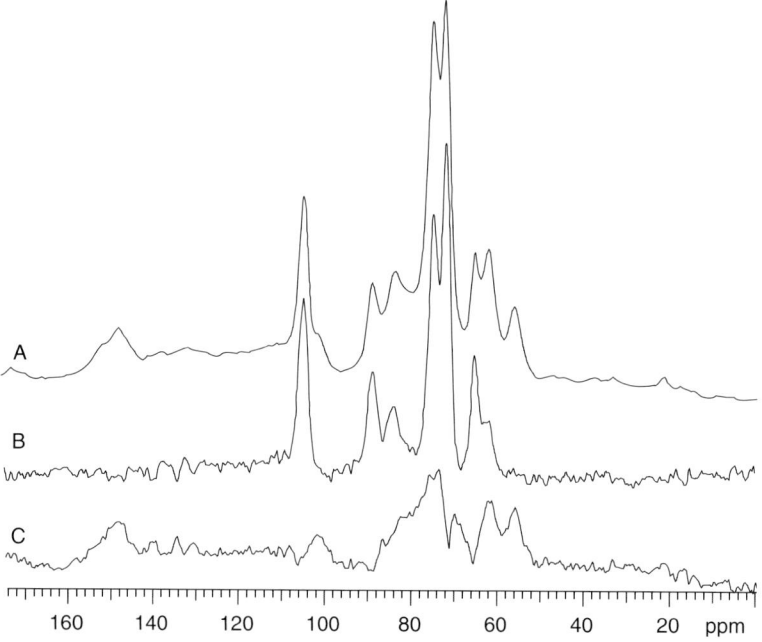

Figure 13.2 The ^{13}C CPMAS NMR spectrum of (A) pine wood, and the separated subspectra of (B) cellulose, and (C) lignin–hemicellulose matrix. From Maunu et al. [4], Figures 3–7, copyright 2005 with permission from Appita Inc.

determined reliably. The subspectra of more ordered cellulose and the amorphous matrix are separated on the basis of the different relaxation times when the PSRE method is used. Linear combination procedure is used to obtain the subspectrum of cellulose as subspectrum B) in Figure 13.2.

The crystallinity is determined in our studies from the area of the crystalline (86–92 ppm) cellulose C4-signal, a, and the area of the amorphous (79–86 ppm) cellulose C4-signal, b, according to Teeäär et al. [26] as

$$\text{CrI} = a/(a+b) \times 100\% \tag{13.1}$$

Only the highly ordered cellulose in the interior of the crystallites is considered as crystalline cellulose by this method and the remaining less-ordered cellulose, including the fibril surfaces, is considered as being amorphous [8]. The spin-locking technique has been used to investigate the cellulose crystallinity in various woods [2,3]. Slightly lower crystallinities for hardwoods, 54% (mean value for six woods), than for softwoods, 57% (mean value for five woods), were obtained.

We have studied variations in cellulose morphology by taking samples from different parts of wood stem: from near the pith, from mature wood, and at different heights of the stem [12]. The cellulose crystallinities (CrI) were obtained by the spin-locking technique for three different Scots pine and three different Norway spruce samples. It was observed that the crystallinity of cellulose was the same from the pith to the bark for both woods within the accuracy of the determination. Besides NMRs the samples were studied also by x-ray-scattering measurements. Solid-state NMR was applied to determine the crystallinity of cellulose and x-ray crystallography to determine the crystallinity of wood. For comparison, the crystallinity of wood was also calculated from the NMR crystallinity of cellulose using mass fractions 52.2% and 48% for cellulose in *Pinus sylvestris* and *Picea abies*, respectively. The values of 27% and 25% for Scots pine and Norway spruce were only slightly lower than the crystallinities obtained by wide angle x-ray scattering (WAXS) for the mature wood. Taking into account the accuracy of the estimations, the values agree well enough. It can be concluded from this comparison that the spin-locking technique provides a good way to determine cellulose crystallinity in wood samples [12].

13.4.1.2 Lignin structures in woods

The aromatic carbons of lignin resonate in the ^{13}C CPMAS NMR spectra without nearly any interference from the signals of cellulose and hemicelluloses. The resolution in the spectra, however, remains low (top and bottom spectra in Figure 13.1). The DD technique has proven to be a very useful method when lignin structures, especially those in wood, are studied in more detail [10,11,13,20,21]. In the DD spectra the signals of quarternary aromatic carbons are observable without overlapping of signals from tertiary aromatic carbons. Hardwood lignin contains syringyl (S) as well as guaiacyl (G) units. The spectra of hardwoods show signal at 153 ppm assigned to C3 and C5 carbons and a signal at ~135 ppm assigned to C1 and C4 carbons of the S unit (Figure 13.3a). The G units in softwoods give a signal at ~150 ppm assigned to C3 and C4 carbons, and the signal at ~133 assigned to C1 units and substituted C5 carbons (Figure 13.3b).

The lignin in hardwoods is built up of G and S units in ratio varying from 4:1 to 1:2 for the two monomeric units [27]. The lignin in oak and aspen contains more G units and probably more nonetherified S units than the lignin in birch as revealed by solid-state NMR [5,13,28].

Twenty-five tropical hardwoods from Ghana were examined recently using different spectroscopic techniques in the solid state to explore the chemical structure and phytochemical diversity [29]. DD NMR results showed variations in the S/G proportions in lignin. It was observed that a higher amount of G units correlated with a higher amount of lignin and that this further correlated with better bio-resistance.

13.4.1.3 Chemical structure of thermally modified wood

The utilization of spin-locking and DD methods makes it possible to follow changes that various treatments cause to the chemical structure of wood and its components. In our group we have utilized these solid-state NMR methods to evaluate the thermal modification effects on different woods [10,11,13,20,21].

The conversion of cellulose during thermal modification is observed as a decrease in the signals at 84 and 62 ppm assigned to disordered, amorphous cellulose (Figure 13.4). The PSRE technique and linear combinations of measured spectra were used to isolate subspectra of cellulose and lignin–hemicellulose material as described in the beginning of this chapter. The crystallinity of the cellulose was found to increase with increasing temperature with the CrI being the highest (65%) for pine wood treated at 230°C. This was due to a preferred degradation of the less-ordered molecules during the thermal treatment [11].

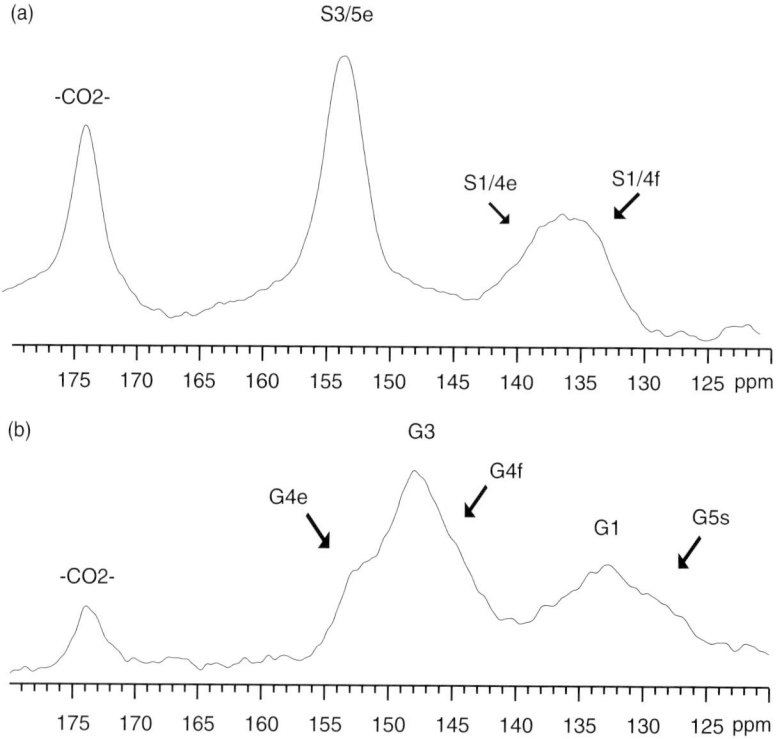

Figure 13.3 The DD spectra of (a) silver birch and (b) scots pine; G refers to guaiacyl units and S to syringyl units; e refers to etherified C4 units, f to free phenolic C4 units and s to substituted structures.

The results show that thermal modification enhances cellulose crystallinity in every studied wood species. The crystallinity indices are higher for the thermally modified softwood samples than for the hardwood samples (Figure 13.5). One relevant factor for the lower cellulose crystallinities of hardwoods than of softwoods studied by this method is the incomplete removal of xylan. The accuracy of the values was estimated to be within ±3% on the basis of spectra processing [11,13].

The relative mass fraction of lignin increased during thermal modification in proportion to the relative mass fraction of carbohydrates, being the highest for the pine wood samples treated at the highest temperature of 230°C. This change was due to the preferential degradation of carbohydrates.

DD measurements were used to follow the changes in lignin structures in more detail. Cleavage of the β-O-4 linkages in lignin is observed in softwoods from the shape of the G C3 and C4 signals at 140–160 ppm (Figure 13.6). The etherified G units in lignin are seen as a shoulder at 153 ppm and free phenolic G units appear as a shoulder at 146 ppm. After thermal modification, softwood lignin contains more free phenolic units than before treatment. The increase in signal height at 128 ppm arises from the C5-substituted structures and indicates an increase in the content of condensed lignin G structures due to the modification [11,13].

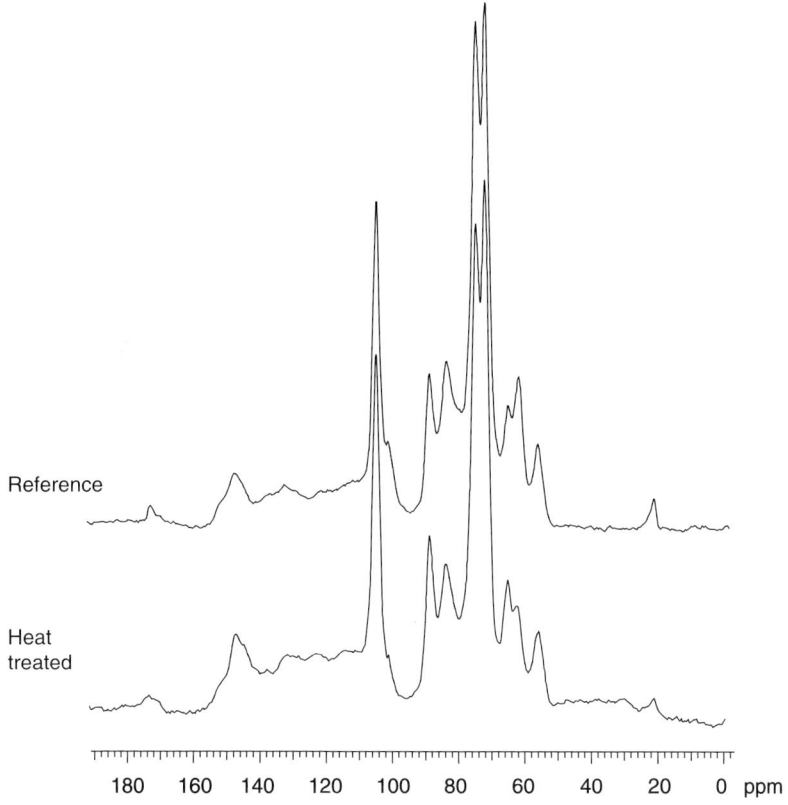

Figure 13.4 The 75 MHz ^{13}C CPMAS spectra of a heat treated wood and its reference sample. From Maunu [1], Figures 5 and 9, copyright 2002 with permission from Elsevier.

For hardwoods the cleavage of the β-O-4 linkages in lignin is observed from the relative intensities of the signal at 153 and 148 ppm [13]. The signal at 153 ppm is assigned to S C3 and C5 units that are etherified at C4 whereas the signal at 148 ppm is assigned to those in free phenolic units and to G C3 and C4. The majority of the S units in hardwoods appear to contain β-O-4 linkages before treatment, whereas after treatment a substantial part of these linkages are cleaved.

Other changes that the thermal modification cause to the wood structure detected on the basis of solid-state NMR measurements were the deacetylation of hemicelluloses and the demethoxylation of lignin. The thermal degradation of wood was shown to begin with the hemicelluloses. The lower thermal stability of hemicelluloses compared to cellulose can be explained by the low amount of crystallinity.

13.4.2 Cellulose fibers

Owing to the capability of measuring samples in their native states, ^{13}C CPMAS NMR can be applied to investigate both the physical and chemical structure of lignocellulosics. The

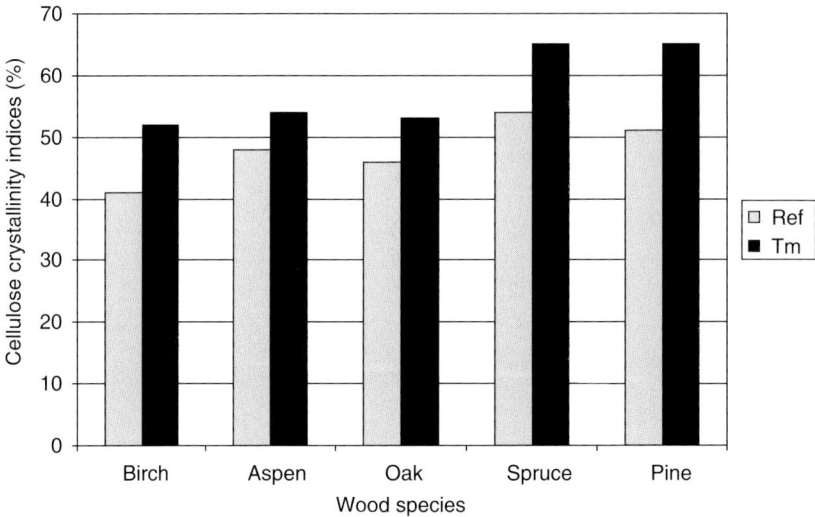

Figure 13.5 Cellulose crystallinity indices (%) for various wood samples; Ref refers to the untreated sample and Tm to thermally modified sample.

delignification reactions during cooking and bleaching lead to structural changes of the components of the pulp fibers and require characterization in the solid phase. Solid-state NMR methods have been used to follow delignification during pulping as well as to study the kinetics of the pulping process. Further, the morphology of cellulose in pulps has been characterized extensively. The structural details of the insoluble residual lignin, resistant to removal during pulping, have been clarified also using solid-state NMR methods. The results are referred to in many reviews and recently also by the present author [1]. In the following sections, the fiber ultrastructure from the viewpoint of cellulose morphology and chemical structure, as well as the condensation in technical lignins, will be discussed based on the results obtained from our laboratory during the past few years.

13.4.2.1 Cellulose morphology in fibers

Structural changes occurring in the morphology of cellulose during various chemical pulping related processes were followed in order to optimize the cooking and bleaching conditions [16,18,23,24,30,31]. According to the ^{13}C CPMAS measurements the crystallinity of cellulose increases during pulping. The CrI values obtained by the ordinary ^{13}C CPMAS measurements are very sensitive to the variations of the lignin and hemicellulose contents. Therefore, the amorphous lignin and hemicelluloses were removed by the PSRE method discussed in Section 13.3. Cellulose crystallinity was calculated from the areas of the crystalline and amorphous C4 signals after deconvolution using the subspectrum from PSRE measurement (Figure 13.7a). When various pulps were studied, a slightly lower degree of cellulose crystallinity was found in birch pulps compared with the corresponding pine pulps [16].

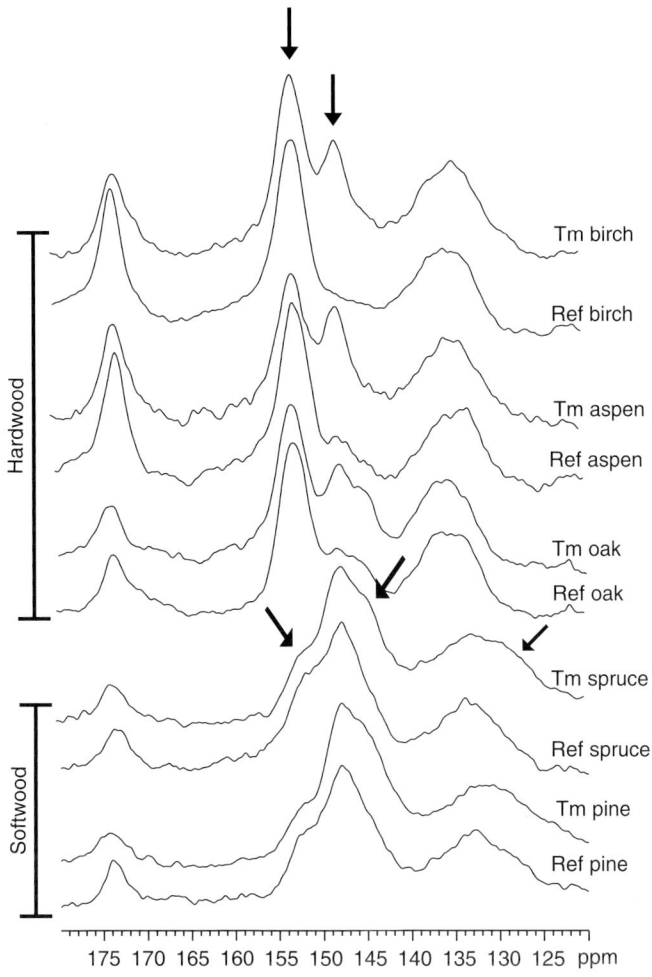

Figure 13.6 The DD spectra of thermally modified (Tm) and unmodified reference (Ref) samples of different wood species; arrows indicate the changes referred to in the text.

The occurrence of various crystalline forms of cellulose can be observed in the original CPMAS spectrum *inter alia* as shoulders in the C4 crystalline signal [7,8]. To determine the relative proportions of cellulose polymorphs, resolution enhancement has to be used in order to succeed with signal deconvolution (Figure 13.7b) [16,22,32]. The increased noise level involved with this method exposes the deconvolution of small signals to errors, and therefore, the determination of cellulose II form is not very accurate. During pulping part of the metastable cellulose, I_α is converted to the more stable cellulose I_β polymorph [18,23,24,30].

The structure of the fiber wall is inhomogeneous and the composition of the fiber surface differs from the inner part of the fiber. To obtain information about the pulping effects on the fiber surface structure, kraft pulp was refined and the fines were separated from the

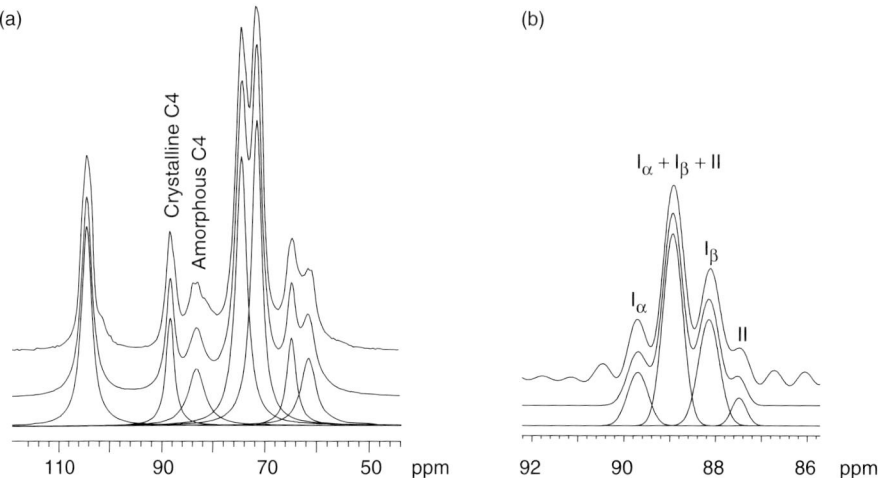

Figure 13.7 The 75 MHz ^{13}C CPMAS spectrum of a softwood pulp sample (as an example) with deconvolution of the spectrum (a) and resolution enhanced crystalline C4 resonance with deconvolution (b). Measured spectra, calculated spectra, and deconvoluted lines are shown from top to bottom, respectively.

long fibers, and various fractions were studied regarding the cellulose morphology [24]. After kraft pulping, cellulose crystallinity was found to be lower in ray cells and on the fiber surface compared with the long-fiber fractions. Only marginal changes in the amounts of the crystalline forms of cellulose were observed during refining. In the fines the relative proportions of I_α and I_β were similar to the corresponding bulk fibers.

The cellulose C4 resonance region in the solid-state ^{13}C CPMAS NMR spectra (80–92 ppm) is very informative when divided into separate resonances by spectral fitting [33–35]. The C4 carbons in the ordered region have been assigned to crystalline cellulose I forms with sharp Lorenzian line shapes, and moreover, these are overlapped by a broad Gaussian line assigned to paracrystalline cellulose with shorter carbon relaxation time. The more disordered region has been divided into two signals from cellulose at accessible fibril surfaces and one signal from cellulose at inaccessible fibril surfaces, all with Gaussian line shapes. Owing to the differences in chemical shifts of the crystalline interiors and the crystallite surfaces, fitted results can be used to determine the average fibril dimensions and further the differences in chemical shifts of accessible and inaccessible fiber surfaces for determination of fibril aggregate dimensions. The method is based on the square cross-section fibril model and aggregated cellulose fibrils model [33]. For successful application of this method, it is necessary, depending on the pulps, to remove hemicelluloses as well as residual lignin from samples before NMR analysis. No major differences were obtained in fibril and aggregate dimensions when isolated cellulose from fines and long fibers were compared [31].

13.4.2.2 Ordered hemicelluloses

For pure cellulose samples the degree of crystallinity can be determined by ordinary ^{13}C CPMAS NMR measurements. However, in chemical pulp samples the remaining

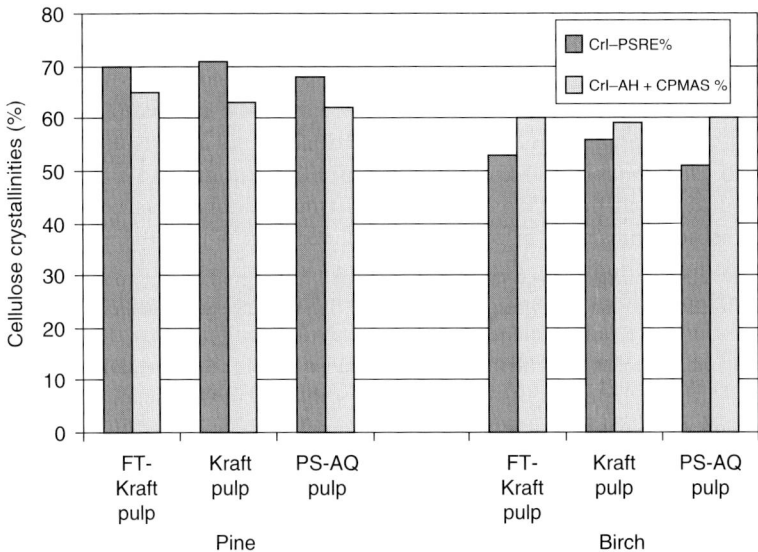

Figure 13.8 Cellulose crystallinities (CrI %) of various chemical pulps determined by the PSRE method, and after 17 h acid hydrolysis and then determined by ordinary CPMAS, respectively.

hemicelluloses and lignin interfere with the determination of cellulose crystallinity. Therefore, to obtain quantitative cellulose spectra these amorphous noncellulosic components must be removed from the samples. This can be done either chemically by $NaClO_2$ delignification and acid (such as hydrochloric acid) hydrolysis (at 100°C) or spectroscopically by the PSRE method described above. To compare these two methods, various pine and birch pulps containing varying amounts of hemicelluloses have been analyzed [14,16,36].

After removal of hemicelluloses by hydrochloric acid hydrolyses at 100°C for 17 h, higher CrI values were obtained in comparison with the unhydrolysed samples. The crystallinities obtained for pine kraft pulps obtained from flow-through (FT) kraft cooking (FT-kraft pulp), kraft cooking (kraft-pulp), and kraft polysulfide–anthraquinone cooking (PS–AQ pulp) after acid hydrolysis ($CrI_{AH+CPMAS}$) were lower than the corresponding crystallinities obtained by PSRE method (CrI_{PSRE}). For birch samples just the opposite was observed (Figure 13.8). The lower values of the pine pulps are at least partly due to the incomplete removal of hemicelluloses by the acid hydrolysis. Another effect of the acid hydrolysis is the dissolution of amorphous cellulose. The higher CrI_{PSRE} values obtained for the pine pulps indicate that the amorphous noncellulosic components may be removed more effectively by the PSRE method than by the acid hydrolysis. We concluded that the PSRE method is a more convenient way to remove the interfering hemicellulose signals from softwood chemical pulps. After the spectroscopic or chemical removal of hemicelluloses, no significant differences in the cellulose crystallinities were observed between pulps obtained by various pulping methods. According to the CrI values, the crystallinity is lower in birch pulps than in the corresponding pine pulps. It has been reported earlier that in hardwoods the degree of crystallinity is slightly lower than in softwoods [2,3,11], and this seems to remain also after pulping.

The CrI$_{PSRE}$ values of birch pulps are even lower than the corresponding CrI$_{AH+CPMAS}$ values. This is a result of the incomplete removal of xylan from the birch pulps by the PSRE method as the xylan signal can still be seen at 82 ppm in the subspectrum of the crystalline component (Figure 13.9a). This indicates similar relaxation behavior of xylan and cellulose, which may be a consequence of a close association and interactions between xylan and cellulose and/or increased relaxation times of xylan due to ordered structure. Likewise, all the glucomannan could not be removed completely by the PSRE method. The C1 signal of the mannose residues can still be seen at 102 ppm in the subspectra of the crystalline components of all the softwood pulps studied (Figure 13.9b). It is therefore likely that some of the glucomannan may as well possess an ordered ultrastructure or interact with cellulose. From this it was concluded that the PSRE method also provides information on the interactions between cellulose and hemicelluloses [14].

13.4.2.3 Extent of lignin condensation

Solid-state ^{13}C CPMAS spectroscopy can be very well utilized in the studies of isolated lignin structures with limited solubility. However, the low resolution in the ordinary CPMAS spectra of lignins limits the investigation of detailed structures, but some structural features can be obtained, and these even more profound when the DD method is used. Condensed aromatic lignin structures have a significant role when the incomplete delignification is considered. In our research projects we have had an opportunity to evaluate the feasibility of the DD method to determine the degree of condensation in various lignin samples [16–19].

Using DD technique the average extent of substitution of the aromatic rings of lignin can be determined. Dipolar interactions between protons and carbons induce a fast dephasing of the signal, and this is more rapid with the tertiary aromatic carbons due to the proximity of the attached protons. Two different methods were used to obtain the degree of condensation and the methods were compared with different model compounds [19].

The ordinary ^{13}C CPMAS spectrum of a kraft pulp residual lignin without any delay and the DD spectrum measured with the 50 µs dephasing delay are shown in Figure 13.10a. The signal A (140–160 ppm) in the lower spectrum is due to the C3 and C4 of the G group whereas the signal B (115–140 ppm) is due to the C1 and the substituted C5 of the G group. The ratio of these signal areas (A/B) provides information on the aromatic substitution and on the degree of condensation [17,18].

The average extent of the substitution can also be determined by monitoring the unequal decay rates of the aromatic tertiary and quarternary carbon signals. This method was in the first instance applied to the analysis of various naturally degraded and fossilized soil and coal samples [37,38]. The natural logarithm of the area of aromatic region is presented as a function of the dipolar dephasing delay time (Figure 13.10b). The total intensity (I_{tot}) of the aromatic area is composed of the intensities of the faster decaying aromatic tertiary carbons (I_a) and the slower decaying quaternary carbons (I_b). The procedure to obtain the number of tertiary carbons in guaiacyl ring and further the extent of condensation was delineated in detail by Liitiä et al. [19]. According to the results obtained on the basis of the model compounds studied, the "d2-array" method was found to be slightly more accurate than the A/B method.

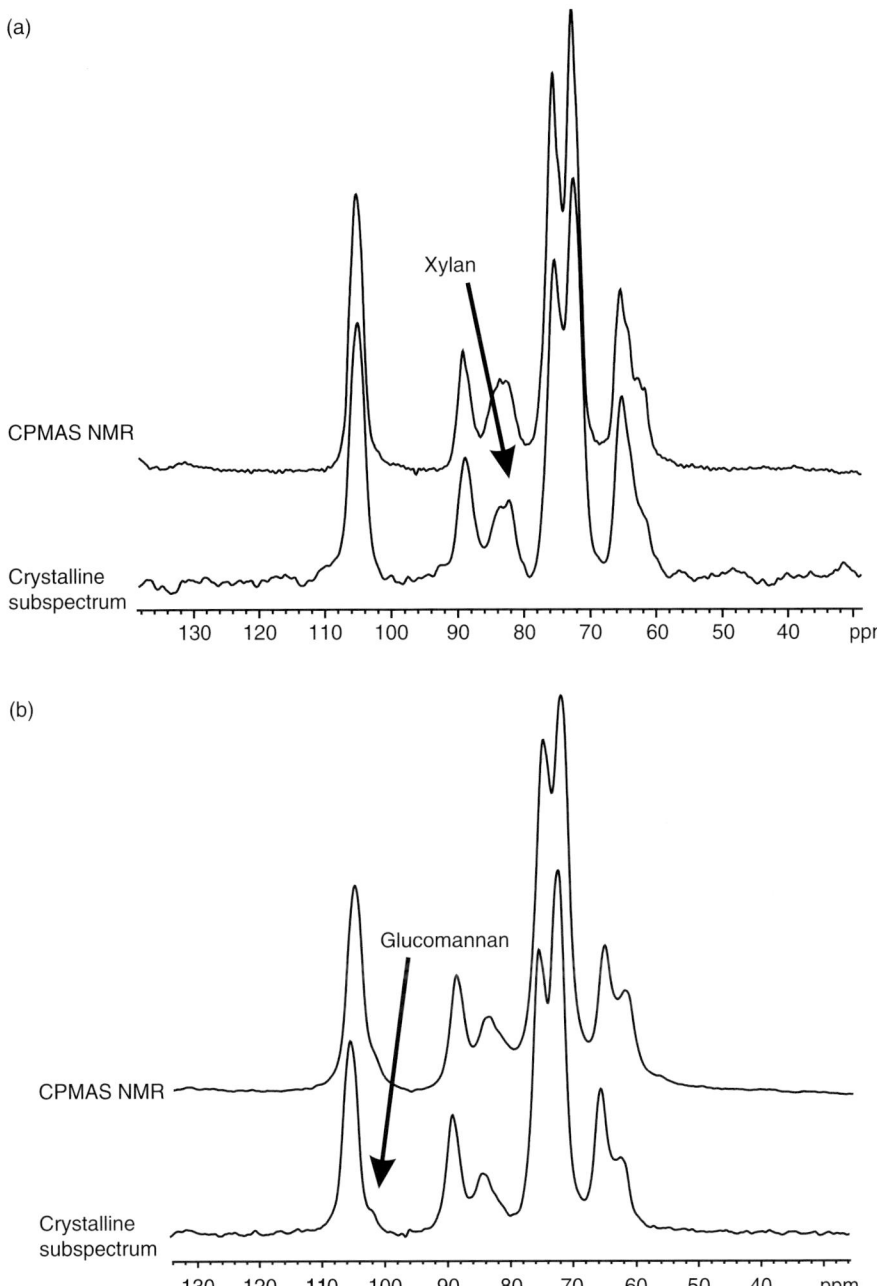

Figure 13.9 The ^{13}C CPMAS NMR spectrum of (a) birch pulp with the subspectrum of the crystalline component and (b) pine pulp with the subspectrum of the crystalline component.

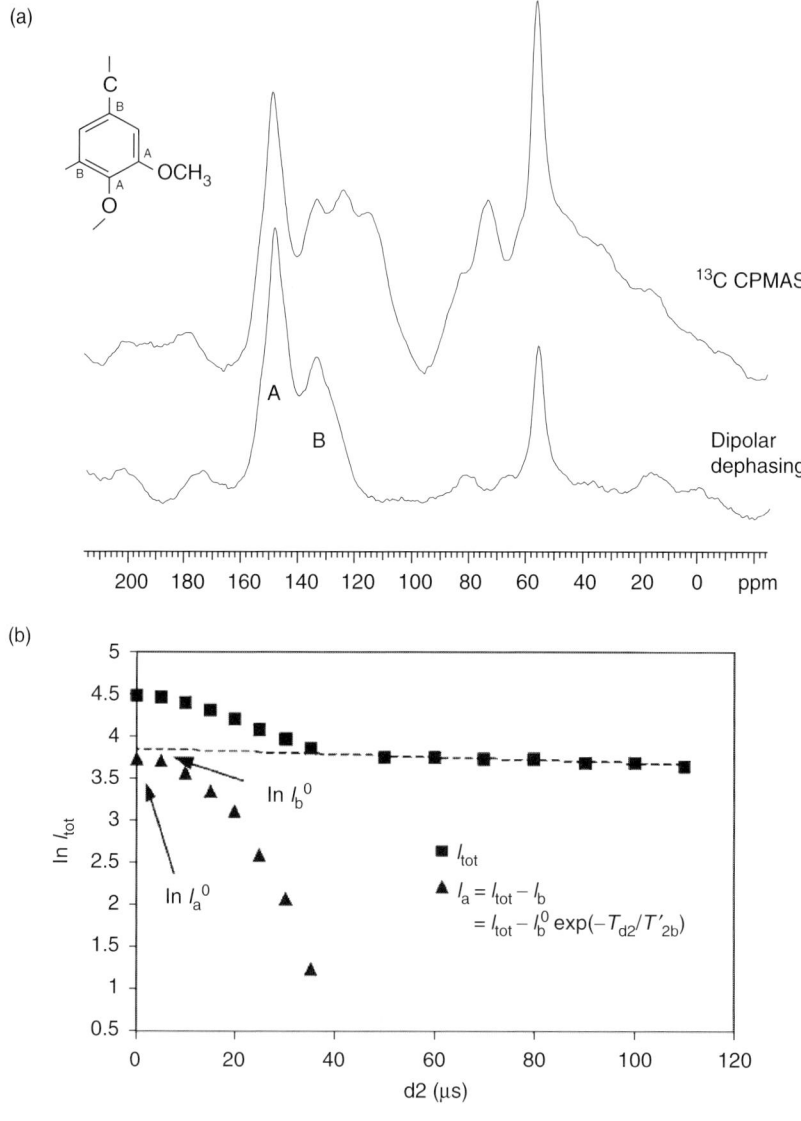

Figure 13.10 (a) The ^{13}C CPMAS and the DD (delay = 50 μs) spectra of a kraft pulp residual lignin; (b) decay of the ln I_{tot} as a function of d2-delay.

The extent of condensation obtained by the d2-array method for native and several technical lignins is represented in Figure 13.11. According to the results, spruce lignin isolated enzymatically from ball-milled wood (MW-E), which is considered to be a better representative of native lignin, was found to be more condensed than pine milled wood lignin (MWL), which might represent the most easily isolated and the least condensed part of lignin. All technical lignins were found to be more condensed than pine MWL. Studies of pine spent liquor lignins (SLLs) dissolved in various stages (60–240 min) of FT cooking, FT-SLL-60 to

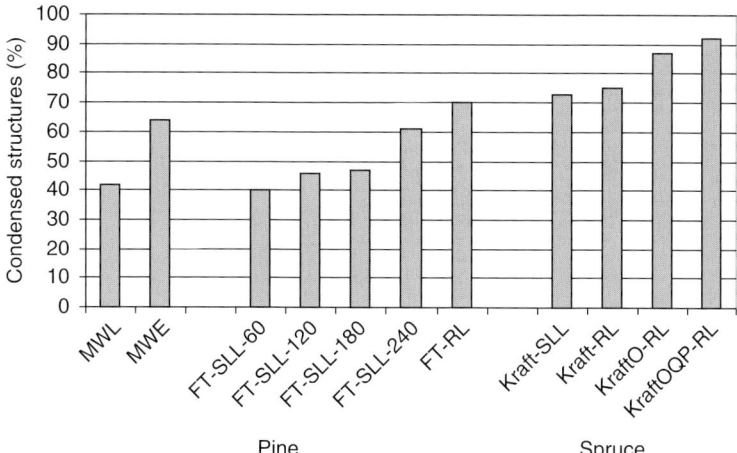

Figure 13.11 The extent of condensed structures in native and technical lignins.

FT-SLL-240, show that the amount of condensed structures increases as the cook proceeds. This indicates that the uncondensed lignin structures are most easily removed in the beginning of the cook. As the cook proceeds more condensed lignin structures are dissolved, the corresponding residual lignin (FT-RL) is the most condensed. The SLL of the conventional spruce kraft pulp is more condensed than the last SLL-fraction of the FT-cook of pine. While assuming that the lignin structural difference between spruce and pine is small, these results indicate that either more condensed lignin structures have been dissolved in the conventional kraft cook or some further condensation has occurred in the black liquor lignin during the conventional cook. In the RLs of the conventional spruce and the FT pine kraft cook, the extent of condensation seems to be quite similar. Oxygen (O) delignification of the kraft pulp was found to further increase the relative amount of the condensed structures in RL. A slight increase in the amount of the condensed structures was observed further after chelation and then peroxide bleaching (QP).

It can be concluded that both A/B and d2-array methods can be used to estimate the degree of condensation in lignin samples and no chemical treatment for the sample preparation is needed. The accuracy of the d2-array method was found slightly better, but the long measurement times required and interference of carbohydrates and all aromatic impurities are the disadvantages of this method. Therefore, the A/B method may be more reliable for samples containing large amounts of carbohydrates or some aromatic impurities.

13.4.3 Derivatives

The applications of high-resolution NMR spectroscopy to the studies of chemically modified cellulose can occasionally be met with difficulties because of the limited solubility of the derivatives in common NMR solvents. The morphological as well as the chemical analysis of solid cellulose derivatives is however possible even though the resolution in the solid-state NMR remains low. A selective derivatization can be followed on the basis of CPMAS spectra, and it can be estimated if the entire initial cellulose is made accessible to the modification reactions. However, successful application of solid-state NMR methods for the characterization of cellulose derivatives requires the control of a variety of parameters [4].

For samples with more diverse chemical structures such as cellulose derivatives, the CP kinetics differs between the aromatic, carboxylic, and aliphatic carbons. The individual carbon intensities depend upon the contact time and relaxation rates. It follows that the comparison of intensities requires the recording of a contact time series together with relaxation evaluations. Signal intensities of a carboxymethyl cellulose (CMC) with varying contact times are presented in Figure 13.12a. During CP the magnetization builds up and

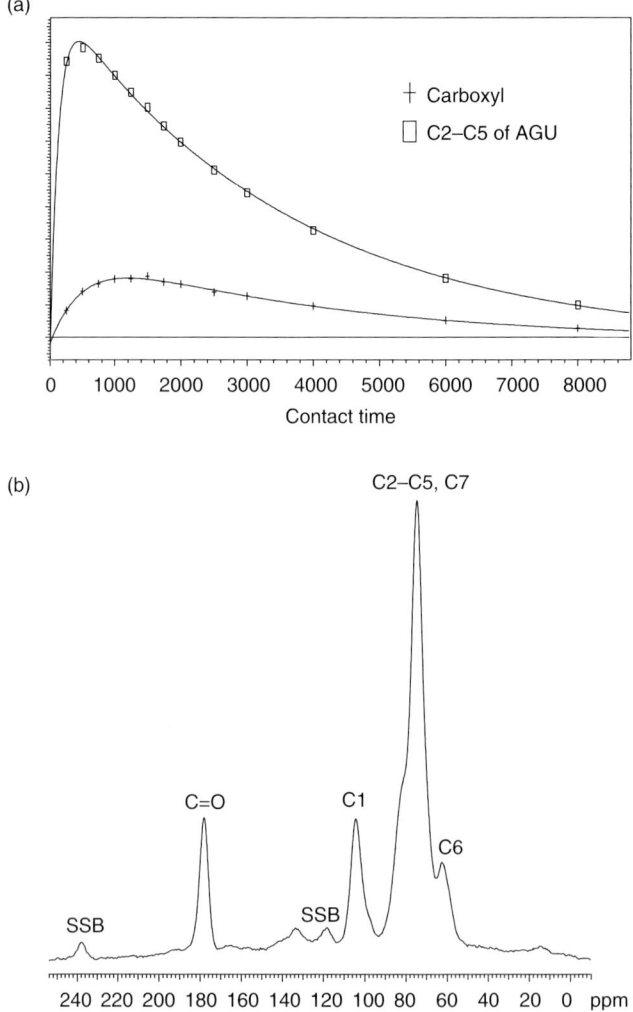

Figure 13.12 (a) Signal intensities of carboxyl carbons and C2—C5 carbons of the anhydroglucose unit (AGU) of CMC as a function of the contact times; (b) the ^{13}C CPMAS NMR spectrum of CMC with signal assignments. From Maunu et al. [4], Figures 3–7, copyright 2005 with permission from Appita Inc.

relaxes with different rates, which requires an extrapolation to contact time = 0 or a selection of a suitable contact time for quantitative comparison of signals [9].

^{13}C CPMAS spectrum measured with 2 ms contact time and 5 s delay between pulses is presented in Figure 13.12b. The appearance of SSB in the spectrum increases the complexity of quantitative analysis because their intensities have to be included in the main signal intensity. The sample rotation rate has to be chosen in a suitable way to prevent signal overlapping and to obtain adequate resolution. Ideal would be if the spinning speed were greater than the static line width expressed in hertz. Unless this condition is attained, SSBs are observed on both sides of the main signal and their intensities have to be included in the main signal intensity. When these are taken into account, the degree of substitution can be evaluated for CMC by comparing the signal intensities of C=O (+SSBs) and the C1 signal of the anhydroglucose unit (AGU).

Because of the complicated wood structure and fiber morphology in wood pulps, the accessibility of cellulose to solvents is limited in many cases. This causes the derivatization of cellulose to be difficult and often results in inhomogeneously substituted cellulose derivatives. If the derivatization reactions are performed in solutions, a more homogeneous product is obtained. However, due to the limited solubility of the starting material, the derivatization processes are usually heterogeneous, resulting in modification mainly in the accessible cellulose domains. The homogeneity of the derivatives can be evaluated spectroscopically by applying the PSRE technique as described above. This method has been used by us in the analysis of a cellulose carbamate sample. The modification is seen as a carbonyl signal at ~158 ppm (Figure 13.13a). The differences in the proton relaxations

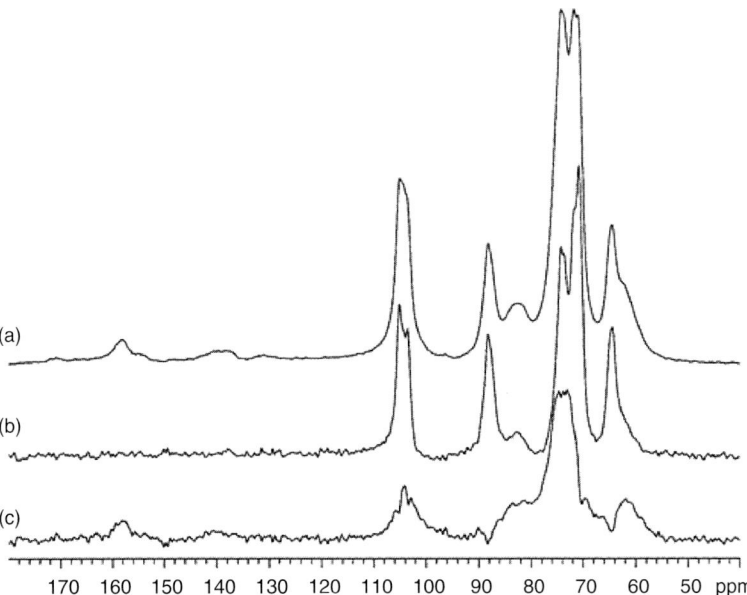

Figure 13.13 The ^{13}C CPMAS NMR spectrum of cellulose carbamate (a) and the separated subspectra of ordered (b) and unordered matrix (c). From Maunu et al. [4], Figures 3–7, copyright 2005 with permission from Appita Inc.

were used to separate the ordered (Figure 13.13b) and unordered phases (Figure 13.13c). As seen in the bottom spectrum the unordered phase contains the carbonyl signal, which leads to the conclusion that the modification has occurred in accessible amorphous phase in this case.

Concluding remarks

Solid-state NMR has the potential to analyze samples in native state and obtain average information of bulk samples. The diversity of NMR techniques, together with the increased knowledge and the development of the hardware and software during recent years, provides many opportunities for characterizing heterogeneous materials. This chapter has demonstrated how advanced ^{13}C CPMAS NMR methods can be applied to obtain morphological as well as chemical information of such complicated substances as woods, cellulose fibers, and cellulose derivatives.

In the PSRE method the differences in the relaxation times ($T_{1\rho H}$) of different spatial domains can be utilized to separate the components into subspectra of their own. The technique is used in this context to separate ordered and unordered matrix from each other. Dipolar interactions between protons and carbons are used to cause dephasing of the proton-attached C signals in the CPMAS spectra and this technique allows the study of the lignin aromatic quaternary carbons more precisely. In addition to the ordinary ^{13}C CPMAS NMR measurements, these methods have been applied to investigate the effects of chemical pulping-related processes on the polymeric components of wood and pulp. Further the changes that a heat treatment process brings in the chemical structure of wood components have been investigated. Solid-state NMR provides also remarkable structural and morphological information of insoluble cellulose derivatives, especially in the studies of the heterogeneous chemical modification of the samples.

References

1. Maunu, S.L. (2002) NMR studies of wood and wood products. *Prog. Nucl. Magn. Reson. Spectrosc.*, 40, 151–174.
2. Newman, R.H. and Hemmingson, J.A. (1990) Determination of the degree of cellulose crystallinity in wood by carbon-13 nuclear magnetic resonance spectroscopy. *Holzforschung*, 44, 351–355.
3. Newman, R.H. (1999) Estimation of the lateral dimensions of cellulose crystallites using ^{13}C NMR signal strengths. *Solid State Nucl. Magn. Reson.*, 15, 21–29.
4. Maunu, S.L., Wikberg H., and Hiltunen, M. (2005) Solid state NMR possibilities in the characterisation of cellulose and its derivatives. In: *13th Proc. Intern. Symp. Wood Fibre Pulping Chem.*, Vol. 2, pp. 349–352. Appita Inc. New Zealand.
5. Manders, W.F. (1987) Solid-state carbon-13 NMR determination of the syringyl/guaiacyl ratio in hardwoods. *Holzforschung*, 41, 13–18.
6. Hawkes, G.E., Smith, C.Z., Utley, J.H.P., Vargas, R.R., and Viertler, H. (1993) A comparison of solution and solid state carbon-13 NMR spectra of lignins and lignin model compounds. *Holzforschung*, 47, 302–312.
7. VanderHart, D.L. and Atalla, R.H. (1984) Studies of microstructure in native celluloses using solid-state carbon-13 NMR. *Macromolecules*, 17, 1465–1472.

8. Atalla, R.H. and VanderHart D.L. (1999) The role of solid-state carbon-13 NMR spectroscopy in studies of the nature of native celluloses. *Solid State Nucl. Magn. Reson.*, 15, 1–19.
9. Käuper, P., Kulicke, W.-M., Horner, S. et al. (1998) Development and evaluation of methods for determining the pattern of functionalization in sodium carboxymethylcelluloses. *Angew. Makromol. Chem.*, 260, 53–63.
10. Wikberg, H. (2005) Advanced solid state NMR spectroscopic techniques in the study of thermally modified wood. Ph.D. thesis, University of Helsinki. Available online at: http://ethesis.helsinki.fi.
11. Sivonen, H., Maunu, S.L., Sundholm, F., Jämsä, S., and Viitaniemi, P. (2002) Magnetic resonance studies of thermally modified wood. *Holzforschung*, 56, 648–654.
12. Andersson, S., Wikberg, H., Pesonen, E., Maunu, S.L., and Serimaa, R. (2004) Studies of crystallinity of scots pine (*Pinus sylvestris* L.) and norway spruce (*Picea abies* (L.) *Karst.*). *Trees*, 18, 346–353.
13. Wikberg, H. and Maunu S.L. (2004) Characterisation of thermally modified hard- and softwoods by ^{13}C CPMAS NMR. *Carbohydr. Polym.*, 58, 461–466.
14. Liitiä, T., Maunu, S.L., Hortling, B., Tamminen, T., Pekkala, O., and Varhimo, A. (2003) Cellulose crystallinity and ordering of hemicelluloses in pine and birch pulps as revealed by solid-state NMR spectroscopic methods. *Cellulose*, 10, 307–316.
15. Liitiä, T., Maunu, S.L., Sipilä, J., and Hortling, B. (2001) Solid state ^{13}C NMR spectrroscopy in residual lignin studies. In: *11th Proc. Intern. Symp. Wood Pulping Chem.*, Vol. I, pp. 289–292, France.
16. Liitiä, T. (2002) Application of modern NMR spectroscopic techniques to structural studies of wood and pulp components. PhD thesis, University of Helsinki. Available online at: http://ethesis.helsinki.fi.
17. Liitiä, T., Maunu, S.L., and Hortling, B. (2000) Solid state NMR studies of residual lignin and its association with carbohydrates. *J. Pulp Paper Sci.*, 26, 323–330.
18. Liitiä, T., Maunu, S.L., and Hortling, B. (2001) Solid state NMR studies on inhomogeneous structure of fibre wall in kraft pulp. *Holzforschung*, 55, 503–510.
19. Liitiä, T., Maunu, S.L., Sipilä, J., and Hortling, B. (2002) Application of solid-state ^{13}C NMR spectroscopy and dipolar dephasing technique to determine the extent of condensation in technical lignins. *Solid State Nucl. Magn. Reson.*, 21, 171–186.
20. Wikberg, H., Maunu, S.L., Jämsä, S., and Viitaniemi, P. (2004) Changes in the structure of thermally modified hard- and softwoods studied by solid state NMR. In: *8th Proc. European Workshop on Lignocellulosics and Pulp*, pp. 29–32. Latvia.
21. Nuopponen, M., Wikberg, H., Vuorinen, T., Maunu, S.L., Jämsä, S., and Viitaniemi, P. (2004) Heat-treated softwood exposed to weathering. *J. Appl. Polym. Sci.*, 91, 2128–2134.
22. Newman, R.H. (1994) Crystalline forms of cellulose in softwoods and hardwoods. *J. Wood Chem. Technol.*, 14, 451–466.
23. Maunu, S.L., Liitiä, T., Kauliomäki, S., Hortling, B., and Sundquist, J. (2000) ^{13}C CPMAS NMR investigations of cellulose polymorps in different pulps. *Cellulose*, 7, 147–159.
24. Liitiä, T., Maunu, S.L., and Hortling, B. (2000) Solid state NMR studies of cellulose crystallinity in fines and bulk fibres separated from refined kraft pulp. *Holzforschung*, 54, 618–624.
25. Wada, M., Okano, T., Sugiyama, J., and Horii, F. (1995) Characterization of tension and normally lignified wood cellulose in *Populus maximowiczii. Cellulose*, 2, 223–233.
26. Teeäär, R., Serimaa, R., and Paakkari, T. (1987) Crystallinity of cellulose, as determined by CP/MAS-NMR and XRD methods. *Polym. Bull.*, 17, 231–237.
27. Sjöström, E. (1993) *Wood Chemistry, Fundamentals and Applications*, 2nd Edn, Academic Press, New York.
28. Hawkes, G.E., Smith, C.Z., Utley, J.H.P., Vargas, R.R., and Viertler, H. (1993) A comparison of solution and solid state carbon-13 NMR spectra of lignins and lignin model compounds. *Holzforschung*, 47, 302–312.

29. Nuopponen, M.H., Wikberg, H.I., Birch, G.M. et al. (2006) Characterization of 25 tropical hardwoods with fourier transform infrared, ultraviolet resonance raman and ^{13}C-NMR cross-polarization/magic-angle spinning spectroscopy. *J. Appl. Polym. Sci.*, 102, 810–819.
30. Liitiä, T., Maunu, S.L., and Hortling, B. (1998) Effects of pulping on crystallinity of cellulose studied by solid state NMR. In: *Cellulosic Pulps, Fibres, and Materials*, (10th *Cellucon '98*, ed. J.F. Kennedy, G.O. Phillips, P.A. Williams, and B. Lönnberg). Woodhead Publishing Limited, England. pp. 39–44.
31. Hult, E.-L., Liitiä, T., Maunu, S.L., Hortling, B., and Iversen, T. (2002) A CP/MAS ^{13}C-NMR study of cellulose structure on the surface of kraft pulp fibers. *Carbohydr. Polym.*, 49, 231–234.
32. Newman, R.H. (1997) Crystalline forms of cellulose in the silver tree fern *Cyathea dealbata*. *Cellulose*, 4, 269–279.
33. Larsson, P.T., Wickholm, K., and Iversen, T. (1997) A CP/MAS carbon-13 NMR investigation of molecular ordering in celluloses. *Carbohydr. Res.*, 302, 19–25.
34. Wickholm, K., Larsson, P.T., and Iversen, T. (1998) Assignment of non-crystalline forms in cellulose I by CP/MAS carbon-13 NMR spectroscopy. *Carbohydr. Res.*, 312, 123–129.
35. Larsson, P.T., Hult, E.-L., Wickholm, K., Petterson, E., and Iversen, T. (1999) CP/MAS carbon-13 NMR spectroscopy applied to structure and interaction studies on cellulose I. *Solid State Nucl. Magn. Reson.*, 15, 31–40.
36. Liitiä, T., Maunu, S.L., Pekkala, O., Varhimo, A., and Hortling, B. (2002) Application of solid-state NMR spectroscopic methods to analyse cellulose crystallinity in pulps with varying hemicellulose contents. In: *Proc. 7th European Workshop on Lignocellulosics and Pulp*, pp. 309–312. Finland.
37. Hatcher, P.G. (1987) Chemical structural studies of natural lignin by dipolar dephasing solid-state ^{13}C nuclear magnetic resonance. *Org. Geochem.*, 11, 31–39.
38. Bates, A.L. and Hatcher, P.G. (1992) Quantitative solid-state ^{13}C nuclear magnetic resonance spectrometric analyses of wood xylem: Effect of increasing carbohydrate content. *Org. Geochem.*, 18, 407–416.

Part III
Characterization of Lignocellulose-Based Composites and Polymer Blends

Chapter 14

Advances in the Characterization of Interfaces of Lignocellulosic Fiber Reinforced Composites

Sherly Annie Paul, Laly A. Pothan, and Sabu Thomas

Abstract

Ecological concern has resulted in a renewed interest in lignocellulosic materials and therefore, issues such as recyclability and environmental safety are becoming increasingly important for the introduction of new materials and products. Natural fibers such as flax, hemp, kenaf, jute, and sisal have a number of techno-economical and ecological advantages over synthetic fibers. The combination of good mechanical and physical properties together with their environmentally friendly character has triggered a number of industrial sectors, notably the automotive industry, to consider these fibers as potential candidates to replace glass fibers in environmentally safe products. A major disadvantage of cellulose fibers is their highly polar nature, which makes them incompatible with nonpolar polymers. The stress transfer at the interface between two different phases is determined by the degree of adhesion. A strong adhesion at the interfaces is needed for an effective transfer of stress and load distribution throughout the interface. The compatibility of hydrophobic polymer and hydrophilic cellulose fibers can be enhanced through the modification of polymer or fiber surface. A clear understanding of the complex nature of cellulose surfaces and interfaces is needed to optimize surface modification procedures and thus to increase the usefulness of lignocelluloses as a constituent of composites. Various treatments of the lignocellulosic surfaces and the characterization techniques have been illustrated. The characterization techniques for lignocellulosic fibers and interfaces that are enumerated in this article are micromechanical, microscopic, spectroscopic, and thermodynamic analysis. Recent interfacial studies of different lignocellulosic fiber reinforced composites have also been cited.

14.1 Introduction

14.1.1 Lignocellulosic fibers and their composites

14.1.1.1 Different types of lignocellulosic fibers

Plant fibers are a composite material designed by nature [1]. Most plant fibers, except cotton are composed of cellulose, hemicelluloses, lignin, waxes, pectins, and some water-soluble compounds [2]. Physical properties of natural fibers are basically influenced by the chemical structure such as cellulose content, degree of polymerization, orientation and crystallinity, which are affected by conditions during growth of plants as well as extraction methods used. There is an enormous amount of variability in plant fiber properties depending upon whether the fibers are taken from which part of the plant, the quality of plant, and location. Different fibers have different lengths and cross-sectional areas and also different defects such as microcompressions, or pits or cracks. Table 14.1 contains the world production of certain commercially important fiber sources [3]. Table 14.2 shows the chemical composition of a few plant fibers [1].

Natural fibers are subdivided based on their origins – plants, animals, or minerals. Plant fibers include bast fibers, leaf fibers, and seed/fruit fibers. Bast consists of a wood core

Table 14.1 Commercially important natural fiber sources.

Fibre source	Species	Origin
Wood	10 000 species	Stem
Cotton lint	*Gossypium* sp.	Fruit
Jute	*Corchorus*	Stem
Flax	*Linum usitatissimum*	Stem
Sisal	*Agave sisilana*	Leaf
Hemp	*Cannabis sativa*	Stem
Coir	*Cocos nucifera*	Fruit
Ramie	*Boehmeria nivea*	Stem
Abaca	*Musa textiles*	Leaf
Sunhemp	*Crorolaria juncea*	Stem
Kenaf	*Hibiscus cannabinus*	Stem
Bamboo	>1250 species	Stem

Table 14.2 Chemical composition of various plants.

Fiber	Cellulose (wt%)	Hemi celluloses (wt%)	Lignin (wt%)	Pectin (wt%)	Moisture content (wt%)	Waxes (wt%)	Microfibrillar angle (deg)
Flax	71	18.6–20.6	2.2	2.3	8–12	1.7	5–10
Hemp	70–74	17.9–22.4	3.7–5.7	0.9	6.2–12	0.8	2–6.2
Jute	61–71.5	13.6–20.4	12–13	0.2	12.5–13.7	0.5	8
Kenaf	45–57	21.5	8–13	3–5	–	–	–
Sisal	66–78	10–14	10–14	10	10–22	2	10–22
Henequen	77.6	4–8	13.1	–	–	–	–
PALF	70–82	–	5–12.7	–	11.8	–	14
Banana	63–64	10	5	–	10–12	–	12–14
Cotton	85–90	5.7	0–1	–	7.85–8.5	0.6	–

surrounded by a stem. Within the stem, there are a number of fiber bundles, each containing individual fiber cells or filaments. Fibers are extracted from stems after a process called retting [3]. Examples include flax, hemp, jute, kenaf, and ramie. Leaf fibers such as sisal, abaca, banana, and henequen are coarser than bast fibers. Cotton is the most common seed fiber. Other examples include coir and oil palm. Other sources of lignocellulosics can be from agricultural residues such as rice hulls from a rice processing plant, sun flower seed hulls from an oil-processing unit, and bagasse from a sugar mill.

14.1.1.2 Properties of lignocellulosic fibers

To compare different kinds of lignocellulosic fibers, knowledge about fiber length and fiber diameter is important. The effective length of the long fibers after processing is based on the type of plant, the processing technology, the precut length of the stalks, and the operation data of the decorticator. The length of the natural fibers varies in a wide range from 10 to 250 mm. An important parameter is the aspect ratio (length/diameter) that has an influence on the mechanical properties of the composite [4]. A high aspect ratio is very important in agro-based fiber composites as it gives an indication of possible high strength properties. This aspect ratio is highly sensitive to attrition during processing (extrusion, injection) [5]. Table 14.3 shows the dimensions and mechanical properties of some common natural fibers [1].

The fiber strength is an important factor in selecting a specific vegetable fiber for a specific application. The mechanical properties are influenced by several external factors. Because of this, harvesting, storage and processing of the fibers have to be controlled as much as possible to eliminate the loss of fiber strength [6]. From Table 14.3 it can be seen that mechanical properties of vegetable fibers vary widely depending on the type of the fibers. The retting process increases the tensile strength of hemp fibers [6]. This means that fibers become stronger and harder, while simultaneously elasticity and elongation decreases. Tensile strength of natural fibers is comparable to that of high-tensile steel which is classified by strength of 450 N mm^{-2} (85 000 Ib in.$^{-2}$) [6]. The modulus of elasticity of natural fibers depends on the type of plants and retting varying from 24 to 60 kN mm^{-2} [6]. It is about one-seventh the modulus of steel. This modulus marks the outstanding tensile stress of

Table 14.3 Mechanical properties and dimensions of some common natural fibers.

Fiber	Tensile strength (MPa)	Young's modulus (GPa)	Elongation at break (%)	Density (g cm^{-3})	Diameter (μm)
Jute	393–800	13–26.5	1.16–1.5	1.3–1.49	25–200
Hemp	690	70	1.6	1.47	25–500
Cotton	287–800	5.5–12.6	7–8	1.5–1.6	12–38
Sisal	468–700	9.4–22	3.7	1.45	50–200
Coir	131–220	4–6	15–40	1.15–1.46	100–460
Kneaf	930	53	1.6	–	–
Flax	345–1500	27.6	2.7–3.2	1.5	40–600
Banana	600–750	28–29	2–5	1.3	80–120

fibers. Because of this, natural fibers are predestined for application in composites and construction components requiring high package crushability, for example, bumpers of cars [6].

Fiber fineness is a measure of the degree of fiber bundles to loosen up into thinner bundles or elementary fibers. Modern decortications methods, for example, the hammer mill with integrated cleaning effect, provide a high fiber fineness of less than 15 tex [6]. Pretreatment of fiber bundles by retting improves the fineness of fibers after processing [6]. Flax and linseed fibers, which have thinner fiber bundles, show a higher fineness after processing. The fineness of hemp fibers meets the requirements of boards, mats, and fleeces (<18 tex). Retted hemp and unretted flax and linseed meet the requirements of heat insulation mats, coarse yarn, and compression moulding [6].

Cellulosic fibers have amorphous and crystalline domains with a high degree of organization. The crystallinity degree depends on the origin of the material. Cotton, flax, ramie, sisal, and banana fibers have high degrees of crystallinity (65–70%), but the crystallinity of regenerated cellulose is only 35–40%. Progressive elimination of the less organized parts leads to fibrils with ever increasing crystallinity, until almost 100% crystallinity leading to whiskers [7]. Crystallinity of cellulose results partially from hydrogen bonding between the cellulosic chains, but some hydrogen bonding also occurs in the amorphous phase [7] although its organization is low. In cellulose there are many hydroxyl groups available for interaction with water by hydrogen bonding. In contrast to glass fibers, where water absorption is important only at the surface, cellulosic fibers interact with water not only on the surface but also in the bulk. The quantity of water absorbed depends on the relative humidity of the confined atmosphere with which the fiber is in equilibrium. The sorption isotherm of cellulosic material depends on the purity of cellulose and the degree of crystallinity. All –OH groups in the amorphous phase are accessible to water, whereas only a small amount of water interacts with the surface –OH groups of the crystalline phase.

14.1.1.3 Lignocellulosic fiber composites

Different plant fibers and wood fibers are found to be interesting reinforcements for rubber, thermoplastics, and thermosets [8–12]. Extensive research work has been carried out by Thomas and coworkers [13–18] on the utility of various natural fibers as reinforcement in plastics and rubbers. The addition of particulate fillers (wood flour) and short fibers (sisal) into an unsaturated polyester (UP) matrix was performed and analyzed by Marcovich *et al.* [19]. The efficiency of the filler treatment was carefully investigated, in particular, esterification with two different anhydrides, maleic anhydride (MAN) and an alkenyl succinic anhydride (ASA). The efficiency of the reactions was assessed by Fourier transform infrared spectroscopy (FTIR), titrimetric techniques, and moisture absorption values. The results showed that esterification improves the wettability of the fillers by the resin so that higher concentrations of filler could be incorporated into the composite. The use of short palm tree lignocellulosic fibers as a reinforcing phase in polyester and epoxy matrices has been reported by Kaddami *et al.* [20]. The morphology and the mechanical properties of the resulting composites were characterized using scanning electron microscopic analysis, differential scanning calorimetry, dynamical mechanical analysis and three-point bending tests. It was shown that the interfacial adhesion was better in the case of epoxy-based composites. To improve interfacial adhesion, the esterification of the lignocellulosic filler in alkaline medium was performed using acetic or MAN. They concluded that such chemical modification, which led to a change in the chemical composition of the filler, only succeeded to improve the mechanical properties of the epoxy-based composites. Another interesting area is that of hybrid composites. The incorporation of two or more fibers within a single matrix is known as hybridization and the resulting material is referred to as hybrid composites. The behavior of hybrid composites is a weighted sum of the individual components in which there is a more favorable balance between the inherent advantages and disadvantages. Hybrid composites were prepared from glass fiber mat and chemically modified coir fiber mat with polyester resin by Rout *et al.* [21]. The comparative study of the water absorption revealed that water uptake was less for glass-hybridized composites. Study of the mechanical properties of sisal/oil palm hybrid fiber-reinforced natural rubber composite was conducted by Jacob *et al.* [22]. Their conclusion was that increasing the concentration of fibers resulted in reduction of the tensile strength, but increased modulus of the composite.

Researchers have reported the results of their studies on green composites of different plant fibers and various biodegradable matrices like poly(lactic acid) (PLA), poly(butylene succinate) (PBS), and soy-based matrices. As growing environmental concerns are making plastic a target of criticism due to their lack of degradability, biodegradable polymers are considered an eco-friendly option to manage waste. They constitute a loosely defined family of polymers that are designed to degrade through the action of living microorganism [23]. The major advantage of green composites is that they are environmentally friendly, fully degradable, and sustainable. Lee *et al.* [24] investigated the effect of lysine-based diisocyanate (LDI) as a coupling agent on the properties of biocomposites from PLA, PBS, and bamboo fibers (BFs). Tensile properties, water resistance, and interfacial adhesion of both the PLA–BF and the PBS–BF composites were improved by the addition of LDI, whereas thermal flow became somewhat difficult due to cross-linking between the polymer matrix and BF. Crystallization temperature and enthalpy in both composites were increased and decreased, respectively, with increasing LDI content. The heat of fusion in both composites was decreased by the addition of LDI, whereas there was no significant change in the melting temperature. Thermal degradation temperature of both composites was lower than those of pure polymer matrix, but the composites with LDI showed higher degradation temperature than those without LDI. Dweib *et al.* [25] reported the studies of natural-fiber-reinforced composites from thermosetting resin developed and manufactured from plant oil (soybean).

Natural-fiber-reinforced composites have yet many challenges to overcome to become largely used as a reliable engineering materials for structural elements. Availability, price, performance, and biodegradability are some of the factors that have catalyzed the urge of using lignocellulosic fibers as reinforcements in polymeric materials. However, the large diversity of lignocellulosic fibers results in high variability in their properties.

14.1.2 *Interfaces of lignocellulosic fiber reinforced composites*

14.1.2.1 *Interface and interphase*

The term interface is defined as a two-dimensional region between the fiber and matrix having properties intermediate between those of fiber and matrix. Matrix molecules can be anchored to the fiber surface by chemical reaction or adsorption, which determine the extent of interfacial adhesion. In certain cases, the interface may be composed of an additional constituent such as a bonding agent or as an interlayer between the two components of the composite. The region separating the bulk polymer from the fibrous reinforcement is of utmost importance to load transfer. This region was originally called the interface but is now viewed as an interphase because of its three-dimensional heterogeneous nature. It is not a distinct phase, as the interphase does not have a clear boundary. It is more accurately viewed as a transition region that possesses neither the properties of the fibers nor those of the matrix. An interphase that is softer than the surrounding polymer would result in lower overall stiffness and strength, but greater resistance to fracture. On the other hand, an interphase that is stiffer than the surrounding polymer would give the composite less fracture resistance but make it very strong and stiff. The nature of the interphase varies with the specific composite system. The interphase is generally thought to be thin ($<5\,\mu m$) with very subtle differences in properties between bulk polymer and interphase. The developments of atomic force microscope and nano-indentation devices have facilitated the investigation of the interphase [26].

14.1.2.2 Types of interfacial bonding

The structure and properties of the fiber–matrix interface play a major role in the mechanical and physical properties of composite materials. The large difference between the elastic properties of the matrix and the fibers have to be communicated through the interface or in other words the stress acting on the matrix are transmitted to the fiber across the interface. These interface effects are seen as a type of adhesion phenomenon and are often interpreted in terms of the surface structure of the bonded material, that is, the surface factors such as wettability, surface free energy, the presence of polar groups on the surface and the surface roughness of the material to be bonded. Several mechanisms have been suggested to account for fiber–matrix interfacial adhesion.

When two electrically neutral surfaces are brought sufficiently close together, there is a physical attraction that is best understood by considering the wetting of solid surfaces by liquids. In the case of two solids being brought together the surface roughness on a micro and atomic scale prevents the surface coming into contact except at isolated points which will lead to weak adhesion. For effective wetting of a fiber surface, the liquid resin must cover every hill and valley of the surface to displace all the air. It is possible to form a bond between two polymer surfaces by the diffusion of the polymer molecules on one surface into the molecular network of the other surface. The bond strength will depend on the amount of molecular entanglement and the number of molecules involved. Interdiffusion may be promoted by the presence of solvents and plasticizing agents and the amount of diffusion will depend on the molecular conformation and constituents involved and the ease of molecular motion.

Forces of attraction occur between two surfaces when one surface carried a net positive charge and the other surface a net negative charge as in the case of acid–base interactions and ionic bonding. The strength of the interface will depend on the charge density. The anionic and cationic species present at the fiber and matrix phases will have an important role in the bonding of the fiber–matrix composites via electrostatic attraction. Introduction of suitable coupling agents at the interface can enhance the bonding through the attraction of cationic functional groups by anionic surface and vice versa. Introduction of silane coupling agent on the glass surface is an example of this type of interaction. In this case, the pH of the silane can be controlled to obtain optimum coupling effect. A chemical bond is formed between a functional group on the fiber surface and an appropriate functional group in the matrix. The strength of the bond depends on the number and type of bonds, and interface failures mostly involve bond breakage. The process of bond formation and breakage are in some form of thermally activated dynamic equilibrium. Mechanical interactions can occur in a number of ways. For example, when a liquid polymer matrix is made to flow on the rough surface of a solid substrate, a "lock and key configuration" results in solidification. Surface roughness can increase the adhesive bond strength by promoting wetting or providing mechanical anchoring sites.

14.1.2.3 Chemical modifications to improve interfacial bonding

Natural fibers are amenable to chemical modification due to the presence of hydroxyl groups. The hydroxyl groups may be involved in the hydrogen bonding within the cellulose molecules thereby reducing the activity towards the matrix. Chemical modifications may activate these groups or can introduce new moieties that can effectively interlock with the matrix. One of the most common methods of chemical modification is mercerization, which is a treatment with sodium hydroxide. Alkali treatment improves the fiber surface adhesive characteristics by removing natural and artificial impurities from the surface. The surface tension and hence the wettability of the mercerized fibers are higher, and this results in better bonding between the fibers and the matrix. Mercerization also leads to fiber fibrillation which increases the effective surface area available for contact with the wet matrix [27].

Coupling agents are widely used to strengthen composites containing fillers and fiber reinforcements. The most common coupling agents are silanes, isocyanates, titanates, MAN, and maleic anhydride-grafted polypropylene (MAA-PP). The chemical composition of these coupling agents allows them to react with the fiber surface to form a bridge of chemical bonds between the fiber and matrix. It is expected that the formation of strong covalent bonds between cellulose fibers and the coupling agents and of weak non-covalent bonds between the thermoplastics and the coupling agents would improve the mechanical properties of fiber-reinforced thermoplastics. Pretreatment of fibers by encapsulated coating with silanes or isocyanates, grafting, and so on, provides better dispersion by reducing the fiber–fiber interaction with the formation of coating on the fiber surface.

Maleated coupling agents are widely used to strengthen composites containing fillers and fiber reinforcements. Interactions between the anhydride groups of maleated coupling agents and the hydroxyl groups of natural fibers can overcome incompatibility problem to increase the tensile and flexural strengths of natural fiber thermoplastic composites. Schematic representation of the reaction of MAPP with natural fibers is represented in Figure 14.1 [28].

Another popular chemical modification is silanation. The bifunctional silane molecules act as a link between the resin and the cellulose fibers by forming a covalent bond with the surface of the fibers through a siloxane bridge while its alkyl (R) groups, for example, bond to the polymer resin via van der Waals interactions. Such a covalent bond and van der Waals interactions gives molecular continuity across the interface region of the composite. A number of factors affect the microstructure of the coupling agent, which in turn controls

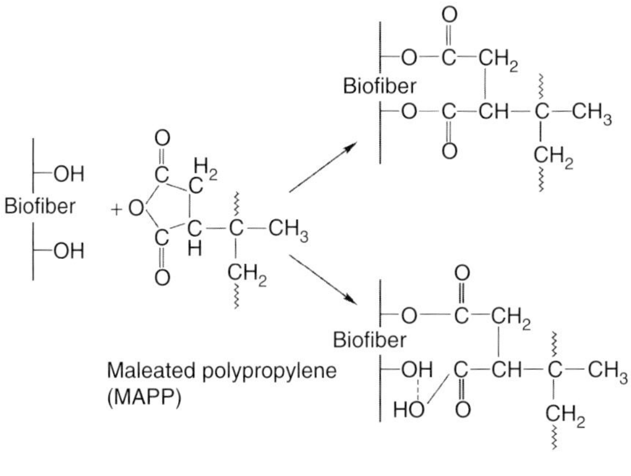

Figure 14.1 Reaction of natural fibers with MAPP. Reprinted with permission from Mohanty *et al.* [28], copyright 2002 with permission from BRILL$_{NV}$.

the mechanical and physical properties of the composites. They are the silane structure including the structure of the R groups, acidity, drying conditions and homogeneity, the topology, and the chemical composition of the fiber surface. Figure 14.2 shows the schematic representation of silane and cellulose reaction [29].

Grafting is an effective method for the modification of natural fibers. Grafting efficiency, proportion and frequency determine the degree of compatibility of cellulose fibers with a polymer matrix. The grafting parameters are influenced by the type and concentration of initiator, the monomer to be grafted, and the reaction conditions [30]. Acetylation is an attractive method of modifying the surface of natural fibers and making it more hydrophobic. The principle of the method is to react the hydroxyl groups of fiber constituents with the acetyl groups. The reaction is catalyzed by bases or acids, and a large number of catalysts including pyridine and sulphuric acid have been used in the past; but the use of catalysts creates many problems. Strong acids cause hydrolysis of cellulose resulting in damage of fiber structure. The probable reaction between the fiber and acetic anhydride can be represented as follows:

Another method of modifying the surface of natural fibers is sizing with fatty acids such as stearic acid. The carboxyl group reacts with hydroxyl groups of the fiber through an esterification reaction, and hence the treatment reduces the number of hydroxyl groups available for bonding with water molecules. Also, the long hydrocarbon chain of stearic acid provides an extra protection from water, as it is itself quite hydrophobic. Another advantage of stearic acid is that it is not susceptible to oxidation at the processing temperatures of natural fiber systems.

14.2 Characterization of the interfaces of lignocellulosic fiber reinforced composites

14.2.1 Micromechanical techniques

The characterization of the interface gives the chemical composition as well as information on the interactions between the fibers and the matrix. Mechanical properties of

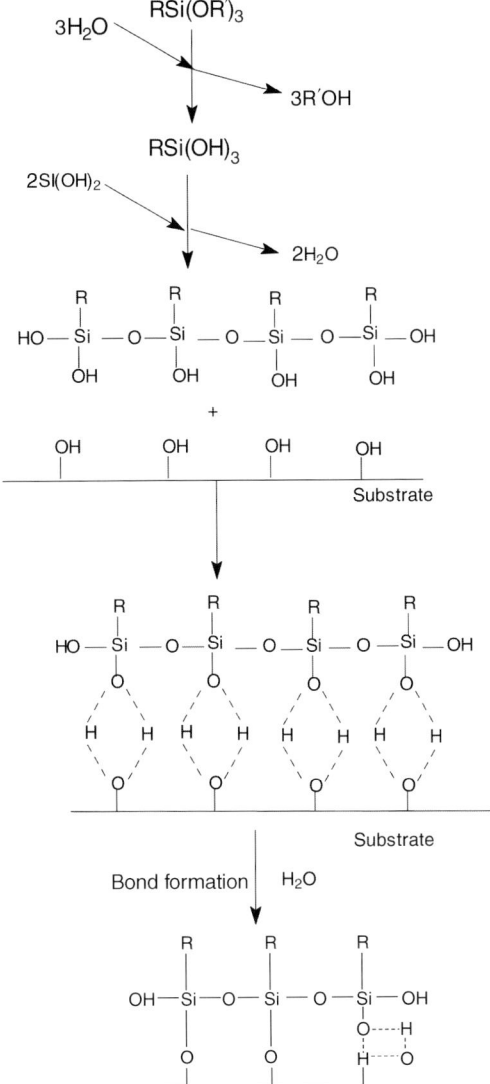

Figure 14.2 Schematic representation of silane and cellulose reaction/interaction. From Pothan *et al.* [29], Figure 3, copyright 2006 with permission from Elsevier.

fiber-reinforced plastics highly depend on the properties of the interface between the fibers and the matrix. Various methods are available for the characterization of the interface. Two popular methods are the microbond test and the single-fiber pull-out test [31]. These are among the most effective and convenient methods to determine the fiber–matrix interfacial properties. In the microbond test, a small droplet of matrix is deposited on a fiber and sheared off by restraining the droplet while the fiber is pulled out. In the single-fiber pull-out test, the end of a fiber is embedded in a larger amount of matrix and pulled out while the matrix is held. In both experiments the peak force P, required to debond the fiber is recorded as a function of droplet length (microbond) or embedded fiber length (pull-out) test. The extent of fiber–matrix interface bonding can also be determined by microindentation/microcompression/fiber push-out test and single-fiber fragmentation test.

Craven et al. [32] evaluated the interface of *Bombyx mori* silkworm silk–epoxy composite by microbond test. Cured resin droplets were sheared from the silk fibers by knife-edged jaws of a microvice. The microbond test has been used in this work due to its suitability for fibers that can carry only low loads. Silk is a strong fiber but its individual fibers are fine and so have a limited load-bearing capacity. The microbond technique involves depositing a microdroplet of resin onto a fiber. After the resin has cured, it is sheared from the fiber by two parallel plates attached to a microvice. They observed that the mean interfacial shear strength (IFSS) of silk–epoxy composite is 15 (±2) MPa. It was found that silk fibers could offer useful reinforcement to an epoxy system due to their high tensile strength and extensibility. In an interesting study, Luo and Netravali [33] characterized the interfacial properties of green composites made from pineapple fibers and poly(hydroxybutyrate-covalerate) (PHBV) resin by means of microbond technique and single-fiber fragmentation. The IFSS value was found to be very low. The IFSS depends mainly on two factors: mechanical locking and chemical bonding. In the case of pineapple fiber–PHBV resin, H-bonding is possible between the ester group on the resin and the OH group on the fibers; however, because of the hydrophobicity of methyl and ethyl pendant groups in the resin, H-bonding probability is low. As a result, IFSS is mainly attributed to high surface irregularity of pineapple fibers and the resulting mechanical interaction. The high viscosity of the resin also seemed to preclude much mechanical bonding. Zafeiropoulos et al. [34] characterized the interface in flax-fiber-reinforced PP composites by means of single-fiber fragmentation test. Flax fiber was modified by means of two surface treatments: acetylation and stearation. The authors observed that acetylation improved the stress transfer efficiency at the interface. Stearic acid treatment was also found to improve the stress transfer efficiency but only for shorter treatment times. They also found that stearation for longer durations deteriorated the interface. This was attributed to a decrease of fiber strength upon treatment for a longer duration and the fact that excess of stearic acid acted more as a lubricant than as a compatibilizer. The interfacial characterization of flax-fiber-reinforced thermoplastic composites was carried out by Stamboulis et al. [35]. They employed the single-fiber pull-out technique. Two types of flax fibers were used, namely, dew retted and upgraded duralin fibers. The duralin fibers were treated for improved moisture resistance. The IFSS of dewretted and upgraded duralin fibers in low-density polypropylene (LDPE), high-density polypropylene (HDPE), and MAN-modified PP were determined. The force displacement curves for all the samples were found to be typical of a brittle fracture mode interface behavior. They also did not observe any improvement of IFSS in the case of upgraded flax-fiber-reinforced composite. Another study utilizing single-fiber pull-out test was attempted by Vandevelde and Kiekens [36]. The authors used dew retted hackled long flax treated with propyltrimethoxy silane, phenyl isocyanate, and MAA-PP. The studies revealed that composite prepared with flax fiber treated with MAA-PP exhibited the highest IFSS. The interfacial adhesion of flax-fiber-reinforced polypropylene (PP) and polypropylene–ethylene propylene diene terpolymer (PP–EPDM) blends were investigated by Manchado et al. [37]. In this study both the matrices were modified with MAN. The single-fiber pull-out test was conducted to investigate the extent of interfacial adhesion. They observed that the addition of small proportions of MAN to the matrices significantly increased the shear strength. The authors were of the opinion that introduction of functional groups in the matrix reduced the interfacial stress concentrations preventing fiber–fiber interactions which are responsible for premature composite failure. Also, in the presence of MAN, the esterification of flax fibers takes place, which increases the surface energy of the fibers to a level closer to that of the matrix. Hence, a better wettability and interfacial adhesion is obtained. In an innovative study [38] interfacial evaluation of the untreated and treated jute and hemp fibers reinforced different matrix polypropylene–maleic anhydride polypropylene copolymer (PP–MAPP) composites was investigated by micromechanical technique combined with acoustic emission (AE) and dynamic contact angle (DCA) measurement. The studies revealed that the alkali-treated fibers raised the surface energy due to the removal of the weak boundary layers, and thus increasing the surface area, whereas the surface energy of silane-treated jute and hemp fibers decreased due to the blocking of high energy sites. MAPP in the PP–MAPP matrix caused the surface energy to increase due to the acid–base sites introduced. Microfailure modes of two natural fiber composites were observed differently due to different tensile strength of the natural fibers, and the results were also confirmed by nondestructive AE analysis indirectly.

14.2.2 Microscopic techniques

Microscopic studies such as optical microscopy, scanning electron microscopy (SEM), transmission electron microscopy (TEM), and atomic force microscopy (AFM) can be used to study the morphological changes on the fiber surface and can predict the extent of mechanical bonding at the interface. The adhesive strength of fiber to various matrices can be determined by AFM studies. AFM has become a technique of major interest in composite science. It is a useful technique to determine the surface roughness of fibers. The effect of silane coupling agent on the morphological properties of luffa fiber (LF)–PP composites were studied by Demira et al. [39]. They used AFM to study the morphology of the fibers and reported that the use of silane coupling agents decreased the surface roughness of the fibers. Figure 14.3 shows the AFM topographic pictures of the untreated and treated LFs. Balnois et al. [40] used AFM method to investigate the effect of chemical treatments on the surface microstructure and adhesion properties of flax fibers. The studies revealed that after chemical treatments, the fiber surface appear to be less heterogeneous in topology and smoother and no significant roughness difference was found between the different treatments.

In an innovative study, Michaeli et al. [41] characterized the interfacial adhesion between flax fibers and UP resin with different microscopic methods. The techniques employed were SEM, TEM, and AFM. The authors used two methods to incorporate the coupling agent: a fiber finish process in which the butyl titanate coupling agent was directly added to the fibers and a resin additive process where the coupling agent was mixed with the resin. From the SEM studies a very clear boundary layer on the fibers that were coated with the coupling agent was observed. For the fibers with the titanate agent mixed into the matrix, no clear boundary line was observed. TEM studies revealed that butyl titanate had a greater adhesion effect in the resin additive process than in the fiber finish process. AFM images supported SEM and TEM findings. Furthermore, the images implied that the boundary layer possessed many small white areas, which differentiated the boundary layer with its phase shift from the fibers and the matrix. The interface of eucalyptus kraft

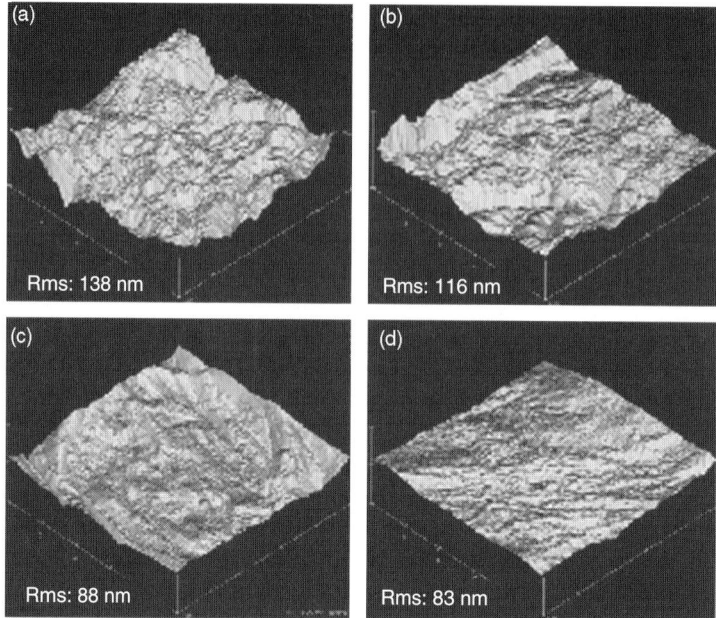

Figure 14.3 AFM topographic pictures of (a) untreated, (b) NaOH-treated, (c) NaOH + (3-aminopropyl)-triethoxysilane (AS)-treated, and (d) NaOH + 3-(trimethoxysilyl)-1-propanethiol (MS), treated luffa fibers (LFs). From Demira et al. [39], Figure 9, copyright 2006 with permission from Elsevier.

fiber-reinforced silicone composites was analyzed by Redendo et al. [42]. The eucalyptus fibers were modified with vinyltriethoxysilane. The interface was studied by field emission SEM (FESEM). In and only in the untreated composites, there was a lack of adhesion between the fibers and the silcone matrix resulting in pull-out. This fracture was found to occur exclusively in the matrix and perpendicularly to the fiber direction. This indicated the silane coupling agent promoted adhesion between the fiber and silicone composite. The usefulness of lignocellulosic waste flours, that is, spruce, olive husk, and paper flours as a source of filler for the preparation of cost-effective and biodegradable polymer matrix composites was studied by Tserki et al. [43]. The biodegradable polyester Bionolle 3020 was used as matrix. Waste flour–matrix interfacial adhesion was promoted by means of flour surface treatment, acetylation and propionylation, and by the addition of MAN-grafted Bionolle as a compatibilizer. The composite materials produced were characterized by means of mechanical property measurements, SEM, water absorption, and biodegradation studies. Compatibilizer addition resulted in materials with improved mechanical properties, while flour treatment with acetic or propionic anhydride significantly reduced the material water uptake. In addition, waste flour incorporation into the polymeric matrix increased its biodegradation rate. Figure 14.4 shows the SEM micrographs of the fracture surfaces of the composites with untreated, acetylated, propionylated paper flours, and compatibilizer added.

Dufresne et al. [44] investigated the ultrastructure and morphology of potato (*Solanum tuberosum* L.) tuber cells by optical, scanning, and transmission electron microscopies. After

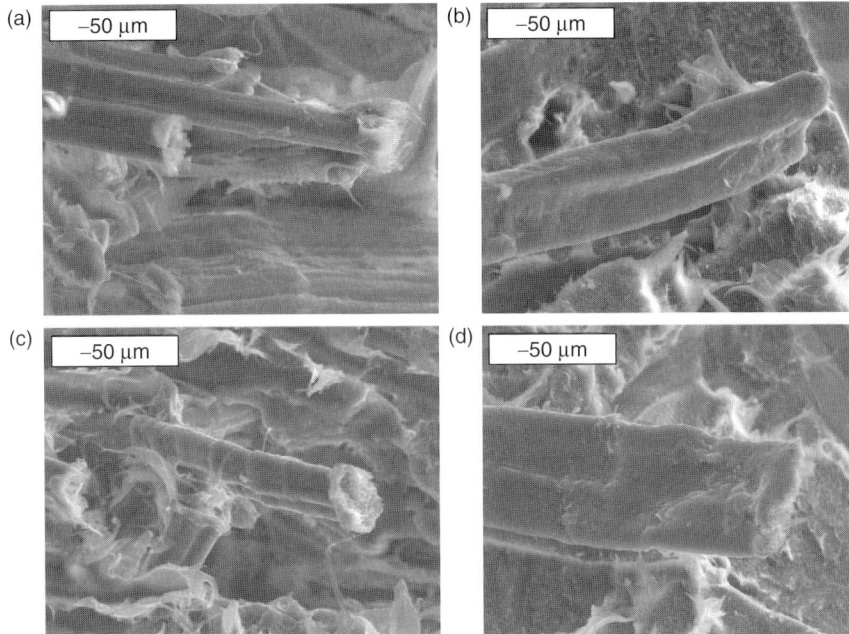

Figure 14.4 SEM micrographs of the fracture surfaces of composites with (a) untreated, (b) acetylated, (c) propionylated paper flours, and (d) compatibilizer addition. From Tserki et al. [43], Figure 4, copyright 2006 with permission from Elsevier.

the removal of starch granules, pectins and hemicelluloses were solubilized under alkaline conditions. Composite materials were processed from this potato cellulose microfibril suspension using gelatinized potato starch as a matrix and glycerol as a plasticizer. After blending and casting, films were obtained by water evaporation. The mechanical properties and water absorption behavior of the resulting films were investigated, and differences were observed depending on the glycerol, cellulose microfibrils, and relative humidity content. Dong et al. [45] studied the effect of preparation conditions on the structures of the cellulose microcrystals from sulphuric acid hydrolysis of cotton fiber. Characterization of the cellulose whiskers were performed using different techniques such as TEM, x-ray and neutron diffraction, nuclear magnetic resonance (NMR), and AFM. The interaction between coupling agent and lubricants in wood–polypropylene composites was conducted by Harper and Wolcott [46]. They studied the influence of the coupling agent and lubricants on the crystallization of polypropylene in the bulk and interphase regions and the subsequent spatial distribution of the additives. Differential scanning calorimetry and polarized light microscopy were used to determine kinetic parameters for the crystallization.

14.2.3 Spectroscopic techniques

Electron spectroscopy for chemical analysis (ESCA) also referred to as x-ray photoelectron spectroscopy (XPS), Fourier transform infrared spectroscopy (FTIR), laser Raman spectroscopy (LRS), NMR, and photoacoustic spectroscopy have been shown to be successful in polymer surface and interface characterization.

XPS, originally developed by Siegbahn and coworkers [47] for the investigation of a solid surface, provides an understanding of both the quality of interfacial bond and the performance of the bond during service. Gellerstedt and Gatenholm [48] characterized succinylated lignocellulosic fibers using XPS. The results showed that carboxylic functionalities were predominantly introduced to the fiber surface. The surface properties of the modified chemical pulp, as measured by XPS, showed that the fiber surfaces were modified to a higher extent than the cell wall. Cai et al. [49] studied the effect of surface grafted ionic groups on the performance of cellulose-fiber-reinforced thermoplastic composites. The elements present on the fiber surface were detected using XPS analysis; the atomic ratio oxygen/carbon (O/C) increased slightly upon the treatment of the fibers. XPS study of the raw banana fibers by Pothan et al. [50] showed the presence of numerous elements on the surface of the fibers (Figure 14.5). Investigation of the surface after alkali treatment showed the removal of most of the elements. Silane treatment was found to introduce considerable amount of silicon on the surface of the fibers. The XPS surface characterization of the argon and air plasma-treated wood fibers carried out by Yuan et al. [51] showed that the air plasma treatment increased the O/C ratio more than that of the argon plasma-treatment. Effect of γ-aminopropyl trimethoxy silane on the performance of jute–polycarbonate composites was reported by Khan and Hassan [52]. The jute surface was modified with γ-aminopropyl trimethoxy silane to improve interfacial adhesion between jute fiber and polycarbonate. The treated and untreated jute fiber surfaces were investigated by XPS. The study indicated that the silane played an important role in forming interfacial bonding with the jute fibers and polycarbonate.

Hristov and Vasileva [53] employed FTIR to examine the interface of wood-fiber-reinforced polypropylene composites. The authors modified polypropylene matrix with maleated polypropylene (PPMA) and poly(butadiene styrene) rubber. They inferred from the FTIR spectra that the coupling agent was located around the wood fibers rather than randomly distributed in the PP matrix. From the spectra of the treated composite they concluded that the compatibilizer was attached to the wood fibers either by ester or hydrogen bonds. The effect of different chemical modifications of aspen fibers on the interfacial characteristics of the aspen-fiber-reinforced high-density polyethylene (HDPE) composite was conducted by Colom et al. [54]. The interaction between the aspen fibers and HDPE was improved by the addition of two coupling agents, maleated polyethylene (epolene C-18) and γ-methacryloxy-propyl trimethoxy silane (silane A-174). Interfacial morphology was studied using FTIR. The spectral band at 1635 cm^{-1} corresponds to water absorption. Owing to the hydrophilic character of the lignocellulosic fiber, this band was found to be the highest for untreated composites and the lowest for composite modified with maleated polyethylene. The different patterns of this band were due to the hydrophilic nature of the lignocellulosic fibers and the protective effect produced by the coupling agents. They also observed that silane was a better coupling agent than epolene.

Raman spectroscopy has been extensively utilized for the study of deformation micromechanics of fiber-based composites. The technique relies on the fact that Raman bands corresponding to the vibrational modes of bonds in the polymer fiber shift towards a lower wave number because of molecular stress and resultant strain. This has been used to map stresses along fibers embedded in a resin matrix to determine the interfacial shear

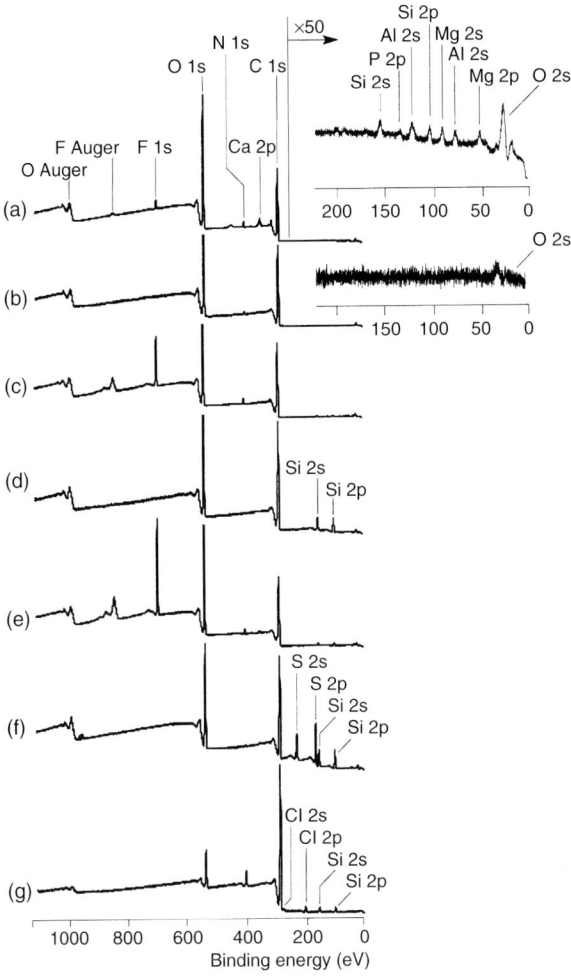

Figure 14.5 Series of XPS survey spectra of raw and chemically treated banana fibers. (a) untreated fibers, (b) fibers after a treatment with 0.25% NaOH, (c) fibers after acetylation, (d) fibers grafted with A-174 (γ-methacryloxy-propyl trimethoxy silane), (e) fibers grafted with fluorosilane, (1H,1H,2H,2H-perfluorooctyl triethoxysilane), (f) fibers grafted with sulfur containing silane, [bis(3-triethoxysilylpropyl)tetrasulfone)], (g) fibers treated with $C_{18}T$ (2,4-dichloro 6-n-octa-decyloxy-s-triazine). The insets are cuts of the survey spectra of untreated (a) and NaOH-treated (b) banana fibers showing the presence or absence of accompanying elements on the sample surfaces. From Pothan et al. [50], Figure 4, copyright 2006 with permission from the American Clinical Society.

stress. Recently, Raman spectroscopy has also been used to investigate the deformation micromechanics of natural and regenerated cellulose fibers. Studies have also been carried out in the case of composites but it was found that bonding across the ends of the fibers gave rise to good stress transfer and it was difficult to evaluate the properties of the interface. Eichom et al. [55] used Raman spectroscopy to follow the composite micromechanics of hemp fibers and epoxy resin microdroplets. The authors observed that the 1095 cm^{-1} cellulose Raman band had been shifted to a lower wave number under the application of strain and stress, indicating molecular deformation. It has also been shown that it is possible to map the stress of a hemp fiber inside an epoxy resin droplet on the surface. Use of this technique has also shown that an interfacial shear stress for epoxy and hemp fibers is comparable to amide–epoxy and glass–epoxy systems but coupled with good adhesion of the fiber ends may lead to low fracture toughness of a natural fiber composite. Solid-state ^{13}C NMR spectroscopy using cross polarization and magic angle spinning (CP/MAS) is useful for characterizing wood-polymer composites since detailed information can be obtained from solid samples. These include the composition, glass transition temperature, melting transition, percent crystallinity, and the number and type of crystalline phases [56]. In an innovative study [57] the composites formed by sugarcane bagasse and thermoplastic polymers, such as PP, PE, and ethylene-co-vinyl acetate (EVA), have been analysed by high-resolution, solid-state, CP/MAS ^{13}C NMR spectroscopy with varying contact times and proton spin-lattice relaxation times in the rotating frame. The NMR responses showed that these techniques can be used to observe the degree of compatibility and homogeneity of different polymer composites. Preparation and characterization of EVA–sisal fiber composites was reported by Malunka et al. [58]. When the samples were prepared in the presence of dicumyl peroxide, cross-linking of EVA as well as grafting between EVA and the sisal fibers appeared to take place. SEM, FTIR, surface free energy, and gel content analyses strongly indicated grafting of EVA onto sisal fibers under the composite preparation conditions, even in the absence of the peroxide. But the grafting mechanism could not be confirmed from solid-state ^{13}C NMR analysis. Physicochemical and ^{13}C NMR characterization of different treatment sequences on kenaf bast fibers has been reported by Keshk et al. [59].

Another interesting technique for characterizing fibers is solvatochromism. Surface polarity of the lignocellulose fibers and related materials can be measured by solvatochromism using Ultraviolet-visible (UV-Vis) spectroscopy of adsorbed solvatochromic probes on the fibers and using the linear solvation–energy relationship. Fischer et al. [60] studied the surface polarity of native cellulose using genuine solvatochromic dyes. Spange et al. [61] studied the α, β, and π^* parameters of the native celluloses, carboxy methyl celluloses, cellulose tosylates, and other derivatives with different degrees of substitution, and found that α depends on both the amount and the acidity of the accessible surface acidic groups. Pothan et al. [62] reported the determination of polarity parameters of chemically modified banana fibers by solvatochromic technique and reported that different silanes, NaOH, and long alkyl groups used to modify the cellulose fiber surface had changed the hydrogen bond donating ability of the fibers. UV-vis absorption spectra of iron dye on the untreated and silane-treated banana fibers [62] is shown in Figure 14.6.

14.2.4 Thermodynamic and other techniques

The frequently used thermodynamic methods for the characterization of lignocellulosic fiber reinforced polymers are wettability study, inverse gas chromatography (IGC), and zeta potential measurements. Contact angles provide useful information on wettability, the improvement of which is the aim for the modification of fiber surfaces in many cases. The degree to which liquids wet a fiber determines how easily the liquid can penetrate fiber

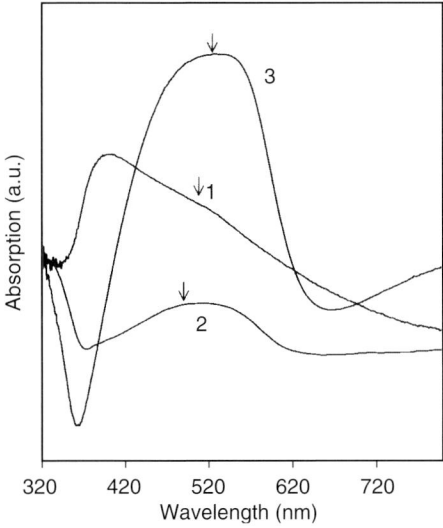

Figure 14.6 UV-vis absorption spectra absorption spectra of iron dye on untreated and silane-treated banana fibers. 1: untreated, 2: 0.25% NaOH + A-174, and 3: 0.5% NaOH + A-151. A-174 = γ-methacryloxy-propyl trimethoxy silane, A-151 = vinyl triethoxysilane. From Pothan et al. [62], copyright 2000 with permission from John Wiley & Sons, Inc.

assemblage. The interface can also be characterized qualitatively by other methods such as dynamic mechanical analysis and stress relaxation technique. Swelling techniques have been used to assess the level of interfacial adhesion in the case of fiber-reinforced elastomer composites.

X-ray diffraction method is a powerful tool for detecting the stress on the incorporated fiber in the composite under load *in situ* and nondestructively. X-ray diffraction studies of the stress transfer of kenaf-reinforced poly(L-lactic acid) (PLLA) composite was performed by Nishino et al. [63]. The stress on the incorporated fibers in the composite under transverse load was monitored *in situ* and nondestructively using x-ray diffraction. The outer applied stress was found to be well transferred to the incorporated kenaf fibers through the PLLA matrix, which suggested a strong interaction between the fibers and the matrix. In addition, it was also revealed that a silane coupling treatment of the kenaf fibers was effective for the improvement of interfacial adhesion. Morphological aspects and chemical characterization of piassava (*Attalea funifera*) fibers were reported by Almeida et al. [64]. Chemical composition study revealed that piassava were lignin rich fibers, 48.4 wt%. X-ray diffraction showed that cellulose I was their main crystalline constituent. Natural fiber surface properties can be examined using contact angle measurements [65]. Shen et al. [66] employed the capillary rise technique to study the surface energy of BF. Based on the obtained contact angles, they determined the γ_S^d of bamboo to be 42.3 mJ m^{-2} and found that bamboo was significantly more basic ($\gamma_S^- > \gamma_S^+$)γ than cotton linter samples. Rong et al. [67] evaluated the surface properties of sisal fibers by the capillary rise technique using the Washburn analysis. They obtained γ_S^d of 18.8 mJ m^{-2} for the untreated fibers. Cantero et al. [68] by employing the same technique found γ_S^d for natural flax to be 23.0 mJ m^{-2}. The surface energies of the natural fibers can be investigated using IGC. The surface energy of highly crystalline cellulose has been reported to be between 60 and 66 mJ/m^2. William and Douglas [69] used DCA analysis and IGC to probe the surface-chemical changes in wood pulp fibers during recycling. The DCA measurements revealed that the overall effect of recycling was an increase in the nonpolar (dispersive) component and a corresponding decrease in the surface free energy of the polar component, hence resulting in a total surface free energy that remained essentially unaltered. The DCA experiment also showed that virgin fibers lost both their electron accepting ($\gamma_S^+ \gamma$ and their electron-donating ($\gamma_S^- \gamma$ characteristics when converted to papers. Upon rehydration, the fibers recovered some surface acidity ($\gamma_S^+ \gamma$ but surface basicity (γ_S^-)γ continued to decrease. The changes in polar surface free energy correlated well with the changes in hydroxyl number determined independently using the acetylation method. But IGC could not detect changes in the dispersive component of the surface free energy induced by recycling. The acid–base changes in the IGC measurements were also indistinguishable between virgin fibers and recycled fibers.

The zeta potential technique has been found to be useful in characterizing lignocellulosic fibers [70]. The presence of acidic or basic dissociable surface functional groups can be estimated by measuring the pH dependence of the zeta potential. The measurement of the pH dependence of the zeta potential results in the qualitative measurement of the acidity or basicity of solid surfaces. Schematic sketch of the electrokinetic analyzer is given in Figure 14.7. The characterization of solid surfaces by electrokinetic phenomena was reported by Grundke et al. [71]. Stana-Kleinschek and Ribitsch [72] discussed the electrokinetic properties of processed cellulose fibers. Pothan et al. [73] studied the influence of chemical treatments on the electrokinetic properties of banana fibers. Figure 14.8 shows the pH dependence of zeta potential on the raw and silane-treated banana fibers.

Dynamic mechanical analysis has been used to analyze the interfacial adhesion in chemically modified sisal-fiber-reinforced polypropylene composites by Joseph et al. [74]. The dynamic modulus value and the damping parameter, used to quantify interfacial interaction

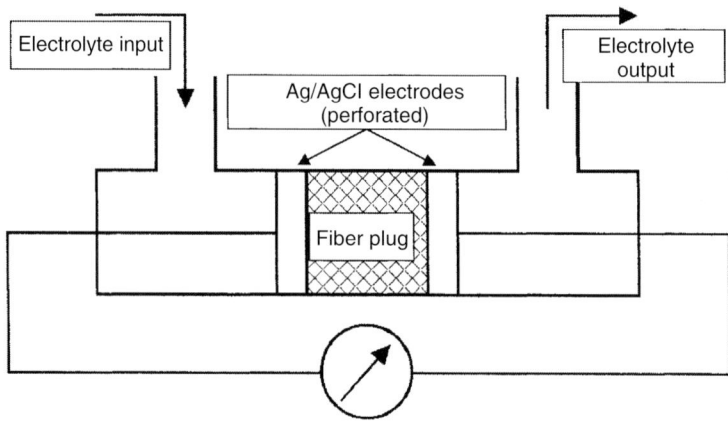

Figure 14.7 Special cell used for the investigation of zeta (ζ) potential. From Pothan et al. [73], Figure 2, copyright 2002 with permission from BRILL$_{NV}$.

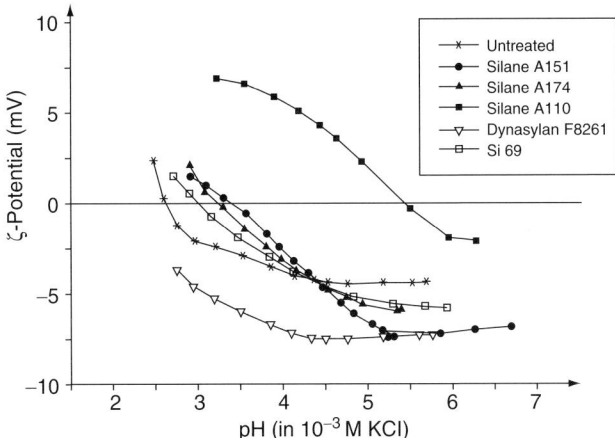

Figure 14.8 pH dependence of zeta potential on the raw and silane-treated fibers. From Pothan et al. [73], Figure 17, copyright 2002 with permission from BRILL$_{NV}$.

in composites were investigated with special reference to the effect of temperature and frequency. Increased dynamic modulus values and low damping value showed the improved interactions between the fibers and the matrix Stress relaxation experiments could also be used as an efficient method for characterizing the interface. In a series of stress relaxation tests carried out by Geethamma et al. [75], it was observed that the nature of the relaxation curve depended on the nature of the interface.

The interfacial adhesion in fiber-reinforced elastomer composites can be analyzed by swelling technique. Jacob et al. [22] investigated the interfacial bonding in sisal/oil palm hybrid fiber-reinforced natural rubber composites by means of anisotropic swelling measurements. The authors observed that the presence of fibers restricted the entry of solvent. Swelling was found to be at a minimum for composites containing a strong interface while the presence of a weak interface enabled the entry of solvent.

Conclusion

New environmental regulations and societal concerns have triggered the search for new materials, products and processes that are compatible with the environment. The use of lignocellulosic fibers derived from annually renewable resources as a reinforcing phase in polymeric matrix composites provides positive environmental benefits with respect to ultimate disposability and raw material use. One of the promising applications of natural fibers is in polymer composites that can be moulded into a variety of flat and complex shaped components by exploiting their unique reinforcing potential. Lignocellulosic composites have a bright future, yet their acceptability on a large scale has remained limited due to various reasons. In this chapter, different types of lignocellulosic fibers, their properties and the use of these fibers as reinforcement in polymer matrices have been described. Effective chemical modification of the surface of lignocellulosic fibers with the aim of using them in polymer matrices can be carried out following a number of procedures. The chemical modification of fibers leads to the formation of composites with good strength. The interface is an area that has generated tremendous interest. The properties of composites depend on the response of the interface. Different types of interfacial bonding and methods to improve interfacial bonding have also been discussed. Several methods for the characterization of the interface have also been described. The industrial application of natural fibers requires making high quality fibers on a continuous basis in large quantities at competitive prices. Conventional processing technologies cannot meet the strict demands of the modern industries. Consequently, new technologies have to be developed in order to successfully build efficient process plants for natural fibers.

References

1. Bismarck, A.; Mishra, S.; Lampke, T., Plant Fibers as Reinforcement for Green Composites; In: Mohanty, A.K.; Misra, M.; Drzal, L.T. (eds); *Natural Fibers, Biopolymers and Biocomposites*; CRC Press; New York; 2005, pp. 37–108.
2. Bledzki, A.K.; Gassan, J., Composites Reinforced with Cellulose-Based Fibers; *Prog. Polym. Sci.* 1999, 24, 22–32.
3. Bolton, J., The Potential of Plant Fibres as Crops for Industrial Use; *Outlook Agri.* 1995, 24, 85–89.
4. Joseph, S; Jacob, M.; Thomas, S., Natural Fiber – Rubber Composites and their Applications; In: Mohanty, A.K.; Misra, M.; Drzal, L.T. (eds); *Natural Fibers, Biopolymers and Biocomposites*; CRC Press; New York; 2005, pp. 435–475.
5. Dalvag, H.; Klason, C.; Stromvall, H.E., The Efficiency of Cellulosic Fillers in Common Thermoplastics; *Int. J. Polymer. Mater.* 1985, 11, 9–38.
6. Munder, F.; Furll, C.; Hempel, H., Processing of Bast Fiber Plants for Industrial Applications; In: Mohanty, A.K.; Misra, M.; Drzal, L.T. (eds); *Natural Fibers, Biocomposites and Biopolymers*; CRC Press; New York; 2005, pp. 109–140.
7. Herrera-Franco, P.J.; Valadez-Gonzalez, A., Fiber–Matrix Adhesion in Natural Fiber Composites; In: Mohanty, A.K.; Misra, M.; Drzal, L.T. (eds); *Natural Fibers, Biocomposites and Biopolymers*; CRC Press; New York; 2005, pp. 177–230.
8. Zou, Y.; Wang, L.; Zhang, H.; Qian, Z.; Mou, L.; Wang, J.; Liu, X., Stabilization and Mechanical Properties of Biodegradable Aliphatic Polyesteramide and its Filled Composites; *Polym. Degrad. Stab.* 2004, 83, 87–92.
9. Aziz, S.H.; Ansell, M.P., The Effect of Alkalization and Fiber Alignment on the Mechanical and Thermal Properties of Kenaf and Hemp Bast Fiber Composites: Part 1 – Polyester Resin Matrix; *Compos. Sci. Technol.* 2004, 64, 1219–1230.
10. Dash, B.N; Rana, A.K; Mishra, H.K.; Nayak, S.K; Mishra, S.C.; Tripathy, S.S., Novel Low-Cost Jute–Polyester Composites, Part-1, Processing and Mechanical Properties, and SEM Analysis; *Polym. Compos.* 1999, 20, 62–68.
11. Qiu, W.; Endo, T.; Takahir, T., Interfacial Interactions of a Novel Mechanochemical Composite of Cellulose with Maleated Polypropylene, *J. Appl. Polym. Sci.* 2004, 94, 1326–1338.
12. Ismail, H.; Jaffri, R.M.; Rozman, H.D., Oil Palm Wood Flour Filled Natural Rubber Composites: Fatigue and Hysteresis Behaviour; *Polym. Int.* 2000, 49, 618–622.
13. George, J.; Bhagavan, S.S.; Thomas, S., Effects of Environment on the Properties of Low-Density Polyethylene Composites Reinforced with Pineapple-Leaf Fiber; *Compos. Sci. Technol.* 1998, 58, 1471–1485.
14. Joseph, K.; Thomas, S., Effect of Ageing on the Physical and Mechanical Properties of Sisal-Fiber–Reinforced Polyethylene Composites; *Compos. Sci. Technol.* 1995, 53, 99–110.

15. Joseph, S.; Sreekala, M.S.; Oommen, Z.; Koshy, P.; Thomas, S., A Comparison of the Mechanical Properties of Phenol Formaldehyde Composites Reinforced with Banana Fibers and Glass Fibers; *Compos. Sci. Technol.* 2002, 62, 1857–1862.
16. Geethamma, V.G.; Mathew, K.T.; LakshimiNarayanan, R.; Thomas, S., Composites of Short Coir Fibers and Natural Rubber: Effect of Chemical Modification, Loading and Orientation of Fiber; *Polymer* 1998, 39, 1483–1491.
17. Kumar, R.P.; Nair, K.C.M.; Thomas, S.; Schit, S.C.; Ramamurthy, K. Morphology and Melt Rheological Behaviour of Short-Sisal-Fiber-Reinforced SBR Composites. *Compos. Sci. Technol.* 2000, 60, 1737–1751.
18. Pothan, L.A.; George, J.; Thomas, S., Effect of Fiber Surface Treatments on the Fiber–Matrix Adhesion in Banana Fiber Reinforced Polyester Composites; *Compos. Interfaces* 2002, 9, 335–353.
19. Marcovich, E.; Reboredo, M.M.; Aranguren, M.I., Lignocellulosic Materials and Unsaturated Polyester Matrix; *Compos. Interfaces* 2005, 12, 3–24.
20. Kaddami, H.; Dufresne, A.; Khelifi, B.; Bendahou, A.; Taourirte, M.; Raihane, M.; Issartel, N.; Sautereau, H.; Gerard, J.-F.; Sami, N., Short Palm Tree Fibers-Thermoset Matrices Composites; *Composites Part A* 2006, 37, 1413–1422.
21. Rout, J.; Misra, M.; Tripathy, S.S.; Nayak, S.K.; Mohanty, A.K., The Influence of Fiber Surface Modification on the Mechanical Properties of Coir-polyester Composites; *Compos. Sci. Technol.* 2001, 61, 1303–1312.
22. Jacob, M.; Thomas, S.; Varughese, K.T., Mechanical Properties of Sisal/Oil Palm Hybrid Fiber Reinforced Natural Rubber Composites; *Compos. Sci. Technol.* 2004, 64, 955–965.
23. Hodzic, A., Bacterial Polyester-based Biocomposites: A Review; In: Mohanty, A.K.; Misra, M.; Drzal, L.T. (eds); *Natural Fibers, Biopolymers and Biocomposites*; CRC Press; New York; 2005, pp. 597–616.
24. Lee, S.-H.; Wang, S., Biodegradable Polymers/Bamboo Fiber Biocomposite with Bio-based Coupling Agent; *Composites Part A* 2006, 37, 80–91.
25. Dweib, M.A.; O'Donnell, A.; Wool, R.P.; Hu, B.; Shenton III, H.W., Houses Using Soy Oil and Natural Fibers Biocomposites; In: Mohanty, A.K.; Misra, M.; Drzal, L.T. (eds); *Natural Fibers, Biopolymers, and Biocomposites*; CRC Press; Boca Raton, FL; 2005, pp. 751–773.
26. Downing, T.D.; Kumar, R.; Cross, W.M.; Kjerengtroen, L.; Kellar, J.J., Determining the Interphase Thickness and Properties in Polymer Matrix Composites Using Phase Imaging Atomic Force Microscope and Nano-indentation; *J. Adhes. Sci. Technol.* 2000, 14, 1801–1812.
27. Cao, Y.; Shibata, S.; Fukumoto, I., Mechanical Properties of Biodegradable Composites Reinforced with Bagasse Fiber Before and After Alkali Treatment; *Composites Part A* 2006, 37, 423–429.
28. Mohanty, A.K.; Drzal, A.T.; Misra, M., Engineered Natural Fiber Reinforced Polypropylene Composites: Influence of Surface Modifications and Novel Powder Impregnation Processing; *J. Adhes. Sci. Technol.* 2002, 16, 999–1015.
29. Pothan, L.A.; Thomas, S.; Groeninckx, G., The Role of Fiber/Matrix Interactions on the Dynamic Mechanical Properties of Chemically Modified Banana Fiber/Polyester Composites; *Composites Part A* 2006, 37, 1260–1269.
30. Escamilla, G.C.; Trugillo, G.R.; Franco, E.P.J.H.; Mendizabal, E.; Puig, J.E., Preparation and Characterization of Henequen Cellulose Grafted with Methyl Methacrylate and its Application in Composites; *J. Polym. Sci.* 1997, 66, 339–345.
31. Nairn, J.A.; Liu, C.-H.; Mendels, D.-A.; Zhandarov, S., Fracture Mechanics Analysis of the Single-Fiber Pull-out Test and the Microbond Test Including The Effects of Friction and Thermal Stresses; *Proc. 16th Ann. Tech. Conf. Am. Soc. Composites 2001*, VPI, Blacksburg VA, September 9–12, 2001.
32. Craven, J.P.; Cripps, R.; Viney, C., Evaluating the Silk/Epoxy Interface by Means of the Microbond Test; *Composites Part A* 2000, 31, 653–660.

33. Luo, S.; Netravali, A.N., Characterization of Henequen Fibers and the Henequen Fiber/Poly(hydroxybutyrate-co-hydroxyvalerate) Interface; *J. Adhes. Sci. Technol.* 2001, 4, 423–437.
34. Zafeiropoulos, N.E.; Baillie, C.A.; Hodgkinson, J.M., Engineering and Characterisation of the Interface in Flax Fiber/Polypropylene Composite Materials. Part II. The Effect of Surface Treatments on the Interface; *Composites Part A* 2002, 33, 1185–1190.
35. Stamboulis, A.; Baillie, C.; Schulz, E., Interfacial Characterisation of Flax Fibre-Thermoplastic Polymer Composites by the Pull-out Test; *Angewandte Makromolekulare Chemie* 1999, 272, 117–120.
36. VandeVelde, K.; Kiekens, P., Influence of Fiber Surface Characteristics on the Flax/Polypropylene Interface; *J. Thermoplast. Compos. Mater.* 2001, 14, 244–260.
37. Manchado, M.A.L.; Arroyo, M.; Biagiotti, J.; Kenny, J.M., Enhancement of Mechanical Properties and Interfacial Adhesion of PP/EPDM/Flax Fiber Composites Using Maleic Anhydride as a Compatibilizer; *J. Appl. Polym. Sci.* 2003, 90, 2170–2178.
38. Park, J.-M.; Quang, S.T.; Hwang, B.-S.; DeVries, K.L., Interfacial Evaluation of Modified Jute and Hemp Fibers/Polypropylene (PP)-Maleic Anhydride Polypropylene Copolymers (PP-MAPP) Composites Using Micromechanical Technique and Nondestructive Acoustic Emission; *Compos. Sci. Technol.* 2006, 66, 2686–2699.
39. Demira, H.; Atikler, U.; Balkose, D.; Tihmihoglu, F., The Effect of Fiber Surface Treatments on the Tensile and Water Sorption Properties of Polypropylene-Luffa Fiber Composites; *Composites Part A* 2006, 37, 447–456.
40. Balnois, E.; Busnel, F.; Baley, C.; Grohens, Y., An AFM Study of the Effect of Chemical Treatments on the Surface Microstructure and Adhesion Properties of Flax Fibers; *Compos. Interfaces* 2007. Available online.
41. Michaeli, W.; Munker, M.; Krumpholz, T., Characterisation of the Fibre/Matrix Adhesion with Different Microscopic Analysing Methods on Natural Fibre-reinforced Thermosets; *Macromol. Mat. Eng.* 2000, 284/285, 25–29.
42. Redendo, S.U.A.; Goncalves, M.C.; Yoshida, I.V.P.J., Eucalyptus Kraft Pulp Fibers as an Alternative Reinforcement of Silicon Composites 11. Thermal, Morphological and Mechanical Properties of the Composites; *J. Appl. Polym. Sci.* 2003, 89, 3739–3746.
43. Tserki, V.; Matzinos, P.; Panayiotou, C., Novel Biodegradable Composites Based on Treated Lignocellulosic Waste Flour as Filler Part II. Development of Biodegradable Composites Using Treated and Compatibilized Waste Flour; *Composites Part A* 2006, 37, 1231–1238.
44. Dufresne, A.; Dupeyre, D.; Vignon, M.R., Cellulose Microfibrils from Potato Tuber Cells: Processing and Characterization of Starch-cellulose Microfibril Composites; *J. Appl. Polym. Sci.* 2000, 14, 2080–2092.
45. Dong, X.M.; Revol, J.F; Gray, D.G., Effect of Microcrystalline Preparation Conditions on the Formation of Colloid Crystals of Cellulose; *Cellulose* 1998, 5, 19–32.
46. Harper, D.; Wolcott, M., Interaction between Coupling Agent and Lubricants in Wood-Polypropylene Composites; *Composites Part A* 2004, 35, 385–394.
47. Siegbahn, K.; Hamrin, K.; Hedman, J.; Johansson, G.; Bergmark, T.; Karlsson, S.E.; Lindgren, I.; Lindbert, B., *ESCA: Atomic, Molecular and Solid State Structure by Means of Electron Spectroscopy*; Almquist and Wiksells, Uppsala, Sweden, 1967.
48. Gellerstedt, F.; Gatenholm, P., Surface Properties of Lignocellulosic Fibers Bearing Carboxylic Groups; *Cellulose* 1999, 6, 103–121.
49. Cai, X.; Riedl, B.; Ait-Kadi, A., Effect of Surface Grafted Ionic Groups on Performance of Cellulose Fiber-reinforced Thermoplastic Composites; *J. Polym. Sci. B Polym. Phys.* 2003, 41, 2022–2032.
50. Pothan, L.A.; Simon, F.; Spange, S.; Thomas, S., XPS Studies of Chemically Modified Banana Fibers; *Biomacromolecules* 2006, 7, 892–898.

51. Yuan, X.; Jayaraman, K.; Bhattacharyya, D., Effects of Plasma Treatment I in Enhancing the Performance of Wood-fiber Polypropylene Composites; *Composites Part A* 2004, 35, 1363–1374.
52. Khan, M.A.; Hassan, M.M., Effect of γ-Aminopropyl Trimethoxy Silane on the Performance of Jute Polycarbonate Composites; *J. Appl. Polym. Sci.* 2006, 100, 4142–4154.
53. Hristov, V.; Vasileva, S., Dynamic Mechanical and Thermal Properties of Modified Poly(propylene) Wood Fiber Composites; *Macromol. Mat. Eng.* 2003, 288, 798–806.
54. Colom, X.; Carrasco, F.; Pages, P.; Canavatc, J., Effects of Different Treatments on the Interface of HDPE/Lignocellulosic Fiber Composites; *Compos. Sci. Technol.* 2003, 63, 161–169.
55. Eichorn, S.J.; Sirichaisit, J.; Young, R.J., Deformation Mechanisms in Cellulose Fibers, Paper and Wood; *J. Mater. Sci.* 2001, 36, 3129–3135.
56. Gerstein, B.C., High-Resolution NMR Spectrometry of Solids; *Anal Chem.* 1983, 55, Part 2 899 A.
57. Stael, G.C.; Ines, M.; Tavares, B., Solid-State Carbon ^{13}NMR Study of Material Composite Based on Bagasse and Thermoplastic Polymers; *J. Appl. Polym. Sci.* 2001, 82, 2150–2154.
58. Malunka, M.E.; Luyt, A.S.; Krump, H., Preparation and Characterization of EVA-Sisal Fiber Composites; *J. Appl. Polym. Sci.* 2006, 100, 2, 1607–1617.
59. Keshk, S.; Suwinarti, W.; Sameshima, K., Physicochemical Characterization of Different Treatment Sequences on Kenaf Bast Fiber; *Carbohydr. Polym.* 2006, 65, 202–206.
60. Fischer, K.; Spange, S.; Fischer, S.; Bellmann, C.; Adams, J., Probing the Surface Polarity of Native Celluloses Using Genuine Solvatochromic Dyes; *Cellulose* 2002, 9, 31–40.
61. Spange, S.; .Fischer, K.; Prause, S.; Heinze, T., Empirical Polarity Parameters of Cellulose and Related Materials; *Cellulose* 2003, 10, 201–212.
62. Pothan, L.A.; Zimmermann, Y.; Thomas, S.; Spange, S., Determination of Polarity Parameters of Chemically Modified Cellulose Fibers by Means of the Solvatochromic Technique; *J. Polym. Sci. B Polym. Phys.* 2000, 38, 2546.
63. Nishino, T.; Hirao, K.; Kotera, M., X-ray Diffraction Studies on Stress Transfer of Kenaf Reinforced Poly(L-lactic acid) Composite; *Composites Part A* 2006, 37(12), 2269–2273.
64. Almeida, J.R.M.; Aquino, R.C.M.P.; Monteiro, S., Tensile Mechanical Properties, Morphological Aspects and Chemical Characterization of Piassava (*Attalea funifera*) Fibers; *Composites Part A* 2006, 37, 1473–1479.
65. Pasquini, D.; Belgacem, M.N.; Gandini, A.; Curvelo, A.A., Surface Esterification of Cellulose Fibers: Characterization by DRIFT and Contact Angle Measurements; *J. Colloid Interface Sci.* 2006, 295, 79–83.
66. Shen, Q.; Gao, Y.L.D.-S., Chen, Y., Surface Properties of Bamboo Fiber and a Comparison with Cotton Linter Fibers; *Colloids Surf. B Biointerfaces* 2004, 35, 193–195.
67. Rong, M.Z.; Zhang, M.Q.; Liu, Y.; Yan, H.M.; Yang, G.C.; Zeng, H.M., Interfacial Interaction in Sisal/Epoxy Composites and its Influence on Impact Performance; *Polym. Compos.* 2002, 23, 182–192.
68. Cantero, G.; Arbelaiz, A.; Llano-Ponte, R.; Mondragon, I., Effects of Fiber Treatment on Wettability and Mechanical Behaviour of Flax/Polypropylene Composites; *Compos. Sci. Technol.* 2003, 63, 1247–1254.
69. William, T.T.; Douglas, G., Contact Angle and IGC Measurements Probing Surface-Chemical Changes in the Recycling of Wood Pulp Fibers; *J. Adhes. Sci. Technol.* 2001, 15, 223–241.
70. Bismarck, A.; Mohanty, A.K.; Askargorta, I.A.; Czapla, S.; Misra, M.; Hinrichsen, G.; Springer J., Surface Characterization of Natural Fibers; Surface Properties and the Water Up-Take Behavior of Modified Sisal and Coir Fibers; *Green Chem.* 2001, 3, 100–107.
71. Grundke, K.; Jacobasch, H.-J.; Simon, F.; Schneider, S.T., Physico-Chemical Properties of Surface Modified-Polymers; *J. Adhes. Sci. Technol.* 1995, 3, 327–350.
72. Stana-Kleinschek, K.; Ribitsch, V., Electrokinetic Properties of Processed Cellulose Fibers; *Colloid Surf. A* 1998, 127–138.

73. Pothan, L.A.; Bellman, C.; Kailas, L.; Thomas, S., Influence of Chemical Treatments on the Electrokinetic Properties of Cellulose Fibers; *J. Adhes. Sci. Technol.* 2002, 16, 2, 157–178.
74. Joseph, P.V.; Mathew, G.; Joseph, K.; Groeninckx, G.; Thomas, S., Dynamic Mechanical Properties of Short Sisal Fiber Reinforced Polypropylene Composites; *Composites Part A* 2003, 34, 275–290.
75. Geethamma, V.G.; Pothen, L.A.; Rhao, B.; Neelakantan, N.R.; Thomas, S., Tensile Stress Relaxation of Short-Coir-Fiber-Reinforced Natural Rubber Composites; *J. Appl. Polym. Sci.* 2004, 94, 96–104; *Composites Part A* 2003, 34, 75–290.

Chapter 15
Thermal and Mechanical Analysis of Lignocellulose-Based Biocomposites

Hyoe Hatakeyama, Takashi Nanbo, and Tatsuko Hatakeyama

Abstract

Biocomposites composed of a kraft lignin–ethylene glycol-type polyol polyurethane matrix and a biofiller or an inorganic filler have been characterized by thermogravimetry (TG), derivative thermogravimetry (DTG), and mechanical analysis. Depending on the filler type and content, up to three thermal degradation temperatures are observed and the mass residues at 450–550°C range from ∼30% to >90%. Densely filled composites (with apparent density of ∼1.2–1.8 g cm^{-3}) having good mechanical performance (with bending strength ≥25 MPa) can be obtained by proper selection of the molecular weight of the ethylene glycol-type polyol, the type of the filler, and, most importantly, the content of the filler.

15.1 Introduction

In the present polymer industry, attention is being drawn to polymers that are compatible with the natural environment. For the development of environmentally compatible polymers, it should be recognized that nature has already constructed a variety of polymeric materials that are being used or can be readily transformed to usable products in our daily life. Major plant components such as polysaccharides and lignin contain hydroxyl groups that can be used as reactive sites for the synthesis of polymers such as polyurethanes.

Lignin, a by-product of the chemical pulping industry, is mostly burnt as a fuel, even though it is one of the most promising natural resources. Most industrial lignins are obtained from the kraft and sulfite pulping processes. As lignins are natural polymers with random cross-linking of their substructural phenyl propane units, their physical and chemical properties are highly dependent on their isolation processes. The higher order structure of lignin is fundamentally amorphous with two or three of its phenylpropanoid monomers, p-coumaryl alcohol, coniferyl alcohol, and/or synapyl alcohol cross-linked to produce a three-dimensional lignin polymer via a radical coupling process during its biosynthesis. Lignin and polysaccharides have been found to be useful raw materials for the preparation of lignin- and polysaccharides-based films, polyesters, polyurethanes, and polyblends [1–23].

Biocomposites obtained from various natural resources have been utilized from prehistoric times. Clay mixed with straw has been used as a strong construction material. The

Table 15.1 Classification of biocomposites.

Group	Matrix	Filler
I	Epoxy resin [22,23] or polyolefine	Cellulose fiber or powder, or wood meal
II	Bio-urethane [24,25]	$BaSO_4$, $CaCO_3$, $Al(OH)_3$, talc or sand
III	Bio-urethane	Cellulose fiber or powder, wood meal, empty fruit bunches, or coffee grounds

durability of biocomposites prepared using elaborate manual methods is higher than that of modern composites consisting of glass fiber and epoxy resin. That this is true is illustrated by the fact that a long wall constructed in 2 BC in the Taklimakan desert still remains. Although various definitions exist concerning composites [24], biocomposites should consist of two or more mechanically separable biocomponents. Biocomposites prepared under controlled conditions are expected to possess improved properties over their individual components.

Biocomposites are more and more in demand not only because the supply of petroleum is limited, but also because environmentally compatible properties are now required for many products. To meet such a demand, a large variety and quantity of renewable resources need to be adapted to the industrial scale of modern production systems. Bioresources obtained as by-products of agriculture-related industries have received much attention, because the amount of the materials can easily be estimated and unique properties can often be obtained. The pulp and paper industry, the sugar industry, and the food and coffee industries produce a large amount of residual by-products that have not been fully utilized and can potentially serve as key components of biocomposites.

Depending on the source of their components, biocomposites can be classified into three groups: with one of the following components, (I) filler components are biomaterials; (II) matrix components are biomaterials; and (III) the whole composites are fabricated from biomaterials. Table 15.1 shows a general classification of biocomposites. In early biocomposite research, attention was paid to biofillers. The major difficulty of synthetic polymer matrix–biofiller composites came from the fact that the surface compatibility of the two materials was insufficient and the composites were delaminated. In this chapter, biocomposites belonging to groups II and III are discussed based on recent results obtained in our laboratory.

15.2 Sample preparation and measurements

Kraft lignin (KL)-based polyurethane (PU) matrices and composites were prepared as follows: A KL (33 g), kindly provided by Westvaco Asia Co. Ltd., was first dissolved in 67 g of diethylene glycol (DEG), 67 g of triethylene glycol (TEG), and 67 g of polyethylene glycol (PEG) with molar mass of $\sim$200, respectively, to give three polyol solutions containing KL, designated as KLD, KLT, and KLP. To each of the polyol solutions was added poly(phenylene-methylene) polyisocyanate (MDI) (polymerized diphenylmethane diisocyanate) (150 g). The mixture was vigorously stirred to give each of the PU matrices designated as KLDPU,

KLTPU, and KLPPU, respectively. To prepare the KL-based PU composites, each of the polyol solutions was mixed with a known amount of the biofiller, an industrially ground wood meal (60–80 mesh) of Japanese cedar kindly provided by Taiyo Chemicals Co., Ltd., or an inorganic filler such as $BaSO_4$, $Al(OH)_3$, $CaCO_3$ or talc before the addition of MDI.

Measurements of the thermal and mechanical properties of the prepared samples were carried out as follows: Thermogravimetry (TG) was performed using a Seiko TG 220 at a heating rate of $20°C\,min^{-1}$ in a nitrogen atmosphere. Derivative thermogravimetric (DTG) curves were obtained by differential calculation of TG curves. Mass residues (MR) measured at 450°C, 500°C, and 550°C were designated as m_{450}, m_{500}, and m_{550}, respectively. m_{20} refers to the mass residue at 20°C which was the amount of the material used for the TG experiments. Apparent density [ρ = weight (M)/apparent volume (V), $g\,cm^{-3}$] was measured on samples of 15.0 (length) × 3.0 (width) × 1.0 (thickness) cm using a Mitsutoyo ABS digital solar caliper and an electric balance (Shimadzu balance EB-4300DVW). The average caliper values of each part of a sample were measured at three different spots and used to calculate the apparent volume (V). The mass (M) of each sample was weighed using the Shimadzu balance.

Bending strength was measured on samples of the same size as those for ρ measurement on a Shimadzu Autograph AG-IS. Applied stress was varied in order to control the rate in a range from 1.0×10^{-3} to $1.0 \times 10^{-2}\,m\,min^{-1}$. Bending strength ($\sigma$, MPa) was calculated according to JIS A1408 [26]. Modulus of elasticity (E, GPa) was determined using the initial stage of stress–strain curve at 25°C according to Equation (15.1) (JIS Z 2101) [27].

$$E = (\Delta P \times L)/(\Delta L \times A) \tag{15.1}$$

where P is the difference of the upper and the lower limit bending strength (kN) in the linear part of the bending curve; L is the distance between the two points; ΔL is the strain difference between the upper and lower limit and A is the cross-section area (m^2).

15.3 Thermal and mechanical properties of PU composites with wood meal as the bio-filler

Figure 15.1 shows the TG and the DTG curves of the KLPPU composites with wood meal as the filler. Both the TG and the DTG curves show the presence of two thermal degradation temperatures (T_d's), DT_{d1} and DT_{d2}. DT_{d2} seems to be specific to the degradation of the wood meal, because the DT_{d2} peak becomes prominent when the wood meal contents in the PU composites are over 70%, and because only the DT_{d2} peak is present when the wood meal content is 100%.

Figure 15.2 shows the change of MR at 450°C, 500°C, and 550°C of the wood meal–KLPPU composites with increasing wood meal content. Both the KLPPU matrix and the wood meal undergo significant decomposition at ≥450°C. The decomposition between 450°C and 550°C changes only slightly with the wood meal content due likely to the already significant decomposition of the composites at ∼450°C.

Figure 15.3 shows the relationship between the apparent density (ρ) and the wood meal content in wood meal–KLD, wood meal–KLT, and wood meal–KLPPU composites, respectively. The ρ value reaches a maximum value at a wood meal content of ∼50–70%. There is

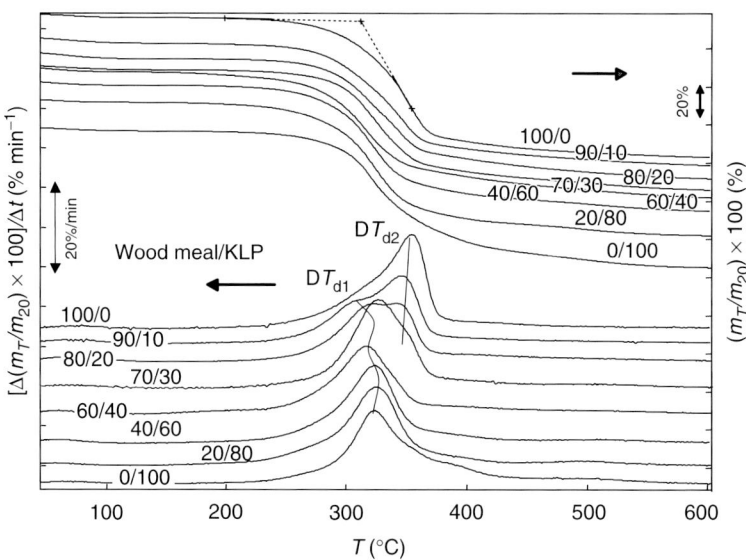

Figure 15.1 TG and DTG curves of the wood meal-KLPPU composites; numerals are the values of wood meal/KLP %.

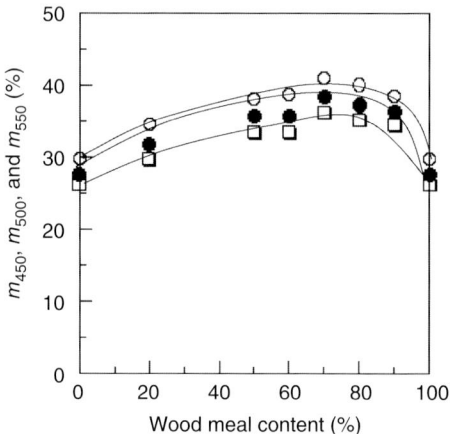

Figure 15.2 Mass residues at 450°C, 500°C, and 550°C vs wood meal content of the wood meal-KLPPU composites; ○ = m_{450}, ● = m_{500}, and □ = m_{550}.

little difference among the three composites at the same wood meal content, showing that the effect of molecular weight of the ethylene glycol-type polyol is not prominent in such composites.

The bending strength (σ) of the wood meal–KLD, wood meal–KLT, and wood meal–KLPPU composites initially increases with increasing wood meal content, reaches a maximum value between a wood meal content of ~50–70%, and then decreases (Figure 15.4). The σ values of the composites, however, depend to a certain extent on the molecular weight

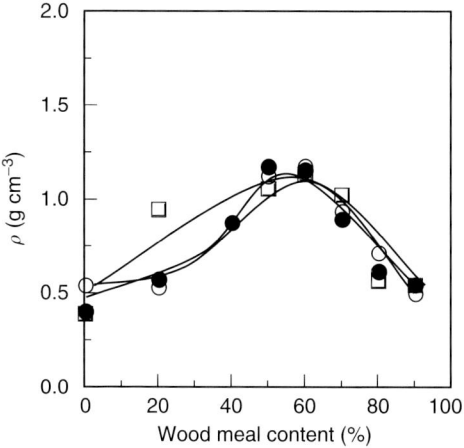

Figure 15.3 Apparent density (ρ) vs wood meal content of the wood meal-KLD (O), -KLT (●), and -KLP (□) PU composites.

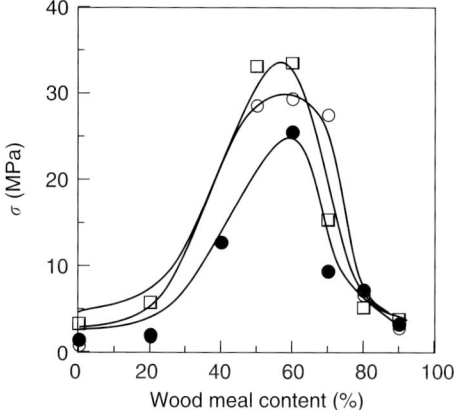

Figure 15.4 Bending strength (σ) vs wood meal content of the wood meal-KLD (O), -KLT (●), and -KLP (□) PU composites.

of the ethylene glycol-type polyol (DEG, TEG, or PEG) used along with the KL to make the PU matrix (KLD-, KLT- or KLPPU). The maximum (σ) value varies from ~25 to 35 MPa, depending on the ethylene glycol-type polyol and the wood meal content used for the preparation of the composite. These results indicate that densely filled composites having good mechanical performance can be obtained using a bio-KL-based PU as the matrix and a wood meal bio-filler with proper control of the structure of the matrix and the amount of the filler used.

The modulus of elasticity (E) of the composites calculated according to Equation (15.1) described in Section 15.2 also increases initially with the wood meal content, reaches a maximum value at a wood meal content of ~50–70%, and then decreases (Figure 15.5).

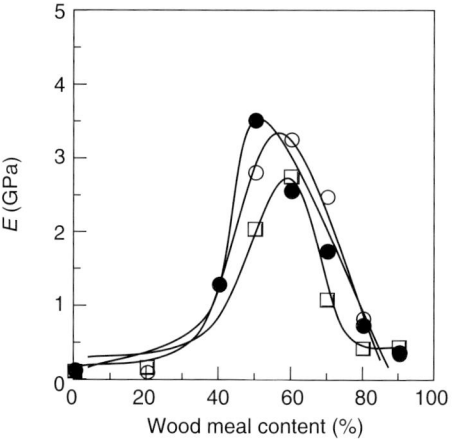

Figure 15.5 Modulus of elasticity (*E*) vs wood meal content of the wood meal-KLD (○), -KLT (●), and -KLP (□) PU composites.

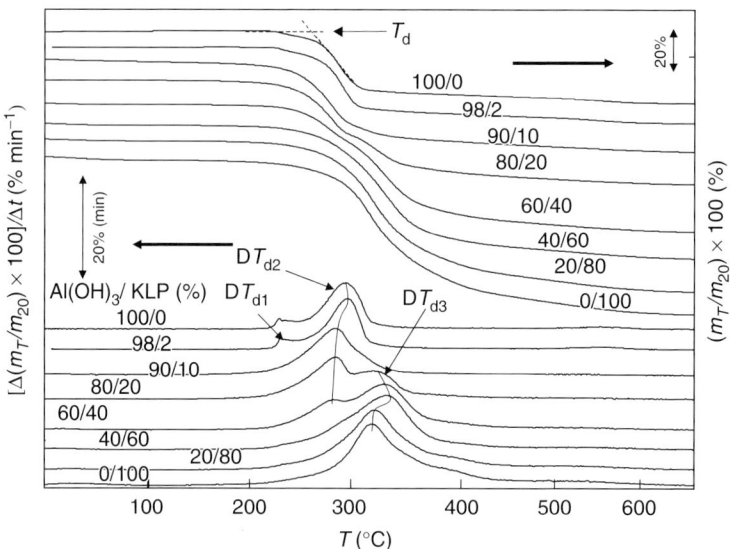

Figure 15.6 TG and DTG curves of the KLPPU composites with Al(OH)$_3$ as the filler; numerals are the values of Al(OH)$_3$/KLP %.

15.4 Thermal and mechanical properties of PU composites with inorganic fillers

Figures 15.6 and 15.7 show the TG and the DTG curves of KLPPU composites with Al(OH)$_3$ and CaCO$_3$ as the filler, respectively, at various filler contents. The composites with Al(OH)$_3$ as the inorganic filler decompose in three stages at ∼230°C (DT_{d1}), 260°C (DT_{d2}), and

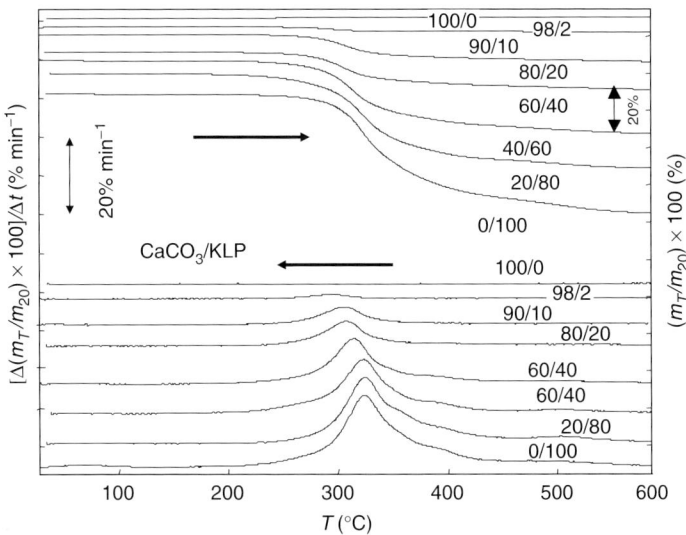

Figure 15.7 TG and DTG curves of the KLPPU composites with $CaCO_3$ as the filler; numerals are the values of $CaCO_3$/KLP %.

330°C (DT_{d3}) (Figure 15.6). The first decomposition observed at ∼230°C, DT_{d1}, is likely the thermal degradation of $Al(OH)_3$ that corresponds to the decomposition of the hydrated water. The second decomposition observed at 260°C, DT_{d2}, is attributed to the dissociation of the isocyanate and the phenolic hydroxyl groups of lignin [25]. The third decomposition observed at 330°C, DT_{d3}, can be attributed to the decomposition of the PEG chain [25]. The TG and DTG curves of the composites with $CaCO_3$ (or talc, data not shown) as the inorganic filler show only one decomposition at ∼300°C ascribable to the decomposition of the KLPPU matrix due to the fact that the inorganic filler ($CaCO_3$ or talc) decomposes at >700°C [19].

Figure 15.8a and b show the plots of the mass residue of the KLPPU composites at 450°C (m_{450}), 500°C (m_{500}), and 550°C (m_{550}) vs the content of $Al(OH)_3$ and $CaCO_3$, respectively. The amount of mass residue ranges from 30% to 70% in the KLPPU composites with $Al(OH)_3$ as the filler and is independent on the three decomposition temperatures studied (Figure 15.8a). The majority of the residue is the aluminum compound with a portion of the decomposition of $Al(OH)_3$ occurring due to the decomposition of the hydrated water. At 100% $Al(OH)_3$ content, the mass residue is ∼70%. For the PU composites with $CaCO_3$ as the filler, the amount of mass residue is also independent on the three decomposition temperatures studied (Figure 15.8b). However, the mass residue at any given content of $CaCO_3$ is higher than that at the same content of $Al(OH)_3$ with ∼100% mass residue at 100% $CaCO_3$ content compared to ∼70% mass residue at 100% $Al(OH)_3$ content. This is because $CaCO_3$, the major or sole composition of the residue, does not decompose at a temperature lower than 550°C. The KLPPU composites with talc as the filler gave a similar result to that of the PU composites with $CaCO_3$ as the filler.

Figure 15.9a and b show the plots of the apparent density (ρ) vs $Al(OH)_3$ and $CaCO_3$ contents, respectively, of the KLDPU, KLTPU, and KLPPU composites. As clearly seen from

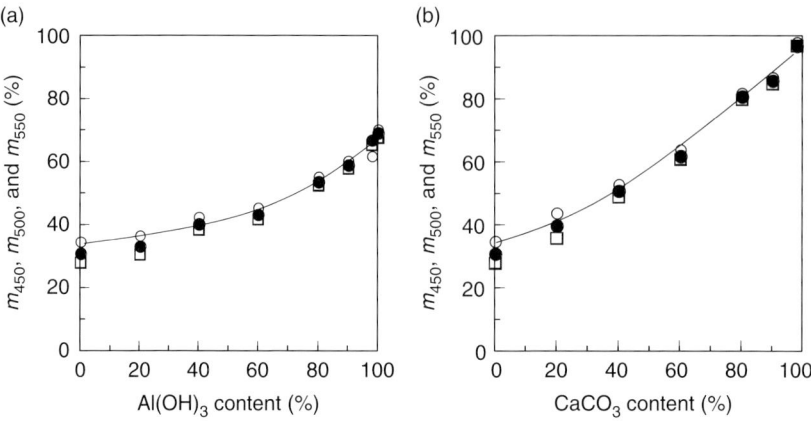

Figure 15.8 (a) Mass residues at 450°C, 500°C, and 550°C vs Al(OH)$_3$ content of the KLPPU composites with Al(OH)$_3$ as the filler; O = m_{450}, ● = m_{500} and □ = m_{550}. (b) Mass residues at 450°C, 500°C, and 550°C vs CaCO$_3$ content of the KLPPU composites with CaCO$_3$ as the filler; O = m_{450}, ● = m_{500} and □ = m_{550}.

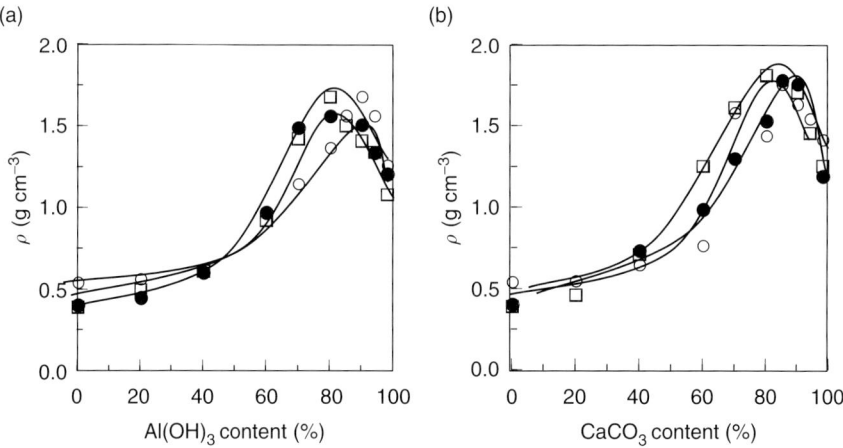

Figure 15.9 (a) Apparent density (ρ) vs Al(OH)$_3$ content of the KLD (O), KLT (●), and KLP (□) PU composites with Al(OH)$_3$ as the filler. (b) Apparent density (ρ) vs CaCO$_3$ content of the KLD (O), KLT (●), and KLP (□) PU composites with CaCO$_3$ as the filler.

the figures, no significant differences between Al(OH)$_3$ and CaCO$_3$ fillers are observed in the trend of the change of ρ with the filler content except for the ρ values and particularly the maximum ρ values of the composites. The maximum ρ values of the composites with CaCO$_3$ as the filler are higher than those with Al(OH)$_3$ as the filler due to the higher density of CaCO$_3$ than Al(OH)$_3$.

Figure 15.10 shows the plots of the apparent density (ρ) vs the filler content of the KLPPU composites with Al(OH)$_3$, CaCO$_3$, and talc as the fillers, respectively. As seen from

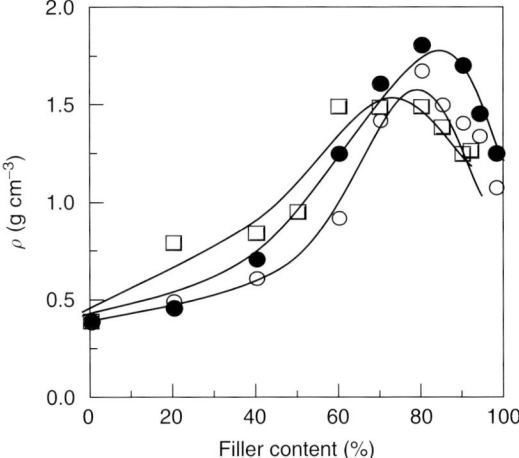

Figure 15.10 Apparent density (ρ) vs filler content of the KLPPU composites with Al(OH)$_3$ (○), CaCO$_3$ (●), and talc (□) as the fillers, respectively.

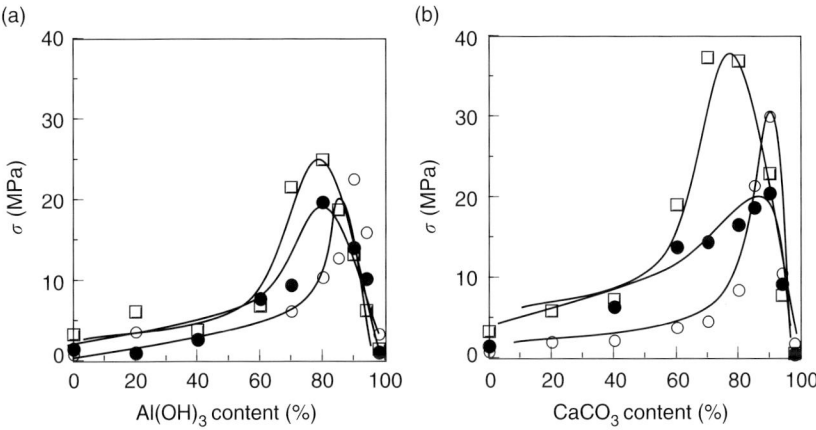

Figure 15.11 (a) Bending strength (σ) vs Al(OH)$_3$ content of the KLD (○), KLT (●), and KLP (□) PU composites with Al(OH)$_3$ as the filler. (b) Bending strength (σ) vs CaCO$_3$ content of the KLD (○), KLT (●), and KLP (□) PU composites with Ca(CO)$_3$ as the filler.

the figure, the trend of the change in ρ of the KLPPU composites with the filler content is similar. However, the ρ values including the maximum ρ values are dependent on the type/density of the filler. In addition, the optimal amount of the filler to give the maximum ρ value differs from one filler to another.

Figure 15.11a and b show the plots of the bending strength (σ) vs Al(OH)$_3$ and CaCO$_3$ contents, respectively, of the KLDPU, KLTPU, and KLPPU composites. In general, the σ values initially increase slightly with the filler contents, then increase rapidly to reach maximum values, and finally decrease rapidly with further increase of the filler contents.

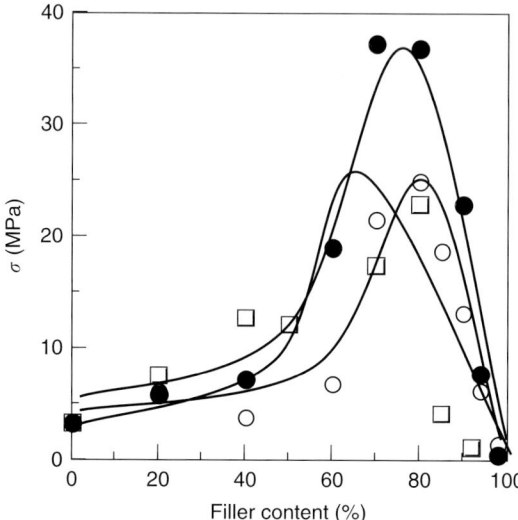

Figure 15.12 Bending strength (σ) vs filler content of the KLPPU composites with Al(OH)$_3$ (O), CaCO$_3$ (●), and talc (□) as the fillers, respectively.

The σ values, particularly the maximum σ values, depend on both the PU matrix and on the filler type; the highest maximum σ value is achieved for the KLPPU composite with CaCO$_3$ as the filler (Figure 15.11b).

Figure 15.12 shows the plots of σ vs filler content of the KLPPU composites with Al(OH)$_3$, CaCO$_3$, and talc as the fillers, respectively. As shown in the figure, the maximum σ value of the composite with CaCO$_3$ as the filler is higher than that of the composite with Al(OH)$_3$ or talc as the filler. In addition, there exists an optimum content of the inorganic filler at which the biocomposites possess the highest bending strength of ≥ 25 MPa. Such high values of bending strength render these composites sufficient for practical applications. The results in Figure 15.12 and those in Figure 15.11b show that the inorganic filler type is an important factor and so is the type of polyol used in preparing the KL-polyol PU matrix.

Modulus of elasticity (E) of the KLDPU, KLTPU, and KLPPU complexes with Al(OH)$_3$ or CaCO$_3$ as the filler all shows an maximum value at an optimal filler content (Figure 15.13a and b) similar to what is observed with bending strength. However, with Al(OH)$_3$ as the filler the highest E value is obtained for the KLDPU matrix while with CaCO$_3$ as the filler the highest E value is obtained for the KLPPU matrix. The effect of the molecular weight of the ethylene glycol-type polyol on the maximum E value is different than that on the maximum σ value where KLPPU matrix gives a higher maximum σ value than the KLTPU or KLDPU matrix whether the filler is Al(OH)$_3$ or CaCO$_3$.

Figure 15.14 shows the plots of E vs filler content of the KLPPU composites with Al(OH)$_3$, CaCO$_3$, and talc as the fillers, respectively. The maximum E value of the composite with an inorganic filler is dependent on the kind of the inorganic filler and the content of the filler. For the KLPPU composites, CaCO$_3$ filler gives a higher maximum E value than Al(OH)$_3$ or talc. These results and those in Figure 15.13a and b show that the inorganic filler type

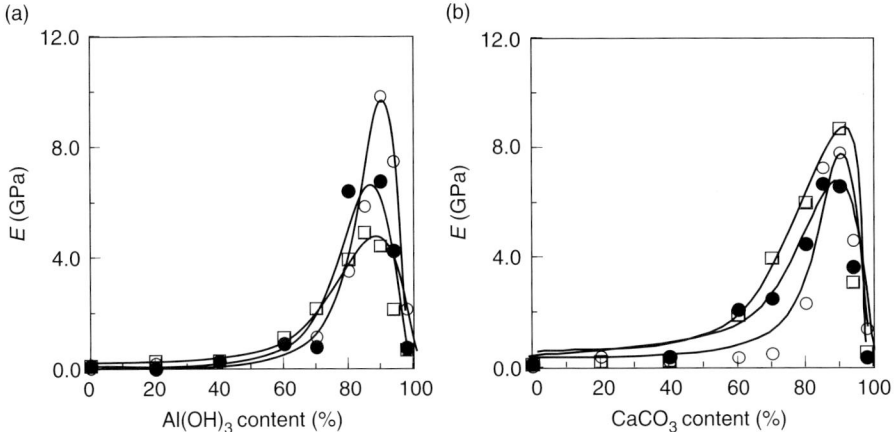

Figure 15.13 (a) Modulus of elasticity (E) vs Al(OH)$_3$ content of the KLD (O), KLT (●), and KLP (□) PU composites with Al(OH)$_3$ as the filler. (b) Modulus of elasticity (E) vs CaCO$_3$ content of the KLD (O), KLT (●), and KLP (□) PU composites with CaCO$_3$ as the filler.

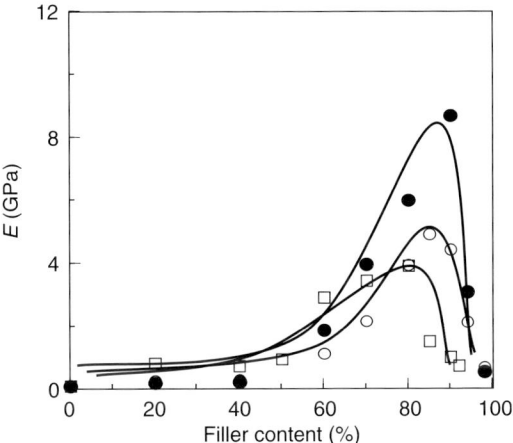

Figure 15.14 Modulus of elasticity (E) vs filler content of the KLPPU composites with Al(OH)$_3$ (O), CaCO$_3$ (●), and talc (□) as the fillers, respectively.

is an important factor and so is the type of polyol used in preparing the KL-polyol–PU matrix.

References

1. Baumberger, S., Starch-lignin Films. In: Hu, T.Q. (ed.); Chemical Modification, Properties, and Usage of Lignin; Kluwer Academic/Plenum Publishers; New York, 2002; pp. 1–19.

2. Hatakeyama, H., Polyurethanes Containing Lignin. In: Hu, T.Q. (ed.); Chemical Modification, Properties, and Usage of Lignin; Kluwer Academic/Plenum Publishers; New York, 2002; pp. 41–56.
3. Gandini, A.; Belgacem, M.N.; Guo, Z.X.; Montanari, S., Lignins as Macromonomers for Polyesters and Polyurethanes. In: Hu, T.Q. (ed.); Chemical Modification, Properties, and Usage of Lignin; Kluwer Academic/Plenum Publishers; New York, 2002; pp. 57–80.
4. Feldman, D., Lignin and Its Polyblends – A Review. In: Hu, T.Q. (ed.); Chemical Modification, Properties, and Usage of Lignin; Kluwer Academic/Plenum Publishers; New York, 2002; pp. 81–99.
5. Nägele, H.; Pfitzer, J.; Nägele, E.; Inone, E.R., Arboform – A Thermoplastic Processable Material from Lignin and Natural Fibers. In: Hu, T.Q. (ed.); Chemical Modification, Properties, and Usage of Lignin; Kluwer Academic/Plenum Publishers; New York, 2002; pp. 101–119.
6. Hatakeyama, H., Thermal Analysis of Environmentally Compatible Polymers Containing Plant Components in the Main Chain; *J. Therm. Anal. Cal.* 2002, 70, 755–759.
7. Hatakeyama, H.; Asano, Y.; Hatakeyama, T., Biobased Polymeric Materials. In: Chellini, E.; Solario, R. (eds); Biodegradable Polymers and Plastics; Kluwer Academic/Plenum Publishers, New York, 2003; pp. 103–119.
8. Zetterlund, P.; Hirose, S.; Hatakeyama, T.; Hatakeyama, H.; Albertsson, A.C., Thermal and Mechanical Properties of Polyurethanes Derived from Mono- and Disaccharides; *Polym. Inter.* 1997, 42, 1–8.
9. Hatakeyama, H.; Kobashigawa, K.; Hirose, S.; Hatakeyama, T., Synthesis and Physical Properties of Polyurethanes from Saccharide-based Polycaprolactones; *Macromol. Symp.* 1998, 130, 127–138.
10. Hatakeyama, T.; Tokashiki, T.; Hatakeyama, H., Thermal Properties of Polyurethanes Derived from Molasses Before and After Biodegradation; *Macromol. Symp.* 1998, 130, 139–150.
11. Asano, Y.; Hatakeyama, H.; Hirose, S.; Hatakeyama, T., Preparation and Physical Properties of Saccharide-based Polyurethane Foams. In: Kennedy, J.F.; Philips, G.O.; Williams P.A.; Hatakeyama, H. (eds); Recent Advances in Environmentally Compatible Polymers; Woodhead Publishing Ltd.; Cambridge, 2001; pp. 241–246.
12. Yoshida, H.; Mörck, R.; Kringstad, K.P.; Hatakeyama, H., Kraft Lignin in Polyurethanes. II. Effects of the Molecular Weight of Kraft Lignin on the Properties of Polyurethanes from a Kraft Lignin-Polyether Triol-Polymeric MDI System; *J. Appl. Polym. Sci.* 1990, 40, 1819–1832.
13. Reimann, A.; Mörck, R.; Hirohisa, Y.; Hatakeyama, H.; Kringstad, K.P., Kraft Lignin in Polyurethanes. III. Effects of the Molecular Weight of PEG on the Properties of Polyurethanes from a Kraft Lignin-PEG-MDI System; *J. Appl. Polym. Sci.* 1990, 41, 39–50.
14. Nakamura, K.; Mörck, R.; Reimann, A.; Kringstad, K.P.; Hatakeyama, H., Mechanical Properties of Solvolysis Lignin-derived Polyurethanes; *Poly. Adv. Technol.* 1991, 2, 41–47.
15. Nakamura, K.; Hatakeyama, T.; Hatakeyama, H., Thermal Properties of Solvolysis Lignin-derived Polyurethanes; *Polym. Adv. Technol.* 1992, 3, 151–155.
16. Hirose, S.; Nakamura, K.; Hatakeyama, H.; Meadows, J.; Williams, P.A.; Phillips, G.O., Preparation and Mechanical Properties of Polyurethane Foams from Lignocellulose Dissolved in Polyethylene Glycol. In: Kennedy, J.F.; Williams, P.A.; Phillips, G.O. (eds); Cellulosics: Chemical, Biochemical and Materials; Ellis Horwood Limited; Chichester, 1993; pp. 317–331.
17. Nakamura, K.; Hatakeyama, H.; Meadows, J.; Williams, P.A.; Phillips, G.O., Mechanical Properties of Polyurethane Foams Derived from Eucalyptus Kraft Lignin. In: Kennedy, J.F.; Williams, P.A.; Phillips, G.O. (eds); Cellulosics: Chemical, Biochemical and Materials; Ellis Horwood Limited; Chichester, 1993; pp. 333–340.
18. Hatakeyama, H.; Hirose, S.; Nakamura, K.; Hatakeyama, T., New Types of Polyurethanes Derived from Lignocellulose and Saccharides. In: Kennedy, J.F.; Williams, P.A.; Phillips, G.O. (eds); Cellulosics: Chemical, Biochemical and Materials; Ellis Horwood Limited; Chichester, 1993; pp. 525–536.
19. Hatakeyama, T.; Liu, Z., Handbook of Thermal Analysis; John Wiley & Sons; Chichester, 1998; pp. 275, 296, and 325.

20. Hirose, S.; Kobayashi, M.; Kimura, H.; Hatakeyama, H., Synthesis of Lignin-based Polyester-epoxy Resins; In: Kennedy, J.F.; Phillips, G.O.; Williams, P.A.; Hatakeyama, H. (eds); Recent Advances in Environmentally Compatible Polymers; Woodhead Publishing Ltd.; Cambridge, 2001; pp. 73–78.
21. Hirose, S.; Hatakeyama, T.; Hatakeyama, H., Glass Transition and Thermal Decomposition of Epoxy Resins from the Carboxylic Acid System Consisting of Ester-Carboxylic Acid Derivatives of Alcoholysis Lignin and Ethylene Glycol with Various Dicarboxylic Acids; *Thermochim. Acta* 2005, 431, 76–80.
22. Hatakeyama, H.; Tanamachi, N.; Matsumura, H.; Hirose, S.; Hatakeyama, T., Biobased Polyurethane Composite Foams with Inorganic Fillers Studied by Thermogravimetry; *Thermochim. Acta* 2005, 431, 155–160.
23. Hatakeyama, H.; Nakayachi, A.; Hatakeyama, T., Thermal and Mechanical Properties of Polyurethane-based Geocomposites Derived from Lignin and Molasses; *Composites A Appl. Sci. Manufac.* 2005, 36(5), 698–704.
24. Hull, D., An Introduction to Composites Materials; Cambridge University Press; Cambridge, 1982.
25. Saunders, S.H.; Frisch, K.C., Polyurethanes, Chemistry and Technology, Part I Chemistry; Interscience Publishers; New York, 1962, pp. 103–111.
26. The Japanese Industrial Standard, JIS A 1408:2001, Test Method of Bending and Impact for Building Boards.
27. The Japanese Industrial Standard, JIS Z 2101:1994, Methods of Test for Woods.

Chapter 16
New Insights into the Mechanisms of Compatibilization in Wood–Plastic Composites

Kaichang Li and John Ziqiang Lu

Abstract

Compatibilization between hydrophilic wood and hydrophobic thermoplastics is critical for improving the interfacial adhesion in the resultant wood–plastic composites (WPCs). The interfacial adhesion has a significant impact on the final mechanical properties of WPCs. In this chapter, compatibilization mechanisms of WPCs with and without a compatibilizer, increase in the strength and stiffness of WPCs through strengthening and stiffening wood, and the effect of hydrophobic interaction on the WPC strength are reviewed with a focus on topics related to polypropylene and polyethylene thermoplastic matrices and references to the analytical methods used for studying the mechanisms.

16.1 Introduction

In the past decade, wood–plastic composites (WPCs) have been one of the most rapidly growing composite materials. WPCs have been extensively used for automotive, building products, furniture, packaging materials, and other applications [1,2]. WPCs maintain a strong momentum for further growth in the 21st century [2–4].

Commonly used thermoplastics for WPCs include polypropylene (PP), polyethylene (PE), polyvinyl chloride, and polystyrene, with PE being the most commonly used one [5]. Wood is usually used in the forms of flour and fibers in WPCs [5]. The majority of WPCs are produced through an extrusion process, while a small volume of them is produced through a compression or injection molding process. In addition to wood and a thermoplastic, various additives such as lubricants, compatibilizer (also called coupling agent), and pigment may be added in the production of WPCs [5]. A number of literature reviews of WPCs have been published [1,5–12].

Wood is hydrophilic and thermoplastics are hydrophobic. The interfacial adhesion between wood and a thermoplastic in a WPC is typically weak. Stress cannot be effectively transferred from the thermoplastic matrix to the reinforcing wood. Mechanical properties of WPCs are thus typically low. Extensive efforts have been devoted to improvement of the interfacial adhesion, that is, the compatibilization, between wood and a thermoplastic

[6,10,13–20]. Because PE and PP are the predominant thermoplastics for WPCs, this review will focus on various compatibilization mechanisms for PE- and PP-based WPCs with references to the analytical methods used for studying the mechanisms.

16.2 Compatibilization mechanisms for WPCs without a compatibilizer

A compatibilizer can significantly improve the interfacial adhesion between wood and a thermoplastic, thus greatly enhancing the strength and stiffness of WPCs. However, the use of a compatibilizer adds cost to WPCs. At present, a compatibilizer is not commonly used in the commercial production of WPCs. As hydrophilic wood is not compatible with a hydrophobic thermoplastic, a question arises regarding how WPCs without a compatibilizer can maintain decent strengths for various applications.

PP was investigated as a hot melt wood adhesive for making plywood [21]. The examination of PP-bonded plywood with scanning electron microscopy (SEM) revealed that PP penetrated into lumens of wood cells and filled gaps between wood fibers [21]. More specifically, molten PP penetrated into some vessels away from the glueline. It also penetrated deeply into the wood fibers, ray parenchyma cells, and intercellular canals. PP penetrated not only into cell lumens but also into pit cavities. It easily formed mushroom-like projections on vessels and filled bordered pit cavities. It was concluded that the anchoring effect of PP that had penetrated into the interior of wood played an essential role in the adhesion [21].

With a combination of optical microscopy and SEM techniques, studies on PP-wood veneer laminates (models of wood–PP composites) also revealed that PP considerably penetrated vessel lumens, ray cells, pits, and intercellular spaces of wood [22,23]. On the wood–PP interfaces, the vessel lumens, ray cells, pits, and intercellular spaces were partly blocked by PP [22,23]. Some penetrated PP experienced plastic deformation during a peel fracture [22,23].

PP was used as an adhesive to bond glass and wood together [24]. The wood–PP interface was investigated through etching wood away using a combination of chromic acid, nitric, and sulfuric acids [24]. SEM characterization of the PP interface revealed that PP penetrated the vascular tissues, and conformed to the features of the wood [24]. Recently SEM characterization also showed that high-density PE (HDPE) penetrated deep into fiber cracks, pit cavities, and other voids in the wood–HDPE composites [25].

All these results indicated that a thermoplastic penetrated wood in either wood/plastic laminate or molded/extruded WPCs. Wood is porous and has many gaps and cracks among fibers. In the production of WPCs, a thermoplastic is melted and then compounded with wood. It is understandable that some of the molten thermoplastic is forced into cracks, gaps, and lumens of wood during compounding. Once WPCs are made and cooled down, the penetrated thermoplastic remains in wood and forms mechanical interlocks with wood. Mechanical interlocks appear to be the predominant forces that hold wood and the thermoplastic together.

Numerous studies using differential scanning calorimetry (DSC), polarizing microscopy equipped with a hot stage, dynamic mechanical analysis (DMA), dynamic mechanical thermoanalysis (DMTA), and other methods have revealed that wood and other natural fibers can induce crystallization and transcrystallization of PP and PE in WPCs [26–33]. Crystallization of PP phase was dominated by the gradual formation of spherulites [28] on the wood fibers embedded in the PP matrix. The average spherulite diameter of PP in WPCs was normally smaller than that of pure PP [27,28]. Incorporation of fillers into thermoplastic matrices generally increases the crystallization temperature and accelerates the nucleation of thermoplastics, thus improving its crystallization rate in WPCs [27,29,32,34].

Extensive evidences show that inserting cellulose/wood fibers accelerates transcrystalline growth of PP matrix around the fiber surface [26–33]. The PP spherulites produce a cylindrical layer of columnar crystalline structures around the fibers. These oriented aggregations of lamellae are arranged along the fiber laterally and their growth is usually restricted in the radial direction only [29]. As a nucleating agent, the wood/cellulose fibers inserted in the PP matrix not only induce the formation of transcrystalline layers but also increase the crystallization rate of the PP matrix [26–28,32,33].

Crystalline structures of thermoplastics are stronger and stiffer than their amorphous structures. The crystalline structures of a penetrated thermoplastic can thus significantly strengthen the mechanical interlocks between the thermoplastic and wood. Moreover, the stiff transcrystalline layers around the fibers can improve the shear stress transfer between the fibers and the thermoplastic matrix. The transcrystallization on the fiber is expected to further increase the interfacial adhesion. As a result, crystallization and transcrystallization of PP or PE effectively enhance the final mechanical properties of WPCs.

16.3 Increasing the strength and stiffness of WPCs through strengthening and stiffening wood with wood adhesives

The WPCs without any additives typically have lower strength than their corresponding thermoplastics [34]. Such WPCs cannot meet property requirements of many applications. To improve the strength and stiffness of WPCs, some WPC manufacturers add wood adhesives such as powder phenol-formaldehyde (PF) resin in the production of WPCs.

PF resin and polymeric diphenylmethane diisocyanate (PMDI) resin are commonly used adhesives for making wood composites such as plywood, particleboard, and oriented strandboard [35,36].

SEM, DMA and water-uptake studies revealed that PF resin significantly increased the strength and stiffness and reduced the water-uptake rate of the PF-treated WPCs [37,38]. The water-uptake rate of the PF-treated WPCs was measured under damp conditions by soaking the injection-molded WPC samples in water at ambient temperature for a period of 80 days [37]. The water-uptake rate of each sample was then measured on days 18, 44, and 80, respectively. Moisture contents of WPCs varied with the soaking time in water and the PF concentration in WPCs; untreated WPCs had a higher water-uptake rate than PF-treated WPCs on day 80 [37]. The 30 wt% PF-treated WPCs showed the lowest water-uptake rate (∼0.5 wt%) [37].

It was demonstrated that ethyl isocyanate, 1,6-hexamethylene diisocyanate, toluene 2,4-diisocyanate, and PMDI enhanced the strength and stiffness of wood–HDPE and wood–LLDPE (linear low-density PE) composites [39]. Among these four isocyanates, PMDI was the most effective, and ethyl isocyanate was the least effective [39].

Several other studies using mechanical testing methods also confirmed that PMDI significantly increased both the strength and stiffness of wood–HDPE composites [13,20,40–42]. PMDI resin is moisture sensitive. It was found that re-drying of wood flour prior to compounding with a plastic further increased the modulus of rupture (MOR) and the modulus of elasticity (MOE) of the resulting WPCs [41,42]. When HDPE was first mixed with PMDI, and then with wood flour, the MOR and MOE of resulting WPCs were much higher than those when wood flour was first mixed with PMDI and then with HDPE [41,42]. It is still poorly understood why the mixing order makes such a big difference.

It is well known that PF resin and PMDI are not good adhesives for bonding PP or PE together. In other words, PF resin and PMDI cannot form strong interfacial adhesion between wood and PE (or PP). The question is why PF resin and PMDI can significantly improve the strength and stiffness of wood–PE/PP composites.

As discussed previously, there are many fines and loose fibers on wood surfaces, and there are many gaps and cracks inside wood. These gaps and cracks become the weak parts when stress is applied to WPCs. There is no doubt that PF resin, and PMDI can consolidate fines and loose fibers and bridge the gaps and cracks of wood, thus strengthening and stiffening wood. A number of studies using mechanical testing methods revealed that the strength and stiffness of WPCs increased when the stiffness and strength of wood flour/fibers increased [15,43–46]. Strengthening and stiffening of wood by these wood adhesives are expected to greatly contribute to the improvement of the strength and stiffness of WPCs by these wood adhesives. Treatment of wood with these wood adhesives can significantly increase the hydrophobicity of wood, thus improving the wettability of wood by PP or PE. The improved wettability can facilitate the coverage of PP or PE on wood surfaces and the penetration of PP or PE into the interior of wood, thus improving the mechanical interlocks between wood and PP or PE. The improved wettability can also improve the dispersion of wood in the PP or PE matrix, thus improving the uniformity of WPC properties. The improved hydrophobicity of wood can improve hydrophobic interactions between wood and PP or PE. Contributions of the hydrophobic interactions to the strength and stiffness of WPCs are expected to be small because hydrophobic forces are typically very weak. At present, the relative contributions of wood-strengthening/ wood-stiffening, improved mechanical interlocks, improved dispersion, and hydrophobic interactions on the overall properties of WPCs are not well understood.

16.4 Effect of hydrophobic interaction on the WPC strengths

Several studies using infrared (IR) and SEM characterization and mechanical testing methods revealed that chemical modifications of wood and other natural fibers with acetic anhydride, stearic acid, and maleic anhydride improved the strength of the resulting WPCs [37,47–51]. Such chemical treatments increased the hydrophobicity of wood and other natural fibers, thus improving their wettability and hydrophobic interactions with thermoplastics. The chemical treatments can also improve the penetration of thermoplastics into wood and the dispersion of wood in the thermoplastic matrix during the preparation of WPCs. In contrast to the treatment of wood with wood adhesives, these chemical treatments cannot strengthen and stiffen wood fibers. Therefore, hydrophobic interactions and mechanical interlocking are expected to be the main compatibilization mechanisms for these chemical-treated, wood-based WPCs. In general, these chemical treatments are not very effective in improving the strength and stiffness of WPCs.

Numerous silane coupling agents with a basic structure of R–Si(OR′)$_3$ have been extensively studied and widely used for silica-filled composite materials because they can form covalent linkages with silica [52]. The R group in R–Si(OR′)$_3$ has to vary with the matrix materials for achieving maximum compatibilization effects [52]. For example, silane coupling agents containing a mercapto group (–SH) such as HSCH$_2$CH$_2$CH$_2$Si(OCH$_2$CH$_3$)$_3$ are typically used for silica-filled rubber composites because the –SH group can covalently link to rubber via a vulcanization process [52]. Silane coupling agents containing unsaturated C=C bonds such as 3-(methacryloyloxy)propyltrimethoxysilane, CH$_2$=C(CH$_3$)COOCH$_2$CH$_2$CH$_2$Si(OCH$_3$)$_3$, are typically used in glass-fiber-reinforced or silica-filled unsaturated polyester because the C=C bonds in the silane coupling agents can react with the unsaturated C=C bonds in the polyester via a free radical process to form strong covalent linkages [52]. Because of their ready availability, various silane coupling agents have been investigated for improving the strength and stiffness of WPCs. They include vinyltri(2-methoxyethoxy)silane, 3-(methacryloyloxy)propyltrimethoxysilane, 2-(3,4-epoxy cyclohexyl)ethyltrimethoxysilane, (3-glycidyloxypropyl)trimethoxysilane, and γ-aminopropyltrimethoxysilane [40,53–60].

Silane coupling agents can react with various hydroxyl groups in wood to form covalent linkages although such covalent linkages are susceptible to hydrolysis by water. Silane coupling agents can also be readily hydrolyzed to form silanols, R–Si(OH)$_3$, that can form hydrogen bonding with hydroxyl groups in wood. They can increase the hydrophobicity of wood fibers, and thus the compatibilization mechanisms are similar to those for the chemical treatments of wood fibers with acetic anhydride, stearic acid, and maleic anhydride. More specifically, hydrophobic interactions between wood and thermoplastics and improved mechanical interlocks from improved penetration of thermoplastics into wood are main forces for the interfacial adhesion. However, silane coupling agents cannot form strong covalent linkages with PP or PE unless they contain C=C bonds and are used with a free radical initiator [60]. Silane coupling agents also cannot form entanglement or co-crystalline structures with PP or PE because their hydrocarbon chains are too short. Generally speaking, silane coupling agents are not very effective in improving the strength and stiffness of WPCs.

16.5 Increasing the strength and stiffness of WPCs with a compatibilizer

An effective compatibilizer must be able to bond with wood and plastics. Ideally, a compatibilizer molecule should have two domains: a wood-binding domain and a plastic-binding domain capable of bonding with wood and with plastics, respectively. The domain concept is borrowed from protein chemistry. A domain can be referred to as a small functional group such as a succinic anhydride group in maleic-anhydride-grafted PP (MAPP) (Figure 16.1) or as a polymer with a complex structure. Theoretically, any chemicals/polymers that can bond with wood can be used as a wood-binding domain. However, structural requirements for a plastic-binding domain are extremely strict because thermoplastics, especially PP and PE, are difficult to bond. For PP and PE, formations of entanglements and

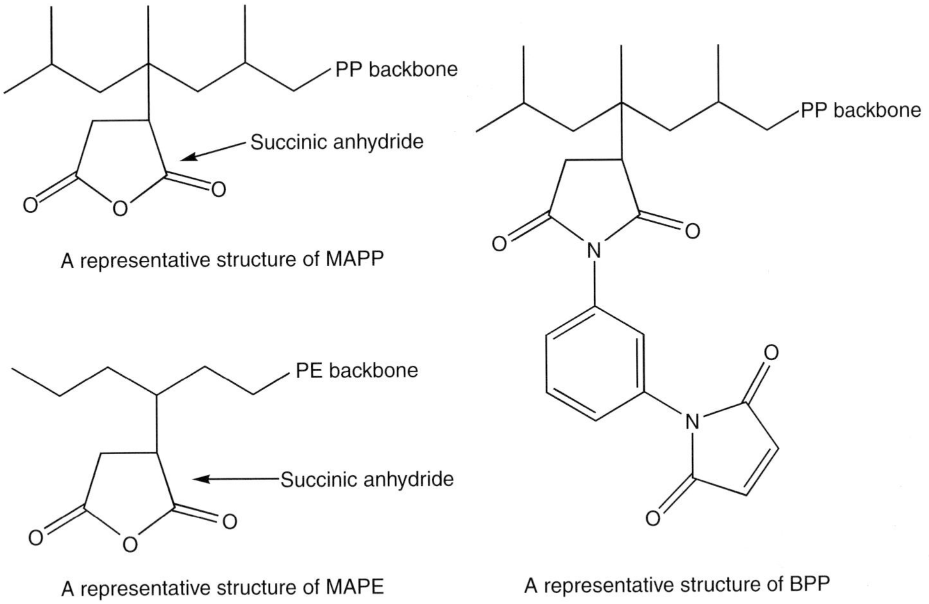

Figure 16.1 Representative structures of maleic-anhydride-grafted polypropylene (MAPP), maleic-anhydride-grafted polyethylene (MAPE), and N,N'-m-phenylene-bismaleimide-grafted polypropylene (BPP).

a co-crystalline structure between PP/PE and a plastic-binding domain are the only effective ways to strongly bond a plastic-binding domain with PP/PE. MAPP is a more effective compatibilizer for wood–PP composites than for wood–PE composites because the PP chains in MAPP cannot form crystalline structures with PE chains. Ideally, a plastic-binding domain should have exactly the same chemical structure as the thermoplastic used in WPCs.

At present, the most extensively studied compatibilizers for WPCs are MAPP and maleic-anhydride-grafted PE (MAPE) (Figure 16.1) [10,14–17,57,58,61,62]. MAPP and MAPE have the same compatibilization mechanisms. The compatibilization mechanisms of MAPP can be summarized as follows: MAPP binds wood through covalent ester linkages and hydrogen bonds. The covalent ester linkages result from the reactions of succinic anhydride ($-C_4H_3O_3$) groups in MAPP with the hydroxyl groups of wood components, and hydrogen bonds are formed between the succinic acid groups (from the hydrolysis of some of the succinic anhydride groups in MAPP or from mono-esterification of the succinic anhydride groups in MAPP with the hydroxyl groups in wood) and the hydroxyl groups in wood. The backbones of MAPP form entanglements and co-crystallization with PP matrix in wood–PP composites. MAPP can significantly improve the strength and stiffness of wood–PE composites through the entanglements between the PP backbones and the PE matrix. However, MAPE is a more effective compatibilizer than MAPP for wood–PE composites because the PE backbones of MAPE can form both entanglements and co-crystallization with the PE matrix. MAPP and MAPE can also significantly improve the hydrophobicity and wettability of wood, thus improving the penetration of PP or PE into wood and the dispersion of wood in the PP or PE matrix. Therefore, hydrophobic interactions between wood and PP/PE and mechanical interlocking also contribute to the improved strength and stiffness of WPCs by MAPP or MAPE.

It was first demonstrated using electron spectroscopy for chemical analysis (ESCA), infrared (IR), contact angle measurement, titrimetric analysis, SEMs and other methods that MAPP bonded with wood through covalent ester linkages [63,64]. ESCA spectra of untreated and MAPP-treated cellulose fibers revealed that the MAPP-treated fibers had a much stronger C—C peak at 285 eV than the untreated fibers. In addition, the O/C and the O/(O—C=O) ratios of the MAPP-treated fibers were lower than those of the untreated fibers by 40% and 50%, respectively [64]. The contact angle of the MAPP-treated fibers was found to be high (up to 140°). These results indicated that MAPP-treated cellulose/wood fibers had hydrophobic properties on the surfaces due to the attachment of the PP tail of MAPP onto the fibers by ester linkages [64].

The covalent ester linkages on the MAPP-treated fibers were confirmed by IR spectroscopy [63,64]. IR spectrum of the neat MAPP showed a characteristic peak of a dicarboxylic acid (from succinic acid) at 1717 cm^{-1} and characteristic peaks of a cyclic anhydride group (from succinic anhydride) at ~1786 and 1862 cm^{-1}, respectively (the peak at 1862 cm^{-1} is usually difficult to detect) [64]. In contrast, the IR spectrum of the MAPP-treated wood fibers showed an absorbance peak of the carbonyl group (—C=O) of ester linkages at ~1740 cm^{-1} [64]. Titrimetric analyses also showed that MAPP-treated fibers had a lower free hydroxyl value, but higher free acid and saponification values than the untreated fibers, indicating the formation of ester linkages between the fibers and MAPP. One succinic anhydride group in MAPP can react with two or one hydroxyl group(s) in wood to form two types of covalent linkages: two ester linkages or one ester linkage with one free carboxylic acid. All experimental results indicated that these two types of covalent linkages both existed [63,64].

However, there was one report showing that no direct evidence of ester linkages between wood fibers and MAPP could be obtained [65]. This could be due to the typically low content of the ester linkages and their embedment in WPCs. The usage of MAPP is typically only a few percent of WPCs, and the amount of succinic anhydride groups in MAPP is typically below 20 wt% of MAPP. Moreover, not all succinic anhydride groups can react with hydroxyl groups in wood to form ester linkages. The amount of the resultant ester linkages is thus very low in WPCs and the linkages are embedded in WPCs. It is not surprising that the ester linkages cannot be detected by IR spectrometry if the WPC samples are not properly treated prior to IR analysis. One way to allow the detection of a low content of the ester linkages is to remove PP and MAPP that are not covalently bonded with wood fibers by extraction of the WPC sample with hot p-xylene using a Soxhlet extractor [63]. This technique has been successfully used to assist in the IR detection of covalent linkages of various compatibilizers onto wood fibers in wood–PP and wood–PE composites [41,42,66].

Under heating, interfusion easily occurs between MAPP and the PP matrix, thus forming polymer chain entanglement [6,30,67]. Dynamic mechanical spectroscopy, polarizing optical microscopy, and other studies show that the MAPP backbones form transcrystalline structures with the PP matrix [22,23,26–33,61,62]. Crystalline PP or PE is stronger and stiffer than amorphous PP or PE. The transcrystallization is expected to further increase the interfacial adhesion.

Some studies suggested that maleated polyolefins hindered the transcrystallization of PP or PE in WPCs [26,33]. MAPP tended to decrease the crystallization temperature of PP in wood–PP composites [26,33]. Moreover, the spherulites in MAPP-treated WPCs were bigger than those in untreated WPCs [26]. It was likely that MAPP around the fibers hindered the molecular chain motion of the PP matrix and retarded its overall crystallization rate [26]. It was also observed by polarizing optical microscopy that MAPP was outside the PP transcrystallization region and was gradually expelled from the fiber surface during crystallization of PP [33].

The compatibilization efficiency of maleated polyolefins is usually dependent upon the acid number (i.e., the amount of succinic anhydride) and the molecular weight of maleated polyolefins [25,61,62,68]. The amount of succinic anhydride is directly correlated with the amount of ester linkages and hydrogen bonds between MAPP/MAPE and wood. The molecular weight is related to the efficiency of forming entanglements and segmental crystalline structures between MAPP/MAPE and the plastic matrix.

Other maleated thermoplastics include a maleic-anhydride-grafted styrene–ethylene–butylene–styrene (MA-SEBS) triblock copolymer [69,70].

N,N'-m-phenylene-bismaleimide-grafted PP (BPP) (see representative structure in Figure 16.1) was demonstrated as an effective compatibilizer for wood–PP composites [71,72]. Vinyltrimethoxysilane-grafted PE, PE–$(CH_2CH_2Si(OCH_3)_3)_n$, was prepared via a free radical grafting process and shown to be an effective compatibilizer for wood–PE composites [73–75].

The compatibilization mechanisms of BPP and PE–$(CH_2CH_2Si(OCH_3)_3)_n$ are similar to those of MAPP and MAPE. The imide group in BPP can react with the hydroxyl groups of wood components to form ester linkages [71,72] with the concurrent formation of an amide. The amide group can form hydrogen bonding with the hydroxyl groups in wood. The imide group in BPP is less reactive than the succinic anhydride group in MAPP. The PP backbone of BPP can form chain entanglements and co-crystalline structures with the PP matrix in wood–PP composites. For PE–$(CH_2CH_2Si(OCH_3)_3)_n$, the hydroxyl groups of wood replace the methoxy groups in PE–$(CH_2CH_2Si(OCH_3)_3)_n$ to release methanol and to form silane–wood covalent linkages, PE–$(CH_2CH_2Si(O-wood)_3)_n$ [73–75]. PE–$(CH_2CH_2Si(OCH_3)_3)_n$ can be readily hydrolyzed to form the corresponding silanols, PE–$(CH_2CH_2Si(OH)_3)_n$, that can form hydrogen bonding with wood [73–75]. The PE backbone of PE–$(CH_2CH_2Si(OCH_3)_3)_n$ can form entanglements and co-crystalline structures with the PE matrix [73–75].

Maleated polyolefins such as MAPP and MAPE are usually synthesized through grafting maleic anhydride onto the backbone of polyolefins via a free radical process. Hence, maleic anhydride groups are randomly attached onto the molecular chain of a polyolefin. Because the grafting reaction is a free radical reaction, it is difficult to control the distribution of the succinic anhydride on the polyolefin backbones. In efforts to overcome the drawbacks of maleated polyolefins, a number of new compatibilizers have been developed and characterized in recent years [19,20,41,42,66,76,77].

A combination of stearic anhydride and polyaminoamide-epichlorohydrin (PAE) resin was superior to PAE, stearic anhydride, or MAPE in terms of enhancing the strength and stiffness of wood–PE composites [18,66]. PAE resin has been widely used as a paper additive to improve the wet strength of paper and is known to be able to strongly bond with wood fibers [18,66]. It was demonstrated by Fourier Transform-Infrared (FT-IR) analysis that stearic anhydride covalently linked with wood via ester linkages [66]. The treatment of wood with the PAE resin facilitated the covalent linkages of stearic anhydride onto wood [66]. The compatibilization mechanism for the combination of stearic anhydride and the PAE resin has been proposed as follows: (1) the PAE resin bonded with wood, thus strengthening and stiffening the wood fibers, which is evidenced by the subtracted FT-IR spectra (e.g., the p-xylene-extracted wood residuals containing PAE as a compatibilizer had an additional relatively strong absorbance peak at 1600–1650 cm^{-1}, representing the carbonyl group (C=O) of the amide linkages in the PAE resin), (2) stearic anhydride formed covalent linkages with the PAE-treated wood fibers, which is manifested by the subtracted FT-IR spectra (e.g., the WPCs treated with 3% stearic anhydride had a relatively strong characteristic peak for the C=O group of esters at ~1740 cm^{-1}), and (3) the hydrocarbon chain of stearic anhydride formed entanglements and segmental crystalline structures with the PE matrix [66].

A combination of PMDI and stearic anhydride or stearic acid was a more effective compatibilizer system than stearic anhydride, stearic acid, PMDI, or MAPE individually [41,42]. Stearic anhydride or stearic acid was covalently bonded with wood. The compatibilization mechanism for the combination of stearic anhydride (or stearic acid) and PMDI was proposed to be similar to that for the combination of stearic anhydride and the PAE resin.

PE–MDI (4,4′-methylenediphenyl diisocyanate) and PE–PMDI block copolymers were prepared and investigated as compatibilizers for wood–PE composites [20]. These block copolymers effectively enhanced the strength and stiffness of the resulting wood–PE composites [20]. These block copolymers are expected to form interfacial bridges between wood and PE because they have a strong wood-binding domain (MDI or PMDI) and an ideal plastic-binding domain (PE). However, they may not be able to strengthen and stiffen wood fibers as effectively as MDI or PMDI alone.

Conclusions

This chapter systematically summarizes various compatibilization mechanisms in WPCs. For WPCs without a compatibilizer, mechanical interlocking between penetrated thermoplastic (PP or PE) and wood fibers is the primary force that holds the thermoplastic and wood fibers together. Transcrystallization of PP or PE on the fiber surface strengthens the mechanical interlocking and facilitates stress transfer from the thermoplastic matrix to the wood fibers.

A compatibilizer can significantly increase the strength and stiffness of the resultant WPCs. All published results indicate that effective compatibilizers all contain two domains, a wood-binding domain able to bond with wood fibers through chemical or physical bonding (e.g., covalent bonding, hydrogen bonding, and van der Waals forces), and a plastic-binding domain capable of bonding with the thermoplastic through transcrystallization and entanglements between the compatibilizer and the thermoplastic matrix. MAPP and MAPE are the most extensively used compabilizers for wood–PP composites and wood–PE composites, respectively. The following compatibilization mechanisms for MAPP and MAPE have been well established: MAPP and MAPE form ester linkages and hydrogen bonds with wood fibers via their MA functional group, and transcrystalline structures and entanglements with the thermoplastic matrices via their PP and PE backbones, thus improving the interfacial adhesion between wood and the thermoplastics. The formation of the ester linkages and the transcrystalline structures has been extensively demonstrated by IR spectrometry, ESCA, SEM, and contact angle measurements.

WPCs can also be stiffened and strengthened with wood adhesives (e.g., PF and PMDI). These synthetic resins cannot effectively improve the interfacial adhesion between the fibers and PP or PE matrix. However, they consolidate and strengthen the fibers through filling and bonding the cracks and gaps in the fibers, thus enhancing the reinforcement effect of the fibers in WPCs. Chemical modification of the fibers with acetic anhydride, stearic acid, or maleic anhydride increases the hydrophobicity of the fibers, thus improving the hydrophobic interactions between the fibers and the thermoplastics, and facilitating the dispersion and penetration of the thermoplastic onto the fibers.

References

1. Youngquist, J.A., Unlikely partners? The marriage of wood and nonwood materials; *Forest Prod. J.* 1995, 45, 25–30.
2. Clemons, C., Wood–plastic composites in the United States – the interfacing of two industries; *Forest Prod. J.* 2002, 52, 10–18.
3. Mapleston, P., It's one hot market for profile extruders; Modern Plastics 2001, June, 49–52.
4. Wolcott, M.P. and Smith, P.M., Opportunities and challenges for wood–plastic composites in structural applications; The Progress in Woodfibre–Plastic Composites Conference, Toronto, Canada, 2004, 1–10.
5. Wolcott, M.P. and Englund, K., A technology review of wood–plastic composites, In Proceedings of the 33rd Washington State University International Particle board/Composites Materials Symposium, 1999, April 13–15, 1999, pp. 103–111.
6. Gauthier, R.; Joly, C.; Coupas, A.C.; Gauthier, H. and Escoubes, M., Interfaces in polyolefin/cellulosic fiber composites: chemical coupling, morphology, correlation with adhesion and aging in moisture; *Polym. Compos.* 1998, 19, 287–300.
7. Eichhorn, S.J.; Baillie, C.A.; Zafeiropoulos, N.; *et al.*, Review: current international research into cellulosic fibres and composites; *J. Mater. Sci.* 2001, 36, 2107–2131.
8. Bledzki, A.K.; Reihmane, S. and Gassan, J., Thermoplastics reinforced with wood fillers: a literature review; *Polym.-Plast. Technol. Eng.* 1998, 37, 451–468.
9. George, J.; Sreekala, M.S. and Thomas, S., A review on interface modification and characterization of natural fiber reinforced plastic composites; *Polym. Eng. Sci.* 2001, 41, 1471–1485.
10. Lu, J.Z.; Wu, Q. and McNabb, H.S., Jr., Chemical coupling in wood fiber and polymer composites: a review of coupling agents and treatments; *Wood Fiber Sci.* 2000, 32, 88–104.
11. Renneckar, S.; Zink-Sharp, A.; Esker, A.; Johnson, R. and Glasser, W., Novel methods for interfacial modification of cellulose-reinforced composites. In: Oksman, K. and Sain, M. (eds); *Cellulose Nanocomposites: Processing, Characterization, and Properties*, 2005; vol. 938, pp. 78–96.
12. Saheb, D.N. and Jog, J.P., Natural fiber polymer composites: a review; *Adv. Polym. Technol.* 1999, 18, 351–363.
13. Raj, R.G.; Kokta, B.V.; Maldas, D. and Daneault, C., Use of wood fibers in thermoplastics. VII. The effect of coupling agents in polyethylene–wood fiber composites; *J. Appl. Polym. Sci.* 1989, 37, 1089–1103.
14. Nitz, H.; Reichert, P.; Romling, H. and Mulhaupt, R., Influence of compatibilizers on the surface hardness, water uptake and the mechanical properties of poly(propylene) wood flour composites prepared by reactive extrusion; *Macromol. Mater. Eng.* 2000, 276/277, 51–58.

15. Bledzki, A.K. and Faruk, O., Wood fibre reinforced polypropylene composites: effect of fibre geometry and coupling agent on physico-mechanical properties; *Appl. Compos. Mat.* 2003, 10, 365–379.
16. Djiporovic, M.; Dingova, E.; Miljkovic, J. and Popov-Pergal, K., the influence of different coupling agents on some properties of polypropylene–wood composites; *Mater. Sci. Forum* 2003, 413, 219–224.
17. Coutinho, F.M.B. and Costa, T.H.S., Performance of polypropylene–wood fiber composites; *Polym. Test.* 1999, 18, 581–587.
18. Geng, Y.; Li, K. and Simonsen, J., Effects of a new compatibilizer system on the flexural properties of wood–polyethylene composites; *J. Appl. Polym. Sci.* 2004, 91, 3667–3672.
19. Zhang, C.; Li, K. and Simonsen, J., Improvement of interfacial adhesion between wood and polypropylene in wood–polypropylene composites; *J. Adhesion Sci. Technol.* 2004, 18, 1603–1612.
20. Zhang, C.; Li, K. and Simonsen, J., Terminally functionalized polyethylenes as compatibilizers for wood–polyethylene composites; *Polym. Eng. Sci.* 2006, 46, 108–113.
21. Goto, T.; Saiki, H. and Onishi, H., Studies on wood gluing XIII: gluability and scanning electron microscopic study of wood–polypropylene bonding; *Wood Sci. Technol.* 1982, 16, 293–303.
22. Kolosick, P.C.; Myers, G.E. and Koutsky, J.A., Polypropylene crystallization on maleated polypropylene-treated wood surface: effects on interfacial adhesion in wood polypropylene composites. In: Rowell, R.M.; Laufenberg, T.L. and Rowell, J.K. (eds); *Materials Interactions Relevant to Recycling of Wood-based Materials*; Materials Research Society, Pittsburgh, PA. 1992; vol. 266, pp. 137–154.
23. Kolosick, P.C.; Myers, G.E. and Koutsky, J.A., Bonding mechanisms between polypropylene and wood: coupling agent and crystalline effects; The 1st Conference on Woodfiber–Plastic Composites, Madison, WI, 1993, pp. 15–19.
24. Smith, M.J.; Dai, H. and Ramani, K., Wood-thermoplastic adhesive interface – method of characterization and results; *Int. J. Adhes. Adhesives* 2002, 22, 197–204.
25. Lu, J.Z.; Negulescu, I.I. and Wu, Q., Maleated wood-fiber/high-density-polyethylene composites: Coupling mechanisms and interfacial characterization; *Comp. Interfaces* 2005, 12, 125–140.
26. Hristov, V. and Vasileva, S., Dynamic mechanical and thermal properties of modified poly(propylene) wood fiber composites; *Macromol. Mater. Eng.* 2003, 288, 798–806.
27. Zafeiropoulos, N.E.; Baillie, C.A. and Matthews, F.L., A study of transcrystallinity and its effect on the interface in flax fibre reinforced composite materials; *Composites Part A* 2001, 32, 525–543.
28. Wolcott, M.P.; Yin, S. and Rials, T.G., Using dynamic mechanical spectroscopy to monitor the crystallization of PP/MAPP blends in the presence of wood; *Comp. Interfaces* 2000, 7, 3–12.
29. Sanadi, A.R. and Caulfield, D.F., Transcrystalline interphases in natural fiber–PP composites: effect of coupling agent; *Comp. Interfaces* 2000, 7, 31–43.
30. Gatenholm, P.; Felix, J.; Klason, C. and Kubàt, J., Cellulose–polymer composites with improved properties. In: Salamone, J.C. and Riffle, J.S. (eds); *Contemporary Topics in Polymer Science: Advances in New Materials*; Plenum Press, New York. 1992; vol. 7, pp. 75–82.
31. Lenes, M. and Gregersen, Ø.W., Effect of surface chemistry and topography of sulphite fibers on the transcrystallinity of polypropylene; *Cellulose* 2006, 13, 345–355.
32. Amash, A. and Zugenmaier, P., Study on cellulose and xylan filled polypropylene composites; *Polym. Bull.* 1998, 40, 251–258.
33. Son, S.-J.; Lee, Y.-M. and Im, S.-S., Transcrystalline morphology and mechanical properties in polypropylene composites containing cellulose treated with sodium hydroxide and cellulose; *J. Mater. Sci.* 2000, 35, 5767–5778.
34. Lu, J.Z.; Doyle, T.W. and Li, K., Preparation and characterization of wood–(nylon 12) composites; *J. Appl. Polym. Sci.* 2007, 103, 270–276.
35. Sellers, T., Jr., Wood adhesive innovations and applications in North America; *Forest Prod. J.* 2001, 51, 12–22.

36. Johns, W.E., Isocyanate as wood binders: a review; *J. Adhesion* 1982, 15, 59–67.
37. Chtourou, H.; Riedl, B. and Ait-Kadi, A., Reinforcement of recycled polyolefins with wood fibers; *J. Reinf. Plast. Comp.* 1992, 11, 372–394.
38. Simonsen, J. and Rials, T.G., Enhancing the interfacial bond strength of lignocellulosic fiber dispersed in synthetic polymer matrices. In: Rowell, R.M.; Laufenberg, T.L. and Rowell, J.K. (eds); *Materials Interactions Relevant to Recycling of Wood-base Materials*; Materials Research Society, Pittsburgh, PA. 1992; vol. 266, pp. 105–111.
39. Raj, R.G.; Kokta, B.V.; Maldas, D. and Daneault, C., Use of wood fibers in thermoplastic composites. VI. Isocyanate as a bonding agent for polyethylene–wood fiber composites; *Polym. Comp.* 1988, 9, 404–411.
40. Raj, R.; Kokta, B.V.; Grouleau, G. and Daneault, C., The influence of coupling agents on mechanical properties of composites containing cellulosic fillers; *Polym. Plast. Technol. Eng.* 1990, 29, 339–353.
41. Geng, Y.; Li, K. and Simonsen, J., A combination of poly(diphenylmethane diisocyanate) and stearic anhydride as a novel compatibilizer for wood–polyethylene composites; *J. Adhesion Sci. Technol.* 2005, 19, 987–1001.
42. Geng, Y.; Li, K. and Simonsen, J., A commercially viable compatibilizer system for wood–polyethylene composites; *J. Adhesion Sci. Technol.* 2005, 19, 1363–1373.
43. Andersons, J.; Sparnins, E. and Joffe, R., Stiffness and strength of flax fiber/polymer matrix composites; *Polym. Comp.* 2006, 27, 221–229.
44. Shaler, S.M., Mechanics of the interface in discontinuous wood fiber composites; The 1st Woodfiber–Plastic Composites Conference, Madison, WI, 1993, pp. 9–14.
45. Mamunya, Y.; Zanoaga, M.; Myshak, V.; Tanasa, F.; Lebedev, E.; Grigoras, C. and Semynog, V., Structure and properties of polymer–wood composites based on an aliphatic copolyamide and secondary polyethylenes; *J. Appl. Polym. Sci.* 2006, 101, 1710–1700.
46. Neagu, R.C.; Gamstedt, E.K. and Berthold, F., Stiffness contribution of various wood fibers to composite materials; *J. Comp. Mater.* 2006, 40, 663–699.
47. Joseph, K.; Thomas, S. and Pavithran, C., Effect of chemical treatment on the tensile properties of short sisal fiber-reinforced polyethylene composites; *Polymer* 1996, 37, 5139–5149.
48. Khalil, H.P.S.A.; Ismail, H.; Rozman, H.D. and Ahmad, M.N., The effect of acetylation on interfacial shear strength between plant fibers and various matrices; *Eur. Polym. J.* 2001, 37, 1037–1045.
49. Rowell, R.M., Acetylation of wood: journey from analytical technique to commercial reality; *Forest Prod. J.* 2006, 56, 4–12.
50. Tserki, V.; Zafeiropoulos, N.; Simon, F.; and Panayiotou, C., A study of the effect of acetylation and propionylation surface treatments on natural fibers; *Composites Part A* 2005, 36, 1110–1118.
51. Kalaprasad, G.; Francis, B.; Thomas, S.; et al., Effect of fibre length and chemical modifications on the tensile properties of intimately mixed short sisal/glass hybrid fibre reinforced low density polyethylene composites; *Polym. Int.* 2004, 53, 1624–1638.
52. Plueddemann, E.P., *Silane Coupling Agents*, 2nd edition, Plenum Press, New York, 2004.
53. Bataille, P.; Richard, L. and Sapieha, S., Effects of cellulose fibers in polypropylene composites; *Polym. Comp.* 1989, 10, 103–108.
54. Beshay, A.D.; Kokta, B.V. and Daneault, C., Use of wood fibers in the thermoplastic composites. II: Polyethylene; *Polym. Comp.* 1985, 6, 250–257.
55. Felix, J. and Gatenholm, P., Controlled interactions in cellulose–polymer composites. I. Effect on mechanical properties; *Polym. Comp.* 1993, 14, 449–457.
56. Maldas, D.; Kokta, B.V.; Raj, R. and Daneault, C., Improvement of the mechanical properties of sawdust wood fiber–polystyrene composites by chemical treatment; *Polymer* 1988, 29, 1255–1265.

57. Maldas, D.; Kokta, B.V. and Daneault, C., Influence of coupling agents and treatments on the mechanical properties of cellulose–fiber polystyrene composites; *J. Appl. Polym. Sci.* 1989, 37, 751–775.
58. Raj, R.; Kokta, B.V.; Maldas, D. and Daneault, C., Use of wood fibers in thermoplastics. VII. The effect of coupling agents in polyethylene–wood fiber composites; *J. Appl. Polym. Sci.* 1989, 37, 1089–1103.
59. Xanthos, M., Processing conditions and coupling agents effects in polypropylene/wood flour composites; *Plast. Rubber Process. Appl.* 1983, 3, 223–228.
60. Kuan, C.-F.; Kuan, H.-C.; Ma, C.-C.M. and Huang, C.-M., Mechanical, thermal and morphological properties of water-crosslinked wood flour reinforced linear low-density polyethylene composites; *Composites Part A* 2006, 37, 1696–1707.
61. Olsen, D.J., Effectiveness of maleated polypropylenes as coupling agents for wood flour/polypropylene composites; The 49th Annual Technical Conference – Society of Plastics Engineers, Montreal, Canada, 1991, 1886–1891.
62. Snijder, M. and Bos, H., Reinforcement of polypropylene by annual plant fibers: optimisation of the coupling agent efficiency; *Compos. Interfaces* 2000, 7, 69–79.
63. Kishi, H.; Yoshioka, M.; Yamanoi, A. and Shiraishi, N., Composites of wood and polypropylenes I; *Mokuzai Gakkaishi* 1988, 34, 133–139.
64. Felix, J.M. and Gatenholm, P., The nature of adhesion in composites of modified cellulose fibers and polypropylene; *J. Appl. Polym. Sci.* 1991, 42, 609–620.
65. Kazayawoko, M.; Balatinecz, J.J.; Woodhams, R.T. and Law, S., Effect of ester linkages on the mechanical properties of wood fiber–polypropylene composites; *J. Reinf. Plast. Comp.* 1997, 16, 1383–1406.
66. Geng, Y.; Li, K. and Simonsen, J., Further investigation of polyaminoamide-epichlorohydrin/stearic anhydride compatibilizer system for wood–polyethylene composites; *J. Appl. Polym. Sci.* 2006, 99, 712–718.
67. Felix, J.M. and Gatenholm, P., Formation of entanglements at brushlike interfaces in cellulose–polymer composites; *J. Appl. Polym. Sci.* 1993, 50, 699–708.
68. Lu, J.Z.; Wu, Q. and Negulescu, I.I., Wood fiber/high density polyethylene composites: coupling agent performance; *J. Appl. Polym. Sci.* 2005, 96, 93–102.
69. Oksman, K.; Lindberg, H. and Holmgren, A., The nature and location of SEBS-MA compatibilizer in polyethylene–wood flour composites; *J. Appl. Polym. Sci.* 1998, 69, 201–209.
70. Oksman, K. and Lindberg, H., Influence of thermoplastic elastomers on adhesion in polyethylene–wood flour composites; *J. Appl. Polym. Sci.* 1998, 68, 1845–1855.
71. Sain, M.M. and Kokta, B.V., Polyolefin–wood filler composites. I. Performance of *m*-phenylene bismaleimide-modified wood fiber in polypropylene composites; *J. Appl. Polym. Sci.* 1994, 54, 1545–1559.
72. Sain, M.M.; Kokta, B.V. and Maldas, D., Effect of reactive additives on the performance of cellulose fiber-filled polypropylene composites; *J. Adhesion Sci. Technol.* 1993, 7, 49–61.
73. Bengtsson, M.; Gatenholm, P. and Oksman, K., The effect of crosslinking on the properties of polyethylene/wood flour composites; *Compos. Sci. Technol.* 2005, 65, 1468–1479.
74. Bengtsson, M. and Oksman, K., The use of silane technology in crosslinking polyethylene/wood flour composites; *Compos. Part A* 2006, 37, 752–765.
75. Bengtsson, M. and Oksman, K., Silane crosslinked wood plastic composites: processing and properties; *Compos. Sci. Technol.* 2006, 66, 2177–2186.
76. Geng, Y.; Li, K. and Simonsen, J., Effect of a new compatibilizer system on the flexural properties of wood–polyethylene composites; *J. Appl. Polym. Sci.* 2004, 91, 3667–3672.
77. Saputra, H.A.; Simonsen, J. and Li, K., Effects of compatibilizers on the flexural properties of grass straw–polyethylene composites; *J. Biobased Mat. Bioenergy* 2007, 1(1), 137–142.

Chapter 17
X-Ray Powder Diffraction Analyses of Kraft Lignin-Based Thermoplastic Polymer Blends

Yi-ru Chen and Simo Sarkanen

Abstract

Plasticization of methylated kraft lignin-based polymeric materials by miscible low glass transition temperature (T_g) aliphatic polyesters affects the x-ray powder diffraction pattern in a counterintuitive manner. The d-spacing at the point of maximum intensity in the characteristic amorphous halo is displaced towards smaller values. A reduction in molecular weight of the kraft lignin components alone engenders a similar trend. These effects can be understood by describing the diffuse scattering from methylated kraft lignin preparations as a sum of two overlapping Lorentzian component peaks centered at $d = \sim 4.1$ Å and $d = \sim 5.1$ Å. Comparison with the crystal structures of selected di- and trilignols indicates that these two peaks encompass d-spacings between aromatic rings which are, respectively, in roughly parallel-but-offset and edge-on orientations relative to one another. As would be expected, the latter peak is broadened much more than the former when a miscible low-T_g polymer is blended into a methylated kraft lignin-based material.

17.1 Introduction

17.1.1 X-Ray diffraction patterns from lignin preparations

The first wide-angle x-ray powder diffraction patterns of any consequence obtained from simple lignin derivatives were published in 1974 [1]. The "milled wood lignin" and "dioxane lignin" were prepared by extracting ball-milled cedar wood meal with aqueous 96% dioxane and refluxing dioxane containing 0.4% concentrated hydrochloric acid, respectively; a portion of the dioxane lignin was also methylated. The diffuse haloes observed were characteristic of amorphous polymers [1]; no attempt was made to distinguish which atoms or groups might be separated by the various inter- and intramolecular distances that could give rise to the broad overlapping peaks of diffracted intensity vs 2θ (where θ is the angle of diffraction). Nevertheless, the variations with temperature of the d-spacings at the discernible outer peak maxima were used to estimate transition temperatures for the three lignin preparations that were close to the glass transition temperatures (T_gs) determined by differential scanning calorimetry [1].

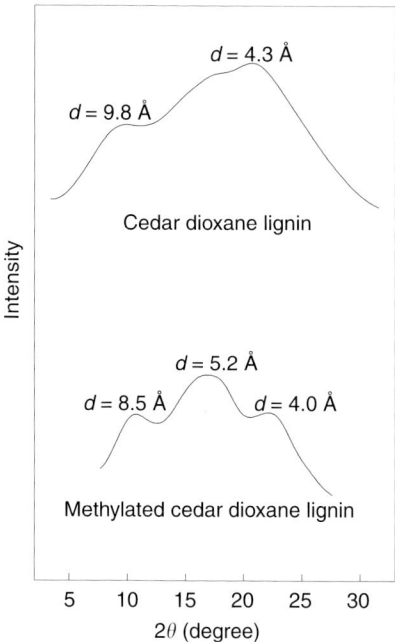

Figure 17.1 Wide-angle x-ray diffraction patterns from cedar dioxane lignin and its methylated derivative, with *d*-spacings given at the broad peak maxima. Data from Reference 2.

Similar analyses of very similar lignin samples published 8 years later yielded virtually identical results [2] from data, however, that were not exactly the same (Figure 17.1). The authors commented upon the partial resolution of the outer 4.3 Å halo of the dioxane lignin into two diffuse peaks centered at 4.0 and 5.2 Å as a result of methylating the lignin derivative. They suggested that hydroxyl group methylation might expand some of the intermolecular distances in the lignin domains, but they offered no interpretation about which separations might have contributed to the distinguishable haloes in the x-ray diffraction patterns of the amorphous lignin preparations [2]. This had to wait for another 23 years [3].

Actually, the enhancement in resolution of the diffuse 4.0 and 5.2 Å peaks, engendered in the x-ray diffraction pattern by methylating the dioxane lignin, has important implications. It suggests that hydrogen-bonding and the other intermolecular forces between the individual lignin components tend to favor significantly different arrangements of the interacting species. Furthermore, it indicates that the two kinds of intermolecular interactions have comparable effects as far as their relative strengths are concerned. This has, indeed, been substantiated by MP2/6-31G(d) molecular orbital calculations [4].

17.1.2 *Amorphous polymeric materials*

Interestingly, the year 1974, which witnessed the publication of the first x-ray diffraction patterns from simple lignin derivatives [1], also heralded the organization of a symposium on the Physical Structure of the Amorphous State at the 168th National Meeting of the

American Chemical Society [5]. There, P.J. Flory declared that progress towards an understanding of the physical properties of polymeric materials requires knowledge of the spatial conformations and intermolecular packing of polymer chains in the amorphous state. He cautioned that ostensible difficulties in packing flexible macromolecules to bulk density would not be alleviated by partial ordering [6]. Nevertheless, even though there is no long-range three-dimensional order in amorphous polymeric materials, dense packing of macromolecular chain segments entails some correlation between their relative positions and orientations [7]. The resulting anisotropy in short-range order is embodied in the autocorrelation function of the electron density distribution in physical space and the corresponding scattered intensity distributions in reciprocal space. Information about molecular structure is more directly accessible in the radial distribution function which can be derived from the electron density distribution, but greater clarity in regard to intermolecular arrangements may be gleaned from the scattered intensity distributions [7].

Empirically, the diffuse scattering from atactic polystyrene is adequately described as a sum of two broad Lorentzian peaks [8] centered at d-spacings of 8.9 Å ($2\theta = 10°$) and 4.6 Å ($2\theta = 19°$). According to a radial distribution function analysis, 87% of the amorphous outer peak centered at 4.6 Å results both from intramolecular distances between benzene ring and chain atoms, and also from intermolecular separations between the aromatic rings [9]. Above T_g, the 8.9 Å d-spacing at the maximum of the inner halo was reported to increase with temperature [10] more than that of the outer one; the lower-angle peak was therefore attributed primarily to intermolecular separation distances [8]. Despite some inconsistencies in regard to the temperature sensitivity of the 8.9 Å polystyrene peak [1], the foregoing interpretation largely conformed with the views of the authors responsible for the first x-ray diffraction studies of simple lignin derivatives [2].

Generally speaking, the diffuse scattered intensity in the x-ray diffraction pattern from an amorphous polymeric material is quite difficult to interpret, but in some instances the amorphous domains in semicrystalline polymers may be more amenable to explicit examination. This is illustrated for poly(ethylene terephthalate) (PET) in Figure 17.2, where the respective x-ray diffraction patterns from a highly crystalline oligomeric powder and a typical annealed polymer film are analytically resolved into crystalline and amorphous peaks. At least two Lorentzian functions (centered at $2\theta = 17.5°$, $d = 1/0.20 = 5.0$ Å and $2\theta = 23.5°$, $d = 1/0.26 = 3.8$ Å, respectively) are necessary to fit the diffuse scattering from the amorphous PET domains in both samples [11].

With oriented amorphous PET fibers, the intensities of the two amorphous haloes, as they vary with azimuthal angle, reach their maximum values at the equator [11]. Therefore, the diffuse 3.8 and 5.0 Å peaks have been related to distances along the two directions transverse to the polymer chain axis. Indeed, since the 3.8 Å peak is very close to the intense 100 crystalline PET reflection [3.4 Å in Figure 17.2(a)], it was assigned to intermolecular spacings normal to aligned aromatic rings that are rotationally correlated on adjacent chains. Moreover, as the 5.0 Å halo is centered at the same 2θ value (namely, 17.5°) as the 010 reflection [Figure 17.2(a)], intermolecular separations in the planes of the aromatic rings have been thought to contribute significantly to this amorphous peak [11]. There is an important corollary to describing the diffuse scattering from semicrystalline PET in terms of spacings along two crystallographic directions; such an interpretation implies that incipient order could exist even in the amorphous domains. The question arises as to whether a comparable situation may be present in simple lignin derivatives.

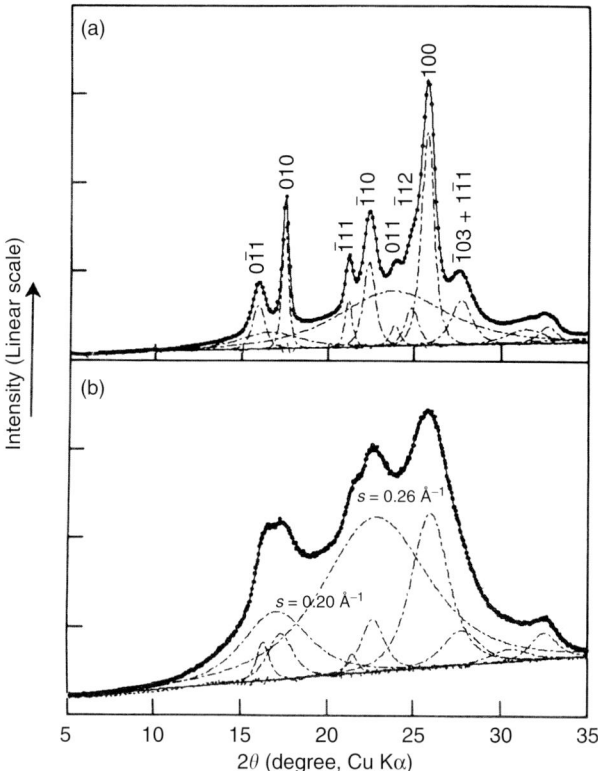

Figure 17.2 X-Ray diffraction diagrams for (a) highly crystalline low-molecular-weight PET powder, and (b) typical annealed polymeric PET film. Solid line over data points indicates sum of resolved crystalline and amorphous peaks represented by dashed lines. From Murthy et al. [11], Figure 1, copyright 1991 with permission from the American Chemical Society.

17.2 Kraft lignin-based polymeric materials

17.2.1 X-Ray powder diffraction patterns from paucidisperse kraft lignin fractions

Decisions about which lignin samples are worthy of characterization by wide-angle x-ray diffraction are based on a variety of considerations. One concern might depend on how closely the lignin preparation reflects the native biopolymer from which it is derived. Another criterion might involve the actual availability of a particular lignin derivative with promising mechanical properties for producing useful polymeric materials. This second qualification would point to the by-product lignins from the kraft pulping process as the lignin derivatives of choice. Kraft lignins are interesting in that they have been formed under quite severe chemical conditions (e.g., 2 h exposure to aqueous 45 g L^{-1} of NaOH and 12 g L^{-1} of Na$_2$S at 170°C, as far as the delignification of softwood chips is concerned), during which many of the substructures along the macromolecular lignin chains may be cleaved or otherwise modified, and new covalent bonds can be formed between separate molecular components [12].

Nevertheless, kraft lignins do not possess configurations that have been irretrievably scrambled in relation to the native biopolymer. This is evident from the selectivity inherent in the effects of the powerful noncovalent interactions that prevail between the discrete kraft lignin components. Despite the extent to which these individual molecular species might have been modified, they can still associate in their thousands with one another in remarkably specific ways to form well-defined supramacromolecular complexes possessing hydrodynamic volumes above that of 20-million-molecular-weight polystyrene [13,14]. The resulting molecular weight distributions of acetylated and methylated kraft lignin derivatives are clearly multimodal in form, not unimodal as they would be if the huge associated entities were randomly assembled.

Intermolecular association between lignin components is governed not by hydrogen bonding but rather by strong noncovalent interactions between aromatic rings that can be equivalent in their effects to at least two or three hydrogen bonds [4]. Yet, there are no indications that crystalline domains exist in kraft lignin preparations (nor other kinds of lignin derivatives). This assertion is based upon the kinds of experimental observations that are exemplified in Figure 17.3, which depicts x-ray powder diffraction patterns from a series

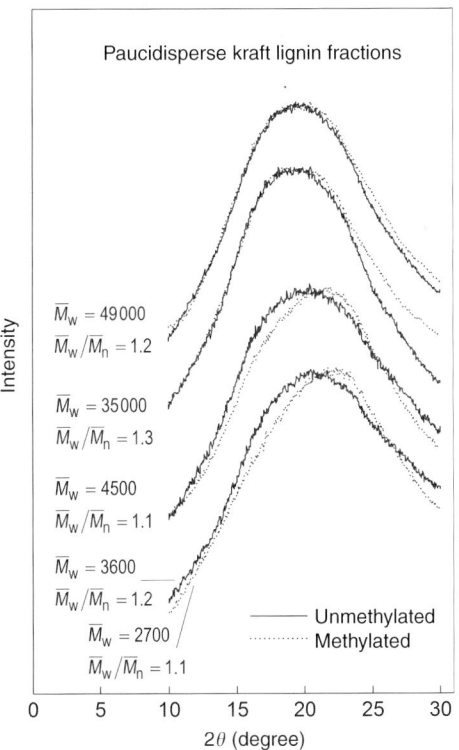

Figure 17.3 Wide-angle x-ray diffraction patterns from underivatized and methylated paucidisperse kraft lignin fractions. Data from Reference 3 and Y. Li and S. Sarkanen, unpublished results.

of paucidisperse kraft lignin fractions with successively decreasing molecular weights. The maximum in the intensity of the amorphous halo is displaced toward smaller d-spacings (larger scattering angles) as the molecular weight is reduced, and the effect is enhanced when the kraft lignin components have been methylated. Moreover, above the diffuse peak maxima, the difference in scattered intensity between the methylated and unmethylated (underivatized) higher-molecular-weight kraft lignin fractions increases with scattering angle (Figure 17.3). Thus, in contrast to what previous claims would contend [2], hydroxyl group methylation does not seem to expand any of the inter- or intramolecular separation distances in softwood kraft lignin preparations.

17.2.2 Alkylated kraft lignin-based thermoplastic blends with low-T_g polymers

In the US, kraft lignins are produced in huge quantities, amounting to ~45 million metric tons annually according to year 2000 estimates [15]. Thus, in terms of availability, these lignin derivatives would be suitable starting materials for manufacturing lignin-based plastics and other value-added by-products of converting lignocellulose into paper.

In this connection, the year 1997 was a watershed in the field of lignin-based polymeric materials. A full-length paper was published about thermoplastics with promising mechanical properties (tensile strength extending to 26 MPa and Young's modulus reaching 1.5 GPa) that were composed of 85% (w/w) kraft lignin [16]. Moreover, the same year saw a preliminary account [17] on ethylated and methylated 100% kraft lignin-based polymeric materials also with fairly encouraging tensile strengths (~15 MPa) and moduli (~1.3 GPa). Both reports were unprecedented, and significant improvements in mechanical behavior were soon to be realized by alkylating the kraft lignin preparations more completely [18]. These new thermoplastics had broken the barrier to becoming genuinely "lignin-based", in that, during the two previous decades, it had typically not been possible to exceed 30–45% limits to the incorporation of simple lignin derivatives in cohesive polymeric materials [19].

The first formulations to accommodate very high levels of any lignin derivative in promising polymeric materials were achieved with homogeneous blends of 85% (w/w) underivatized kraft lignin containing 12.6% poly(vinyl acetate), 1.6% diethylene glycol dibenzoate, and 0.8% indene [16]. Hereby, it had become feasible to fabricate a polymeric material composed predominantly of a common lignin derivative, but the utility of these new thermoplastics was limited by their tendency to dissolve partially in aqueous alkaline solutions. Their primary value resided in what they revealed about the possible states of the individual molecular components in lignin-based polymeric materials. A particular oddity in the effect of composition on their mechanical properties clearly pointed to a functional distinction between individual lignin components that are and are not incorporated into supramacromolecular complexes [14,16].

Soon it became apparent that ethylated and methylated kraft lignin derivatives could be used on their own to produce polymeric materials with tensile properties similar to those of polystyrene [18]. In contrast to kraft lignins that have not been alkylated, these simple kraft lignin derivatives can be plasticized by miscible low-T_g polymers such as aliphatic main-chain polyesters or poly(ethylene glycol) [3,20]. The plasticizing effects of poly(trimethylene glutarate) and poly(ethylene glycol), as they are blended in progressively

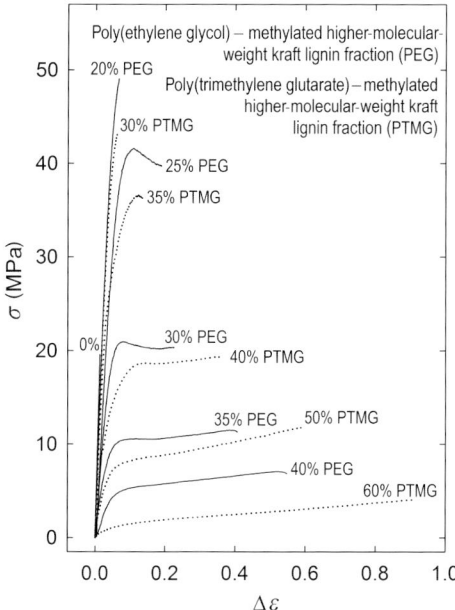

Figure 17.4 Progressive plasticization of methylated polydisperse higher-molecular-weight kraft lignin fraction in blends with poly(ethylene glycol) and poly(trimethylene glutarate). Data from References 3 and 20.

increasing quantities with a methylated polydisperse higher-molecular-weight kraft lignin fraction ($\bar{M}_w = 2.3 \times 10^4$, $\bar{M}_w/\bar{M}_n = 2.6$), are compared in Figure 17.4. The proportion (35%) of poly(trimethylene glutarate) required to reach the threshold of plasticization is significantly larger than that (<25%) of poly(ethylene glycol). The difference arises from the strengths of the intermolecular interactions between the individual alkylated kraft lignin components and the low-T_g polymer chains. When the interactions are stronger, the associated supramacromolecular kraft lignin complexes [13,14] in the polymeric material are (counterproductively) dismantled to a greater extent than when these intermolecular forces are weaker. As a result, more low-T_g polymer is required in the composition of the blend at the plasticization threshold.

The relative magnitudes of the intermolecular interactions between the alkylated kraft lignin and low-T_g polymer are reflected in the concavity of the T_g–composition curve for (homogeneous) blends of the two macromolecular components [3,14]. Thus, as shown in Figure 17.5, the T_g of the poly(ethylene glycol) blends increases very little as the content of the methylated kraft lignin fraction varies from 0% to 65% (w/w). On the other hand, the T_g of the corresponding blends with poly(trimethylene glutarate) increases by more than 60°C over the same composition range. Analyses of the data using the Gordon–Taylor formalism [21] provide direct confirmation of the qualitative impressions created by the concavity of the T_g–composition curves [3,20]. The other miscible low-T_g polymers tested fell within the range between these two bounds. For example, ~30% poly(1,4-butylene adipate) was required to exceed the plasticization threshold characteristic of its blends with the methylated higher-molecular-weight kraft lignin fraction [3].

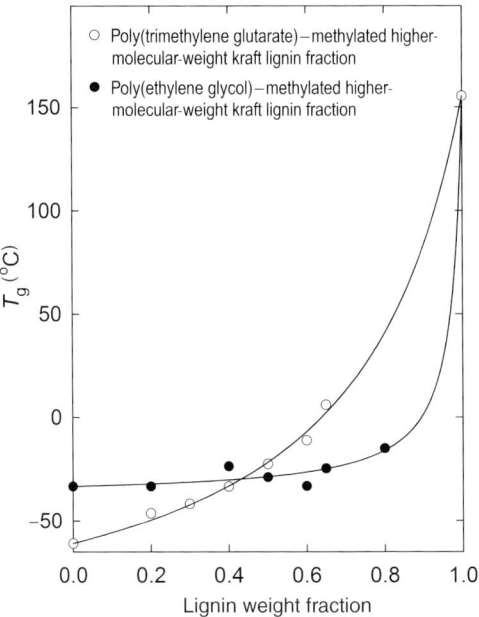

Figure 17.5 Variation of T_g with composition for blends of methylated polydisperse higher-molecular-weight kraft lignin fraction with poly(trimethylene glutarate) and poly(ethylene glycol). Data from References 3 and 20.

These miscible low-T_g aliphatic polyesters are often semicrystalline; the point at which the characteristic crystalline peaks appear in the x-ray diffraction patterns from their blends with the methylated kraft lignin fraction occurs at a polyester content of ∼55% and >70% for poly(butylene adipate) (Figure 17.6) and poly(trimethylene glutarate), respectively [20]. Thus, the methylated kraft lignin components have a greater tendency to disrupt the crystalline domains of poly(trimethylene glutarate) than those of poly(butylene adipate) because they interact with the former more strongly. Similar conclusions are obtained from analyses of the T_g–composition curves (e.g., Figure 17.5) for the corresponding blends [3,20] using the classical Gordon–Taylor formalism [21]. It is worth pointing out that the composition of the alkylated kraft lignin preparation itself affects how much plastic deformation the thermoplastic blend can withstand before fracture. Beyond the plasticization threshold, the ultimate strains borne by the alkylated kraft lignin-based materials are extended by the presence of low-molecular-weight lignin components, which hereby synergistically enhance the action of the miscible low-T_g polymers as plasticizers in these blends [3,19].

17.2.3 X-Ray powder diffraction patterns from methylated kraft lignin-based plastics

A prominent effect is observed in the x-ray powder diffraction pattern when miscible low-T_g aliphatic polyesters are blended in progressively greater quantities with methylated kraft lignin-based polymeric materials [3]. The d-spacing at the point of maximum intensity in

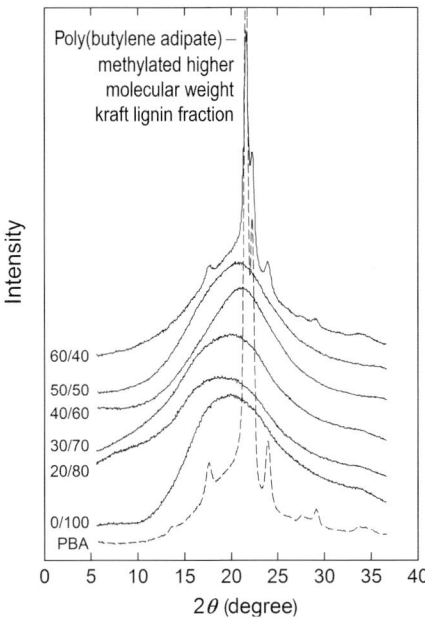

Figure 17.6 Wide-angle x-ray diffraction patterns from blends of methylated polydisperse higher-molecular-weight kraft lignin fraction with poly(1,4-butylene adipate) (PBA). From Li *et al.* [20], Figure 10, copyright 2002 with permission from the American Chemical Society.

the characteristic amorphous halo is displaced towards smaller values, that is, the corresponding scattering angle increases (Figure 17.7). This counterintuitive trend is encountered whether the intermolecular interactions between the methylated kraft lignin components and the low-T_g plasticizing polymer are strong or weak. Any attempt to interpret such a result must contend with the fact that the polyester itself will contribute appreciably, if less strongly than the aromatic lignin components, to the diffuse scattered intensity. Thus, it should be easier to understand the cause(s) of the effect by searching for a context where an analogous trend occurs without the introduction of any foreign blend components. Indeed, a similar result is simply obtained with a series of paucidisperse methylated kraft lignin fractions of successively decreasing molecular weight without introducing any (miscible) low-T_g polymer (Figure 17.3). It is perhaps fitting that a comparable trend is also observed in this case: alkylated low-molecular-weight kraft lignin components synergistically enhance the amount of plastic deformation that an alkylated kraft lignin-based polymeric material beyond the plasticization threshold can bear, prior to fracture (*vide supra*; [3]).

The amorphous scattering (Figure 17.3) from the paucidisperse methylated kraft lignin fractions in the range $2\theta = 5 - 35°$ (Cu Kα) can be described in terms of sums of two Lorentzian component peaks [3] centered at $2\theta = 16.9°$, $d = 5.25$ Å and $2\theta = 22.1 \pm 0.6°$, $d = 4.0 \pm 0.1$ Å, respectively (Figure 17.8). These two peak maxima occur at the same d-spacings as observed much earlier [2] with a methylated cedar dioxane lignin preparation (Figure 17.1), although the paucidisperse methylated kraft lignin fractions showed no sign of the (less intense) inner halo at $d = 8.5$ Å.

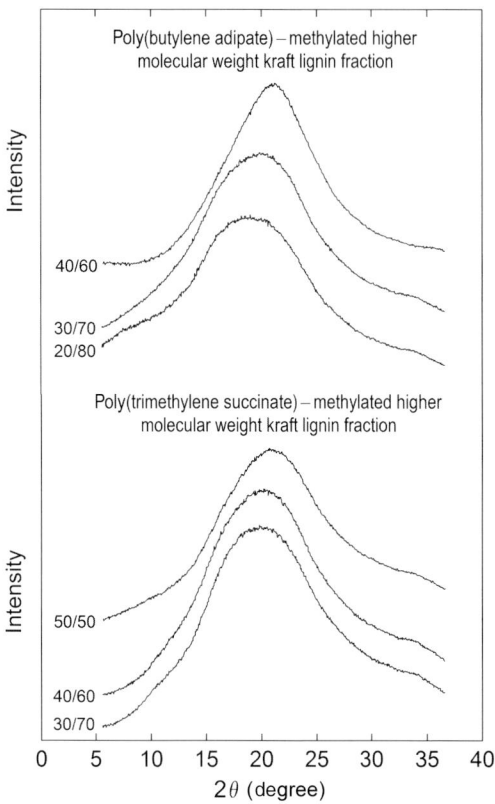

Figure 17.7 Wide-angle x-ray diffraction diagrams from blends of methylated polydisperse higher-molecular-weight kraft lignin fraction with poly(butylene adipate) and poly(trimethylene succinate). From Li et al. [3], Figure 3, copyright 2005 with permission from the American Chemical Society.

Unlike partially crystalline polymers such as PET (Figure 17.2), lignins and lignin derivatives do not exhibit any detectable degree of crystallinity in the solid state. Thus, the origins of the Lorentzian component peaks with maxima at $2\theta = \sim17°$ and $22°$ that make up the amorphous haloes from methylated kraft lignin fractions (Figure 17.8) and methylated cedar dioxane lignin (Figure 17.1) cannot be deduced by comparison with the x-ray diffraction patterns engendered by crystalline domains in these kinds of materials (cf. Figure 17.2). The predominant contributions to the diffuse scattered intensity are expected to arise from the distributions of spacings between the aromatic rings [3]. The relative positions and orientations of these moieties may be distributed about the various geometries that are evident in the crystal structures of dimeric and trimeric lignin model compounds. At one extreme, nonbonded orbital interactions could be important when the perpendicular separation distances between cofacial aromatic rings are small (≤ 3.4 Å) [22]. The crystal structures of divanillyltetrahydrofuran [23] and a 5-5′,8-O-4″-linked trilignol derivative [24] may embody circumstances where such interactions could occur. In contrast, when nonbonded orbital interaction energies cannot overcome the effects of π-electron repulsion, face-to-face

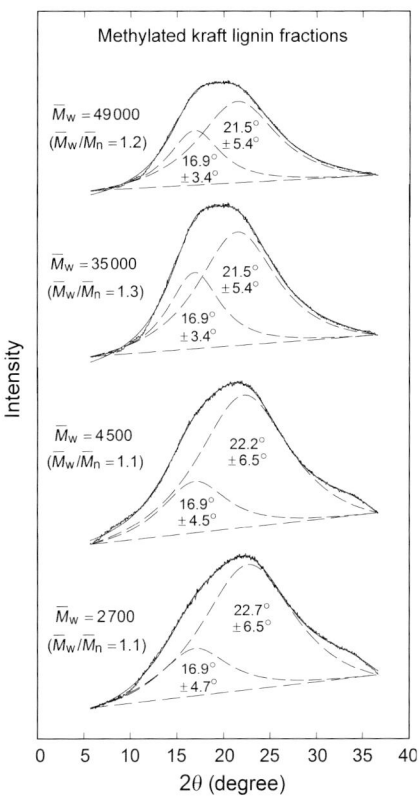

Figure 17.8 Fits of two principal Lorentzian functions to diffuse scattered intensity in the x-ray diffraction patterns from methylated paucidisperse kraft lignin fractions. From Li *et al.* [3], Figure 4, copyright 2005 with permission from the American Chemical Society.

juxtaposition of the aromatic rings will be replaced by a continuum of possible geometries varying between edge-on and parallel offset orientations [25]. The latter has been observed at ∼4.6 Å intermolecular separations in the crystal structure of a phenolic 8-5′-linked dilignol [26], while an arrangement approximating the former has been found at ∼5.3 Å intermolecular separation distances in the crystal structure of a nonphenolic 8-O-4′-linked dilignol [27].

Accordingly, the *d*-spacings between roughly parallel aromatic rings in di- and trimeric lignin model compounds vary quite broadly (±0.6 Å) about 4.0 Å, while those between aromatic rings that are approximately edge-on with respect to one another are centered around 5.3 Å. These values are essentially identical not only to those of the two most prominent diffuse peaks in the x-ray diffraction pattern from methylated cedar dioxane lignin (Figure 17.1) but also to the maxima of the two Lorentzian component peaks describing the amorphous haloes emanating from paucidisperse methylated kraft lignin fractions (Figure 17.8). Of course, the packing of polymer chains in the amorphous state has requirements that differ from that of small molecules in a crystal lattice. Yet, as with polystyrene, liquid styrene exhibits an amorphous halo centered at 4.6 Å [8], although the

8.9-Å subsidiary maximum in the diffuse scattering obtained from the polymer is absent. Such a remarkable correlation between the results from a monomer and the corresponding polymer suggest that the x-ray diffraction patterns of di- and trilignols could provide important insights for interpreting those of polymeric lignin derivatives. Thus, the (component) peak maxima at 4.0 Å and 5.2 Å in the amorphous haloes from the methylated dioxane and kraft lignin preparations (Figures 17.1 and 17.8) may provisionally be assigned, at least in part, to parallel and edge-on arrangements of the aromatic rings, respectively [3].

Interestingly, the ∼5.2 Å d-spacing at the maximum of the Lorentzian component peak that includes the scattering from edge-on arrangements of the aromatic rings does not vary with the molecular weight of the methylated paucidisperse kraft lignin fractions; however, the peak itself is ∼30% narrower for the higher-molecular-weight species (Figure 17.8). This is presumably caused by the fact that most of the higher-molecular-weight kraft lignin components are incorporated into supramacromolecular associated complexes where the range of permissible intermolecular registration is more restricted. The other Lorentzian component peak with a maximum corresponding to a d-spacing of 4.1 Å encompasses the diffuse scattering from more-or-less parallel aromatic rings in the methylated kraft lignin complexes (Figure 17.8). This peak is broadened somewhat (by ∼20%) while undergoing a shift to slightly smaller d-spacings (3.9–4.0 Å at its maximum intensity) for the lower-molecular-weight kraft lignin components; the latter presumably allow more freedom in regard to relative positioning of the aromatic rings.

The methylated polydisperse higher-molecular-weight kraft lignin fraction, upon which the materials in Figures 17.4–17.6 are based, contains associated complexes in a more complete state of assembly than the methylated paucidisperse fractions in Figure 17.8. The reason is that thousands of individual components with a fairly broad range of molecular weight are required to form the huge supramacromolecular assemblies in a well-defined manner. Most of the discrete species that were not incorporated into associated complexes were removed during the preparation of the higher-molecular-weight polydisperse kraft lignin fraction [3]. The effect of this is evident in the x-ray diffraction pattern from the methylated fraction (Figure 17.9) in the Lorentzian component peak ($2\theta = 17.4 \pm 2.4°$) that embraces the scattering from aromatic rings oriented edge-on with respect to one another. The component peak width is ∼30% narrower than that from the methylated paucidisperse high-molecular-weight kraft lignin fractions, and the d-spacing at the corresponding peak maximum is displaced to a slightly lower value (5.1 Å). However, the other Lorentzian component peak ($2\theta = 21.5 \pm 5.0°$), which includes the diffuse scattering from parallel aromatic rings (Figure 17.9), remains centered at the same d-spacing (4.1 Å) as found with the methylated paucidisperse high-molecular-weight kraft lignin fractions, even though its width is ∼10% narrower [3].

Miscible aliphatic polyesters have a substantial impact upon the x-ray diffraction patterns from alkylated kraft lignin preparations as they form homogeneous blends with these materials (Figure 17.7). For example, blending of 30–40% poly(trimethylene succinate) with the methylated polydisperse higher-molecular-weight kraft lignin fraction broadens the two Lorentzian component peaks centered at d-spacings of 4.1 and 5.1 Å in the amorphous halo by ∼10% and 60%, respectively (Figure 17.9). The effects of adding the low-T_g polymer are not strictly systematic; the integration of the reflections to which the poly(trimethylene succinate) chains contribute in the x-ray diffraction pattern of the blend occurs in a manner that cannot be easily predicted. Nevertheless, the distribution of separation distances

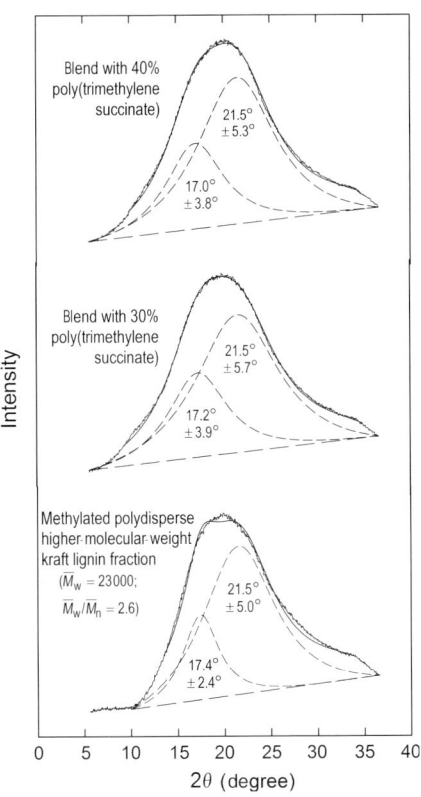

Figure 17.9 Fits of two principal Lorentzian functions to diffuse scattered intensity in the x-ray diffraction patterns from the methylated polydisperse higher-molecular-weight kraft lignin fraction and its blends with poly(trimethylene succinate). From Li et al. [3], Figure 5, copyright 2005 with permission from the American Chemical Society.

between edge-on aromatic rings is clearly broadened much more than that between parallel aromatic rings in the methylated kraft lignin components [3]. Such a difference would be expected if the interactions between the former (which are separated to a greater extent from one another) are weaker than those between the latter. Thus, the effects on the x-ray diffraction pattern of adding a low-T_g polymer (Figure 17.9) are qualitatively similar to those of reducing the molecular weight of the kraft lignin components (Figure 17.8). This is in harmony with the finding that the lower-molecular-weight components in alkylated kraft lignin-based polymeric materials enhance the impact of miscible aliphatic polyesters upon them as plasticizers [3].

Concluding remarks

In 2006, the analysis of x-ray powder diffraction patterns from lignin-based polymeric materials was still in its infancy. The inherent difficulty lies in the fact that no crystalline domains have ever been detected in simple lignin derivatives. Consequently, progress in gleaning useful insights from the amorphous haloes to which lignin preparations give rise can only be made through extensive comparisons of systematic trends in the diffuse scattering from many series of lignin derivatives, fractions, and blends. Evaluation of the x-ray powder diffraction patterns from methylated kraft lignin-based materials in terms of two overlapping Lorentzian component peaks represents a promising beginning. One peak encompasses scattering from cofacial aromatic rings that are roughly parallel but offset from one another; the other arises, in part, from aromatic rings that are more-or-less edge-on with respect to each other. Accordingly, a basis has been proposed for understanding how these Lorentzian peaks are affected by the molecular weight of the kraft lignin components and plasticization with miscible low-T_g polymers. It will be interesting to see how well this framework will accommodate further trends from different families of lignin-based polymeric materials in the future.

Acknowledgments

The work carried out at the University of Minnesota was supported by the U.S. Department of Agriculture (Grant No. 98-35103-6730), the U.S. Environmental Protection Agency through the National Center for Clean Industrial and Treatment Technologies, the Vincent Johnson Lignin Research Fund, and the Minnesota Agricultural Experiment Station (Project No. 43-68, maintained by Hatch Funds).

References

1. Hatakeyama, T.; Hatakeyama, H., X-ray Studies on Lignin and Its Model Polymers; *Rep. Prog. Polym. Phys. Jpn.* 1974, 17, 711–712.
2. Hatakeyama, T.; Hatakeyama, H., Temperature Dependence of X-ray Diffractograms of Amorphous Lignins and Polystyrenes; *Polymer* 1982, 23, 475–477.
3. Li, Y.; Sarkanen, S., Miscible Blends of Kraft Lignin Derivatives with Low-T_g Polymers; *Macromolecules* 2005, 38, 2296–2306.
4. Sarkanen, S.; Chen, Y-r., Towards a Mechanism for Macromolecular Lignin Replication; *59th Appita Proc.* 2005, 2, 407–414.
5. Allen, G.; Petrie, S.E.B. (eds), *Physical Structure of the Amorphous State*; Marcel Dekker; New York, *J. Macromol. Sci. Phys.* 1976, B12, 1-301
6. Flory, P.J., Theoretical Predictions on the Configurations of Polymer Chains in the Amorphous State; *J. Macromol. Sci. Phys.* 1976, B12, 1–11.
7. Ruland, W., The Structure of Amorphous Solids; *Pure Appl. Chem.* 1969, 18, 489–515.
8. Murthy, N.S.; Minor, H.; Bednarczyk, C.; Krimm, S., Structure of the Amorphous Phase in Oriented Polymers; *Macromolecules* 1993, 26, 1712–1721.
9. Wecker, S.M.; Davidson, T.; Cohen, J.B., A Structural Study of Glassy Polystyrene; *J. Mater. Sci.* 1972, 7, 1249–1259.
10. Kilian, H.-G.; Boueke, K., Röntgenographische Strukturanalyse von amorphem Polystyrol, VI; *J. Polym. Sci.* 1962, 58, 311–333.
11. Murthy, N.S.; Correale, S.T.; Minor, H., Structure of the Amorphous Phase in Crystallizable Polymers: Poly(ethylene terephthalate); *Macromolecules* 1991, 24, 1185–1189.

12. Gierer, J., Chemical Aspects of Kraft Pulping; *Wood Sci. Technol.* 1980, 14, 241–266.
13. Dutta, S.; Sarkanen, S., A New Emphasis in Strategies for Developing Lignin-Based Plastics. In: Caulfield, D.F.; Passaretti, J.D.; Sobczynski, S.F. (eds); *Materials Interactions Relevant to the Pulp, Paper and Wood Industries*; Materials Research Society; Pittsburgh, *MRS Symp. Proc.* 1990, 197, 31–39.
14. Li, Y.; Sarkanen, S., Biodegradable Kraft Lignin-based Thermoplastics. In: Chiellini, E.; Solaro, R. (eds); *Biodegradable Polymers and Plastics*; Kluwer Academic/Plenum Publishers; New York, 2003; pp. 121–139.
15. *Biobased Industrial Products: Research and Commercialization Priorities*; Commission on Life Sciences, National Research Council; Washington, D.C., 2000; p. 82.
16. Li, Y.; Mlynár, J.; Sarkanen, S., The First 85% Kraft-Lignin-Based Thermoplastics; *J. Polym. Sci. B Polym. Phys.* 1997, 35, 1899–1910.
17. Li, Y.; Sarkanen, S., The First Alkylated 95–100% Kraft Lignin Based Plastics. In: *Proc. 9th Intern. Symp. Wood Pulping Chem.*; CPPA; Montreal, PQ, Canada, 1997; 63, pp. 1–6.
18. Li, Y.; Sarkanen, S., Thermoplastics with Very High Lignin Contents. In: Glasser, W.G.; Northey, R.A.; Schultz, T.P. (eds); *Lignin – Historical, Biological, and Materials Perspectives*; American Chemical Society; Washington, D.C., *ACS Symp. Ser.* 2000, 742, 351–366.
19. Chen, Y.-r.; Sarkanen, S., From the Macromolecular Behavior of Lignin Components to the Mechanical Properties of Lignin-Based Plastics; *Cellul. Chem. Technol.* 2006, 40, 149–163 and references therein.
20. Li, Y.; Sarkanen, S., Alkylated Kraft Lignin-Based Thermoplastic Blends with Aliphatic Polyesters; *Macromolecules* 2002, 35, 9707–9715.
21. Gordon, M.; Taylor, J.S., Ideal Copolymers and the Second-Order Transitions of Synthetic Rubbers. I. Noncrystalline Copolymers; *J. Appl. Chem.* 1952, 2, 493–500.
22. Morokuma, K., Why Do Molecules Interact? The Origin of Electron Donor–Acceptor Complexes, Hydrogen Bonding and Proton Affinity; *Acc. Chem. Res.* 1977, 10, 294–300.
23. Lundquist, K.; Stomberg, R., On the Occurrence of Structural Elements of the Lignan Type (β–β Structures) in Lignins: The Crystal Structures of (+)-Pinoresinol and (±)-*trans*-3,4-Divanillyltetrahydrofuran; *Holzforschung* 1988, 42, 375–384.
24. Roblin, J.-P.; Duran, H.; Duran, E.; Gorrichon, L.; Donnadieu, B., X-Ray Structure of a Trimeric 5,5′-biaryl/*erythro*-β-O-4-ether Lignin Model: Evidence for Through-Space Weak Interactions; *Chem.-Eur. J.* 2000, 6, 1229–1235.
25. Hunter, C.A.; Sanders, J.K.M., The Nature of π–π Interactions; *J. Am. Chem. Soc.* 1990, 112, 5525–5534.
26. Stomberg, R.; Lundquist, K., The Crystal Structure of *trans*-2,3-Dihydro-2-(4-hydroxy-3-methoxyphenyl)-3-hydroxymethyl-7-methoxybenzofuran; *Acta Chem. Scand. B* 1987, 41, 304–309.
27. Stomberg, R.; Lundquist, K., Stereochemical Assignment of the Diastereomers of 1-(3,4-Dimethoxyphenyl)-2-(2-methoxyphenoxy)-1,3-propanediol from X-Ray Analysis; *Acta Chem. Scand. A* 1986, 40, 705–710.

Chapter 18
DSC and DMA of ECs and EC–MC Blends

Shizuka Horita, Tatsuko Hatakeyama, and Hyoe Hatakeyama

Abstract

Ethylcelluloses (ECs) crossed-linked with hexamethylene diurethane and blends of EC and methylcellulose (MC) have been characterized by differential scanning calorimetry (DSC) and dynamic mechanical analysis (DMA). The main chain motion of the cross-linked EC, detectable as a change in glass transition temperature (T_g) by DSC and in α-dispersion by DMA, is affected by the molecular mass and degree of substitution (DS) of the EC and the extent of cross-linking in a complex manner. The effect of molecular mass disappears at high extents of cross-linking. At the same molecular mass, DS significantly affects the molecular motion, particularly at higher extents of cross-linking. In EC–MC blends, the MC component acts as a hard segment in the rubbery state where molecular motion of the EC chains is restricted.

18.1 Introduction

Ethylcellulose (EC) prepared from industrial processes is categorized into two groups, one with a degree of ethylation in a range from 0.8 to 1.3 per one glucose unit and the other with a degree of ethylation from 1.8 to 2.6. EC of the former group is water soluble and used as a thickener and adhesive. The latter group of EC is insoluble in water, soluble in many organic solvents, and used in paints and plastic sheets [1]. Like other cellulose derivatives [2,3], EC is known to form liquid crystal in concentrated acetic acid or dichloroacetic acid solution [4].

EC with a high degree of ethylation has film-forming ability, which is an important factor in developing new applications for cellulose derivatives. Although EC has many desirable characteristics such as film formation, transparency, and appropriate mechanical properties, its water-retention capability is limited. To prepare stable films having a certain degree of hydrophilicity from EC, we have considered the following approaches: (1) the introduction of soft segments as a cross-linker and (2) molecular blends with other cellulose derivatives such as methylcellulose (MC). The chemical structures of EC and MC are shown in Figure 18.1.

Previously, we have investigated the cross-linking of water-soluble polysaccharides via a urethane unit in aqueous media and the physical properties of the chemically crossed-linked

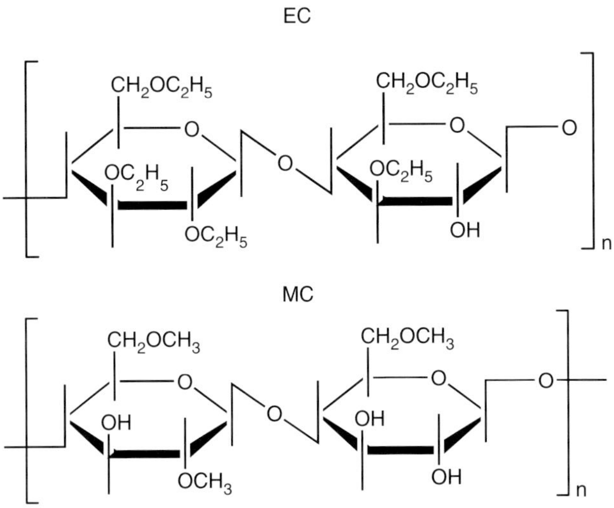

Figure 18.1 Chemical structures of ethylcellulose (EC) and methylcellulose (MC).

hydrogels [5,6]. Chemically cross-linked hydrogels derived from polysaccharides are stable in a wide range of temperatures, when compared with physical gels.

Blends of different types of polysaccharides have been widely investigated to obtain new biocompatible materials having unique features [7–13]. Aqueous solution of MC is known to form thermoreversible hydrogels by heating via hydrophobic molecular interaction [14–16]. In this chapter, we describe our viscoelastic studies and thermal measurements of the molecular motions of EC chemically cross-linked via a diurethane unit in an organic solvent and of EC–MC blends. We chose MC as the other component in the EC blends, because its chemical structure is similar to that of EC but it is more hydrophilic.

18.2 Experimental

18.2.1 Preparation of diurethane cross-linked EC films with various NCO/OH ratios

EC powders with different molecular mass and degree of substitution (DS) were obtained from General Science Co. (Tokyo, Japan). They were EC with DS of 2.43 and viscosities of 4, 10, 22, and 100 cp, respectively, and EC with viscosity of 100 cp and DS of 2.29. The diurethane cross-linked EC films with various NCO/OH ratios were prepared as follows: (1) EC powder was dissolved in chloroform (concentration = 6.25%), maintained at 50°C and stirred until a homogeneous solution was obtained; (2) a small amount of dibutyltin dilaulate was added to the above solution as a catalyst; (3) hexamethylene diisocyanate (HDI) (Asahi Kasei Co., Tokyo, Japan) was dissolved in chloroform and stirred at room temperature (∼25°C) until a homogeneous solution was obtained; (4) EC and HDI solutions were mixed and stirred at room temperature at a known ratio to give NCO/OH molar ratio of 0 (without any diurethane cross-linking), 0.2, 0.4, 0.6, 0.8, and 1.0, respectively; (5) after stirring, the mixed solution was maintained at ~25°C for several minutes, and the solution was extended on a glass plate for one night; and (6) the obtained films were stored in a desiccator. The schematic chemical structure of the diurethane crossed-linked EC from the reaction of the hydroxyl (OH) groups in EC and the diisocyanate group in HDI is shown in Figure 18.2.

18.2.2   Preparation of films from EC–MC blends

The MC powder with a known molecular mass (viscosity = 100 cp) and DS (= 1.80) was obtained from Nacalai Tesque, Inc. (Kyoto, Japan). Films from EC–MC blends were prepared as follows: (1) MC powder was dissolved in chloroform (concentration = 6.25%) maintained at ~25°C and stirred until a homogeneous solution was obtained; (2) homogeneous EC solution in chloroform (concentration = 6.25%) was prepared and maintained at ~25°C; (3) MC and EC solutions were mixed under stirring at various blending ratios. MC content of EC–MC blends was defined as follow:

$$\text{MC Content} = \frac{m_{MC}}{(m_{MC} + m_{EC})} \times 100(\%) \tag{18.1}$$

where m_{MC} and m_{EC} are masses of MC and EC, respectively; and (4) the solutions were extended on a glass plate for one night and the films prepared were stored in a desiccator.

18.2.3   Measurements

18.2.3.1   Viscoelastic measurements

A Seiko dynamic mechanical analyzer (DMA) DMS210 equipped with a tension mode probe was used. Temperature was in a range from −120°C to 200°C. Heating rate was 1.0°C min^{-1} and sample size was 20 mm (length) × 10 mm (width) × ~0.1 mm (thickness). Frequency was varied from 0.5 to 10 Hz. Measurement was carried out in air [17,18]. The dynamic modulus (real part of the complex modulus) (E'), the dynamic loss modulus (imaginary part of the complex modulus) (E''), and the loss tangent (tan δ) were simultaneously obtained as functions of temperature and frequency. The tan δ is defined as follows [19]:

$$\tan \delta = \frac{E''}{E'} \tag{18.2}$$

Activation energy was calculated using peak temperatures of tan δ curve at each frequency.

18.2.3.2   Differential scanning calorimetry (DSC)

A Seiko Instrument DSC 220C equipped with a cooling apparatus was used. Scanning rate was 10°C min^{-1} and nitrogen flow rate was 30 mL min^{-1}. The temperature was varied as follows: (1) the sample was cooled from 40°C to −150°C at cooling rate of 10°C min^{-1}; (2) the sample was maintained at −150°C for 5 min and heated at 10°C min^{-1} to 180°C;

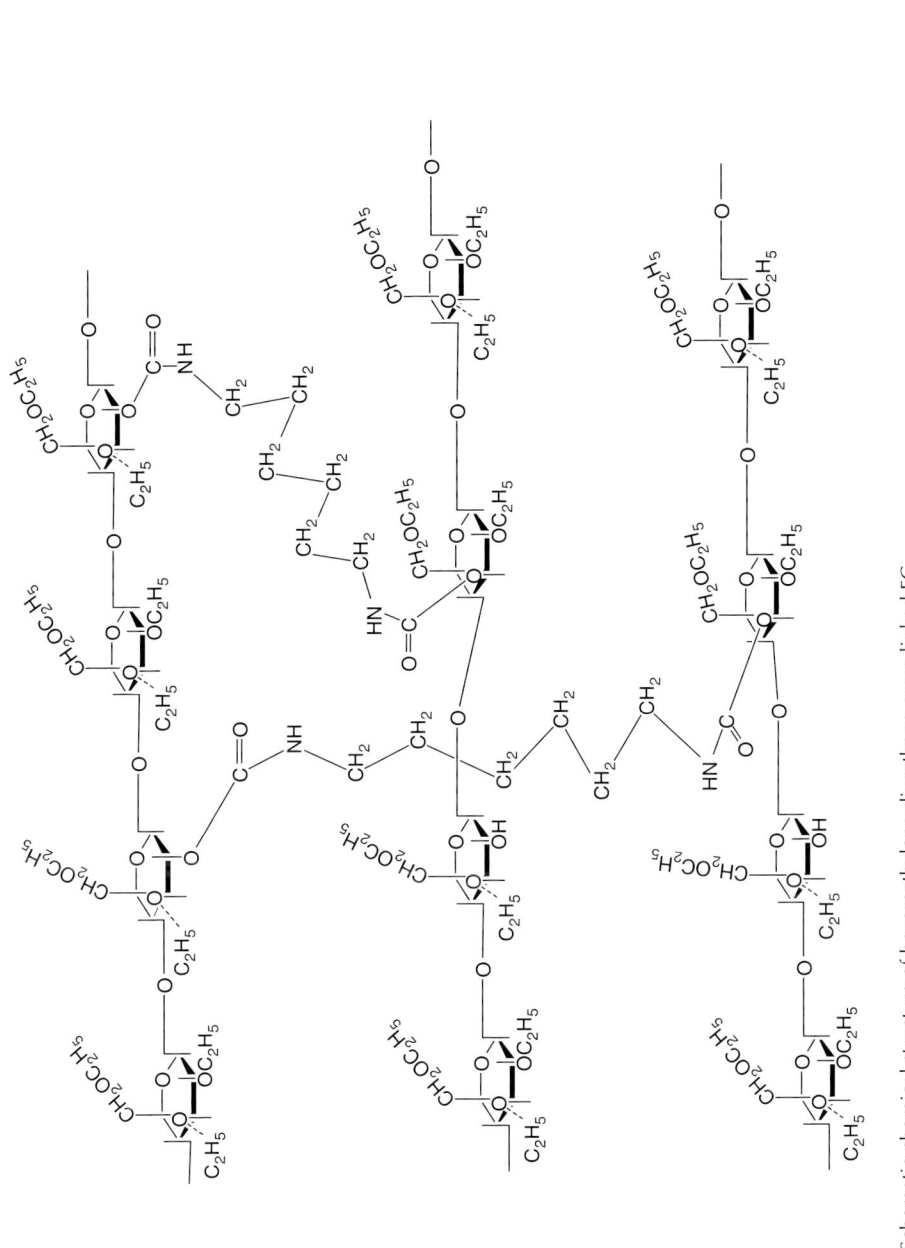

Figure 18.2 Schematic chemical structure of hexamethylene diurethane cross-linked EC.

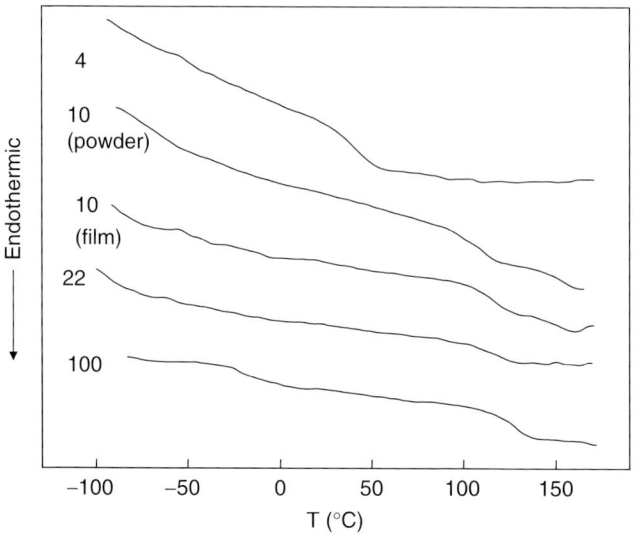

Figure 18.3 DSC heating curves of four EC films and one EC powder from EC with DS of 2.43 and various viscosities; numerals in the figure are values of viscosity (cp).

and (3) the process (1) with cooling from 180°C (instead of 40°C) to −150°C and the process (2) were repeated. The data obtained by the heating run from (3) was mainly used for the analysis [18].

18.3 Effect of inter-molecular hexamethylene diurethane linkage on molecular motion of EC

Figure 18.3 shows the DSC heating curves of four control (with NCO/OH ratio = 0, i.e. without any cross-linking) EC films with various molecular mass (viscosities) and a EC powder with viscosity of 10 cp. Heat capacity gap (ΔC_p) due to glass transition was observed for all the samples. It is well known that molecular motion of polysaccharide is restricted via inter-molecular hydrogen bonding and glass transition is hardly detected by DSC [2,18]. The molecular motion of EC main chain is observable due to the effect of long side-chains; however, the value of ΔC_p is not large (from 0.02 to 0.06 J g^{-1} K^{-1}) compared with that of synthetic polymers (0.2–0.4 J g^{-1} K^{-1}) [20]. Although several small heat capacity gaps were observed in the DSC curves of the samples with viscosity of 10 cp (Figure 18.3), glass transition temperature (T_g) value was determined from the largest gap taking into consideration of the sensitivity of DSC. The values of T_g for the EC films and powder with DS = 2.43 defined and obtained from Figure 18.3 were then plotted against the viscosities (Figure 18.4). T_g value of the sample increased rapidly from viscosity of 4 to 10 cp but gradually from viscosity of 10–100 cp. The effect of DS (DS 2.43 vs 2.29) for the EC films with a viscosity of 100 cp on T_g was not statistically significant (data not shown).

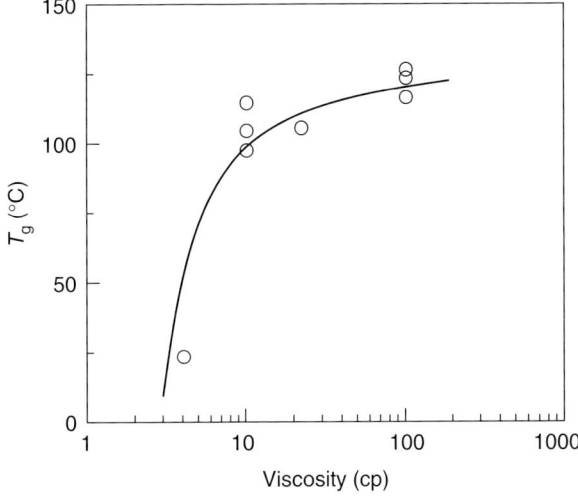

Figure 18.4 Relationship between T_g of the EC films and the viscosity of the EC (DS = 2.43).

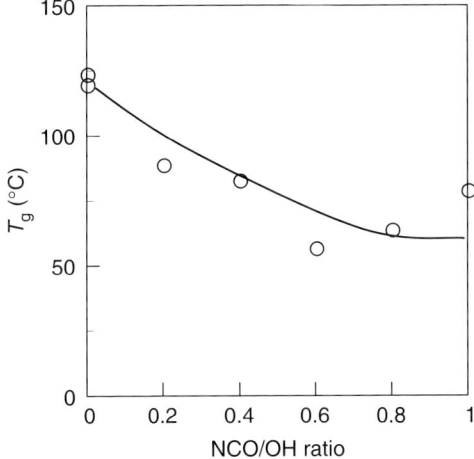

Figure 18.5 Relationship between T_g and NCO/OH ratio of the EC (viscosity = 100 cp and DS = 2.43) films; NCO/OH = 0 is for the control EC film without any cross-linking.

Hexamethylene diurethane cross-linked EC films were analyzed by DSC and T_g values were obtained. The heat capacity gap at T_g of these chemically cross-linked EC films became smaller and T_g was scarcely defined for samples with a high NCO/OH ratio. It is possible that molecular enhancement diminishes via cross-linking. Figure 18.5 shows the relationship between T_g and NCO/OH ratio of the cross-linked EC (viscosity = 100 cp and DS = 2.43) films (NCO/OH = 0 for the control EC film without cross-linking). T_g decreases with

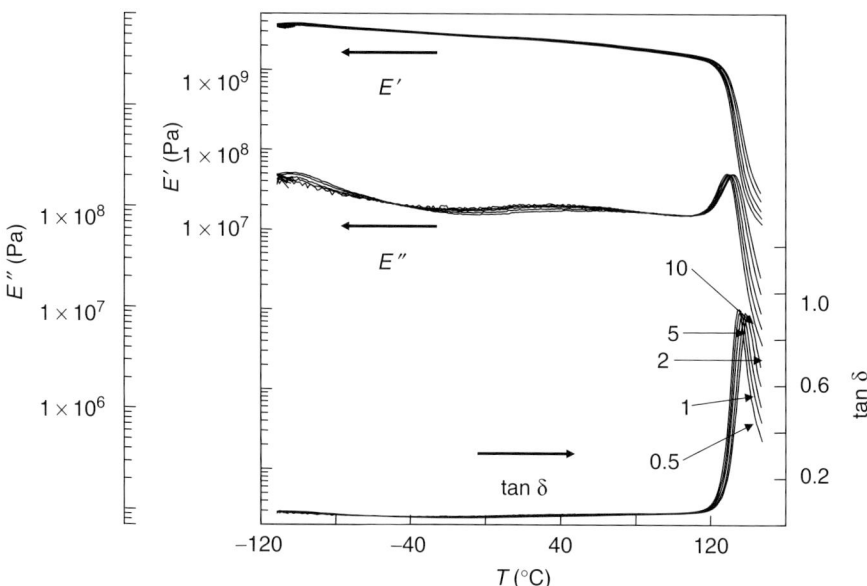

Figure 18.6 E', E'', and tan δ curves of the EC (viscosity = 100 cp and DS = 2.43) films measured at various frequencies; heating rate = 1.0°C min^{-1}, numerals in the figure are values of the frequency.

increasing NCO/OH ratio. This result seems to be unusual because in general cross-linking restricts free rotation of molecular chains, which would increase T_g. However, as described in the latter section, the T_g data obtained by DSC agree well with those obtained by DMA, and are attributed to the fact that the hexamethylene diurethane cross-linker introduced to the EC films works as a soft segment.

Figure 18.6 shows the dynamic modulus (E') (also called elastic or storage modulus, and associated with the stiffness of the material), dynamic loss modulus (E'') (damping term determining the dissipation of energy as heat) and the loss tangent (tan δ) (dissipation factor, tan δ = E''/E') of the control EC (viscosity = 100 cp, DS = 2.43) films measured at various frequencies. E' slightly decreased at ~20°C and steeply decreased at ~130°C. E'' shows a broad peak from 0°C to 90°C (β-dispersion) and a distinct and sharp peak at ~130°C (α-dispersion). The peak of tan δ due to α-dispersion was observed at ~135°C. The β-dispersion was not distinct in the tan δ curve. When temperature/viscosity was increased/decreased, tan δ peak was significantly depressed, suggesting that the effect of cross-linking was significant for short molecular chains.

Figure 18.7 shows the E' and tan δ curves of the control EC (viscosity = 100 cp, DS = 2.43) and the diurethane cross-linked EC films with various NCO/OH ratios measured at 10 Hz. As clearly seen from the figure, the temperature at which E' starts to rapidly decrease shifts to the low temperature side with increasing NCO/OH ratio (i.e., increasing cross-linking). In addition, E' at the rubbery state can clearly be observed for the cross-linked samples, but not for the sample without cross-linking. The peak temperature of tan δ also shifts to the low temperature side while the peak height and width of tan δ markedly decreases and broadens, respectively, with increasing NCO/OH ratio.

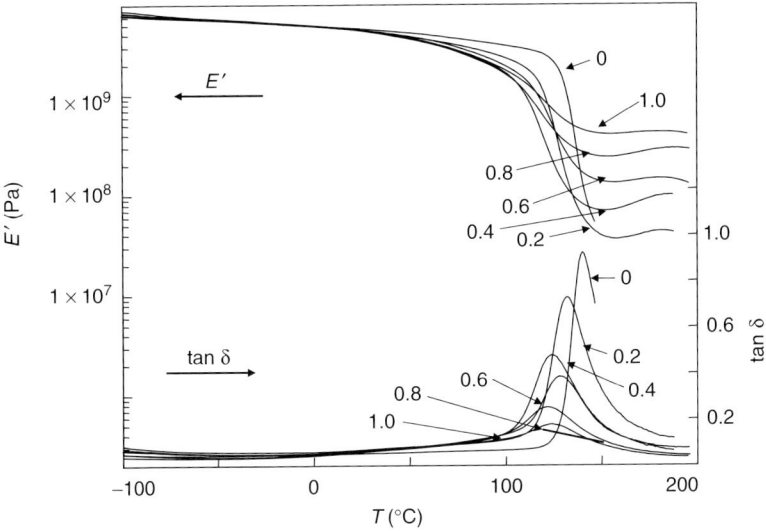

Figure 18.7 E' and tan δ curves of the control EC film (viscosity = 100 cp and DS = 2.43, NCO/OH = 0) and the cross-linked EC films with NCO/OH ratios of 0.2–1.0; heating rate = $1.0°C\,min^{-1}$, frequency = 10 Hz, numerals in the figure are the NCO/OH ratios.

Viscoelastic measurements were carried out for two more series of control and diurethane cross-linked EC films from EC powder with different molecular mass (viscosity = 10 cp, DS = 2.43) and DS (viscosity = 100 cp, DS = 2.29). Figure 18.8 shows the relationships between the peak temperature of tan δ of the EC films with different molecular mass (curve I vs curve II) or DS (curve I vs curve III) and the NCO/OH ratio. At the same DS, the tan δ peak temperature for the EC films with viscosity of 100 cp (curve I) and 10 cp (curve II) initially decreases and increases, respectively, with increasing NCO/OH ratio, but then reaches and remains at a similar and constant value. Such a disappearance of the effect of molecular mass at higher NCO/OH ratios was likely due to the increasing restriction of the molecular motion by cross-linking. As shown in the figure, the effect of DS is more profound; the peak temperature of tan δ increases with increasing NCO/OH ratio for the EC with DS = 2.29 (curve III), in contrast to the decrease of tan δ peak temperature for the EC with DS = 2.43 at the same viscosity value (curve I). It is possible that when DS slightly varies, the number and/or the position of cross-linking changes, and thus the molecular motion is significantly affected. The role of long alkyl-chains such as the hexamethylene unit in our studies is thought to act in two manners according to the number of cross-linking; one role is to act as a soft segment to enhance molecular motion, and the other is to act as a bridge to restrict the mobility of the main chain.

The features of the tan δ peak as a function of NCO/OH ratio for the various EC films are shown in Figure 18.9. The heights of the tan δ peaks for the EC films prepared from EC with different molecular mass or DS all decrease with increasing NCO/OH ratio while the peak widths all increase with increasing NCO/OH ratio. These results suggest that the distribution of relaxation times caused by inhomogeneous structure spreads via cross-linking of the molecular chains.

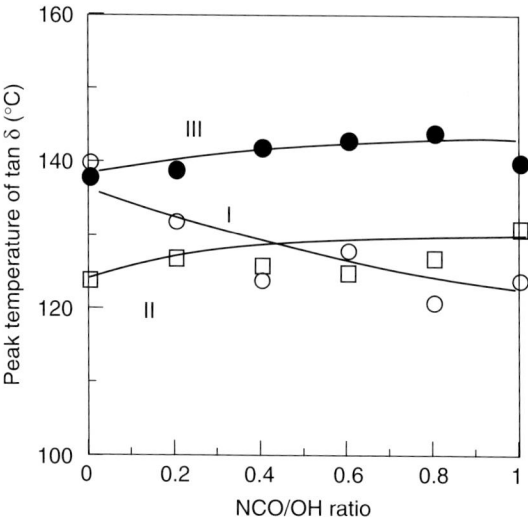

Figure 18.8 Relationships between the tan δ peak temperature and the NCO/OH ratio of the EC films; curve I: viscosity = 100 cp, DS = 2.43, curve II: viscosity = 10 cp, DS = 2.43, and curve III: viscosity = 100 cp, DS = 2.29.

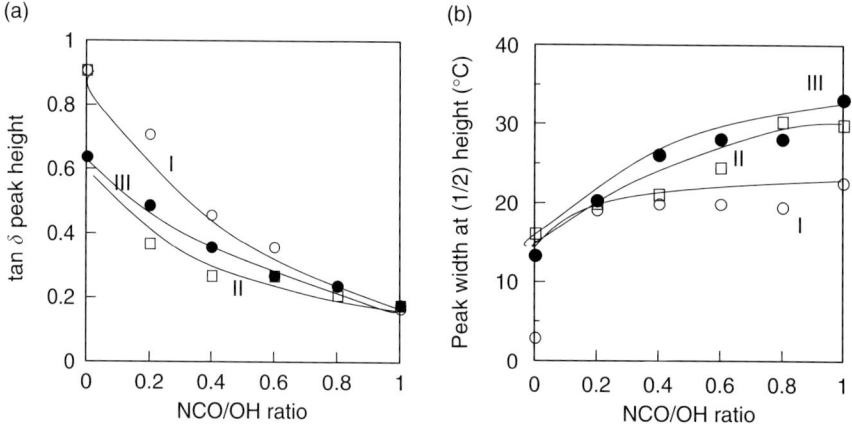

Figure 18.9 Relationships between the peak height (a) and width (b) of the tan δ peak and NCO/OH ratio of the EC films; curve I: viscosity = 100 cp, DS = 2.43, curve II: viscosity = 10 cp, DS = 2.43, and curve III: viscosity = 100 cp, DS = 2.29.

Figure 18.10 shows the plots between E' values at $-100°C$, at a temperature at which E' starts to rapidly decrease (start of the α-dispersion) and at 180°C (rubbery state), respectively, vs NCO/OH ratio for the EC films with the same DS = 2.43 but different molecular mass. As shown in this figure, E' at a temperature at which α-dispersion starts decreases slightly with increasing NCO/OH ratio from 1.4×10^9 to 1.1×10^9 Pa for the

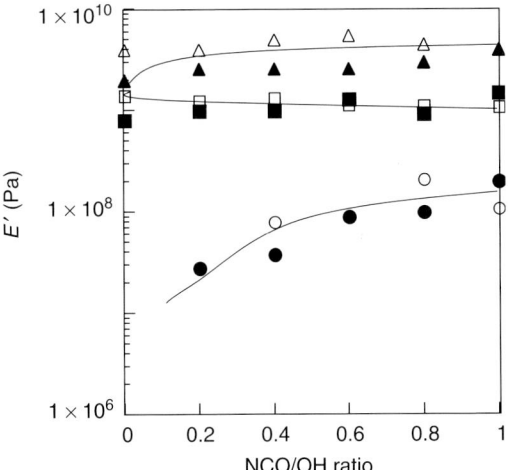

Figure 18.10 Plots between E' values at $-100°C$ (△, ▲), at a temperature at which E' starts to rapidly decrease (□, ■), and at 180°C (rubbery state) (○, ●), respectively, vs NCO/OH ratio; △, □ and ○: EC with viscosity of 100 cp and DS of 2.43, ▲, ■ and ●: EC with viscosity of 10 cp and DS of 2.43.

EC films with a viscosity of 100 cp. This corresponds to the shifts of E'' and tan δ peak to the low temperature side. In contrast, E' at the rubbery state increases with increasing NCO/OH ratio, showing that by intermolecular linking, the rubbery flow of the system is severely restricted.

As shown in Figure 18.6, in a temperature range from $-120°C$ to $160°C$, two dispersions, α and β, can be observed. From the frequency dependency of each peak temperature, activation energy (ΔE) of each molecular motion can be calculated based on Arrhenius-type equation [Equation (18.3)].

$$f = A \exp(-\Delta E/RT) \tag{18.3}$$

where f is the frequency, T the absolute temperature, and R the gas constant. Figure 18.11 shows the plots of ΔE of the α-dispersion (ΔE_α) and β-dispersion (ΔE_β) vs NCO/OH ratio. ΔE_α slightly decreases with increasing NCO/OH ratio, but ΔE_β maintains a constant value. The values of ΔE_α are larger than those of synthetic polymer [21]. This may be due to the fact that large, rigid glucopyranose rings exist in the molecular chain. The β-dispersion observed as a broad and small peak at $\sim 50°C$ is attributed to the rotation of the ethoxyl group when ΔE_β value is taken into consideration.

18.4 Effect of MC blending on molecular motion of EC

When MC (viscosity = 100 cp, DS = 1.80) is blended with EC (viscosity = 100 cp, DS = 2.43), smooth film could be obtained in a MC content lower than 60%. For blends with MC content ≥80%, the films disintegrated when they were separated from the glass

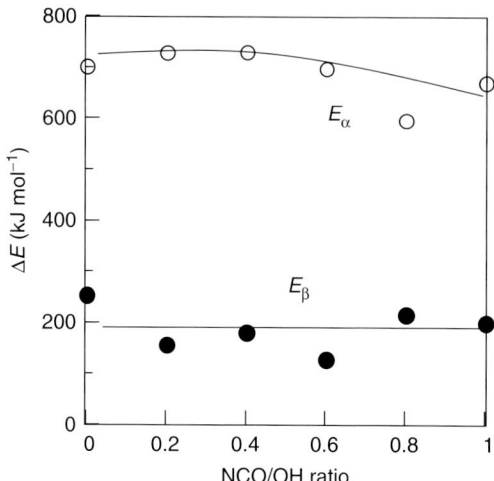

Figure 18.11 Plots of the activation energy of α-dispersion (E_α) and β-dispersion (E_β) vs NCO/OH ratio of the EC (viscosity = 100, DS = 2.43) films.

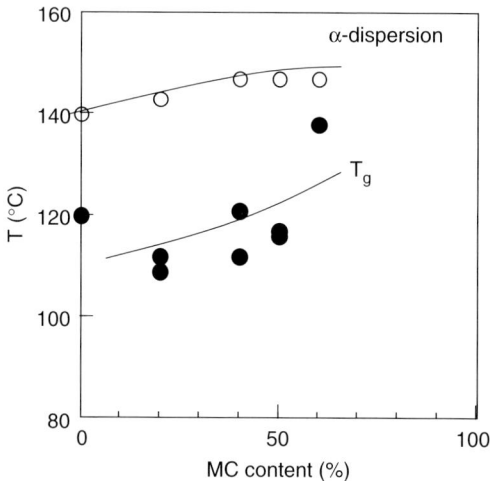

Figure 18.12 Plots of T_g measured by DSC and temperature of α-dispersion measured by DMA vs MC content of the EC/MC blends.

plate. Both T_g and temperature of the α-dispersion increase with increasing MC content (Figure 18.12) due likely to the higher rigidity of the MC molecular chain than EC chain, as shown in Figure 18.1.

E' at $-100°C$, E' at a temperature at which the α-dispersion starts and at the rubbery state (180°C) increase with increasing MC content as shown in Figure 18.13. The fact that

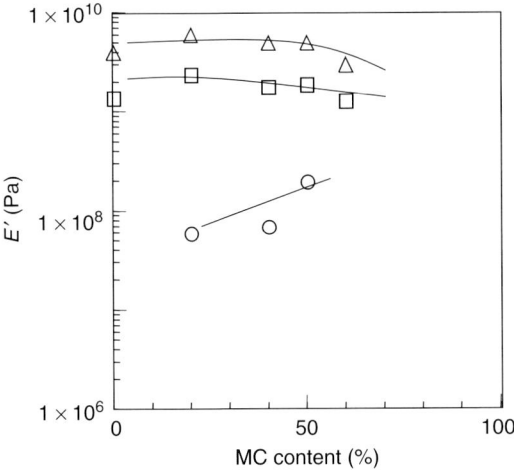

Figure 18.13 Plots of E' values at $-100°C$ (△), at a temperature at which E' starts to rapidly decrease (□), and at $180°C$ (rubbery state) (○) vs MC content of the EC/MC blends.

the rubbery state can be observed by blending EC with MC suggests that MC works as a physical bridge between the EC molecular chains.

References

1. Onda, Y.; Hayakawa, K., *Cellulose Handbook*; Asakura Publisher; Tokyo, 2000; pp. 485–489
2. Hatakeyama, H.; Hatakeyama, T., Interaction between Water and Hydrophilic Polymers; *Thermochim. Acta* 1998, 308, 3–22.
3. Desbrieres, J.; Hirrien M.; Ross-Murphy, S.B., Thermogelation of Methylcellulose: Rheological Considerations; *Polymer* 2000, 41, 2451–2461.
4. Zugenmaier, P.; Haurand, P., Structural and Rheological Investigations on the Lyotropic Liquid-crystalline System-O-Ethylcellulose-Acetic Acid–Dichloroacetic Acid; *Carbohydr. Res.* 1987, 160, 369–380.
5. Hatakeyama, H.; Asano, Y.; Hatakeyama T.; Kennedy J.F., Thermal Properties of Hyaluronic Acid-based Polyurethane Derivatives Associated with Water; In: Kennedy, J.F.; Phillips, G.O.; Williams, P.A.; Hascall, V.C. (eds); *Hyaluronan, Vol 1; Chemical, Biochemical and Biological Aspects*; Woodhead; Cambridge, 2002; pp. 313–322.
6. Hatakeyama T.; Hatakeyama H., Phase Transition of Sodium Hyaluronate Hylan and Polyurethanes Derived from Hyaluronic Acid in the Presence of Water; In: Kennedy, J.F.; Phillips, G.O.; Williams, P.A.; Hascall, V.C. (eds); *Hyaluronan, Vol 1; Chemical, Biochemical and Biological Aspects*; Woodhead; Cambridge, 2002; pp. 323–328.
7. Amici, E.; Clark, A.H.; Normand, V.; Johnson, N.B., Interpenetrating Network Formation in Agarose-sodium Gellan Gel Composites; *Carbohydr. Polym.* 2006, 46, 383–391.
8. Li, C-Y.; Tomasik P.; Zaleska, H.; Liaw, S-C.; Lai, V.M.-F., Carboxymethylcellulose-gelatin Complexes; *Carbohydr. Polym.* 2002, 50, 19–26.

9. Roscaa, C.; Poppa, M.I.; Lisaa, G.; Chitanub, G.C., Interaction of Chitosan with Natural or Synthetic Anionic Polyelectrolytes. 1. The Chitosan–Carboxymethylcellulose Complex; *Carbohydr. Polym.* 2005, 62, 35–41.
10. Nakamura, K.; Kinoshita, E.; Hatakeyama T.; Hatakeyama, H., Thermal Properties of Alginic Acid-polylysine Molecular Composites; In: Nishinari, K. (ed.); *Hydrocolloids-Part 1*; Elsevier Science B.V.; Amsterdam, 2000; pp. 189–195.
11. Lee, S.J.; Kim, S.S.; Lee, Y.M., Interpenetrating Polymer Network Hydrogels Based on Poly(ethylene glycol) Macromer and Chitosan; *Carbohydr. Polym.* 2000, 41, 197–205.
12. Takahashi, M.; Iijima, M.; Kimura, K.; Hatakeyama T.; Hatakeyama H., Thermal and Viscoelastic Properties of Xanthan Gum/Chitosan Complexes in Aqueous Solutions; *J. Therm. Anal. Calorim.* 2006, 85, 669–674.
13. Hirrien, M.; Chevillard, C.; Desbrières, C.J.; Axelos M.A.V.; Rinaudo, M., Thermogelation of Methylcellulose; New Evidence for Understanding the Gelation Mechanism; *Polymer* 1998, 39, 6251–6259.
14. Desbrieres, J.; Hirrien M.; Ross-Murphy, S.B., Thermogelation of Methylcellulose: Rheological Considerations; *Polymer* 2000, 41, 2451–2461.
15. Nishinari, K.; Hofmann, K.; Moritaka, E.; Kohyama; H.; Nishinari, K., Gel-sol Transition of Methylcellulose; *Macromol. Chem. Phys.* 1997, 198, 1217–1226.
16. Sarkar, N.; Walker, L.C., Hydration-dehydration Properties of Methylcellulose and Hyroxypropylmethylcellulose; *Carbohydr. Polym.* 1995, 27, 177–185.
17. Hatakeyama, T.; Quinn, F.X., *Thermal Analysis, Fundamentals and Applications to Polymer Science*, 2nd edition; John Wiley; Chichester, 1999.
18. Hatakeyama, T.; Hatakeyama, H., *Thermal Properties of Green Polymers and Biocomposites*; Kluwer Academic Publishers; Dordrecht, 2004.
19. Van Krevelen, D.W., *Properties of Polymers, their Correlation with Chemical Structure; their Numerical Estimation and Prediction from Additive Group Contributions*, 3rd edition; Elsevier; Amsterdam, 1997; pp. 389–390.
20. Hatakeyama, T.; Hatakeyama, H., Effect of Chemical Structure of Amorphous Polymers on Heat Capacity Difference at Glass Transition Temperature; *Thermochim. Acta* 1995, 267, 249–257.
21. Hatakeyama T.; Li, Z. (eds), *Handbook of Thermal Analysis*; John Wiley; Chichester, 1999.

Chapter 19
DSC and AFM Studies of Chemically Cross-Linked Sodium Cellulose Sulfate Hydrogels

Toru Onishi, Hyoe Hatakeyama, and Tatsuko Hatakeyama

Abstract

Sodium cellulose sulfate (NaCS) and chemically cross-linked NaCS hydrogels have been characterized by differential scanning calorimetry (DSC) and atomic force microscopy (AFM). DSC studies show that the molecular motion of NaCS chains is remarkably enhanced in the presence of water but is restricted by cross-linking. The maximum amount of nonfreezing water plays an important role in the molecular motion in both the NaCS and the cross-linked systems. AFM studies demonstrate that the entangled molecular bundles of NaCS become flat by cross-linking and that the cross-linked NaCS molecular chains form a large and relaxed network structure.

19.1 Introduction

Aqueous solutions of cellulose derivatives are known to show unique functional properties. For example, aqueous solutions of methylcellulose form thermoreversible hydrogels and those of hydroxypropylcellulose produce lyotropic liquid crystals [1–4]. Polyelectrolyte cellulose derivatives such as sodium carboxymethylcellulose (NaCMC) and sodium cellulose sulfate (NaCS) do not form hydrogels but thermotropic lyotropic liquid crystals in the presence of water [4–8]. Previously we reported that NaCS molecules self-assemble and ally in one direction, sandwiching sodium ions within a certain range of water content and temperature [9]. The intermolecular distance of NaCS molecules increases with increasing temperature or water content to a point at which NaCS molecules start to arrange in a random way. CMC having various counter cations forms lyotropic and thermotropic liquid crystals in the presence of water [10], and hydrogels that are induced by the metal cations [11].

NaCS molecules self-assemble in the presence of water via breaking of hydrogen bonding under certain conditions; however, the assembled structure could not be maintained when water content is increased. For the studies and development of functional properties of NaCS it is important to control the higher-order structure of NaCS molecules in the presence of water because the crucial role water plays in such a system. Chemical cross-linking is a method to obtain a stable structure regardless of water content. In this chapter, we describe the chemical cross-linking of NaCS, and our studies of the water–NaCS interaction and the morphological features of the cross-linked NaCS using differential scanning calorimetry (DSC) and atomic force microscopy (AFM), respectively.

19.2 Experimental

19.2.1 Preparation of hydrogels from cross-linked NaCS

The NaCS powder (molecular mass $= 2.9 \times 10^5$ and the degree of substitution per C_6 unit, DS $\simeq 1.0$) was obtained from Acros Organics (New Jersey, USA). The cross-linked NaCS hydrogel samples were prepared as follows: (1) NaCS powder was dissolved in distilled water (concentration = 12.0–26.0%), maintained at 50°C, and stirred until homogeneous solution was obtained; (2) water-soluble hexamethylene diisocyanate (HDI) (Asahi Kasei Co., Tokyo, Japan) was dissolved in water (concentration = 24.5 %) and stirred at 50°C until homogeneous solution was obtained; (3) a small amount of dibutyltin dilaurate was added to the HDI solution as a catalyst; (4) NaCS and HDI aqueous solutions were mixed and stirred at a predetermined ratio to give NCO/OH molar ratio of 0.0–0.6; (5) the mixed solution was maintained at 50°C for several minutes, and then kept in a desiccator for one night; (6) the gels obtained (abbreviated as NaCS for the control sample with NCO/OH ratio of 0.0, and as NaCS–PU for samples with NCO/OH ratio >0.0) were kept in a refrigerator. The schematic chemical structure of NaCS–PU is shown in Figure 19.1.

19.2.2 Differential scanning calorimetry

A Seiko Instrument DSC 220C equipped with cooling apparatus was used. Scanning rate was 10°C min^{-1} and nitrogen flow rate was 30 mL min^{-1}. The temperature was varied as follows: (1) the sample was cooled from 40°C to −150°C at a cooling rate of 10°C min^{-1}; (2) the sample was maintained at −150°C for 5 min and heated at 10°C min^{-1} to 80°C; and (3) step (1) with cooling from 80 (instead of 40) to −150°C and step (2) were repeated. The data obtained by the second heating run was mainly used for the analysis.

Water content of the sample was defined in Equation (19.1):

$$W_c = m_{\text{water}}/m_{\text{dry sample}} \tag{19.1}$$

where m_{water} is the mass of water in the sample and $m_{\text{dry sample}}$ is the mass of the dry sample.

W_c of each hydrogel sample was controlled as follows: (1) the sample was placed in an aluminum sample pan the surface of which had previously been treated so as not to react with water, and (2) water was evaporated until a predetermined mass was reached, after which the sample was hermetically sealed using an automatic sealer.

Nonfreezing water content (W_{nf}), the content of water showing no phase transition, was calculated according to Equation (19.2) [4,6]:

$$W_{\text{nf}} = \frac{[1 - ((\Delta H_m)/334)]}{m_{\text{dry gel}}} \tag{19.2}$$

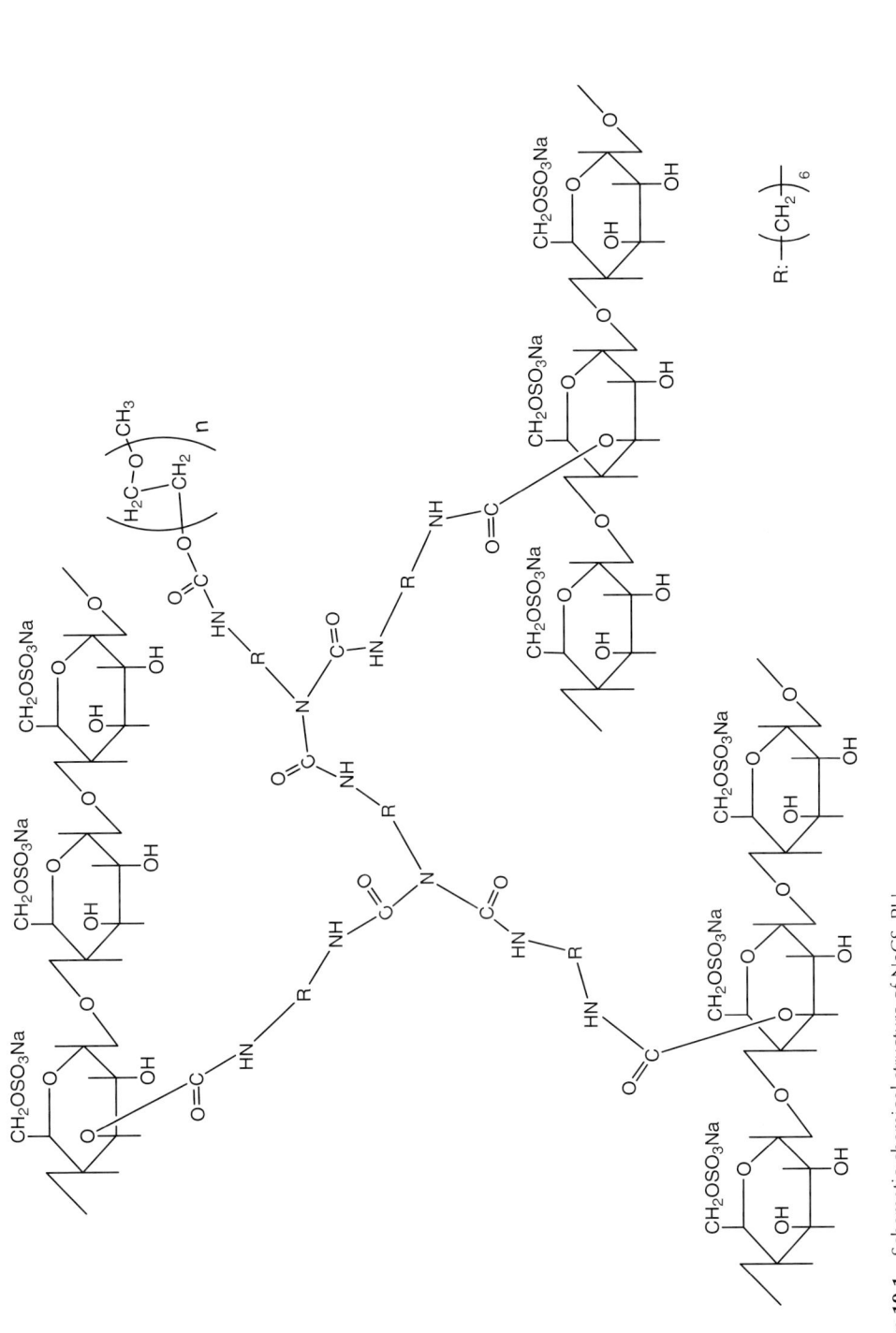

Figure 19.1 Schematic chemical structure of NaCS–PU.

where ΔH_m is the enthalpy calculated from the melting peak of the gels from the DSC heating curves, the value 334 (J g^{-1}) is the enthalpy of the melting of water, and $m_{dry\ gel}$ is the mass of the dry gel.

19.2.3 Atomic force microscopy

Both NaCS and NaCS–PU were dissolved in distilled water to give 1.0 wt% solution. The solution was then diluted to 0.01% or lower concentrations when necessary depending on the measurement conditions. The solution, the microfilter, the syringe, and syringe needle were maintained at 40°C for 30 min. Adhesive tape was attached to the mica surface and detached. One drop of the solution was extended on the newly cleaved surface thus obtained.

A Seiko Instruments, Tokyo, Japan, scanning probe microscope, SPA400, was used. The frequency was 1.0 Hz and the measurements were carried out using the tapping mode. The value of measured width of the sample was calibrated taking into consideration the geometrical shape of the cantilever; the size of cantilever tip is ordinarily larger than that of the samples. Accordingly, apparent width (W) of each sample is larger than the real width as shown in Figure 19.2, in which the cross-section of a sample is assumed either rectangle or circle. The real width was calculated by the equations shown in the figure caption of Figure 19.2. In this study, by examining the molecular shape, the real width was calibrated assuming circular cross-section.

19.3 Phase diagram of NaCS–water and NaCS–PU–water systems

Figure 19.3a and b show the DSC heating curves of NaCS–water and NaCS–PU–water hydrogels with NCO/OH molar ratio of 0.1, respectively, at various water contents. The baseline deflection due to the glass transition can be observed for all the NaCS–water systems (Figure 19.3a). For the NaCS–water system with $W_c = 0.58$, an exothermic peak

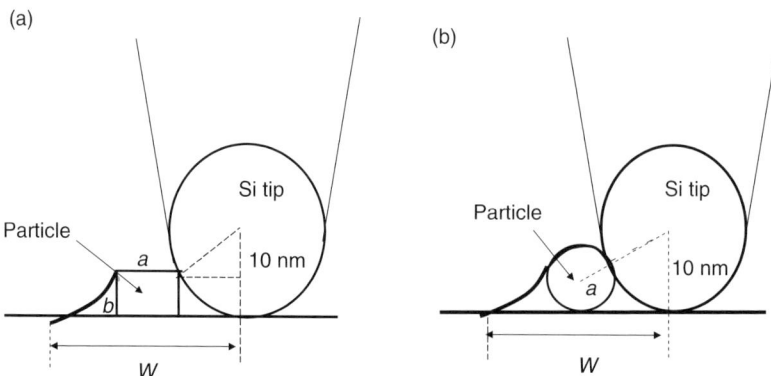

Figure 19.2 Locus of cantilever when the cross-section of sample is assumed either rectangle or circle; (a) rectangle (width a, height b) apparent width (W) $W = 2\sqrt{20b - b^2} + a$, (b) circle (radius a) $W = 4\sqrt{10a}$.

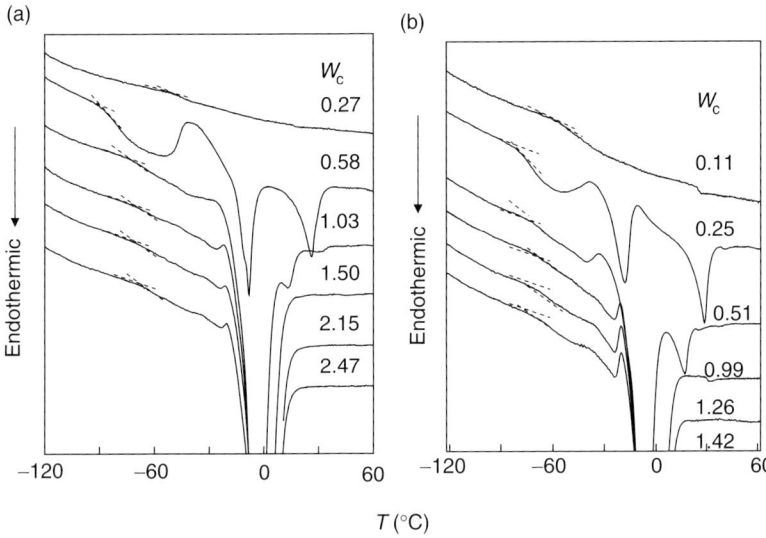

Figure 19.3 (a) DSC heating curves of NaCS–water system and (b) NaCS–PU–water system with NCO/OH ratio of 0.1; numerals shown in the figures are water contents (W_c) in g/g.

due to cold crystallization, and two endothermic peaks due to melting and liquid crystal transitions at progressively higher temperatures were also observed. A shoulder peak can be detected on the low-temperature side of the melting peak. With increasing W_c ($W_c \geq 1.03$) glass transition shifts to the high-temperature side and liquid crystallization temperature shifts to the low-temperature side. Similar DSC heating patterns are observed for the NaCS–PU–water systems, although cold-crystallization, melting, and liquid crystallization occur at W_c values lower than those of NaCS–water systems (Figure 19.3b).

From the DSC heating curves of NaCS and NaCS–PU hydrogels with various NCO/OH ratios at various water contents, transition temperatures were obtained and phase diagrams established. Figure 19.4a shows the representative phase diagrams of NaCS and NaCS–PU hydrogels with NCO/OH ratio of 0.2, while Figure 19.4b shows the schematic phase diagrams of the two systems. As shown in Figure 19.4a, the glass transition temperature (T_g) decreased initially with increasing W_c, reached a minimum value then increased and leveled off at a characteristic temperature. The lowest T_g and the W_c at which the lowest T_g is obtained are defined as $T_{g\ min}$ and $W_{cT_{g\ min}}$, respectively, while the temperature at which T_g levels off is defined as $T_{g\ level\ off}$ (Figure 19.4b). As discussed in our previous report [12], T_g markedly decreases with increasing W_c in the low W_c range due to the catalytic breaking of the intermolecular bonding by the water molecules. DSC results suggest that fewer than several water molecules per one repeating unit of the NaCS molecules are involved in the above event.

Changes of the cold-crystallization temperature (T_{cc}), the low-temperature side shoulder peak (T_{ml}), the main melting peak temperature (T_{mh}), and the liquid crystal transition temperature (T_{lc}) are shown schematically in Figure 19.4b. T_{mh} increases gradually with increasing W_c and then levels off at a characteristic value defined as $T_{mh\ level\ off}$.

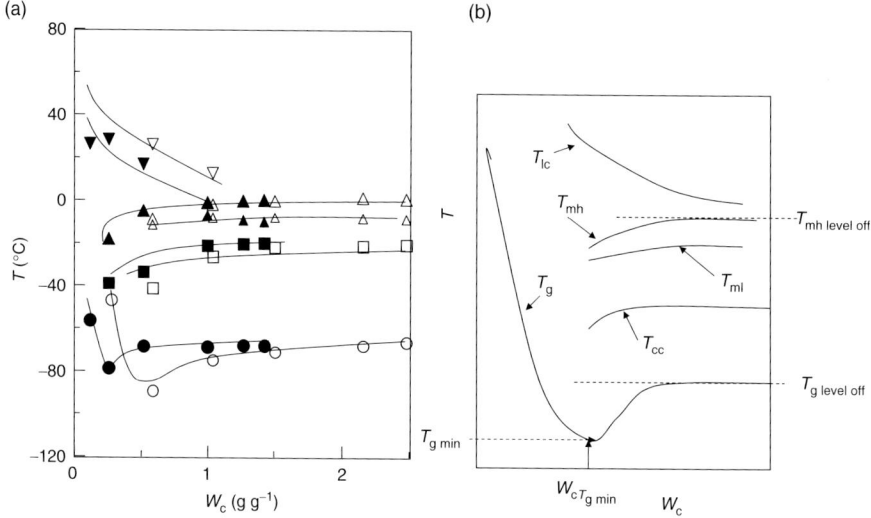

Figure 19.4 (a) Representative phase diagrams of NaCS (unfilled symbols such as ○ and □) and NaCS–PU with NCO/OH ratio of 0.2 (filled symbols such as ● and ■), and (b) schematic phase diagrams showing various characteristic temperatures vs water contents.

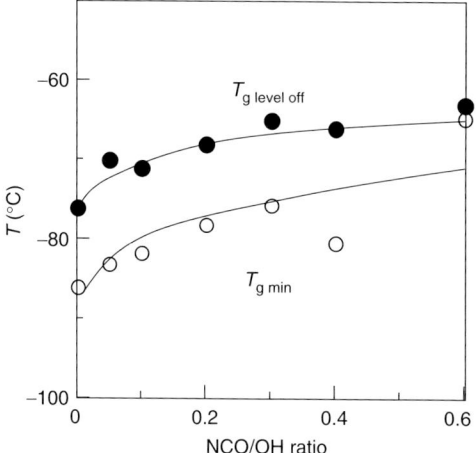

Figure 19.5 Plots of $T_{g\ min}$ and $T_{g\ level\ off}$ vs NCO/OH ratio; data at NCO/OH = 0.0 are from the NaCS–water system and those at NCO/OH = 0.05–0.6 are from the NaCS–PU–water systems.

Figure 19.5 shows the plots of $T_{g\ min}$ and $T_{g\ levels\ off}$ vs NCO/OH ratio. Both $T_{g\ min}$ and $T_{g\ level\ off}$ increase with increasing NCO/OH ratio. The effect of cross-linking extent clearly indicates that molecular motion of the main chain is restricted by cross-linking even though the network structure of the NaCS–PU hydrogels is not extensive.

Figure 19.6 shows the relationship between the nonfreezing water content (W_{nf}) calculated from Equation (19.2) and W_c for the NaCS and the NaCS–PU hydrogels. W_{nf} values

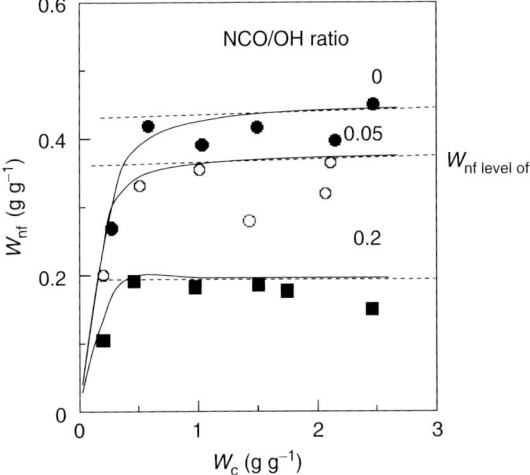

Figure 19.6 Relationship between nonfreezing water content and water content of NaCS (NCO/OH = 0) and NaCS–PU hydrogels with NCO/OH = 0.05–0.2.

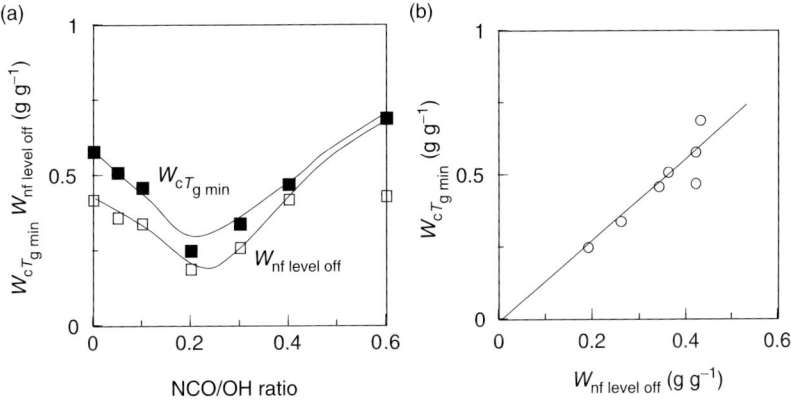

Figure 19.7 (a) Plots of $W_{nf\ level\ off}$ and $W_{cT_{g\ min}}$ vs NCO/OH ratio, and (b) plot of $W_{nf\ level\ off}$ vs $W_{cT_{g\ min}}$.

increase initially with increasing W_c and then levels off. The value of W_c at which W_{nf} levels off is specific for each sample depending on the chemical and higher-order structure. In our previous studies [13], it was shown that W_{nf} values depended on the number of hydrophilic groups such as hydroxyl, carboxylate, and sulfate groups in the repeating unit of the linear polysaccharides. The value of $W_{nf\ level\ off}$ that is the maximum value of W_{nf} decreased with an increase of NCO/OH ratio from 0 to 0.2 (Figure 19.6), but increased with a further increase of NCO/OH ratio (see also Figure 19.7a).

The plots of $W_{nf\ level\ off}$ and $W_{cT_{g\ min}}$ vs NCO/OH ratio are shown in Figure 19.7a. Both the $W_{nf\ level\ off}$ and the $W_{cT_{g\ min}}$ values decreased initially with increasing NCO/OH

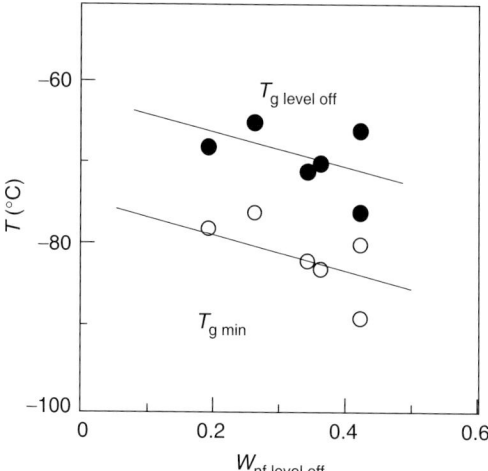

Figure 19.8 Plots of $T_{g\ level\ off}$ and $T_{g\ min}$ vs $W_{nf\ level\ off}$ for the NaCS–water and the NaCS–PU–water systems.

ratio, reached their respective minimum values, and then increased. The values of $W_c T_{g\ min}$ increased linearly with increasing values of $W_{nf\ level\ off}$ (Figure 19.7b). Both the cold crystallization temperature and the melting temperature of the hydrogels reached constant values when the W_c of the sample reached the characteristic value at which W_{nf} leveled off (data not shown).

Both the $T_{g\ level\ off}$ and the $T_{g\ min}$ appear to be linearly correlated to the $W_{nf\ level\ off}$ for the NaCS–water and the NaCS–PU–water systems (Figure 19.8). This result shows that molecular motion is enhanced at a low temperature in the presence of a large amount of nonfreezing water.

19.4 Atomic force micrographs of NaCS and NaCS–PU hydrogel

Atomic force micrographs of NaCS (Figure 19.9) and NaCS–PU hydrogel (NCO/OH = 0.2) (Figure 19.10) were taken under the conditions described in Section 19.2. The micrograph of NaCS shows that its molecular chains are entangled at the concentration of 0.01 wt%. The height and width of each bundle structure were ~3.6 and ~14 nm, respectively. As described in the Section 19.2, the width was calibrated taking into consideration the shape of the cantilever. Both measured and calibrated sizes of the bundles for the NaCS and NaCS–PU samples are listed in Table 19.1. Based on the height and the width of each bundle structure and the consideration of the molecular structure of cellulose obtained by x-ray diffractometry [14], it is estimated that for four NaCS molecules pile and more than ten molecular chains co-aggregate with each other.

When a more diluted solution (<0.01 wt%) of NaCS was used, the effect of the solution flow on the mica surface was observed. Each molecular bundle straightened and the

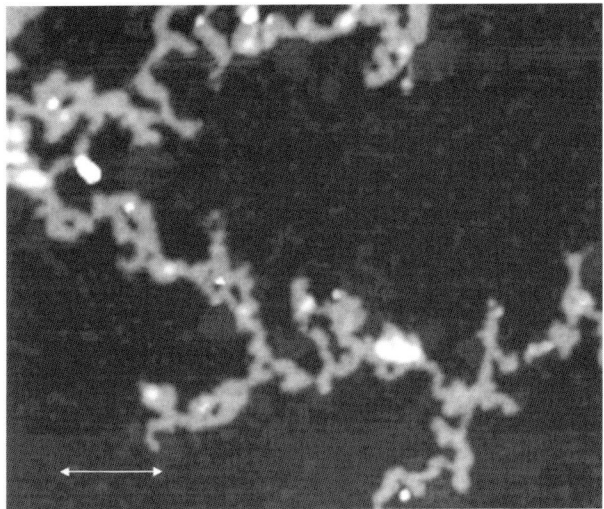

Figure 19.9 Atomic force micrograph of NaCS at a concentration of 0.01 wt%; the arrow length is 0.5 μm.

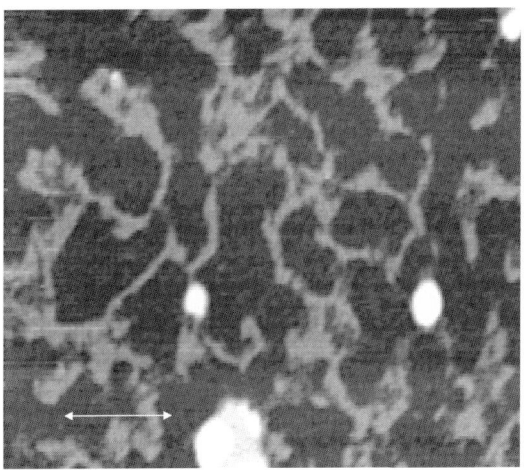

Figure 19.10 Atomic force micrograph of NaCS–PU (NCO/OH = 0.2) at a concentration of 0.001 wt%; the arrow length is 0.5 μm.

Table 19.1 Height and width of the molecular bundles of NaCS and NaCS–PU (NCO/OH = 0.2) determined by AFM.

Sample	Height (nm)		Observed width (nm)		Calibrated width (nm)	
	Average	SD	Average	SD	Average	SD
NaCS	3.6	0.4	47.4	6.1	14.2	3.6
NaCS–PU	1.0	0.04	50.4	6.8	16.1	4.3

molecular chains were oriented to the flowing direction. It is likely that the liquid crystalline structure is formed in such NaCS–water systems [4]. The height and the width of the calculated molecular bundles for the more diluted solutions of NaCS were similar to those shown in Figure 19.9. It is thought that the molecular entanglement loosens by dilution and that the molecular chains form a straight structure and are aligned in one direction.

For the NaCS–PU hydrogel, a network structure of entangled chains can clearly be seen (Figure 19.10). The width of the bundle structure of the NaCS–PU hydrogel was similar to that of NaCS, but the height of the bundle was remarkably reduced (Table 19.1).

From the above AFM results, it is reasonable to conclude that the bundle structure of the NaCS molecules was altered during their reaction with HDI molecules and that flat bundles were formed by cross-linking. The AFM images indicate that the structure of the NaCS–PU molecular chains form a large and relaxed network structure.

References

1. Desbrieres, J.; Hirrien, M.; Rinaudo, M., A Relation between the Condition of Modification and the Properties of Cellulose Derivatives: Thermogelation of Methylcellulose; In: Heinze, T.; Glasser, W.G. (eds), *Cellulose Derivatives, Modification, Characterization and Nanostructures*; ACS Symposium Series 688, ACS; Washington, D.C., 1998; pp. 332–348.
2. Desbrieres, J.; Hirrien, M.; Rinaudo, M.A., Calorimetric Study of Methylcellulose Gelation; *Carbohydr. Polym.* 1998, 37, 145–152.
3. Kondo, T., Hydrogen Bonds, Cellulose and Cellulose Derivatives; In: Dumitriu, S. (ed.), *Polysaccharides, Structural Diversity and Functional Versatility*, 2nd Edition; Marcel Dekker; New York, 2005; pp. 69–98.
4. Hatakeyama, T.; Hatakeyama, H., *Thermal Properties of Green Polymers and Biocomposites*; Kluwer Academic Publishers; Dordrecht, 2004.
5. Hatakeyama, T.; Naoi, S.; Hatakeyama, H., Liquid Crystallization of Glassy Guar Gum with Water; *Thermochim. Acta* 2004, 416, 121–127.
6. Hatakeyama, H.; Hatakeyama, T., Interaction between Water and Hydrophilic Polymers; *Thermochim. Acta* 1998, 308, 3–22.
7. Hatakeyama, T.; Hatakeyama, H., Molecular Relaxation of Cellulosic Polyelectrolytes with Water; In: Glasser, W.G.; Hatakeyama, H. (eds), *Viscoelasticity of Biomaterials*; ACS Symposium Series 489, ACS; Washington, D.C., 1992; pp. 329–340.
8. Hatakeyama, H.; Hatakeyama, T., Nuclear Magnetic Relaxation Studies of Water–Cellulose and Water–Sodium Cellulose Sulfate System; In: Kennedy, J.F.; Phillips, G.O.; Williams, P.A. (eds), *Cellulose: Structural and Functional Aspects*; Ellis Horwood; Chichster, 1990; pp. 131–136.
9. Hatakeyama, T.; Yoshida, H.; Hatakeyama, H., The Liquid Crystalline State of Water–Sodium Cellulose Sulphate Systems Studied by DSC and WAX; *Thermochim. Acta* 1995, 266, 343–354.
10. Nakamura, K.; Hatakeyama, T.; Hatakeyama, H., DSC Studies on Monovalent and Divalent Cation Salts of Carboxymethylcellulose in Highly Concentrated Aqueous Solutions; In: Kennedy, J.F.; Phillips, G.O.; Williams, P.A. (eds), *Wood and Cellulosics*; Ellis Horwood Ltd.; Chichester, 1987; pp. 97–103.
11. Kamide, K.; Yasuda, K.; Okajima, K., ^{13}C NMR Study on Gelation of Aqueous Carboxymethylcellulose with Total Degree of Substitution of 0.39 Solution Induced by Metal Cations; *Polym. J.* 1988, 20, 259–268.
12. Hatakeyama, T.; Yoshida, H.; Hatakeyama, H., A Differential Scanning Calorimetry Study of the Phase Transition of the Water-Sodium Cellulose Sulfate System; *Polymer* 1987, 28, 1282–1286.

13. Hatakeyama, T.; Hatakeyama, H.; Nakamura, K., Non-freezing Water Content of Mono- and Divalent Cation Salts of Polyelectrolyte–Water Systems Studied by DSC; *Thermochim. Acta* 1995, 253, 137–148.
14. Pérez, S.; Mazeau, K., Conforamtion, Structure and Morphologies of Celluloses; In: Dumitriu, S. (ed.), *Polysaccharides, Structural Diversity and Functional Versatility*, 2nd Edition; Marcel Dekker; New York, 2005; pp. 41–68.

Chapter 20
Microscopic Examination of Cellulose Whiskers and Their Nanocomposites

Ingvild Kvien and Kristiina Oksman Niska

Abstract

Microscopic examination of cellulose nanowhiskers (CNW) and their nanocomposites is described in this chapter. Field emission scanning electron microscope (FESEM) is shown to be a convenient method for the detection of the presence of possible larger agglomerates in the composites. More detailed information on the size and distribution of CNW can be obtained using transmission electron microscope (TEM) and atomic force microscope (AFM). TEM is capable of accessing the bulk structure of CNW, but it may underestimate the length of CNW in the composites. AFM could be a powerful alternative to TEM, but it has only limited access to the bulk structure.

20.1 Introduction

Nanocomposites are a relatively new generation of composite materials where at least one of the constituent phases has one dimension <100 nm [1]. This new family of composites is reported to exhibit remarkable improvements of material properties when compared to conventional composite materials [2,3]. In this chapter we focus on polymer-based nanocomposites where both the continuous and reinforcing phases are bio-based. The utilization of cellulose nanowhiskers (CNW) as reinforcement in nanocomposites has attracted significant attention during the last decade [4,5]. The term whisker refers to the needle-like structure of cellulose monocrystals. These crystals, linked by amorphous regions, build up cellulose microfibrils in, for example, the wood cell wall. The hierarchic structure of different types of reinforcement obtained from wood is shown in Figure 20.1. The first three pictures from the left to the right show scanning electron microscope (SEM) images of wood with annual growth rings, a cross-section of wood cell walls and cellulose fibers. The last two pictures are transmission electron microscope (TEM) images of cellulose microfibrils and nanowhiskers that build up the wood cell wall. Both are utilized as cellulose nanoreinforcements. Microfibrils contain both amorphous and crystalline regions, while whiskers consist of monocrystals. The interest to utilize CNW as reinforcement is due to their renewable nature, abundance, large specific surface area, and good mechanical properties [6]. The large specific area of the CNW leads to increased interaction with

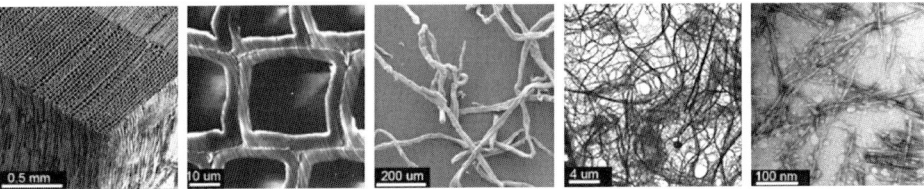

Figure 20.1 The hierarchic structure of wood. From left: wood with annual growth rings, wood cell walls, cellulose fibers, cellulose microfibrils and nanowhiskers.

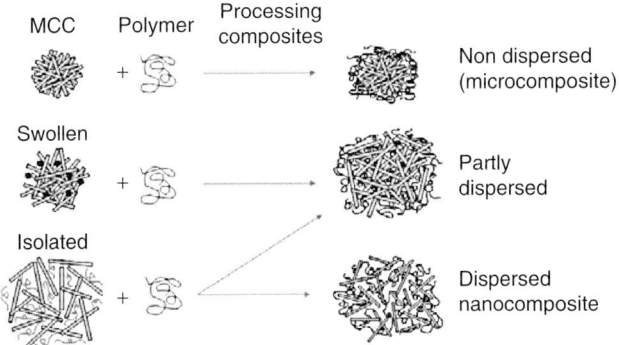

Figure 20.2 Possible structures of nanocomposites based on cellulose nanowhiskers obtained from MCC.

the matrix polymer on molecular level, which will lead to materials with new properties. Because of their small size they will not create large stress concentrations in the polymer matrix [1]. The small size also increases the probability of structural perfection and will in this way be a more efficient reinforcement compared to their micro counterparts. CNW are today only produced on lab-scale from different sources such as wood [7], tunicin [8], ramie [9], cotton [10], wheat straw [11], bacterial cellulose [12], and sugar beet [13]. The size of the CNW depends on the source, and is, for example, ~5 nm in width and 200 nm in length for whiskers from wood [14]. CNW are not commercially available, but they canbe isolated from microcrystalline cellulose (MCC) [7], which is widely used in food industry and as a binder in tablets and capsules [15]. MCC is prepared by hot-treating cellulose from wood with strong mineral acids, vigorous agitation of the slurry and spray drying [16]. Strong hydrogen bonding between the individual cellulose whiskers produced promotes re-aggregation during the drying procedures [15]. Thus, the MCC produced consists of aggregated bundles of whiskers. By using MCC as a starting material for the production of CNW the tedious processing steps by means of purification, bleaching, fibrillation and hydrolysis are reduced. Prior to nanocomposite processing the MCC can either be swelled in an appropriate organic medium or treated by acid hydrolysis to produce a water suspension of isolated CNW [17]. The possible structures of nanocomposites are shown in a schematic drawing (Figure 20.2). In the first case, MCC is not swelled and the polymer is unable to penetrate into the MCC particle and a nanostructure is not obtained. In the second case when MCC is swelled, the polymer is expected to penetrate between the cellulose whiskers and form a partly dispersed nanostructure (comparable to intercalation of layered silicates [2]). In the third case when the whiskers are isolated before blending with the polymer, one can expect two different structures; dispersed whiskers or partly dispersed whiskers because of re-aggregation of the CNW in the polymer.

A few studies have been reported on the preparation of nanocomposites by incorporating CNW into biopolymers such as starch [8,9], cellulose acetate butyrate (CAB) [12,18] or polylactic acid (PLA) [19,20]. CNW are hydrophilic in nature and are reported to produce well-dispersed nanocomposites with starch as matrix [8,9]. For nanocomposites with less hydrophilic matrixes, the CNW tend to aggregate and therefore cannot be regarded as nanocomposites [19,20]. To obtain true nanocomposites and achieve significant increase in material properties, the whiskers should be well separated and evenly distributed in the matrix material. Different processing methods aided with a variety of chemicals (compatibilizers, surfactants, etc.) have been explored to produce nanocomposites [4]. To know how these various processing routes affect the distribution of the nanoparticles in the matrix, it is essential that detailed structural examination of the nanocomposites be performed.

20.2 Characterization methods for nanostructures

The structures of polymer nanocomposites are traditionally characterized by a combination of TEM and wide-angle x-ray diffraction (WAXD) [2]. This combination is, however, suitable only for layered silicate-based nanocomposites because of the ordered stacking of the silicate layers. This regular stacking can be detected by a diffraction peak in the x-ray diffractogram of the pure silicate. When the polymer chains are penetrating in between the layers, the distance will increase, which leads to a shift in the diffraction peak corresponding to the increased height (d in Bragg's equation). For cellulose, only the three-dimensional (3D) arrangement of the cellulose chains in the crystallites are detectable in WAXD and no peaks corresponding to the stacking of the crystallites can be observed. The WAXD method, therefore, cannot be utilized to determine the structure of CNW-based nanocomposites. For the determination of CNW-based nanostructures, different microscopy techniques need to be utilized. As biobased nanocomposites are in general nonconductive, soft, and water-sensitive materials, and consist of low atomic number elements, both sample preparation and instrumentation studies are challenging. For example, the use of electron microscopes will in particular require special attention to electron dose, contrast, and methods to assess the bulk structure without significantly affecting the morphology of the sample.

TEM has frequently been applied to study CNW. Some studies of CNW have also utilized the atomic force microscope (AFM). However, for structure determination of CNW nanocomposites, the conventional SEM has often been utilized [4]. The resolution of a conventional SEM is, however, limited compared to AFM and TEM, and detailed information of the distribution of CNW in the matrix is difficult to obtain. Also, the conductive coating will possibly cover the finer details or broaden the nanostructures. There are, however, SEM microscopes with a so-called field emission gun (FESEM), which has comparable resolution to TEM even at very low voltages and makes it possible to observe organic materials without conductive coating. In environmental SEM (ESEM) gas molecules are present in the specimen chamber, which helps preventing accumulation of electrical charges in the specimen. New SEMs that combine the environmental operating conditions with a field

Table 20.1 Maximum resolution of the various microscopes.

Technique	SEM	FESEM	TEM	AFM
Resolution	5 nm [21]	1 nm [21]	0.2 nm [22]	1 nm (x,y) [34], 0.1 nm (z) [29]

emission gun are now available and are showing promises for the analysis of biopolymer-based nanocomposites. Table 20.1 lists the maximum resolution of the various microscope techniques described in this chapter.

In the following sections, the principles of field emission SEM (FESEM), TEM, and AFM are presented. Examples of different sample preparation methods for the structure analyses of biobased nanocomposites and results from structure characterization of these materials by the different microscopes are given.

20.3 Field emission scanning electron microscope

FESEMs generate electron probes with a higher brightness than conventional SEMs (100 times brighter) [21] and can thus provide high resolution even at low voltages. FESEM is used for surface analysis of various materials. An electron beam is focused down on the specimen and the signals reflected back are collected by detectors to form the image. Different signals are generated when the electron beam hits the specimen. Three important and most used signals are the backscattered electrons, secondary electrons, and x-rays [22]. Backscattered electrons are elastically scattered primary beam electrons [23]. These electrons can give compositional contrast because the scattered fraction depends on the atomic number of the specimen. They are also used for topography imaging and to yield diffraction contrast. Backscattered electrons have high energy, and they can come from depths of 1 μm or more [22]. Secondary electrons have low energy and come from the top few nanometers of the material [22]. These electrons are used to generate topography images. Secondary electrons are generated because of the interaction between primary electrons and weakly bound valence electrons [22]. More secondary electrons are generated if the specimen is tilted because more of the interaction volume is near the surface [21].

20.3.1 Sample preparation

Polymers are generally insulators and a charge is easily built up in the specimen, which can cause bright spots in the image, movement of the sample, and poor signal output [22]. Charging may be decreased by applying a thin conductive layer and assuring contact between the coating and the sample holder. A thin conductive layer will also increase the emission of secondary electrons [24]. There are several methods for applying conductive coatings, and different metal types can be used [24]. Thick gold coatings tend to be granular, cracked, and nonuniform, whereas Au–Pd and platinum are less likely to be resolved [24]. Carbon coatings are applied to specimens when chemical imaging and x-ray analysis are performed because heavy metal coating can influence the information obtained [22]. Carbon does not

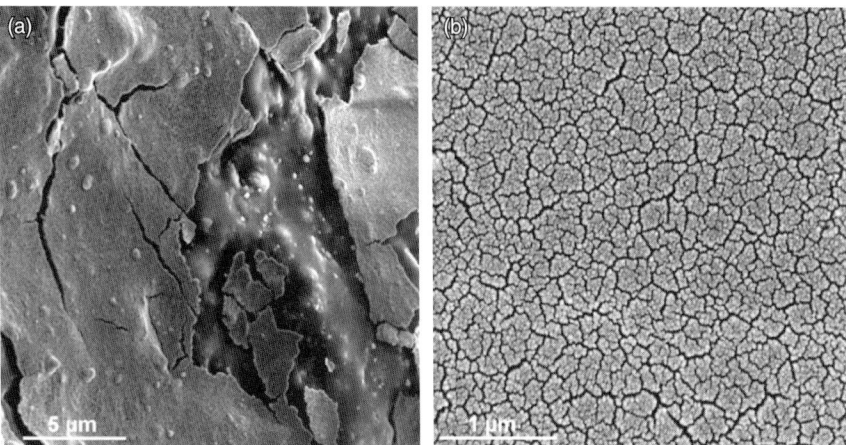

Figure 20.3 FESEM pictures of a nanocomposite showing (a) the platinum coating and the underlying structure and (b) cracked and resolved platinum coating.

provide much electron emission, and often metal coatings must be applied for imaging. However, when analysing nanostructured materials, the coating may cover the finer details, and therefore, it will be more reliable to analyze uncoated samples. Figure 20.3a illustrates that the conductive coating may cover finer details of nanostructured materials. The figure shows a FESEM picture of a bio-nanocomposite sample that is only partly covered by a platinum layer and from which the underlying structure can be observed. Figure 20.3b shows that the FESEM even at relatively low magnifications is able to resolve the platinum coating, which may interfere with the interpretation of the structure. The conductive coating appears to be more easily cracked on smooth surfaces than on rough surfaces. When no conducting layer is applied, it is possible to avoid charging by operating the FESEM at very low acceleration voltages. The resolution of a SEM is, however, proportional to the accelerating voltage and it can be difficult to obtain sufficient resolution for nanostructured materials.

20.3.2 The structure of CNW and nanocomposites

Figure 20.4 shows FESEM pictures of CNW. The sample was prepared by drying a droplet of a water suspension of CNW on a copper grid covered with a holey carbon film for support. The black areas in the pictures are the holes in the carbon film. The picture shown in Figure 20.4a was obtained using backscattered electrons to image the CNW and uranyl acetate to stain the CNW and enhance the contrast. Backscattered electrons are sensitive to the atomic number contrast obtained by the staining, and therefore the presence and shapes of the whiskers were defined through the heavy elements surrounding the whiskers. However, low contrast and resolution made it difficult to discern the whiskers from the carbon foil and to measure the whisker dimensions. Secondary electron imaging was applied to achieve better resolution, but it was still difficult to clearly discern individual whiskers from agglomerated structures (Figure 20.4b).

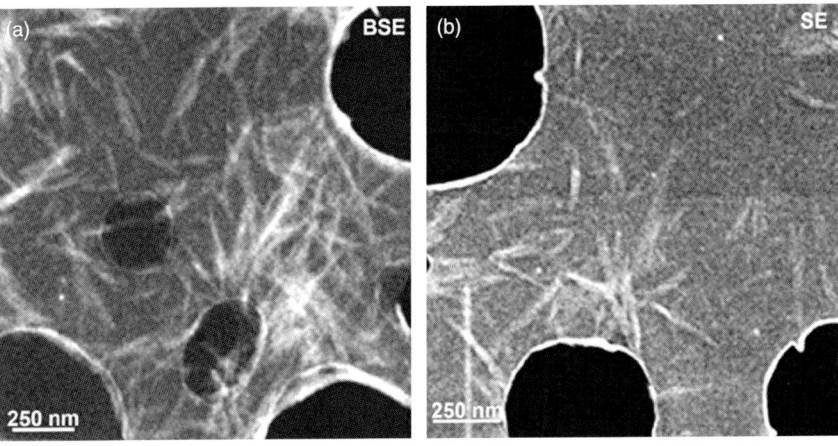

Figure 20.4 Cellulose nanowhiskers observed in FESEM using (a) back-scattered electrons (BSE) and (b) secondary electrons (SE).

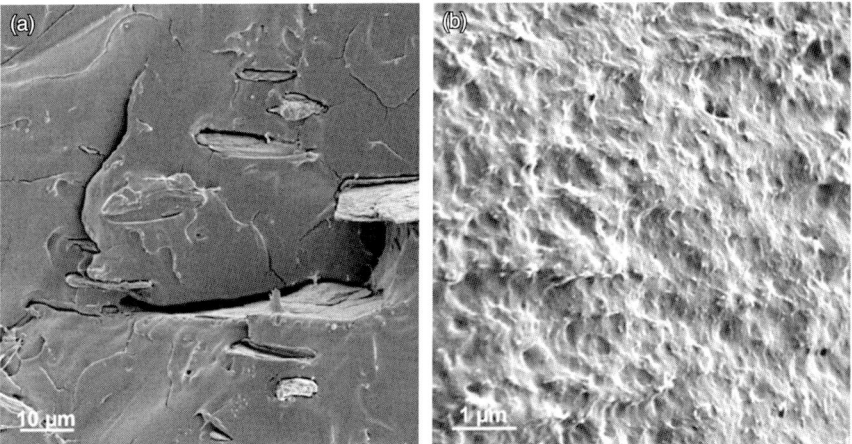

Figure 20.5 (a) The structure of PLA/CNW composite showing large agglomerates of CNW and (b) the structure of uncoated CAB/CNW showing well dispersed CNW.

Figure 20.5 shows examples of two different structures obtained after CNW-bionanocomposite processing. Figure 20.5a shows the structure of a PLA–CNW composite. This sample was fractured in liquid nitrogen and coated with platinum to avoid charging in the electron beam. The cellulose whiskers were agglomerated in the PLA matrix, which was easily detectable in FESEM. This material was therefore a conventional composite, and new strategies to obtain a better distribution of the CNW had to be found. Figure 20.5b shows the structure of a CNW nanocomposite with CAB as matrix. This sample was also fractured, but was not coated. It was still possible to obtain reasonable resolution using a voltage of 1 kV in the FESEM. In this case the CNW was better dispersed in the matrix. There were no big agglomerates detectable in the CAB matrix. It was, however, possible to see small spherical particles in the matrix that might be cross-sections of small clusters of cellulose whiskers or possibly additives in the matrix. In this case a further investigation of the material in AFM or TEM would give more information on the distribution of CNW in the matrix.

FESEM is a very convenient method to verify the presence of possible larger aggregates in the composite, and can thus be an important and quick first step in the analysis of the nanocomposite structures. Lack of aggregates in a sample could be an indication of good dispersion of the cellulose whiskers.

20.4 Transmission electron microscope

TEM is used for analysis of the bulk structure of materials. The principle of TEM is that high-energy electrons are transmitted through an ultrathin section of the specimen and the image is formed when the electrons hit a photo film below the specimen. The image is formed due to scattering of the electrons by the specimen. Bright field (BF) is an imaging mode where an objective aperture is inserted in such a way that the direct unscattered electrons form the image [22]. Areas in the specimen that scatter electrons weakly will therefore appear bright in BF TEM. Regions in the specimen that are thicker or of higher density will scatter more strongly and will appear darker in the image because highly scattered electrons are stopped by the objective aperture [21].

In TEM, there are three basic contrast mechanisms that may all contribute to the image formed: (1) diffraction contrast, (2) mass–thickness contrast, and (3) phase contrast [22]. Figure 20.6 shows schematically the mass–thickness effect. In the thin area almost all electrons are passing through the sample resulting in a brighter image. The thicker part of the same material will scatter more electrons, resulting in a darker image. The gray area illustrates higher density from heavy metal staining which will result in even higher scattering and give the darkest image. The cellulose-based nanocomposites are composed of low-atomic-number elements and therefore scatter electrons weakly, giving poor contrast in the TEM. For these materials, the mass–thickness contrast mechanism can be exploited

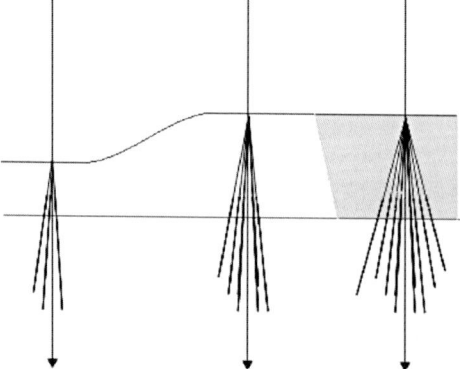

Figure 20.6 A schematic illustration of the mass–thickness contrast; the gray area is stained by a heavy metal and thus the electrons are more scattered.

by deliberately staining the thin specimen with a heavy metal that will highlight specific features of interest [21]. Staining involves the incorporation of high-atomic-number atoms into the polymer to increase the density and thus enhance contrast [23]. In the so-called negative staining, the shape of small particles is shown by staining the regions surrounding the particles rather than the particles themselves [22]. Uranyl acetate is widely and routinely used as a staining medium and is reported to produce high contrast and to be very suitable for high-resolution work [25].

20.4.1 Sample preparation

For the preparation of CNW samples for TEM analyses, a droplet of a highly diluted CNW water suspension can be placed on a carbon-coated copper grid and then dried. The whiskers may be analyzed directly in the TEM using diffraction contrast because they are highly crystalline, or they can be stained by, for example, uranyl acetate to enhance the contrast. Sample preparation of the nanocomposites is a much more time-consuming procedure and there is no single method that will fit all materials. Two methods for sample preparation of nanocomposites are ultramicrotomy and freeze-etching.

20.4.1.1 Ultramicrotomy

Ultramicrotomy is a preparation method where ultrathin (<100 nm) sections of the sample is prepared for TEM study [22]. This is one of the most widely used methods for the preparation of polymeric samples. This method allows direct observation of the actual structure in a bulk material. The steps involved in sample preparation for ultramicrotomy are as follows: (1) specimen mounting, (2) embedding in a resin and curing, (3) trimming and sectioning, and (4) post-staining. To obtain good sectioning it is important that the hardness of the embedding media matches the hardness of the sample. Epoxies are in general the most used embedding media because they are the most stable in the electron beam [22]. Ultramicrotome sections are obtained by first trimming the sample with a glass knife and then cutting the sample with a diamond knife. Figure 20.7 shows a schematic drawing of ultramicrotomy. The resultant slices are collected in a liquid-filled trough and then mounted on grids. The sections are on the order of 50–70 nm in thickness to allow the electron beam

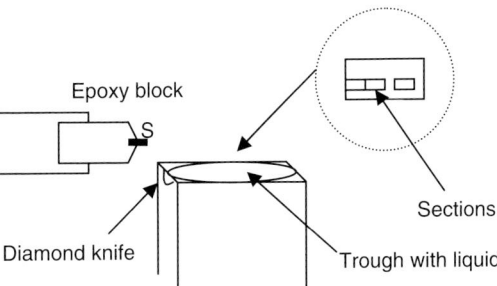

Figure 20.7 The principle of ultramicrotomy. The specimen (S) is embedded in an epoxy block and moved steadily past a diamond knife. Thin sections are floated off the knife into the liquid filled trough.

to transmit through the sample and to obtain sufficient signal to produce an image. It is important to slice the sample to an even thickness to avoid artifacts from mass–thickness contrast.

Polymers that have a glass transition temperature below room temperature can be too soft for ultramicrotomy at room temperature [22]. Cryo-ultramicrotomy is a sectioning method that is performed at low temperatures. The advantages of this technique are that embedding is not required and that soft polymers and water-soluble polymers can be sectioned. Disadvantages are the lengthy time needed, difficulty to collect the sections, and frost build-up [22].

20.4.1.2 Freeze-etching

The freeze-etching method is an alternative to cryo-ultramicrotomy for water-soluble or soft polymers and it is also useful for beam-sensitive materials. The advantage of the freeze-etching technique is that there is no need for chemical fixation or staining of the sample. The freeze-etching method involves five steps: rapid freezing, freeze fracturing, etching, shadowing and replication, and replica cleaning by dissolving the specimen [26]. For rapid freezing, the sample is mounted on a support and then rapidly immersed in propane at its melting point ($-189.6°C$) [26]. The sample is then placed in a liquid nitrogen-cooled chamber and cut with a knife. The freshly cleaved surface is then etched (vacuum sublimation of ice) in the presence of a coldtrap to prevent condensation of water vapor on the surface of the specimen. The surface is shadowed with platinum at an angle of 20–45° [22] and a replication is made with a thin layer of carbon. The replica is floated off the sample, washed and gathered onto TEM grids. Replicas are a copy of the surface characteristics of the original specimen and are very stable in the electron beam.

20.4.2 The structure of CNW and nanocomposites

Figure 20.8 shows the TEM pictures of CNW isolated from MCC using two different methods. Both samples were stained with uranyl acetate to enhance the contrast, although the CNW are highly crystalline and would therefore contribute to diffraction contrast. Figure 20.8a presents CNW isolated from MCC by sulfuric acid treatment. Proper treatment of cellulose with sulfuric acid will lead to esterification of hydroxyl groups by sulfate ions and therefore give a negatively charged surface, which will act as repulsive forces between the whiskers and thereby give a stable colloid suspension [27]. The CNW appeared to be quite well dispersed after drying, and it was possible to observe individual whiskers in TEM. The whiskers were measured to be 5 ± 2 nm in width and 210 ± 75 nm in length [14]. Figure 20.8b shows CNW isolated from MCC by hydrochloric acid treatment. In this case the whiskers were highly agglomerated and it was difficult to discern individual whiskers. This may reflect a lack of surface charges as opposed to the sulfuric-acid-treated CNW. Compared to FESEM examination of CNW, TEM, aided with staining to enhance the contrast, allowed for a more detailed examination of the individual cellulose whiskers.

TEM analysis of the CNW nanocomposites can be challenging for several reasons, depending on the matrix used. Challenges can be the sample preparation due to a water-sensitive or soft matrix, beam sensitivity, and lack of contrast between the whiskers and the matrix. Three examples of TEM examination of CNW nanocomposites are given in

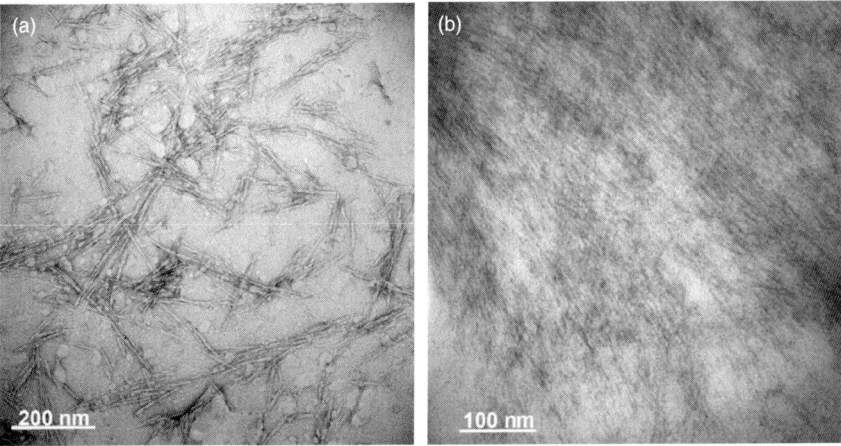

Figure 20.8 TEM pictures of CNW isolated from MCC by (a) sulfuric-acid-treatment and (b) hydrochloric acid treatment.

Figure 20.9. All nanocomposites were produced with whiskers isolated from sulfuric-acid-treated MCC. Figure 20.9a shows the structure of an ultramicrotomed CNW nanocomposite with PLA. The sample was stained with uranyl acetate and the TEM was operated at 100 kV. In this case the staining was not restricted to the close vicinity of the whiskers but was present more like a continuous film. In BF images this gave rise to a continuous dark background, and therefore a reasonable contrast between the whiskers and the surroundings was accomplished. Pores developed in the background film, where the underlying carbon film could be observed, having similar contrast to the whiskers.

Figure 20.9b shows the TEM image of a CNW nanocomposite with CAB. This sample was prepared and operated as for the PLA/CNW nanocomposite in Figure 20.9a. In this case the contrast between the CNW and CAB was very poor. Both CNW and CAB are based on cellulose, and it might therefore be difficult to selectively stain the area around the CNW. The stain appeared to be distributed randomly in the nanocomposite. Staining the nanocomposite longer with uranyl acetate did not improve the contrast, but rather introduced more noise in the image. Figure 20.9c shows a replica of a freeze-etched surface of a thermoplastic starch–CNW nanocomposite. Thermoplastic starch has a strong water affinity, and water has to be excluded from the preparation step. Conventional methods for chemical fixation and ultramicrotomy therefore cannot be used, and the sample was prepared by the freeze-etching method described earlier. The CNW were easily detectable in the thermoplastic starch. They seemed to protrude from the starch matrix and appeared to be wide. The contrast between the starch and the whiskers as it appeared in the replica was due to metal shadowing by platinum, and it was therefore difficult to judge whether the whiskers as observed were individual whiskers broadened by the shadowing or agglomerates of whiskers. A drawback of replication is that it is only a copy of the surface characteristics of the original specimen, and therefore artifacts that make the interpretation difficult may be introduced.

Contrary to SEM, TEM examination allows for determination of the whisker length in the matrix. However, the whiskers that can be distinguished in the TEM image in Figure 20.9b

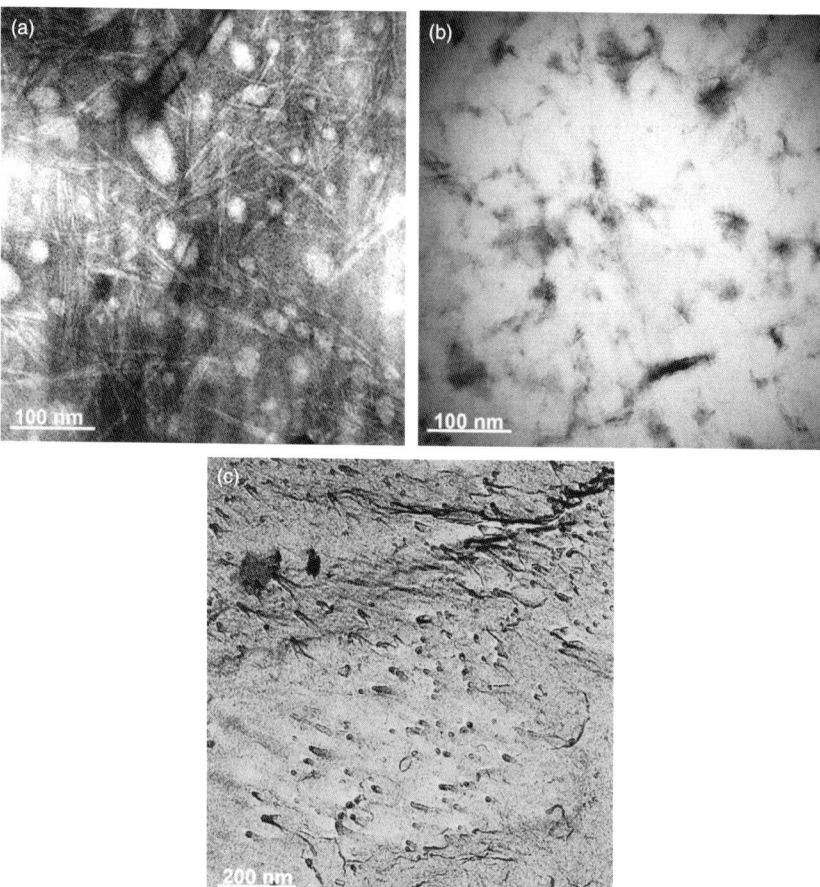

Figure 20.9 TEM pictures of (a) PLA–CNW nanocomposite having good contrast, (b) CAB/CNW nanocomposite having poor contrast, and (c) a replica of freeze-etched surface of thermoplastic starch/CNW nanocomposite.

seemed to be shorter compared to the whiskers in Figure 20.8a. This is probably a result of the sample preparation. When preparing a sample for TEM analysis by ultramicrotomy a sheet with ~50 nm thickness is cut. It is most likely that the whiskers are cut in this procedure, especially if there is a tendency of orientation of the CNW perpendicular to the sections. Thus, TEM examination of the nanocomposites may underestimate the length of the whiskers.

20.5 Atomic force microscope

The AFM is used for topography imaging and can resolve structures in x-, y-, and z-directions as opposed to SEM and TEM [28]. Additional advantages of AFM are that

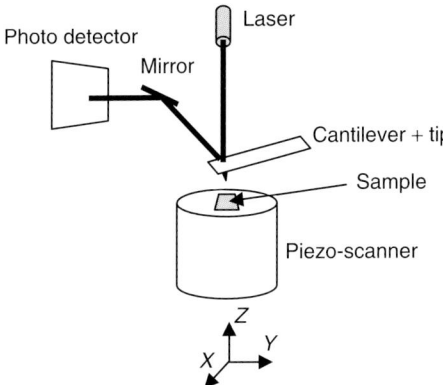

Figure 20.10 Schematic drawing of the atomic force microscope.

none or little pretreatment of the sample is required and that it can be operated in ambient air, liquid, or vacuum [28]. Figure 20.10 shows a schematic diagram of an AFM. A very sharp tip is fixed on the end of a cantilever. Typically, the tip has a radius of curvatures of a few nanometres at the apex and this determines the in-plane resolution [29]. The sample is fixed on a piezoelectric scanner that can move the sample (x-, y-, and z-axis) under the tip. The position of the cantilever is measured using a laser that is reflected on the cantilever and detected by a photodetector [30]. There are three main modes for AFM: contact mode, tapping mode, and noncontact mode. The tapping mode is used to image, for example, surfaces that are easily damaged (soft materials such as polymers) or loosely held to their substrates. The tapping-mode AFM eliminates shear forces present in contact mode [29]. In the tapping-mode AFM, the cantilever is oscillated at a frequency near its resonance (typically a few hundred kHz) [28]. The oscillation is driven by a constant force. The tip is brought toward the sample surface until it begins to touch the surface [28]. The oscillation amplitude is reduced by the contact between the sample and the tip [28]. A feedback control loop of the system maintains this new amplitude constant during the imaging. This is done by the z-component of the scanner. The imaging is obtained by monitoring the z-component of the sample while the tip moves across the surface. The tip is moved across the surface at a relatively slow rate (1 s/scan line) whereas tapping has a very high rate (200–400 kHz) [31]. Very stiff cantilevers with high resonant frequencies are required for the tapping-mode AFM [28]. The tapping-mode AFM is used for topography imaging but can also be extended to obtain phase images, which can be recorded simultaneously with the topographic image [31,32]. The difference in phase between the periodic signal that drives the cantilever and the oscillations of the cantilever are registered [31,32]. A change of phase is associated with a change of the properties of the sample, and therefore, the phase mode is used to detect variations in, for example, composition [32].

20.5.1 Sample preparation

AFM analysis is not carried out in vacuum or using electrons so there are no specific requirements for the sample preparation. For examination of CNW in AFM, a drop of a

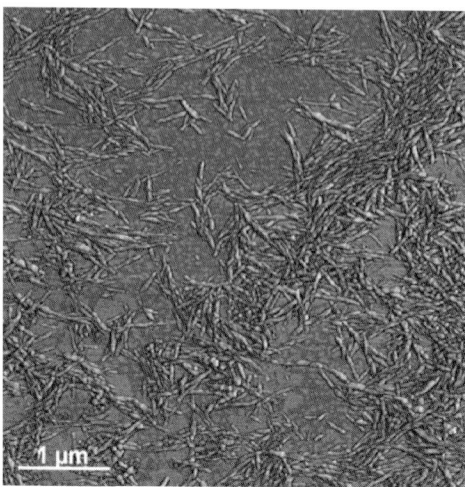

Figure 20.11 AFM phase image of CNW.

diluted CNW water suspension can be placed and dried on a freshly cleaved mica surface to ensure a flat background. For analysis of nanocomposites a flat surface is recommended, and therefore, the nanocomposites can be prepared by ultramicrotomy. For soft materials, the polymers can be smeared during cutting and can cover the underlying structure which will not be accessible in AFM. Therefore, cryo-ultramicrotomy can be beneficial for these materials.

20.5.2 The structure of CNW and nanocomposites

Figure 20.11 shows an AFM phase image of the CNW isolated from MCC by sulfuric acid treatment. These whiskers were dried on a mica surface. The AFM picture was easily obtained without any problems regarding contrast and resolution as for the FESEM and TEM analyses discussed in the earlier sections. However, the shape of the whiskers appeared different than that observed in FESEM and TEM (Figures 20.4 and 20.8). The structures differed from the needle-like shape as observed in TEM. The whiskers appeared significantly broader having a rounded shape. This broadening effect can be explained by the tip used for imaging. In general, the AFM tip has a finite size and shape. As the tip passes over a sample with surface features of comparable size as the tip, the shape of the tip will contribute to the image that is formed [33]. Figure 20.12 illustrates the tip-broadening effect. It was therefore difficult to judge whether the structures observed were individual whiskers or several whiskers agglomerated side by side. The whiskers appeared longer than those from TEM examination, and therefore the structures that appeared to be individual whiskers possibly consisted of several whiskers. Because of the broadening effect it was not possible to determine the width of an individual whisker. It is, however, possible to estimate the thickness of a single whisker in AFM by measuring the height difference between the mica surface and the whiskers because AFM has very good resolution in the z-direction. Line scans across several individual cellulose whiskers showed a $\sim$10–15 nm height difference

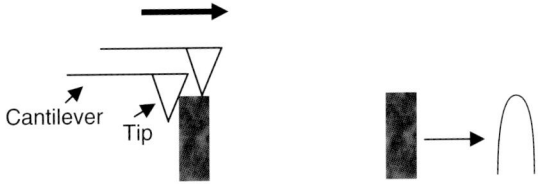

Figure 20.12 Schematic drawing of the tip-broadening effect in AFM.

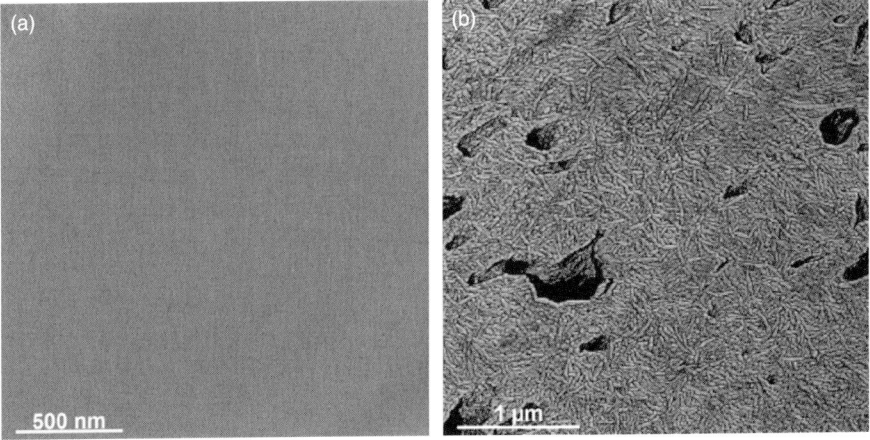

Figure 20.13 AFM phase images of two different PLA–CNW nanocomposites of (a) an ultramicrotomed surface and (b) a cryo-ultramicrotomed surface.

between the mica substrate and the whiskers. From TEM and AFM analyses, the whiskers thus appeared to have almost the same width as thickness.

Figure 20.13 shows AFM phase images of two different PLA–CNW nanocomposites. The sample in Figure 20.13a was prepared by ultramicrotomy. It was not possible to observe any whiskers in the matrix. An explanation for this might be that the soft polymer was covering the whiskers. The nanocomposite in Figure 20.13b was prepared by cryo-ultramicrotomy. The sample was analyzed directly on the cryo-microtomed surface without any further treatment of the sample. In this case the CNW were easily detectable in the PLA matrix. The whiskers partly protruded from the matrix, which may be caused by differences in thermal expansion during heating at cryogenic temperatures. The darker areas in the picture are holes in the polymer matrix. Again the whiskers appeared broader than those from TEM analysis, and a high-resolution tip should therefore be used for analysis of CNW nanocomposites. AFM imaging has the necessary resolution capabilities without the need for staining. AFM could therefore be a powerful alternative to TEM in materials where contrast between the whiskers and the matrix is limited and the beam sensitivity is an issue. The AFM technique described here is, however, a surface technique, and therefore has only limited access to the bulk structure.

Conclusions

For the structure determination of CNW and their bio-nanocomposites, different microscopy methods can be utilized to obtain information both on micro- and nanoscales. The sample preparation and instrumentation of bio-nanocomposites have been found generally to be challenging because they are nonconductive, soft, and water sensitive materials and consist of low-atomic-number elements.

FESEM was shown to be a very convenient method to detect the presence of possible larger agglomerates in the composites and can therefore be an important and easy first step in the analysis of the nanocomposite structure. Lack of aggregates in a sample indicates well-distributed CNW in the matrix.

More detailed information on the distribution of CNW can be obtained using TEM and AFM. In TEM it is possible to access the bulk structure of the material, but the whiskers might be cut in the sample preparation step, and therefore TEM examination of the nanocomposites may underestimate the length of the whiskers. TEM analysis of the CNW nanocomposites can be challenging due to the difficulty in sample preparation of a water sensitive or soft matrix, the beam sensitivity and lack of contrast between the whiskers and the matrix. AFM could be a powerful alternative to TEM because it has sufficient resolution capabilities without the need of staining, and because beam sensitivity is not an issue. The AFM technique described here is, however, a surface technique and therefore has only limited access to the bulk structure.

Acknowledgments

Cargill Dow LLC, Minneapolis, USA, is acknowledged for providing the Nature Works™ PLA polymer and Borregaard AS, Norway, is acknowledged for the microcrystalline cellulose. Norwegian Research Council under the NANOMAT program is acknowledged for financial support of this work. A special thanks to Bjørn Steinar Tanem, SINTEF Materials and Chemistry, for the FESEM pictures of CNW and for all the help with the microscopy work. The Laboratory of Biomass Morphogenesis and Information at the Research Institute for Sustainable Humanosphere (RISH), Kyoto University, Japan, and in particular Professor Junji Sugiyama and Chiori Ito are acknowledged for all their help with the freeze-etching technique.

References

1. Ajayan, P.M.; Schadler, L.S.; Braun, P.V., *Nanocomposite Science and Technology*, 1st Edition; Wiley VCH Verlag GmbH & Co; Weinheim; 2003.
2. Ray, S.S.; Okamoto, M., Polymer/layered silicate nanocomposites: a review from preparation to processing; *Prog. Polym. Sci.* 2003, 28, 1539–1641.
3. Garces, J.M.; Moll, D.J.; Bicerano, J.; Fibiger, R.F.; McLeod, D.G., Polymeric nanocomposites for automotive applications; *Adv. Mater.* 2000, 12, 1835–1839.
4. Azizi Samir, M.A.S.; Alloin, F.; Dufresne, A., Review of recent research into cellulosic whiskers, their properties and their application in nanocomposite fields; *Biomacromolecules* 2005, 6, 612–626.

5. Dufresne, A., Comparing the mechanical properties of high performance polymer nanocomposites from biological sources; *J. Nanosci. Nanotechnol.* 2006, 6, 322–330.
6. Tashiro, K.; Kobayashi, M., Theoretical evaluation of three-dimensional elastic constants of native and regenerated celluloses: role of hydrogen bonds; *Polymer* 1991, 32, 1516–1526.
7. Bondeson, D.; Mathew, A.; Oksman, K., Optimization of the isolation of nanocrystals from microcrystalline cellulose by acid hydrolysis; *Cellulose* 2006, 13, 171–180.
8. Anglés, M.N.; Dufresne, A., Plasticized starch/tunicin whiskers nanocomposites. 1. Structural analysis; *Macromolecules* 2000, 33, 8344–8353.
9. Lu, Y.; Weng, L.; Cao, X., Morphological, thermal and mechanical properties of ramie crystallites-reinforced plasticized starch biocomposites; *Carbohydr. Polym.* 2006, 63, 198–204.
10. Heux, L.; Chauve, G.; Bonini, C., Nonflocculating and chiral-nematic self-ordering of cellulose microcrystals suspensions in nonpolar solvents; *Langmuir* 2000, 16, 8210–8212.
11. Helbert, W.; Cavaillé, J.Y.; Dufresne, A., Thermoplastic nanocomposites filled with wheat straw cellulose whiskers. Part I: Processing and mechanical behavior; *Polym. Compos.* 1996, 17, 604–611.
12. Grunert, M.; Winter, W.T., Nanocomposites of cellulose acetate butyrate reinforced with cellulose nanocrystals; *J. Polym. Environ.* 2002, 10, 27–30.
13. Azizi Samir, M.A.S.; Alloin, F.; Paillet, M.; Dufresne, A., Tangling effect in fibrillated cellulose reinforced nanocomposites; *Macromolecules* 2004, 37, 4313–4316.
14. Kvien, I.; Tanem, B.S.; Oksman, K., Characterization of cellulose whiskers and their nanocomposites by atomic force and electron microscopy; *Biomacromolecules* 2005, 6, 3160–3165.
15. Levis, S. R.; Deasy, P.B., Production and evaluation of size reduced grades of microcrystalline cellulose; *Int. J. Pharm.* 2001, 213, 13–24.
16. Battista, O.A.; Smith, P.A., Microcrystalline cellulose; *J. Ind. Eng. Chem.* 1962, 54, 20–29.
17. Bondeson, D.; Kvien, I.; Oksman, K., Strategies for preparation of cellulose whiskers from microcrystalline cellulose as reinforcement in nanocomposites. In: Oksman, K.; Sain, M. (eds); *Cellulose Nanocomposites Processing Characterization and Properties*; ACS Symposium Series 938; Washington, D.C., 2006; pp. 10–25.
18. Petersson, L.; Oksman, K., Preparation and properties of biopolymer-based nanocomposite films using microcrystalline cellulose. In: Oksman, K.; Sain M. (eds); *Cellulose Nanocomposites Processing Characterization and Properties*; ACS Symposium Series 938; Washington, D.C., 2006; pp. 132–150.
19. Oksman, K.; Mathew, A.P.; Bondeson, D.; Kvien, I., Manufacturing process of cellulose whiskers/polylactic acid nanocomposite; *Compos. Sci. Technol.* 2006, 66, 2776–2784.
20. Petersson, L.; Oksman, K., Biopolymer based nanocomposites: comparing layered silicates and microcrystalline cellulose as nanoreinforcement; *Compos. Sci. Technol.* 2006, 66, 2187–2196.
21. Goodhew, P.J.; Humphreys, J.; Beanland, R., *Electron Microscopy and Analysis*; Taylor & Francis; London, 2001.
22. Sawyer, L.C.; Grubb, D.T., *Polymer Microscopy*, 1st Edition; Chapman and Hall; London, 1987.
23. Magonov, S.N.; Yerina, N.A., Visualization of nanostructures with atomic force microscopy. In: Yao, N.; Wang, Z.L. (eds); *Handbook of Microscopy for Nanotechnology*; Kluwer Academic Publishers; New York, 2005.
24. Serry, F.M.; Strausser, Y.E.; Elings, J.; Magonov, J.; Thornton, S.J.; Ge, L., Surface characterization using atomic force microscopy; *Surf. Eng.* 1999, 15, 285–290.
25. Goldstein, J.; Newbury, D.; Joy, D. et al., *Scanning Electron Microscopy and X-ray Microanalysis*, 3rd Edition; Kluwer Academic/Plenum Publishers; New York, 2003.
26. Echlin, P., Coating techniques for scanning electron microscopy and x-ray microanalysis; *Scanning Electron Microsc.* 1978, 1, 109–132.
27. Lewis, P.R.; Knight, D.P., Staining methods for sectioned material in practical methods. In: Glauert, A.M. (ed.); *Electron Microscopy*; Elsevier/North-Holland Biomedical Press; Amsterdam, 1977.

28. Willison, J.H.M.; Rowe, A.J., Replica, shadowing and freeze-etching techniques. In: Glauert, A.M. (ed.); *Practical Methods in Electron Microscopy*; Elsevier/North-Holland Biomedical Press; Amsterdam, 1980.
29. Marchessault, R.H.; Morehead, F.F.; Koch, M.J., Hydrodynamic properties of neutral suspensions of cellulose crystallites as related to size and shape; *J. Colloid Sci.* 1961, 16, 327–344.
30. Strausser, Y.E.; Heaton, M.G., Scanning probe microscopy. Technology and recent innovations; *Am. Lab.* 1994, 26, 20,22,24–29.
31. Drake, B.; Prater, C.B.; Weisenhorn, A.L. *et al.*, Imaging crystals, polymers, and processes in water with the atomic force microscope; *Science* 1989, 243, 1586–1589.
32. Schmitz, I.; Schreiner, M.; Friedbacher, G.; Grasserbauer, M., Phase imaging as an extension to tapping mode AFM for the identification of material properties on humidity-sensitive surfaces; *Appl. Surf. Sci.* 1997, 115, 190–198.
33. Babcock, K.L.; Prater, C.B., Applications of scanning probe microscopy Part 5: Phase imaging beyond topography; *Am. Lab.* 1996, 4, 28–30.
34. Markiewicz, P.; Goh, M.C., Simulation of atomic force microscope tip-sample/sample-tip reconstruction; *J. Vac. Sci. Technol. B* 1995, 13, 1115–1118.

Index

abaca, 252, 253
acid(s), 7, 19, 30, 50, 51, 63, 73, 91, 95, 151, 175, 177, 214, 218, 256, 258, 261, 268, 294, 295
   acetic, 110, 112, 316
   alginic, 219, 220
   arachidic, 110, 111
   behenic, 110, 111
   boric, 29
   carboxyl(ic), 89, 178, 181, 197, 199, 294
   chromic, 289
   p-coumaric, 177, 181, 183–5
   dicarboxylic, 294
   dichloroacetic, 316
   fatty, 4, 6, 91, 101, 104, 107–13, 115, 175, 258
   ferulic, 177, 183, 184
   heptadecanoic, 110, 111
   hexenuronic, 27
   hydrochloric, 239, 301, 348, 349
   hydroxy-, 177–9, 181, 183–5
   Lewis, 50, 51
   lignoceric, 110–12
   linoleic, 110–12
   mineral, 341
   nitric, 289
   oleic, 41, 42, 104, 106, 110, 115
   palmitic, 110–12
   peracetic, 70, 173
   polyglucuronic, 218
   stearic, 41, 110–12, 258, 260, 291, 292, 296, 297
   succinic, 293, 294
   sulfonic, 195–7
   sulphuric (or sulfuric), 258, 263, 289, 348, 349, 352
   uronic, 7
acid hydrolysis, 5, 7, 74, 163, 221, 239, 263, 341
acidity, 149, 258, 266, 268
acidolysis, 149–53, 173, 318, 325

activation energy, 139, 140, 145, 318, 325
   α-dispersion, 326
   β-dispersion, 326
   initial stage of thermal decomposition of cellulose fabric, 142
   initial stage of thermal decomposition of cellulose powder, 142
adsorption isotherm(s), 46, 47, 70
alkali (E), 37, 38, 43, 47–9, 51, 75, 150, 152, 173, 177, 181, 186, 257, 261, 264
   extraction, 75, 177
   soluble lignin, 152
alkaline, 38, 47, 51, 255, 306
   conditions, 36, 52, 162, 263
   extraction, 173
   hydrogen peroxide, 25, 26
   hydrolysis, 163, 165
   peroxide, 74
   pH, 75
   pulping, 162
alkyl ketene dimer (AKD), 81–98
   sized papers, 83, 91, 94
   sized pilot papers, 81, 87
   structure of, 82
alkylated kraft lignin-based polymeric material, 309, 313
anhydroglucose, 244, 245
apparent density (ρ), 275, 277, 279, 281–3
arabinose, 7, 164, 177
aramid(s), 125, 131
   fibers, 131, 132
ash(es), 171, 175, 176, 186
ash content, 172, 175
aspect ratio, 253
aspen, 163, 191, 233, 236, 237
   dissolved kraft lignin, 164
   fibers, 264
   fiber-reinforced high-density polyethylene (HDPE), 264

aspen (*Cont'd.*)
  lignosulfonate from, 192
  wood, 153
astringin, 3, 7, 8
atomic force microscopy (AFM), 36, 37, 39, 41, 52, 115, 116, 261–3, 329, 330, 332, 337, 338, 340, 342, 343, 346, 350–54
  images, 261, 338
  micrographs, 40
  phase contrast images, 39
  tip, 40, 352
atomic force microscopy (AFM) phase image, 40
  of cellulose nanowhiskers (CNW), 352
  of chemithermomechanical pulp (CTMP), 41
  of polylactic acid (PLA)–cellulose nanowhiskers (CNW) nanocomposites, 353

back-scattered electrons (BSE), 343–6
bagasse, 171, 253, 266
bamboo, 252, 255, 267
bark, 3–7, 9–15, 232
bending strength (σ), 275, 277–9, 283, 284
biocomposites, 176, 255, 275, 276, 284
  classification, 276
  durability, 276
biofiller, 275–7, 279
biomass, 61, 76, 148, 171, 186
bio-nanocomposite(s), 344, 345, 354
biorefining, 148, 149
birch, 74, 107, 155, 163, 191, 233, 234, 236, 237, 239
  bleached kraft pulp (BKP), 103
  chemical pulp, 106
  dissolved lignin, 163
  fibers, 264
  kraft pulp(s), 107
  lignin, 159
  pulp(s), 108, 153, 236, 239–41
  residual lignin, 163
  wood, 150
black spruce, 19, 23
bleaching, 20, 25–7, 101, 102, 107, 175, 222, 236, 341
Bragg's equation, 342

carbohydrate(s), 3, 5, 7, 27, 42, 61, 63, 104, 115, 148–53, 163–5, 176, 177, 234, 243
carboxymethyl cellulose (CMC), 73, 218–21, 244, 266, 329
catechin, 4, 8–10
β-D-cellotetraose, 126, 127
cellouronic acid(s) (CUA), 240, 221

degree(s) of polymerization (DP_W and DP_n), 220, 221
  sodium (Na) salt, 218–20
cellulase(s), 60–64, 66–71, 73–6, 153
cellulolytic enzyme lignin (CEL), 150–53
cellulolytic enzymes, 61
cellulolytic hydrolysis, 60, 63
cellulose, 6, 15, 18–20, 22, 23, 25–7, 29, 30, 42, 44, 46, 60–66, 68–76, 82, 92, 104, 106, 108, 115, 121, 125, 126, 128–34, 138, 140, 142, 145, 171–3, 175, 176, 186, 190, 206–10, 212–22, 224, 227–33, 235–40, 243, 245, 251, 252, 254, 257–9, 263, 266, 290, 294, 336, 340–42, 346, 348, 349
  amorphous, 228, 231, 233, 239
  bacterial, 126, 215, 341
  binding domains (CBDs), 63
  crystalline, 229, 231, 232, 238, 268
  crystallites, 128, 134
  deformation of (in), 23, 121, 128
  deformation processes in, 127, 133
  depolymerization of, 62
  derivatives, 206, 207, 209, 212, 220, 224, 227, 228, 243, 244, 246, 316, 329
  disorder in, 128, 130, 133
  dissolution(s) of, 209
  distribution in the cell walls, 23
  enzyme, 73
  fabric, 138–46
  fiber(s), 87, 95, 121, 125, 126, 128–33, 215, 227, 228, 235, 246, 251, 254, 257, 258, 264–6, 268, 276, 290, 294, 340, 341
  hydrolysis, 61, 64–7, 75, 258
  hydroxyl groups, 68, 82, 209, 218, 224
  lifetime of, 140
  mercerized, 218
  microfibrils, 340, 341
  monocrystals, 340
  native, 126, 128–31, 266
  powder, 138, 139, 142, 143
  pulp(s), 27, 206
  Raman images of, 23, 25
  Raman spectrum of, 128
  regenerated, 125, 129–34, 210, 215, 218, 219, 221, 254, 265
  solution(s), 206, 209, 210
  solvent, 212, 224
  source of, 60
  sugars from, 61
  sulfite pulp, 70
  surface(s), 82, 251
  wood, 29
cellulose acetate butyrate (CAB), 342, 345, 349, 350

cellulose crystallinity (crystallinities), 29, 228, 231–4, 236, 238, 239
cellulose ethers, 212
cellulose I, 127, 139, 238, 267
   crystallinity, 17, 29, 30
cellulose Iα, 126, 127, 237, 238
cellulose Iβ, 126, 127, 231, 237, 238
cellulose II, 126, 127, 237
cellulose nanowhiskers (CNW), 340–42, 344–54
cellulose triethylcarbamate, 209
cellulose triphenylcarbamate (CTCs), 209, 222, 223
cellulose tunicate, 207, 210, 214
cellulose whiskers, 263, 340, 341, 345, 346, 348, 352
cellulosics, 123, 125, 138, 139
   deformation process in, 121, 125
   lifetime, 138
   lifetime prediction of, 138
   native, 121
   regenerated, 121
   swelling, 67
chemical microscopy, 101, 102, 115
co-crystallization, 293
coir, 252, 253, 255
cold crystallization, 333
cold crystallization temperature (T_{cc}), 333, 334, 336
compatibilization mechanisms, 288, 289, 291–3, 295, 296
compatibilizer, 260, 262–4, 288, 289, 292–6, 342
   wood–polyethylene (PE) composites, 295, 296
   wood–polypropylene (PP) composites, 295
composite(s), 93, 123, 131, 133, 172, 251, 253, 254, 256, 257, 260–67, 269, 270, 276–82, 284, 288, 340, 345, 346, 354
   aspen-fiber-reinforced high-density polyethylene (HDPE), 264
   cellulose fiber-reinforced thermoplastic, 264
   cellulose fibers, 95
   densely filled, 275, 279
   epoxy-based, 254, 255, 260
   fiber, 253, 254, 261, 266
   fiber and silicone, 262
   fiber-reinforced, 125, 251, 255, 258, 260
   fiber-reinforced elastomer, 267, 269
   fiber-reinforced, natural rubber, 255, 269
   fiber-reinforced silicone, 262
   flax-fiber-reinforced polypropylene (PP), 260
   fracture surface of, 263
   green, 255, 260
   hybrid, 255
   jute-polycarbonate, 264
   kenaf-reinforced poly(L-lactic acid) (PLLA), 267
   kraft lignin (KL)-based, polyurethane (PU), 277
   kraft lignin diethylene glycol polyurethane (KLDPU), 276–85
   kraft lignin polyethylene glycol polyurethane (KLPPU), 276–85
   kraft lignin triethylene glycol polyurethane (KLTPU), 276–85
   lignocellulosic fiber, 254
   lignocellulosic fiber reinforced, 251, 255, 258
   luffa fiber–polypropylene (PP), 261
   micromechanics, 265
   modulus, 255
   natural fiber thermoplastic, 257
   polylactic acid (PLA)-cellulose nanowhiskers (CNW), 345
   polyurethane (PU), 277, 279–83, 285
   processing, 341
   silica-filled (rubber), 292
   sisal-fiber-reinforced polypropylene (PP), 268
   wood, 290
   wood-fiber-reinforced polypropylene (PP), 264
   wood-high density polyethylene (HDPE), 289–91
   wood-linear low density polyethylene (LLDPE), 290
   wood–polyethylene (PE), 291, 293–6
   wood–polymer, 266
   wood–polypropylene (PP), 263, 289, 291, 293, 295, 296
conductometric titration, 46
coniferaldehyde, 23, 25, 26
coniferyl alcohol, 12, 23, 26, 183, 275
coniferyl units, 190
contact angle(s) (θ), 39, 50, 51, 81, 83, 85, 86, 93–7, 266, 294
   of ethylene glycol, 81, 97, 98
   measurement(s), 36, 48, 52, 81, 85, 261, 267, 294, 297
   of paper, 83, 94, 95, 98
   of water, 81, 83, 95
contour length (L), 216
cotton, 67, 72, 73, 126, 139, 213–15, 251–4, 341
   celluloses, 209
   fiber(s), 63, 263
   linters, 73, 215, 267
coumarates, 171
p-coumarates, 179, 183, 184, 186
cryo-ultramicrotomy, 348, 352, 353
crystal structure(s), 126
   β-D-cellotetraose, 126, 127
   cellulose Iα and Iβ, 127

crystal structure(s) (Cont'd.)
  dimeric and trimeric lignin model
    compounds, 310
  divanillyltetrahydrofuran, 310
  nonphenolic 8-O-4′-linked dilignol, 311
  phenolic 8-5′-linked dilignol, 311
  selected di- and trilignols, 301
crystallinity (Crystallinities), 29, 30, 60–62, 64,
    66, 70, 71, 130, 232, 233, 235, 238, 239,
    252, 254, 266, 310
  cellulose, 29, 228, 231–4, 236, 238, 239
  cellulose I, 17, 29, 30
  index (CrI), 227, 229, 232–4, 236
  of regenerated cellulose, 254
  of wood fiber(s), 17, 29
crystallization, 263, 290, 333
  kinetics, 140
  of polyethylene (PE), 290
  of polypropylene (PP), 263, 290, 295
  of polypropylene (PP) in wood–plastic
    composites (WPCs), 290
  rate, 290, 295
  temperature, 255, 290, 295, 333, 336

deconvolution, 236, 238
  signal, 237
  of spectrum, 238
degree of polymerization (DP), 60, 61, 64, 212,
    221, 252
degree of substitution (DS), 316, 320, 323
  carboxymethyl cellulose(s) (CMC(s)), 221,
    222, 245
  cellulose/OKD β-ketoester, 222–4
  ethylcellulose (EC), 316, 317, 320–26
  methylcellulose (MC), 318, 325
  sodium cellulose sulfate (NaCS), 330
derivative thermogravimetry (DTG), 275
derivative thermogravimetry (DTG) curves
  kraft lignin polyethylene glycol polyurethane
    (KLPPU) composites, 277, 280
  wood meal–KLPPU composites, 278
derivatization followed by reductive cleavage
    (DFRC), 156, 158, 159
2,3-dichloro-5,6-dicyano-1,4-benzoquinone
    (DDQ), 157
  oxidation, 159, 163
differential scanning calorimetry (DSC), 254,
    263, 289, 301, 316, 318, 320–22, 326,
    329, 330, 332, 333
differential scanning calorimetry (DSC) heating
    curves, 320, 332
  for ethyl cellulose (EC) films and powder, 320
  of sodium cellulose sulfate (NaCS) and
    NaCS–polyurethane (PU) hydrogels, 333

  of sodium cellulose sulfate (NaCS)–water and
    NaCS–polyurethane (PU)–water
    hydrogels, 332, 333
dihydroconiferyl alcohol, 12
N,N-dimethylacetamide (DMAc), 209, 210
  chemical structure, 209
  see also lithium
    chloride/N,N-dimethylacetamide
    (LiCl/DMAc)
1,3-dimethyl-2-imidazolidinone (DMI), 209,
    210
  chemical structure, 209
  see also lithium chloride/1,3-dimethyl-2-
    imidazolidinone
    (LiCl/DMI)
dipolar decoupling (dephasing) (DD), 227, 228,
    230, 233, 234, 240, 242
dipolar decoupling (dephasing) (DD) spectrum
    (spectra), 233, 237, 240
  of a kraft pulp residual lignin, 242
  of silver birch and Scots pine, 234
direct blue dye, 68
direct orange dye, 68
α-dispersion, 316, 322, 324–6
β-dispersion, 322, 325, 326
dissolution of amorphous cellulose, 239
dissolution of cellulose, 206, 209, 210
dissolution of regenerated cellulose, 210
dithionite bleaching (Y), 36, 37, 43, 47, 48
Douglas-fir (*Pseudotsuga menziesi*), 66, 75
dynamic Absorption Tester (DAT), 85
dynamic contact angle (DCA), 81, 85, 86, 93–7,
    261, 268
dynamic loss modulus (E''), 318, 322, 325
dynamic mechanical analysis (DMA), 267, 268,
    289, 290, 316, 322, 326
dynamic mechanical thermoanalysis (DMTA),
    289
dynamic modulus (E'), 268, 269, 318, 322–7

electron spectroscopy for chemical analysis
    (ESCA), 37, 39, 102, 263, 294, 297
  see also x-ray photoelectron spectroscopy
    (XPS)
electrospray ionization mass spectrometry
    (ESI-MS), 162, 174
elemental analyses, 177
elemental analysis of lignosulfonates, 199
endoglucanase(s), 62
energy dispersive x-ray fluorescence (EDXRF)
    spectrometer, 200
environmental scanning electron microscopy
    (ESEM), 342

enzymatic hydrolysis, 60–62, 64–7, 70, 72, 73, 76, 149–51, 153
enzymatic mild acidolysis lignin (MEAL), 150–54
enzyme(s), 60–64, 75, 76, 153
  activity, 63
  carbohydrate degrading, 61
  cellulase, 62, 64, 66, 71, 73
  cellulolytic, 61
  glycosyl hydrolase (GH), 63
  hydrolysis, 61
  hydrolytic, 61
  plant cell wall degrading, 62
epicatechin, 9, 10
ethanol, 60, 61, 71, 172, 175, 190
ethyl isocyanate, 290
ethylcellulose (EC), 316–18, 321–3, 325–8
  cross-linked, 316, 321, 323
  diurethane cross-linked, 317, 322, 323
  film(s), 317, 320–26
  hexamethylene diurethane cross-linked, 319, 321
  powder, 317, 320, 323
  solution, 318
eucalyptus, 19, 152, 192, 261
  fibers, 262
  lignin, 156
Eucalyptus globulus, 152, 191
  residual lignin, 163
Eucalyptus grandis, 152, 155, 191
  residual lignin, 162
extractives, 4, 19, 41–6, 51, 52, 87, 89, 92, 101–105, 108, 109, 111, 113–15, 171–3, 175, 177, 186, 229, 230
  content, 43, 45, 52, 106, 107, 176
  hexane-extracted, 41
  hydrophobic, 36, 52
  lipophilic, 6, 175
  microscopy, 101
  secondary ions of, 105, 111
  surface, 44, 102, 104, 106–108, 110, 113–16
  surface content of ($\phi_{extractives}$), 43
  surface coverage by, 103, 104, 106–108
  water-soluble, 176

ferulates, 171, 184, 186
fiber(s), 18, 25, 19, 31, 36, 37, 39, 40, 43, 45–8, 50–52, 60, 61, 63, 66–9, 72, 73, 81, 82, 87, 89–92, 101, 102, 105, 106, 107, 113–15, 121, 123, 125–34, 175, 176, 206, 227, 236–8, 245, 251–70, 288–91, 294, 295, 297
  aramid, 131, 132
  cell wall, 18, 22
cellulose, 87, 95, 121, 125, 126, 128–33, 215, 227, 228, 235, 246, 251, 254, 257,
  subitem 258, 264–6, 268, 276, 290, 294, 340, 341
chemical pulp, 27
chemithermomechanical pulp (CTMP), 41, 51
composite(s), 253, 254, 261, 266
crystalline, 124
deformation, 123, 125
glass, 251, 254, 255, 276, 292
kraft, 41, 51
kraft pulp, 27, 66, 69, 83, 128, 130, 131, 134
length, 60, 66, 253, 260
lignocellulosic, 69, 83, 251, 253, 254, 255, 258, 264, 266, 268, 269
mass, 50
mechanical, 38, 50, 52
mechanical pulp, 26, 36, 39, 48, 51, 83
natural, 251–5, 257, 258, 261, 265–70, 291
plant, 251, 252, 254, 255
polyethylene, 125
polyethylene terephthalate (PET), 131, 132, 303
polypropylene, 131
polymer, 123, 125
pressure groundwood (PGW), 39, 40, 51
pulp, 17, 18, 20, 22, 23, 25, 30, 31, 37, 41, 66, 68, 69, 83, 88, 101, 102, 104, 116, 222, 236, 268
reactive group, 82
recycled, 106, 108, 268
regenerated cellulose, 129–33, 215, 265
reinforced composite(s), 125, 251, 255, 258, 260
reinforced elastomer composites, 267, 269
reinforced high-density polyethylene (HDPE) composite, 264
reinforced natural rubber, 269
reinforced plastics, 259
reinforced polymers, 266
reinforced polypropylene (PP) composites, 260, 268
reinforced silicone composites, 262
reinforced thermoplastic composites
reinforced thermoplastics, 257, 260, 264
rupture, 37
strength, 253, 260, 261
surface, 36, 37, 39–42, 45, 46, 48, 51, 52, 67, 82, 88, 89, 97, 98, 101, 102, 105, 108, 114–16, 237, 238, 251, 255–8, 261, 264, 266, 267, 290, 295, 296
swelling of, 38, 67
synthetic, 251
thermomechanical pulp (TMP), 39, 40, 51

fiber(s) (Cont'd.)
  viscose, 132
  volume, 50
  wall(s), 38, 46, 50, 115, 237
  width, 60, 66
  wood, 17, 22, 23, 26, 29, 254, 264, 289–92, 294–6
fiber quality analyzer (FQA), 65
field emission scanning electron microscopy (FESEM), 262, 342, 343 Filler(s), 36, 82–4, 87, 92, 108, 254, 255, 257, 262, 275–7, 279–85, 290
  inorganic, 275, 277, 280, 281, 284
  lignocellulosic, 254
  paper, 106
  surface, 82
  surface area, 81
  wettability, 254
fines, 36, 39, 42, 45, 46, 52, 66, 82, 107, 109, 237, 238, 291
  pulp, 66
  surface, 41, 42, 46
flavonoid(s), 9–11, 14, 15, 19
flavonoid linkage, 8
flax, 251–4, 267
  fiber(s), 260, 261
fluorescence, 17, 18, 22, 23, 25, 27, 28, 30, 124
freeze-etching, 347–9
fructose, 4, 6
fuel, 3, 60, 275

galactose, 7, 164, 177
gas chromatography (GC), 5, 89, 173, 180
gas chromatography mass spectrometry (GCMS), 5, 29, 180
glass transition temperature(s) (T_g(s)), 138, 266, 301, 303, 306–309, 312–14, 316, 320–22, 326, 333, 334, 348
  composition curve(s), 307, 308
  ethylcellulose (EC) films and powder, 320
  plots, 326
glucan, 15
glucose, 4, 6, 7, 13–15, 61, 62, 130, 133, 164, 176, 177, 190, 224, 229, 316
D-glucose, 128, 132
β-glucosidase, 4, 7, 62, 69, 74
glucoside(s), 3, 4, 6–8, 10, 11, 14, 175, 176
guaiacyl (G), 11, 19, 20, 73, 148, 152–7, 160, 162, 163, 171, 177, 181–4, 215, 231, 233–5, 240

halo(es), 302, 303, 309
  amorphous, 301, 303, 306, 309–12, 314
  diffuse, 301
  of dioxane lignin, 302
hardwood(s), 18, 29, 73, 84, 101, 148, 152, 162, 163, 190, 192, 231–5, 237, 239
  bleached kraft pulp, 212–15
  cells, 101
  crystallinities, 234
  fibers, 89
  kraft pulps, 153
  lignin(s), 153, 156, 159, 162, 163, 233
  lignosulfonates, 190–92, 194
  milled wood lignin(s) (MWL(s)), 155, 156, 159
  native lignin(s), 152, 153
  pulp(s), 29, 229
  residual lignin(s), 153
  unbleached chemical pulps, 218
  unbleached kraft pulp(s), 218, 219
hemicellulases, 62
hemicellulose(s), 19, 28, 30, 39, 60, 61, 64, 65, 68, 73, 75, 76, 101, 115, 149, 171, 173, 175–7, 186, 206, 207, 209, 213, 218, 227, 229–33, 235, 236, 238–40, 251, 263
hemp, 251–3
hemp fibers, 253, 254, 261, 266
heteronuclear multiple bond correlation (HMBC), 158, 164
heteronuclear multiple quantum coherence (HMQC), 157, 159
  spectrum, 165
heteronuclear single quantum correlation (HSQC), 3, 5, 8–12, 14, 154, 157–9, 164, 165, 193, 195, 196
heteronuclear single quantum correlation–total correlation spectroscopy (HSQC–TOCSY), 5, 8–10, 12, 158
hexamethylene diisocyanate (HDI) (1,6-hexamethylene diisocyanate), 290 317, 318, 330, 338
high performance liquid chromatography (HPLC), 4, 7, 8
holocellulose, 4, 172, 173, 175, 176, 210, 218
hydrodynamic radius (R_h), 207, 211, 224
hydrodynamic radius of polymers, 216
hydrophobic interaction(s), 288, 291, 292, 294, 297
hydrophobic interaction chromatography (HIC), 189, 197–9
  analysis of lignosulfonates, 197
  chromatogram, 197–9
  chromatogram of lignosulfonates, 197
  column, 197
p-hydroxyphenyl (H), 148, 152, 155, 160, 171, 177, 181–3
hydroxypropyl cellulose, 219, 220, 329

infrared (IR), 22, 30, 41, 121, 124, 125, 291, 294, 297
    Fourier-transform (FT), 141, 151, 157, 171, 174, 178, 179, 254, 263, 264, 266, subitem 295, 296
    near-, 17, 19, 20, 26–30, 124, 157, 158
    spectrum (spectra), 134, 178, 179, 294, 296
interface(s), 115, 251, 255–62, 264, 265, 267, 269, 270
    adhesion, 251
    air-solid, 88
    celluloses and other materials, 133
    flax-fiber-reinforced polypropylene (PP) composites, 260
    fiber-air, 115
    fiber-matrix, 256, 260
    fiber-vacuum, 115
    lignocellulosic fiber reinforced composites, 251, 255, 258
    mechanical bonding, 261
    polymers and polymer fiber-reinforced composites, 125
    strength, 256
    stress transfer, 251
    stress transfer efficiency, 260
    wood fiber-reinforced polypropylene (PP) composites, 264
    wood–polypropylene (PP), 289
interfacial adhesion, 254–6, 261, 267, 268, 288, 290, 292, 294
    fiber-matrix, 256
    fiber-reinforced elastomer composites, 269
    fibers and polypropylene (PP) or polyethylene (PE) matrix, 297
    flax-fiber-reinforced polypropylene (PP), 260
    flax fibers and unsaturated polyester (UP) resin, 261
    jute fiber and polycarbonate, 264
    waste flour-matrix, 262
    wood and a thermoplastic, 288, 289
    wood and polyethylene (PE) (or polypropylene, PP), 291
intrinsic viscosities, 192
intrinsic viscosity detector, 209
inverse gas chromatography (IGC), 50, 266, 268
isorhapontin, 3, 7, 8
isocyanate(s), 257, 281, 290

jute, 251–3, 261
    fiber(s), 264
    polycarbonate composites, 264
    surface, 264

kenaf, 251–3, 266, 267
    fibers, 267
kraft lignin(s) (KL), 160, 163, 275, 276, 279, 304–10, 312–14
    acetylated, 305
    alkylated, 306–308, 312
    dissolved, 164
    ethylated, 306
    methylated, 301, 305–310, 312, 313
    methylated paucidisperse, 305, 311, 312
    molecular weight of, 301
    paucidisperse, 304, 306
    paucidisperse methylated, 309, 311
    polydisperse, 312
    softwood, 306
kraft lignin (KL)-based polyurethane (PU) composites, 276, 277
kraft lignin (KL)-based polyurethane (PU) matrix, 276, 279
kraft lignin (KL)-polyol polyurethane (PU) matrix, 284, 285
kraft lignin-based (polymeric) materials, 301, 304, 306, 308
kraft lignin-based thermoplastic (polymer) blends, 301, 306
kraft lignin-diethylene glycol polyurethane (KLDPU), 276–85
kraft lignin-polyethylene glycol polyurethane (KLPPU), 276–85
kraft lignin-triethylene glycol polyurethane (KLTPU), 276–85

laser(s), 19, 22, 23, 28, 30, 66, 123, 124, 128, 207, 222, 351
    excitation, 17, 22, 23
    frequency, 123
    light scattering, 189, 207, 214
    proton, 17
    radiation, 124
    Raman, 263
leaf sheath, 171–3, 175, 176, 177, 181, 186
    of banana plant, 171
    dioxane lignin(s) (DL(s)), 173, 177–82, 185
    lignin, 171, 178, 182
    milled wood lignin (MWL), 177
leveling-off degree of polymerization (DP) of regenerated celluloses, 221
lignin(s), 3–5, 11–15, 18–20, 22, 23, 25–8, 30, 36, 39, 41–7, 51, 52, 60, 61, 64–6, 70, 73, 75, 76, 101, 102, 104, 106, 108, 115, 148–54, 156–60, 162, 163, 164, 166, 171–86, 192, 193, 227–35, 239, 240, 242, 243, 246, 251, 252, 267, 275, 281, 301–306, 308–10, 312–14
    cellulolytic enzyme, 150

lignin(s) (Cont'd.)
   dioxane, 150, 153, 171, 173, 178, 180, 182, 185, 301, 302, 309–311
   enzymatic mild acidolysis, 150
   eucalyptus, 156
   hardwood, 153, 156, 159, 162, 163, 233
   isolation, 149, 150, 153
   Klason, 5, 7, 173, 177
   milled wood, 42, 150–56, 159, 160, 173, 174, 177, 230, 242, 243, 301
   model(s), 20–22, 156, 162, 310, 311
   molecular weight(s), 179
   native, 152, 153, 162, 163, 242
   residual, 27, 153, 163, 206, 218, 227, 229, 236, 238, 240, 242, 243
   softwood, 108, 156, 234
   sulfonated, 190, 194
   sulfonation, 38
   surface content of (ϕ_{lignin}), 42
   technical, 148, 157, 162, 163, 192, 193, 236, 242, 243
lignin–carbohydrate (LC) bonds, 149, 152, 153, 160, 163–5
lignin–carbohydrate complex(es) (LCC(s)), 148–54, 156, 163–6
   carbohydrate-rich, 151
   lignin-rich, 151
   model, 164
   structure, 149, 151
lignin–carbohydrate (LC) linkages, 149, 152, 163, 165
lignocellulosics, 18, 20, 65, 66, 68, 73, 75, 148, 149, 157, 166, 228, 235
   sources, 253subitem
   swelling of, 67
lignosulfonate(s), 189–201, 203
   calcium, 200–202
   hardwood, 191–3
   molecular weight(s), 190, 192
   softwood, 191–3
linseed, 254
linseed fibers, 254
liquid crystal transition temperature (T_{lc}), 333, 334
lithium chloride/N,N-dimethylacetamide (LiCl/DMAc), 206, 207, 209–216
lithium chloride/1,3-dimethyl-2-imidazolidinone (LiCl/DMI), 206, 207, 209–214, 216–18, 224
Lorentzian, 103
   component peaks, 301, 309–312, 314
   functions, 303, 311, 313
   peaks, 303, 314
lysine-based diisocyanate (LDI), 255

maleated polypropylene (MAPP), 258
maleic anhydride, 254, 261, 291, 292, 295, 297
maleic anhydride-grafted styrene-ethylene-butylene-styrene (MA-SEBS), 295
mannose, 7, 177, 240
maple, 74
mass residue (MR), 139, 141, 275, 277, 278, 281, 282
   plots, 281
mass spectrometry (MS), 5, 101
mechanical tester, 139
mercury porosimetry, 65, 67, 68, 71
methoxyl group (content) analysis, 174, 177, 178
methylated kraft lignin-based plastics, 308
methylated kraft lignin-based (polymeric) materials, 308, 314
methylcellulose (MC), 316–18, 325–7, 329
   powder, 318
methylene blue adsorption, 46
4,4'-methylenediphenyl diisocyanate, 296
microcomposite, 341
microcrystalline cellulose (MCC), 213–15, 341, 348, 349, 352
microfibrillar angle, 252
microstrain(s), 121
microstress(es), 121, 131, 133
middle lamella (ML), 23, 37–9, 46, 152
milled wood enzymatic lignin (MWEL), 150–52
modulus, 132, 253, 255, 318, 322
   complex, 318
   dynamic, 268, 269, 318, 322
   fiber, 131
   rupture, 291
   steel, 253
modulus of elasticity (E), 277, 279, 280, 284, 285, 291
molecular weight(s) (mass) (MM), 3, 4, 6, 15, 68, 72, 101, 151, 153, 189–95, 197–9, 206, 207, 209–211, 213, 215, 217, 218, 220, 223, 224, 295, 304, 306–310, 312, 313, 316–18, 320, 323, 324, 330
   absolute, 207
   cellulose, 217
   cellulose and cellulose derivatives, 206
   cellulose, hemicelluloses and residual lignin, 207
   determination of lignosulfonates, 189
   distribution curve, 179
   distribution of cellulose and cellulose derivatives, 206
   distribution of cellulose, hemicelluloses and residual lignin, 207
   distribution of dioxane lignins, 180

distribution of softwood kraft pulps, 216
distributions of acetylated and methylated kraft lignin derivatives, 305
ethylene glycol-type polyol, 275, 278, 284
kraft lignin, 301, 313, 314
lignin, 179
lignosulfonate, 190, 192
maleated polyolefins, 295
methylated paucidisperse kraft lignin fractions, 312
number-average (M_n), 191, 192, 194, 199, 213, 215, 220, 307
polystyrene, 305
softwood kraft pulps, 216
standards, 189
values of celluloses and pulps, 212
values of cellulose triphenylcarbamate (CTC), 212
weight-average (M_w), 46–8, 174, 180, 190–92, 194, 199, 207, 208, 212, 213, subitem 215, 218, 220, 307
molecular weight(s) (mass) (MM) plots, 216, 217, 220, 223, 224
cellouronic acid sodium salt (CUA-Na), 220
cellulose triphenylcarbamate (CTC), 224
water-soluble cellulose derivatives and alginic acid, 220

nanocomposite(s), 340–44, 349, 350, 352–4
cellulose acetate-butyrate (CAB)-cellulose nanowhiskers (CNW), 350
cellulose-based, 346
cellulose nanowhiskers (CNW), 342, 345, 348, 349, 353, 354
cellulose nanowhiskers (CNW)-based, 342
field emission scanning electron microscopy (FESEM) pictures, 344
layered silicate-based, 342
polylactic acid (PLA)-cellulose nanowhiskers (CNW), 349, 350, 353
polymer, 342
preparation, 342, 347
processing, 341
structure(s), 341, 346, 348, 352, 354
thermoplastic starch-cellulose nanowhiskers (CNW), 349, 350
nitrobenzene oxidation (NBO), 156, 158, 177
nitrogen (N_2) adsorption, 65, 67–71
nitrogen contents, 173
nuclear magnetic resonance (NMR), 3–6, 8–10, 12, 13, 148, 150, 151, 153, 154, 156–60, 162–5, 171, 174, 192–4, 197, 218, 227, 228, 233, 238, 243, 246, 263, 264, 266

carbon-13 (^{13}C), 5, 8, 9, 11–13, 128, 154–9, 162, 163, 165, 174, 177, 182–5, 192, 209, 224, 228, 266
cross-polarization magic angle spinning (CP-MAS), 174, 178–80, 227–30, 233, subitem 235, 238, 241, 244–6, 266
proton (^1H), 8, 12, 157, 158, 174, 184, 185, 192
proton–carbon-13 (^1H–^{13}C) heteronuclear single quantum correlation (HSQC), 3, 5, 8–12, 14, 154, 157–9, 164, 165, 193, 195, 196
phosphorus-31 (^{31}P), 154, 156–9
solid-state, 126, 128, 150, 174, 177, 227–9, 231–3, 235, 236, 243, 244, 246
three-dimensional (3D), 158, 165
total correlation spectroscopy (TOCSY), 5, 8–10, 12, 158
two-dimensional (2D), 5, 8–12, 15, 155, 157, 162, 164, 165, 189

oak, 73, 233, 236, 237
Ozawa–Flynn–Wall method, 139–41, 143
Ozawa plot of cellulose fabric, 142
ozone (Z), 36–8, 40, 43, 47–9, 51, 52

paper(s), 3, 26, 36, 37, 39, 64, 66, 67, 70, 81–90, 92–4, 96–8, 101–103, 106, 108, 113, 116, 175, 176, 206, 222, 262, 263, 268, 276, 306
additive, 295
filter, 63, 106, 209
newsprint, 101, 102, 107, 108, 114
pilot, 81, 83, 84, 86, 87, 89–92, 94–7
precipitated calcium carbonate (PCC)-filled, 81, 84, 87, 89, 92, 93, 98
sized, 81–3, 91, 92, 94, 98
strength, 102, 295
surface, 81–3, 87, 88–91, 94–8, 101, 105, 106, 108, 113–16
surface roughness, 85
thermomechanical pulp (TMP), 108
wettability, 82, 93
Whatman cellulose, 30
papermaking, 36, 69, 82, 87, 91, 101, 102, 108, 114, 116, 172, 176, 222
chemicals, 108
pectin(s), 36, 45, 47, 52, 251, 252, 263
pentosanes, 175, 176
permanganate oxidation (PO), 155, 156, 158, 171, 174, 180–82, 184
peroxide bleaching (P), 26, 36–8, 40, 41, 43, 46–8, 51, 52, 243

phenol-formaldehyde (PF), 290, 297
phenol-formaldehyde (PF) resin, 290, 291
phenyl isocyanate, 260
photoacoustic spectroscopy, 264
photodiode array (PDA), 207, 210, 211, 218
    detector, 206, 207, 218
piceid, 7, 8
pine, 19, 73, 107, 164, 236, 237, 239, 243
    cellulolytic enzyme lignin (CEL), 150, 151
    dissolved lignin, 163
    kraft pulps, 27, 239
    milled wood lignin (MWL), 151, 242
    pulp, 236, 239, 241
    residual lignin, 163
    Scots, 232–4
    spent liquor lignins (SLL), 242
    thermomechanical pulp (TMP) fibers, 39
    wood, 19, 232–4
plasticization, 301, 307–309, 314
polarizing microscopy, 289
poly(2,5(6)-benzoxazole) (ABPBO), 125, 131
polybrene, 46–8
poly(butylene adipate), 308–10
poly(butylene succinate) (PBS), 255
poly(diallyldimethylammonium
    chloride)(PDADMAC), 46–8
polydispersity (polydispersities) (Mw/Mn), 191,
    192, 194, 199, 213, 215, 220
polyelectrolyte(s), 36, 48, 329
polyelectrolyte adsorption, 36, 37, 46, 52
polyelectrolyte titration, 49
polyethylene (PE), 266, 288–97
    fibers, 125
    high-density (HD), 264, 289
    linear low-density (LLD), 290
    maleated, 264
    maleic anhydride-grafted, 293–6
poly(ethylene glycol) (PEG), 72, 276, 306–308
poly(ethylene terephthalate) (PET), 131, 132,
    303, 310
    amorphous, 303
    crystalline, 303
    fibers, 132, 303
    film(s), 131, 304
    powder, 304
polyflavonoid(s), 4, 10, 11
polyflavanol, 3, 13
poly(hydroxybutyrate-covalerate) (PHBV), 260
poly(lactic acid) (PLA), 255, 342, 345, 349, 350,
    353
polymeric diphenylmethane diisocyanate
    (PMDI), 290, 291, 296, 297
polypeptide, 62
poly(phenylene-methylene) polyisocyanate
    (MDI), 276, 277

polypropylene (PP), 125, 260, 261, 263, 264,
    266, 268, 288–97
    fibers, 131
    high-density (HD), 260
    low-density (LD), 260
    maleic anhydride-grafted (MAA)(MAPP),
        257, 260, 261, 292, 293–6
polysaccharide(s), 6, 7, 63, 149, 176, 177, 182,
    183, 186, 189, 220, 275, 316, 317, 320,
    335
polystilbene, 3
poly(trimethylene glutarate), 306–308
polyurethane (PU), 275–7, 281, 283, 284,
    330–38
    composites, 277, 279–83, 285
    matrix (matrices), 276, 279, 284, 285
potentiometric titration, 46
precipitated calcium carbonate (PCC), 81, 83,
    84, 87, 89–91, 97
    filled paper(s), 81, 84, 87, 89, 92, 93, 98
    filled pilot paper(s), 87–91, 97
primary cell walls (layer) (P), 38, 39
primary walls, 38
profilometer, 85
protein(s), 61, 63, 153, 175, 176, 189, 197, 292
protein contents, 173
proton spin relaxation edition (PSRE), 229, 232,
    233, 236, 239, 240, 245, 246
pulp(s), 3, 6, 17, 25–30, 36–40, 42, 43, 46–8,
    66–72, 84, 90, 101, 104–109, 115, 148,
    149, 153, 175, 176, 186, 206, 207, 212–18,
    222, 228, 229, 236, 238–41, 246, 276
    biomechanical, 26
    bleached, 25, 27, 43, 83, 88, 106
    brightness, 25
    cellulose, 27
    charge of, 47, 48
    chemical, 17, 27, 102, 106–108, 110, 149, 153,
        218, 230, 238, 239, 264
    chemithermomechanical (CTMP), 36–8, 41,
        42, 46–9, 51, 52, 107
    deinked, 111, 112
    elemental chlorine free (ECF), 81, 107
    extracted, 42, 43, 104
    fiber(s), 17, 18, 20, 22, 23, 30, 31, 36, 37, 39,
        41, 48, 49, 51, 52, 66, 68, 69, 83, 88,
        subitem 101, 102, 104, 116, 131, 222,
        236, 268
    freeness, 40
    hydrolysis of, 71, 153
    kraft, 19, 27, 28, 41, 44, 52, 66, 69, 72, 83, 92,
        103, 105, 107, 110, 131, 133, 134,
        subitem 153, 164, 207, 210, 212–16, 218,
        219, 237, 239, 240, 242, 243
    lignin from, 153

mechanical, 17, 25, 26, 36, 37, 39, 41–3, 46, 48, 51, 52, 66, 72, 83, 106–8, 113, 114
organosolv, 71
pressure groundwood (PGW), 37–40, 42, 43, 46–9, 51, 52, 107
recycled, 27, 67, 102, 106–108
sulfite, 68, 70, 213, 215
sulfonated, 75
swollen, 67
thermomechanical (TMP), 25, 26, 30, 36–40, 42, 44–9, 51, 52, 107–109, 111
unbleached, 25, 106
wood, 27, 36, 69, 209, 213, 218, 245, 268
yield, 175
pulping, 71, 101, 108, 162, 175, 186, 192, 218, 222, 236, 237, 239
alkaline, 162
biomechanical, 69
chemical, 102, 236, 246, 275
chemithermomechanical (CTMP), 38
kraft, 162, 163, 175, 238, 275, 304
mechanical, 37, 39
soda, 162
sulfite, 189, 275
pyrolysis gas chromatography mass spectrometry (Pyr-GC-MS), 157, 158, 174, 177subitem

quinone(s), 26, 27
p-quinone(s), 25, 26

rachis, 171–3, 175, 176, 177, 181, 186
  of banana plant, 171
  dioxane lignin(s) (DL(s)), 173, 177–82, 185
  lignin, 171, 178
  milled wood lignin (MWL), 177
Raman, 19, 22, 25–30, 121–3, 125, 128–30, 132, 133, 263
  band(s), 17, 20, 22, 23, 25, 28–30, 130–33, 264, 266
  deformation, 123, 125
  effect, 17–19, 121–3, 131
  Fourier-transform (FT), 17, 19, 20, 25–30, 124
  frequency, 20, 124
  image(s) (imaging), 22, 23, 25
  light scattering, 18
  mapping, 25
  microprobe (microscope), 22, 23, 31, 121, 123, 124, 133
  microscopy, 22, 23, 121, 123–5
  resonance, 18, 19, 26, 27
  scattering, 18, 122, 124
  shift, 122, 130
  spectrometer, 123
  spectroscopy, 17, 22, 26, 27, 30, 41, 121, 123–5, 128, 133, 264, 265
  ultraviolet resonance, 19, 26
Raman spectrum (spectra), 19, 20, 25, 124, 125, 128, 130
  of cellulose, 20, 128
  of lignin(s), 20
  of regenerated cellulose fibers, 129
  of wood and pulp fibers, 20, 128subitem
ramie, 126, 252–4, 341
Rayleigh factor (R_θ), 190
refractive index (RI), 190, 207, 210, 212–14, 218, 219, 221, 222
  detector, 174, 190, 211, 213
  increment (dn/dc), 190, 192, 207, 210, 212, 213, 215
  of solvent, 190, 207
resin(s), 19, 29, 42, 101, 254, 256, 257, 260, 261, 264, 297, 347
  diphenylmethane diisocyanate (PMDI), 290, 291
  epoxy, 264, 276
  heartwood, 19
  ion-exchange, 194
  melamine, 29
  melamine-formaldehyde, 29
  phenol-formaldehyde (PF), 290, 291
  phenolic, 29
  polyaminoamide-epichlorohydrin (PAE), 295, 296
  polyester, 255
  poly(hydroxybutyrate-covalerate) (PHBV), 260
  resorcinol, 29
  thermosetting, 255
  urea, 29
  vinyl acetate, 29
  wood, 101
resin acid(s), 4, 6, 101, 108, 109, 111, 113
rhamnose, 7, 177
rice straw, 75, 171
root mean square (RMS), 201, 202, 262
root mean square (RMS) radius (radii), 206–209, 211–13, 215–18, 223, 224
  plots, 212, 213, 215

scanning electron microscopy (SEM), 261, 262, 266, 289–91, 297, 340, 342–4, 349, 350
  micrographs of fracture surfaces of composites, 262, 263
secondary electrons (SE), 343–5

secondary ion mass spectrometry (SIMS), 37, 41
  see also time-of-flight secondary ion mass spectrometry (ToF-SIMS)
Simons' stain (SS), 60, 65, 68, 69, 75
sitosterol, 110, 111
  β-, 4
  oxo-, 108
size exclusion chromatography (SEC), 171, 174, 189, 190, 192, 206, 207, 209–213, 216, 218, 224
  chromatograms, 179
  column(s), 174, 190, 220
  eluent, 213
  molecular weight distribution of dioxane lignins, 180
size exclusion chromatography (SEC) elution pattern(s), 207, 208, 213, 214, 216, 217, 220, 224
  carboxymethyl cellulose(s) (CMC), 221, 222
  cellulose triphenylcarbamate (CTC), 223, 224
  hardwood bleached kraft pulp (HBKP), 212–14
  hardwood unbleached kraft pulp (HUKP), 218, 219
  softwood unbleached kraft pulp (SUKP), 218, 219
  water-soluble cellulose derivatives, 220
size exclusion chromatography multi-angle (laser) light scattering (SEC-MA(L)LS), 189, 206, 207, 215, 216, 218, 220
  photodiode array (PDA), 207
  photodiode array (PDA) analysis of cellulose, 207
  quasi-elastic light scattering (QELS), 207, 216
size exclusion chromatography multi-angle light scattering (SEC-MALS) analysis, 207, 208, 210, 212, 213, 216, 218, 222–4
  carboxymethyl cellulose (CMC), 222
  cellulose, 210, 216
  cellulose triphenylcarbamates (CTCs), 209
  cellouronic acid, 218
  residual lignin, 218
silane(s), 257–9, 261, 264, 266–9, 295
  coupling agent(s), 256, 261, 262, 292
  treatment, 264
sisal, 251–5, 269
sisal fiber(s), 266–8
sodium cellulose sulfate (NaCS), 329, 330, 332–8
  atomic force micrograph, 337
  cross-linked (NaCS-PU), 329–38
  powder, 330
softwood(s), 18, 29, 73, 101, 148, 151, 152, 162, 190, 192, 208, 218, 231–4, 237, 239, 316
  bleached kraft pulp, 105, 216
  bleached sulfite pulp, 213, 215

cells, 101
chemical pulps, 239
chips, 61, 304
crystallinities, 234
holocelluloses, 218
knots, 101
kraft pulp(s), 44, 52, 128, 130, 131, 133, 153, 207, 210, 216
lignin(s), 108, 156, 234
lignosulfonates, 191–3
milled wood lignin (MWL), 155, 159
pulp(s), 84, 148, 229, 238, 240
unbleached chemical pulps, 218, 219
unbleached kraft pulp, 218
solute exclusion, 65, 67, 68, 71–3, 75
solvent extraction, 3–5, 11, 41, 102, 104, 106, 159
specific surface area (SSA), 39, 60, 64, 65, 89, 340
  of fiber, 89
  of filler, 81
spruce (*Picea abies*), 5, 9–12, 14, 15, 23, 25, 38, 46, 107, 237, 243, 262
  bark, 3, 4, 6, 7, 11, 14
  cellulose crystallinity indices, 236
  kraft pulp, 243
  lignin, 106, 242
  lignosulfonates, 191
  milled wood lignin (MWL), 154, 155, 159
  Norway, 3, 232, 233
  fiber, 25
  mechanical pulp, 106
  tannin, 10, 11
  thermomechanical pulp (TMP), 25, 30, 108, 109
  wood(s), 23, 29
starch, 60, 61, 173, 175–7, 342, 349
  cationic, 82
  cellulose nanowhiskers (CNW) nanocomposites, 349, 350
  granules, 263
  potato, 84, 263
  thermoplastic, 349
stilbene(s), 3, 4, 6–8, 10, 11, 13–15
strain(s), 25, 121, 123, 125, 129–33, 264, 266, 277, 308
  external, 121, 128, 129, 133
  internal, 131, 133
  sensitivity, 121, 125, 130–33
stress(es), 25, 121, 123, 125, 129, 133, 266
  external, 121, 123, 125, 129, 130, 133
  sensitivity, 121
stress–strain
  curve(s), 25, 277
  relationship, 131
  response, 134
suberin, 171, 177, 180, 184–6

sucrose, 4, 6
sulfonation, 38, 42, 51, 75
   degree(s), 192, 194, 200
   of lignin, 38
surface charge(s), 46–9, 348
   determined by adsorption of PDADMAC, 48
   determined by polyelectrolyte titration, 49
   mechanical pulps, 46
   pressure groundwood pulp (PGW), 47
surface content
   of extractives, 43
   of lignin, 42
surface energy, 37, 48, 50, 95, 115, 261, 267
   of bamboo fibers (BF), 267
   of highly crystalline cellulose, 268
   of fiber(s), 101, 261
   of mechanical pulp fibers, 48
   of papers, 95, 114
   of silane-treated jute and hemp fibers, 261
surface morphology of pressure groundwood pulp (PGW) fibers, 40
surface morphology of thermomechanical pulp (TMP) fibers, 40
surface roughness, 39, 83, 85, 93, 256, 257, 261
surface tension(s), 50, 70, 81, 93, 95, 97, 257
swelling, 38, 52, 62, 66–8, 71, 75, 83, 267, 269
   agents, 67, 71
   of cellulosics, 67
   in cotton fibers, 63
   fibers, 38, 67
   of mechanical fibers, 52
   of pulps, 67
syringyl (S), 148, 153, 155–7, 160, 162, 163, 171, 177–9, 181–4, 186, 231, 233–5

tannin(s), 3–15, 152, 177
tencel, 129, 131–4
   deformation of, 131
   fiber, 132
tensile, 253, 255, 257, 306
   deformation, 17, 22, 23
   elongation, 139
   modulus, 132
   strength, 63, 139, 143–6, 253, 255, 260, 261, 306
   stress, 253
tensile modulus, 132
terpenoids, 4, 6
thermogravimeter-differential thermal analyzer (TG/DTA), 139
thermogravimetry (TG), 138–40, 275, 277, 281
   Fourier transform infrared spectrometry (FTIR), 141
thermogravimetry (TG) curves
   of cellulose fabric, 141
   of kraft lignin-polyethylene glycol polyurethane (KLPPU) composites, 277, 280
   of wood meal kraft lignin-polyethylene glycol polyurethane (KLPPU) composites, 278
thioacidolysis, 156–8
time-of-flight secondary ion mass spectrometry (ToF-SIMS), 37, 39, 45, 87, 91, 101, 102, 104–106, 108, 109, 111, 115, 116
   spectrum (spectra), 105, 109–113
time-of-flight secondary ion mass spectrometry (ToF-SIMS) imaging, 105, 109, 111, 113
   of a newsprint paper, 114
   of a pulp handsheet, 109
   of bleached kraft pulp (BKP), 111
   of deinked pulp (DIP), 113
toluene 2,4-diisocyanate, 290
transcrystallization, 290, 294–6
   of polyethylene (PE), 290
      on fiber surface, 296
      in wood–plastic composites (WPCs), 295
   of polypropylene (PP), 290, 295
      on fiber surface, 296
      in WPCs, 290, 295
transmission electron microscopy (TEM), 261, 263, 340, 342, 343, 346–50, 352–4
   image of cellulose nanowhiskers (CNW) nanocomposite with cellulose acetate butyrate (CAB), 349, 350
   picture of poly(lactic acid) (PLA)-cellulose nanowhiskers (CNW) nanocomposite, 350
   pictures of cellulose nanowhiskers (CNW), 348, 349
tricarbanilation of cellulose, 209

ultramicrotomy, 347–50, 352, 353
ultraviolet (UV), 19, 29, 30, 157, 159, 171, 174, 178, 179, 181, 190, 218, 219
   spectra, 174, 178
ultraviolet resonance Raman (UVRR), 19, 20, 23, 27, 28
ultraviolet-visible (UV-vis), 26, 218
   absorption spectra, 267
   spectrophotometer, 174
   spectroscopy, 266

viscoelastic measurements, 318, 323
viscose, 67, 129, 133, 220
   fiber, 132

viscosity, 72, 85, 93, 95, 97, 210, 213, 260, 320, 322, 323
    ethylcellulose (EC) 317, 320–26
    intrinsic, 209
    methylcellulose (MC) powder, 318, 325

water content(s) (W_c), 210, 329, 330, 332–6
    nonfreezing (W_{nf}), 330, 334–6
water retention value (WRV), 65, 67, 68, 72
weight-average contour length (L_w), 215, 216
    of cellulose and pulp samples, 215
    of cellulose chains, 214
wettability, 82, 83, 93, 256, 261, 266, 291
    fillers, 254
    mercerized fibers, 257
    paper, 82, 93
    sized papers, 81
    wood, 291, 294
wheat straw, 70, 171, 341
white birch, 67, 71
white pine, 73
white spruce, 28
wide angle x-ray scattering (WAXS), 233
wood–plastic composite(s) (WPC(s)), 288–97
    compatibilization mechanisms, 289, 296
    compatibilizers, 293
    extruded, 289
    maleic-anhydride-grafted polypropylene (MAPP)-treated, 295
    moisture contents, 290
    phenyl-formaldehyde (PF)-treated, 290
    stiffness, 288–92, 294
    strength(s), 288–92, 294

x-ray crystallography, 229, 233
x-ray fluorescence (XRF), 200, 201
    spectroscopic analysis of lignosulfonates, 201
    spectroscopy, 189, 199, 200, 203
    spectrum of a calcium lignosulfonate, 201
x-ray fluorescence spectrum, 202
x-ray photoelectron spectrometer, 84
x-ray photoelectron spectroscopy (XPS), 36, 37, 39, 41–3, 45, 46, 48, 49, 52, 81–3, 87, 96–8, 101–103, 105, 106, 113, 116, 263–5
    depth, 45
    escape depth, 44
    peak, 44
    spectra, 43, 87–90, 103, 106, 115
    see also electron spectroscopy for chemical analysis (ESCA)
x-ray (powder) diffraction, 227, 267, 301, 303, 304, 342
x-ray diffraction diagrams for highly crystalline low-molecular-weight poly(ethylene terephthalate) PET powder and typical annealed polymeric PET film, 304
x-ray (powder) diffraction pattern(s), 301–303, 305, 308, 310, 312, 313
    alkylated kraft lignin preparations, 312
    amorphous lignin preparations, 302
    amorphous polymeric material, 303
    blends of methylated polydisperse higher-molecular-weight kraft lignin fraction with poly(1,4-butylene adipate)(PBA), 309
    blends of methylated polydisperse higher-molecular-weight kraft lignin fraction with poly(1,4-butylene adipate) and poly(trimethylene succinate), 310
    cedar dioxane lignin and its methylated derivative, 302
    di- and trilignols, 312
    lignin-based polymeric materials, 313
    methylated cedar dioxane lignin, 311
    methylated kraft lignin-based materials, 314
    methylated kraft lignin-based plastics, 308
    methylated polydisperse higher-molecular-weight kraft lignin and its blends with poly(trimethylene succinate), 313
    paucidisperse kraft lignin fraction(s), 304, 311
    simple lignin derivatives, 302
    underivatized and methylated paucidisperse kraft lignin fractions, 305
xylan, 41, 106, 176, 186, 234, 240, 241
xylan binding modules, 63
xyloglucan(s), 176, 186
xylose, 7, 164, 176, 177

young's modulus, 131, 132, 139, 145, 253, 306
    fiber, 130, 131

zeta (ζ) potential, 266, 268, 269